Scientific Foundations of Crime Scene Reconstruction

Introducing Method to Mayhem

Scientific Foundations of Crime Scene Reconstruction

Introducing Method to Mayhem

Jon J. Nordby, PhD

CRC Press
Taylor & Francis Group
Boca Raton London New York

CRC Press is an imprint of the
Taylor & Francis Group, an **informa** business

CRC Press
Taylor & Francis Group
6000 Broken Sound Parkway NW, Suite 300
Boca Raton, FL 33487-2742

Printed and bound in India by Replika Press Pvt. Ltd.
Version Date: 20120822

International Standard Book Number: 978-1-4200-5168-1 (Hardback)

Library of Congress Cataloging-in-Publication Data

Nordby, Jon J.
Scientific foundations of crime scene reconstruction : introducing method to mayhem / Jon J. Nordby.
p. cm.
Includes bibliographical references and index.
ISBN 978-1-4200-5168-1 (hbk. : alk. paper)
1. Crime scenes. 2. Forensic sciences. 3. Crime scene searches. 4. Criminal investigations. I. Title.

HV8073.N63 2013
363.25'2--dc23 2012029902

Visit the Taylor & Francis Web site at
http://www.taylorandfrancis.com

and the CRC Press Web site at
http://www.crcpress.com

To family, friends, and colleagues too numerous to mention; and especially for Kim who always helps and always has my back and my folks who always do both.

Contents

Preface

> The value of a college education is not the learning of many facts but the training of the mind to think.
>
> **Albert Einstein, *The New York Times*, May 18, 1921**

This book is not intended to provide a simple set of directions to teach the reader how to produce either shooting homicide reconstructions or nonshooting homicide reconstructions. While decidedly practical, it is also not a recipe for producing crime scene reconstructions based upon some simplistic formula for sorting and developing crime scene data. The fact remains that there is no such list of activities and there are no such sets of formulas for developing crime scene data. It is just not that simple, despite what various authors may have you believe. Nor does this book provide a rigorous philosophical account of the many natures of scientific method as applied to the theoretical *predictive sciences* or to the theoretical *historical sciences.*[*]

This book aims to demonstrate the application of scientific methods familiar to those natural scientists who produce what are often referred to as historical explanations of given natural phenomenon by applying what philosophers of science have called *reconstructive reasoning.*[†] The chapters are designed to display the necessary applications of critical thinking that I take to be essential for the development of sound data and reliable explanations to be built upon. This sometimes means moving into either new or unfamiliar areas. That move requires an intimate familiarity with scientific methods: one must know what makes one approach scientific, and what makes some alternative approach nonsense. After producing a few hundred such reconstructions and applications of the natural sciences, one

[*] The terms *historical sciences* and *predictive sciences* are used to refer to a basic distinction between two separable but related scientific enterprises. The so-called historical sciences apply a logical method to ensure the reliable reconstruction and explanation of given phenomena that have already occurred and whose occurrences have been both observed and documented but which in turn raise questions about their natures and processes. Examples of such "historical" scientific enterprises include developing geological explanations of continental formations, astrophysical explanations of extraterrestrial environments, genetic explanations of developmental abnormalities, medical explanations of given symptoms exhibited by a patient, or epidemiological explanations of diseases spreading among given populations. On the other hand, the so-called predictive sciences apply a logical method to investigate or develop processes to produce results derived by uniquely conjoining various conditions, elements, or processes in order to produce and investigate specific consequences in an experimental setting. Examples of such predictive sciences include developing novel chemical elements with specific nuclear properties, developing vaccines to fight specific illnesses, or the development of physical laws based upon repeated, controlled experiments that produce the relevant expected result. The predictive sciences are the standard model for simple accounts of natural science and are what most people think of when considering the scientific method. However, the scientific method is not some monolithic, one-size-fits-all entity. Indeed, there are many different "scientific methods" used in various fields, distinguished by their various logical enterprises. The scientific method is, rather, a family of methods united by logic.

[†] This term covers, for the most part, what I have called "dead reckoning" in *Dead Reckoning: The Art of Forensic Investigation*, J. Nordby, CRC Press, 1999.

might be a bit more comfortable with such efforts. But it never gets easier, nor does it ever get familiar. There are simply no sets of protocols to cover all aspects of a real case; no recipes to follow and no flowcharts to provide comfort. Edgar Allan Poe makes a similar point in his fictional work *Murders in the Rue Morgue* (p. 142): "But it is in matters beyond the limits of mere rule that the skill of the analyst is evinced. He makes, in silence, a host of observations and inferences. So, perhaps, do his companions; and the difference in the extent of the information obtained lies not so much in the validity of the inference as in the quality of the observation. The necessary knowledge is that of what to observe."

My effort here involves an attempt to display the methods of the natural sciences, with all their simplicity, complexity, and opacity, for all to examine when such methods are applied to death investigation casework. I divide these projects into shooting and non-shooting reconstructions. Despite this classification, I find that all cases seem to typify the so-called *one-off* case; cases so apparently resistant to generalization that they seem to stand alone among the many cases supposedly classified simply as to type, motive, and reconstructive approach.* In all my some 30 years of producing such analyses, I must confess that I have never found cases that fail to typify this one-off classification in one way or another. Yet this one-off nature is by no means unique to the forensic sciences.†

Studying and learning natural sciences reveals that this one-off nature often typifies scientific explanations in the more traditional areas of physics and chemistry, including the necessary predictive feature for future phenomenon as well as those scientific conundrums more historical in nature.‡ One can only imagine Marie Curie or Louis Pasteur seeing radiation or microorganisms for the first time. I doubt very much that they considered these phenomena as merely part of a cloth already sewn. It remained for those coming after the fact to find or invent an appropriate repository to categorize both observations and theories.

This book characterizes scientific methods designed to produce *historical explanations* of complex physical events that are unique. They are obviously unique because the individual homicide victims are unique: they can presumably only die once. However, the obvious interplay of scientific abstraction also should be obvious: the conditions required to die in one specific case may also apply to other sets of conditions required to die in another specific case. The book displays the fundamental (as in foundational) critical thinking

* See, for example, any of the list makers, flowchartists, or case-type classifiers. If such lists, charts, or classifications are to be at all applicable, then their proponents face three possible choices. First, they may apply their tools simply to describe, chart, or classify the existing case. This hopefully provides useful information that may be indirectly applicable to new cases. Second, they may abstract and generalize the tools in order to apply them to new cases. But the abstracted tools then tend to become overly general and therefore are so broad that they seldom provide any useful strategies for a new case. (My favorite example of this temerity is an effort to summarize the scientific method: (1) Formulate hypothesis, (2) Test hypothesis, (3) Revise hypothesis, etc.) Third, to circumvent this problem, they may once again remodel to produce more specific tools. But they then again become overly specific and therefore are, in fact, so narrow that they also seldom become useful for application to a new case. As legendary college football coach Bear Bryant is supposed to have said, "When you pass the ball, only three things can happen, and two of them are bad." Enough said.

† It seems that some writers on crime scene reconstruction or those who call themselves crime scene reconstructionists fail to notice these nuances and repeatedly bash all cases into submission indiscriminately, using the same hammer as if some mantra uniquely applied to all crime scenes and that, by virtue of their exposure to crime and crime scenes over time, they somehow have become experts in all the natural sciences.

‡ In physics, the exact nature of light, for example, as a wave or as a particle, the properties of photons, and so on.

The author with Roy Rogers.

necessary to gather, develop, and reconfigure the data, not to simply and blindly accept and apply sets of data given by others who apply their specific areas of expertise. It is not sufficient simply to apply any unevaluated work product to the case at hand. The critical thinking and the critical intelligence that scientific methods demand require no less than the review and assessment of all the data from the scientific point of view with an eye toward the specific explanatory purpose presented by the case.

Various scientific practitioners have various specific explanatory goals. To the extent that those goals combine to affect any grander purposes at hand, all of the data must be reevaluated and recast for those grander scientific purposes. To do less is to obviate the nature of our scientific methods and to ignore their complete dependence upon the data: reliance on the primacy of the data becomes reliance upon the quality of the observations made and developed to explain this complex set of mysterious consequences that we call a crime scene.

> Nobody sees a flower—
> really it is so small—we haven't
> time—and to see takes time.
> Like to have a friend takes time.
>
> Georgia O'Keefe

Acknowledgments

I have many to thank and to be grateful for; a few deserve special mention here. I thank each of the police officers, the prosecutors, the defense attorneys, the public defenders, the judges, and the other associated attorneys and paralegals, as well as the medical examiners, the crime laboratory personnel, and the independent forensic scientists for their work with these cases. Without these professionals, I would not have had the opportunity or the assets necessary to work the cases discussed in these pages. I am thankful for the opportunity to assess relationships among the natural sciences and the physical evidence at crime scenes.

I also want to thank fellow members of the Excellence in Forensic Science and Forensic Medicine Consortium.* Member Eugene Liscio of AI² Forensic Animation, Imaging & Illustration deserves special thanks. He provides continual computer and photogrammetry work with images and three-dimensional models used for courtroom presentations of shooting and bloodstain pattern cases. We also enjoy working on ongoing research projects.†

I want to thank Consortium members Dr. Selma Schieveld Eikelenboom and Richard Eikelenboom of Independent Forensic Services (IFS) for their undying support and encouragement. They cover DNA, including touch-DNA, bloodstain pattern analysis, and forensic medicine as valued members of the Consortium. They also provide peer reviews of my and others' scientific work. Thanks as well to Scott Verbonus and Brian Orr of Go Evidence, the remaining Consortium members. Their work and research in both fingerprint development and analysis involves state-of-the art vacuum metal deposition (VMD) machines. Currently, about 20 such machines operate worldwide. Go Evidence has two of them. The future holds great promise for our Consortium.

Thanks also to Martha Schoonhoven, IFS, the Netherlands, and Chuck Vaughn (retired), Washington State Patrol Crime Laboratory, for helpful comments and critical reviews. Thanks as well to the late Inés Carl-Haas (1963–2011) for her dedicated casework with Final Analysis Forensics and inspired conversations covering these scientific topics.

Of course my appreciation extends to my associates in the Final Analysis Forensics laboratory. Emily Wold and Michelle Kang, two of our very gifted interns, provide valued assistance in the daily routine of casework. Thanks as well to Peg A. Foley, my invaluable laboratory assistant for over 25 years (or is it 30 years by now), who also doubles as our accounts manager.

* A group composed of independent practitioners in forensic science and forensic medicine jointly pledged to help support and encourage scientific quality and intellectual integrity in the forensic sciences.

† Scientifically robust three-dimensional images that allow for scaled measurements in all directions within the models will soon become the standard of practice in crime scene documentation and reconstruction. Animation or 360-degree pictures just will not do. Images without measurements are like race cars without engines—perhaps they are pretty to look at, but they do not actually do anything that race cars are supposed to do. Better still, the three-dimensional models do not require fancy or expensive equipment like animations or 360-degree photos do—a total station is nice, but in fact any digital camera will do just fine. That is a tool we all have.

I especially want to thank the exceptionally talented manager of Final Analysis Forensics, Kim Nordby, who also doubles as my wife. She not only read this entire manuscript, but she also pointed out each and every one of my errors—sometimes twice. She continually provided thoughtful suggestions for improving the work and gentle nudging to help me overcome the errors.* She is the best darn wife that I have ever had.

Finally, special tribute goes to my editor, Becky Masterman of CRC Press/Taylor & Francis. Her editorial judgment, eye for detail, and mastery of the English language combined with her strong support and steadfast encouragement to make writing for her truly enjoyable. For all of the above and more, my heartfelt thanks. I am truly blessed.

Peg Foley of Final Analysis Forensics and Eugene Liscio of AI² in the Final Analysis Forensics laboratory.

Intern Michelle Kang of Final Analysis Forensics takes aim with a Sig 45, as I observe.

* She would be the first to underscore the fact that any remaining errors are mine alone.

Intern Emily Wold of Final Analysis Forensics fires a 357 revolver.

Dr. Selma Schieveld Eikelenboom and Richard Eikelenboom of Independent Forensic Services (IFS) examine a rifle.

Scott Verbonus of Go Evidence works on the larger VMD machine for latent prints.

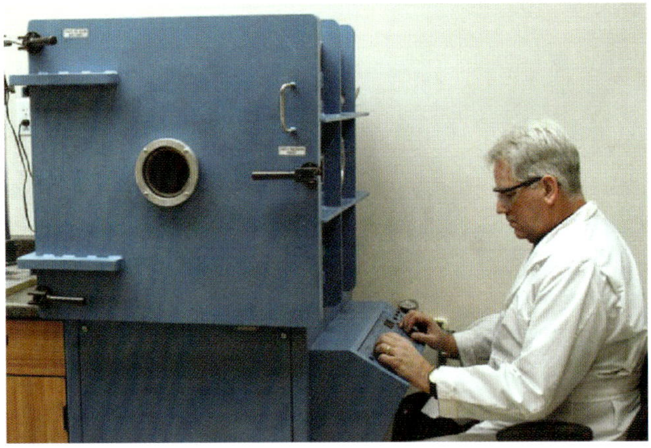

Brian Orr works on the smaller VMD machine. Go Evidence uses these machines to process everything from thermal paper and body bags to car doors.

Jon J. Nordby, Ph.D., D-ABMDI
Final Analysis Forensics
University Place, Washington

Introduction

> Swinging between the Charybdis of the lawyer who knows no math to the Scylla* of the mathematician who misuses his math because he does not have the judgment to select the right model … In other words, we will have to swing between the mistake of listening to the glib nonsense of a lawyer who refuses science and that of applying the flawed theories of some economist who takes his science too seriously.
>
> **Nassim Taleb, *The Black Swan*, p. 206**

No discipline called *crime scene reconstruction* can be found listed among long-established natural sciences such as chemistry, physics, mathematics, biology, geology, or even medicine. A search for its inclusion among social sciences such as psychology or sociology fares no better. In fact, crime scene reconstruction is noteworthy only by its absence from any list of generally accepted scientific disciplines, natural or social. Yet crime scene reconstruction as a discipline has been accepted by the courts, thereby allowing its court-qualified practitioners to provide opinion evidence presumably based upon expertise developed through the study of its principles, their application to data deemed relevant by those principles, and the case experience associated with and developed through the "reconstruction" of many different crime scenes.

By examining the practice, we can almost immediately distinguish the scientific from the nonscientific approaches to crime scene reconstruction. The question is not whether crime scene reconstruction, in general, is a natural science; the question is whether the methods applied in a given crime scene reconstruction are scientific methods. There are practitioners who claim to have years of experience attending crime scenes, and who therefore claim to speak with authority, telling juries and courts greatly entertaining, involved, and persuasive tales detailing the series of complex events that produced the detritus of any given crime scene.† The overwhelming features of this approach include a lack of any relevant measurements, a profound lack of any relevant testing or experiment, and in fact, a lack of any means whatsoever for anyone to approach or identify what may have been the data involved as the foundation of the storytelling. As such, the

* In Greek mythology, Scylla was a monster that lived on one side of a narrow channel, opposite its counterpart Charybdis. The two sides of the strait were within an arrow's range of each other—so close that sailors attempting to avoid Charybdis would pass too close to Scylla and vice versa, both being bad things to do. Scylla was a horrible sea monster with four eyes, six long necks equipped with grisly heads, each of which contained three rows of sharp teeth. Her body consisted of twelve tentacle-like legs, a cat's tail, and four to six dog heads ringing her waist. Charybdis was another sea monster considered to be a shipping hazard in the Strait of Messina. The idiom "between Scylla and Charybdis" has come to mean being between two dangers; choosing either brings harm.

† I have long maintained that arguments from authority based upon experience are simply fallacious: "Merely spending time in the garage doesn't turn a rusty garden rake into a fancy sports car." J.J. Nordby on the role of experience in crime scene reconstruction from lectures on scientific methods, International Association for Identification (IAI), International Association of Arson Investigators (IAAI), and American Academy of Forensic Sciences (AAFS).

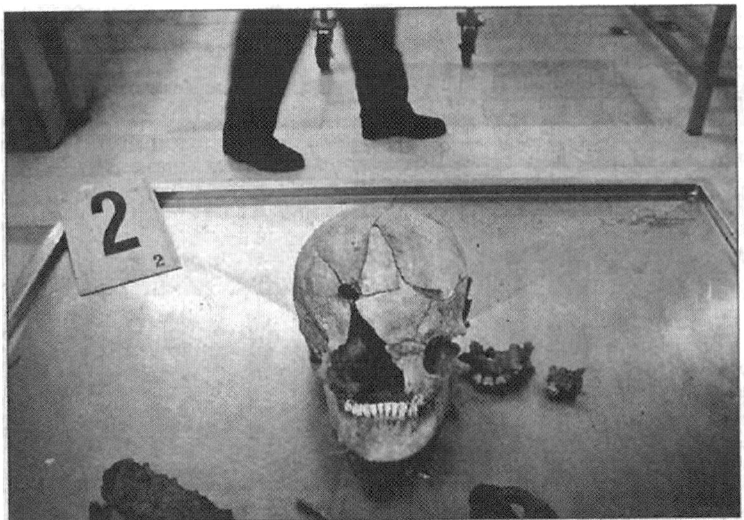

By Damir Sagolj, Reuters

Still no name: A skull of a Muslim man shot with a bullet in the head remains unidentified in Visoko, Bosnia and Herzegovina, which broke from Yugoslavia in 1991.

Somehow one knows that the reconstruction for this case just won't go very well at all. (Photo by Damir Sagolj, Reuters.)

reconstruction is bulletproof. That feature alone belies its extreme remoteness from any truly scientific endeavor.*

Crime scene reconstruction in a scientific sense must also be distinguished from its other perhaps more distant relatives. While both powerful and useful, reconstructions based upon psychology or criminology are not yet accepted by the courts and do not permit practitioners to offer opinion evidence. The latter sense of crime scene reconstruction depends on its cognitive status† upon the disciplines of psychology, criminology, and upon repeated associations among types of data developed through relevant case experience. This sense of crime scene reconstruction attempts to explain psychological characteristics of the criminal. The sense of crime scene reconstruction developed in this book attempts to explain the physical characteristics of the crime scene data, not the psychological characteristics of the perpetrator. Appreciating this distinction helps us to unravel various current practices of crime scene reconstruction.

The effort in this volume is to understand the specific, relevant scientific foundations of crime scene reconstruction as a practice to be accepted by the courts in order to help explain specific characteristics of a crime. Let us briefly consider certain features of crime scene reconstruction as a practice focused upon characteristics of the criminal in order to help understand the scientific foundations of crime scene reconstruction as a practice focused upon characteristics of the crime.

* "In a dying civilization, political prestige is the reward not of the shrewdest diagnostician but of the man with the best bedside manner. It is the decoration conferred on mediocrity by ignorance." Eric Ambler, *A Coffin for Dimitrios*, Knopf, 1939, p. 71.

† Any human activity that makes statements that can be evaluated as either true or false contributes to human knowledge. When organized as an intellectual activity, the activity is called a cognitive discipline. Many areas of human endeavor are not cognitive; music, art, and poetry, for example, are "noncognitive" disciplines. Discussions about music, art, and poetry, however, are cognitive.

Criminal Investigative Analysis and Profiling as Crime Scene Reconstruction

Efforts to "get into the mind of a killer" in order to help solve puzzling murders have proved useful in certain types of cases. This effort has developed increased sophistication in recent years due in part to research conducted through the Federal Bureau of Investigation's Behavioral Sciences Unit. Profiles of the killer are developed through the examination and analysis of various components of the murder including assessing characteristics of the victim, the location or locations and number as well as types of criminal events, and the specific elements of the crime scene(s), including the physical evidence, providing behavioral clues. The primary goal of producing such profiles is not simply to explain a crime, but instead to predict the habits and behaviors of the criminal specifically in order to arrest him.

The courts have been vigilant in restricting the admissibility of such endeavors, leaving this activity exclusively limited to the investigative arena. Despite intellectual advances in psychological theory and recent developments in the discipline of profiling and criminal investigative analysis, the practical import of these disciplines remains solidly investigative. No efforts have been forthcoming solidly to ground these efforts in the realm of the natural sciences.* In fact, both legal and scientific arguments have been developed in an effort to show that such disciplines must remain outside the realm of "science" by their very nature.

According to these arguments, such efforts appear to consist of "mere speculations" and "fanciful conjectures" at best, guesswork; at worst, unsupported prejudice.† The expertise associated with profiling remains, in this view, anecdotal, commonsensical, and theoretically bereft. Yet these intellectual judgments remain far too harsh and tend to misconstrue both the actual and theoretical grounding of these broadly based activities as

* A theoretical natural science typically delegates its explanatory function to a theoretical model fully entailed by a robust scientific theory. Posing questions about the observed world in a mathematical way has been the scientific norm in the West, at least since Galileo. The mathematical modeling process, as an important element of scientific theory, is evident in many examples from biology, physics, and geology. Predator–prey, pollution, evolution, and population sampling models abound in biology. Atomic, pendulum, and equilibrium models infuse theories of physics. These models abstract certain elements at the expense of others and simplify the process of explaining and predicting phenomena. They are idealized heuristic devices whose successes or failures are measured by their predictive and explanatory power. Both the development of so-called scientific approaches to crime scenes and the criticism of such approaches suffer from an apparent failure to understand alternative approaches to some of the most basic issues in scientific explanations that are typical of explanatory efforts among all the natural sciences. In this respect, scientific approaches to crime scenes may not be so different from scientific approaches to atomic properties of the elements or perturbations in planetary orbits after all. However, no such theories yet cover the relevant features of the profiling realm.

† "The Parisian police, so much extolled for acumen, are cunning, but no more. There is no method in their proceedings, beyond the method of the moment. They make a vast parade of measures; but, not infrequently, these are so ill-adapted to the objects proposed. …

"The results attained by them are not infrequently surprising, but, for the most part, are brought about by simple diligence and activity. When these qualities are unavailing, their schemes fail. Vidocq, for example, was a good guesser, and a persevering man. But, without educated thought, he erred continually by the very intensity of his investigations. He impaired his vision by holding the object too close. He might see, perhaps, one or two points with unusual clearness, but in so doing, he, necessarily, lost sight of the matter as a whole. Thus there is such a thing as being too profound." Edger Allan Poe, *Murders in the Rue Morgue*, pp. 152–153. The issues raised about the scientific underpinnings of profiling, then, become equally elusive. (I should think that the Vidocq Society might consider a new name more fitting for the skills of that esteemed group's members.)

well as the practical development of these approaches to the crime scene. While practitioners of profiling and criminal investigative analysis have done a great deal to suggest and pursue productive inroads for more rigorous approaches to crime scenes seen as complex wholes that include psychological properties of the criminal and criminal behaviors, little has been done to pursue and develop the scientific foundations and theoretical justifications that may exist to support and mature such activities.

Method or Methods?

One unique feature of so-called *scientific method* is the distinct difficulty of stating clearly just what it is about some method that makes it "scientific." Sometimes an effort is made to overcome this difficulty by defining a "method" as a clear set of rules—both rules of procedure (what to do) and rules of practice (how to do it). The histories of philosophy, psychology, and technology, for instance, are rife with examples of such attempts, particularly during times when some novel brand of technology comes into popular favor, for instance, the mechanism of the industrial machine age and the reductionism of the computer-driven digital age.* In crime scene reconstruction, this effort to address the issue of method takes the form of a reductionism to rules, flowcharts, or procedures designed to simplify the process so any profoundly disadvantaged Wal-Mart greeter could successfully produce compelling crime scene reconstructions. I do not find such efforts either helpful or practical. Such efforts are best left to tweeters, bloggers, and scrapbookers.

I suggest here that what scientists and philosophers of science call the scientific method is not only a family of logical methods, but it is more a way of being than a way of following some generally acceptable set of rules. The temptation to embrace the latter (particularly by lawyers in courtroom contexts and by supervisory agencies in accreditation contexts) can be explained quite simply: explanations of proper scientific approaches to specific data are constructed *post factum* by an appeal to some set of descriptive rules. However, those rules merely function as a description of some specific processes—they do not and cannot define that process.

That is why attempts to reduce the scientific method to some finite set of rules are fraught with peril. And that is why simply learning some set of rules, even some general higher-order rules, falls short when we attempt to apply them. They are merely descriptive rules, not prescriptive rules. The confusion of descriptive rules with prescriptive rules leads to a great many fallacious attempts to contain, define, and delimit scientific method, or scientific approaches to some discipline, as practiced within some set of simple rules. Counterexamples to any such efforts abound. Scientists, in so far as they are doing science, are anything but simple rule followers.†

* Rene Descartes, with his "ghost in the machine" account of human nature and contemporary efforts to reduce complex human thought processes to sets of rules executable by a computer are two simple examples. Soon such popular developments are replaced by new developments, which then spread among cultures like the most virulent plagues of old.

† We often read and consider such lists of rules when we are asked to characterize the scientific method in our own words. Much of what we end up saying, however, does not withstand any careful intellectual scrutiny if it is offered as a definition. See, for example, any textbook purporting to teach materials associated with a natural science: method = develop hypothesis; test hypothesis; revise hypothesis; retest; produce explanation; go have a beer.

We did not use any rules at all to build our cars back in the day, but that did not stop us from trying. Sometimes the best way to learn is by trying and failing and trying again.

However difficult it may be to capture the essence of scientific method, it can be characterized as a way of being when faced with both recognizing and developing what we call data in order to develop some physical explanation based upon these observations. Whatever the myriad philosophical challenges involved in providing an analysis of the concept scientific method, we can at least agree upon certain necessary conditions that any method must meet in order to be called scientific. What we discuss and develop are some of the necessary conditions for a scientific method. We err if we think that such conditions are also alone sufficient conditions to identify some method's nature as scientific.

"Crime Scene" or "Event" Reconstruction and Scientific Method

A scientific crime scene or event analysis and reconstruction logically links a detailed series of scientific *explanations* in order to provide an understanding of the sequence of events leaving some given set of *data* that may become physical *evidence*. Each explanation is developed, linked, and evaluated by applying a *relevant scientific method* to this available data.[*] This process involves proposing, testing, and evaluating explanatory connections among the physical evidence found to be related to these events. The purpose of the analysis remains to develop and evaluate alternative explanations and finally to determine the *best scientific explanation* of these related events supported by all the available data.

Rules do exist and rules obviously have a place in crime scene management, processing, and reconstruction. However, such rules are not sufficient for producing reliable

[*] I have detailed the nature, scope, and application of the scientific method in my book *Dead Reckoning: The Art of Forensic Detection*, CRC Press, 1999. The book includes concrete examples of the logic essential to the development of scientific explanations by the natural sciences, in general, and the forensic sciences, in particular. I also have explained the nature of the scientific method in my book *Forensic Science: An Introduction to Scientific and Investigative Techniques*, 3rd ed., edited with Stuart James, CRC Press, 2009.

This death scene became someone's garbage can at that person's professional expense. (The press is always handy to capture the event.) I am the one putting on gloves by the open door. My associate received a Velcro mitten the following day.

scientific explanations. Their violation, of course, always has a way of going bad to cause further problems.

Caveats and Required Components

The methods of the natural sciences depend, for their validity, in part, upon being a certain sort of person when one is applying them. Although that may sound odd, the requirement is nothing less than accepting a commitment to follow the evidence wherever it may lead, and to do so however much the result becomes personally distasteful, damaging, or otherwise emotionally unacceptable.* That is one reason why the bulletproof reconstructionist who makes no scientific effort to recognize or to develop data but who simply tells the jury a good story lacks the necessary character in the most fundamental terms to be a natural scientist.

* So much for the claim that the practice of natural science is value-neutral in its fundamental bearing and practice: The data are and must remain in fact neutral, but having the character and the personal integrity to follow that data in order to develop the most reliable scientific explanation, and then to present that explanation with all supporting reasons (show your work as clearly as possible) while inviting reasoned criticism by peers, is anything but value-neutral. The reason that the storyteller's work is bulletproof, that the storyteller has no work to show by way of support other than to say "believe me because I am an expert" is in fact the fatal bullet that should slay the beast immediately if not sooner. That such testimony is admitted into court remains a profound embarrassment to our legal and criminal justice system.

A good story, or even a fanciful bulletproof tale, is seldom bothered by any data collection.

In addition, such analyses must rely upon all the available physical evidence and the analyst's unbiased application of logical scientific methods as we will come to see them. Data may include statements, reports, diagrams and photographs, as well as the physical evidence, but the focus for our purposes remains the physical evidence. The statements of witnesses and interested parties are significant only when tested against the developed evidence. The ability to apply proper logical methods develops in the analyst through relevant education, training, and experience, but involves skills not unfamiliar to the average person.* Thus, jurors, attorneys, and the court should be able to understand the logic behind the reconstruction despite the technical sciences used. In this sense, the testifying forensic scientist by necessity becomes a teacher. This role further depends upon having the character to present the material in an unbiased manner, following the evidence wherever it goes.†

Such analyses may also face logical limitations that must be recognized and explained by the scientist as part of the reconstruction. It remains important to be humble and acknowledge the limitations of the scientific approach to the evidence. *No reconstruction*

* A simple example of this logic may suffice. Suppose that a cadaver dog alerts on an area of disturbed ground next to a building. This does not mean that a human body is buried beneath the ground at this location. What it does mean is that we must apply scientific methods to investigate that area: dig and see if we have a dead dog, cat, bird, or human. The scientific relevance is to provide an initial hypothesis to test. The process does not involve discovering complex probabilistic relationships allowing us to rank alternative explanations numerically. Instead, the process involves discovering supporting or refuting evidence. Thus, it is mistaken to say that the alerting dog shows that one dead thing is more likely than some other dead thing. It is only correct to say that we now have some ideas on how to develop the evidence (i.e., dig). It means nothing without follow-up testing. Also please note that any explanation offered before testing remains unsupported and unscientific guessing. (Given the dog's alerting, what tests could this suggest besides digging?)

† "If the law has made you a witness, remain a man of science: you have no victim to avenge, no guilty or innocent person to ruin or save. You must bear testimony within the limits of science."—Dr. Paul Brouardel, French pioneer in the field of legal medicine.

can explain every element of an event. Many sequences may escape scientific detection, or if detected, may provide no logical grounds for inclusion or no known role in the events under study.

Such analyses must also remain open logically to the discovery of new data that may prove to be relevant to the developing explanation of events and, as such, may supply new evidence. In that circumstance, a scientist must always reserve the right to consider this new evidence and to reassess the development of the reconstruction in its new light. Such analyses must always remain data driven.

As we know, and have stated, there is obviously no natural science called crime scene reconstruction. As should be obvious, in that sense (of listing sciences), crime scene reconstruction is not a "science."[*] For this reason, it is useful to more clearly distinguish *scientific* from *unscientific* crime scene reconstruction. The scientific method or methods of some sort must form the heart of crime scene reconstruction for such activities to be scientific.[†] The focus must remain on the methods of the natural sciences and the reliability of the results must be engendered by this so-called scientific method.[‡]

Scientific versus Unscientific Crime Scene Reconstructions

The methods applied to produce crime scene reconstructions may be either scientific or unscientific.[§] To withstand the rigors of both scientific and legal challenges, crime scene reconstruction in criminal cases must be based upon scientific methods. Consider some basic features of scientific crime scene reconstruction in contrast with some basic features of unscientific crime scene reconstruction.

To be scientific, the method applied must be scientifically *reliable*. To be reliable, in this sense, means more than it gets the right answer more often than it gets the wrong answer. It means, in addition, that reasons that are developed by the method provide support for the answer. This method may involve either *experiments* or *tests*. Part of the experimental or testing process involves identifying appropriate *controls*, usually *positive controls* and *negative controls*, each identified and defined in terms of its relationship to the data being investigated.

[*] Then, if we follow that the same type of classification requirement for being scientific, neither is optometry, ophthalmology, archeology, anthropology, entomology, genetics, and so on.

[†] I discuss this point in "Science Is as Science Does: The Question of Reliable Methodologies in 'Real Science,'" *Shepard's Expert and Scientific Evidence Quarterly*, vol. 2, no. 3, winter 1995.

[‡] See "Is Forensic Taphonomy Scientific?" Jon J. Nordby, in *Forensic Taphonomy: Volume II*, W. D. Haglund, and M. H. Sorg, editors, CRC Press, 2002, Chapter 2, pp. 31–42. Also see "Misplaced Method in the Science of Murder," Jon J. Nordby, in *Scientific Method: Applications in Failure Investigation and Forensic Science*, Randall K. Noon, editor, CRC Press, 2009, pp. 155–172, Chapter 7.

[§] In this context "unscientific" means the same thing as "nonscientific"; its necessary feature involves eschewing logical scientific methods of testing and developing data to evaluate conclusions. No other value, either positive or negative, attaches to this term. In this sense of the term, any activity that does not rely upon scientific methods applied to develop reasons to support a conclusion becomes unscientific. Many human activities are quite properly unscientific in that they do not require testing or data development because they are not rational or cognitive activities. (That is, they do not add to human knowledge.) Traditional examples offered by philosophers include the fields of creative art such as painting or many other inspired expressions of beauty such as poetry. Examples often include human emotions such as love, which often in fact fly in the face of reason. However spelled out, one common feature of such unscientific noncognitive activities remains the inability to articulate rational steps (reasons) taken to reach some result. Although art historians may attempt to explain art after the fact, the point remains that the artist inspired to produce it need not engage in any such activity.

A method's scientific reliability can be measured by a process of *validation*. The validation of a method ensures that the method addresses all relevant data types and produces reasonably repeatable or appropriately uniform results when the same data and the same method are applied—even if applied by different practitioners. This validation is one form of *testing* commonplace in the natural sciences. This underscores the importance of *peer review* in the natural sciences. Science progresses through a community of scholars.

Such validations often result in the calculation of *error rates*, which are usually defined in terms of a range of acceptable results allowed by the method (these are not the same as mistakes made by the practitioner). The error rate of a reliable method can be determined experimentally by validating its results in terms of their *appropriate accuracy* and their *appropriate precision.** In the sense intended here, error rates are not mistakes at all but are really ranges of *acceptable variance* for the result due to the nature of the method as well as the nature of the data.

The method applied must be data driven by *observable data* that anyone can detect without some sort of sixth sense, or psychic ability. The method cannot rely upon such special subjective powers inherent in one individual but unavailable to another trained investigator; the evidence cannot be invisible to those lacking some "special power" or other. This is why the testimony of psychics, for example, cannot be counted as scientific testimony. And this is also why simple claims to have "great experience" remain insufficient as proof for explanations when they are not supported by rigorous independent scientific reasoning. Clairvoyance relies upon special subjective "powers" unavailable to ordinary mortals. For this reason, a scientific result must be based upon systematic evaluations of observable data rather than incidental observations that remain anecdotal in character and subjective in nature, lacking the explanatory logical structure of these scientific methods.

Nor can scientific methods simply be defined in terms of reaching true conclusions; anyone can simply guess at conclusions without using any method at all and still might get the "right" answers. Such answers, even if they turned out to be correct, would be reached randomly in the sense that no reliable method licensed them. For example, the noted psychic and clairvoyant Edgar Cayce, while in a sleep-induced trance, made many startling predictions, which turned out to be true. Although we may marvel at such a phenomenon, in no way can this approach be characterized as scientific regardless of the truth of Cayce's predictions. The limitation that prevents us from calling Cayce's predictions scientific remains to be the lack of any identifiable method, specifically the lack of a scientific method that provides step-by-step reasons,† in turn supporting the conclusions reached. For Cayce, the conclusion simply appears.‡ The fact that he got many correct answers misses

* Accuracy is best conceived as the proximity of the derived or measured result to the true result (sometimes given by theory or proved by what is known to be true), whereas precision is best conceived as the proximity of one result to the next result based upon the same data and supplied by the same method. Accuracy can sometimes be thought of as systematic or methodological variance, whereas precision can be thought of as random or nonsystematic variance apart from the method itself.

† Note that these reasons are not step-by-step rules about how to do anything but are justifications for what is done.

‡ The unscientific approaches evident through documentation of the forensic work done by many law enforcement-based crime laboratories have been publicized by the National Academy of Sciences. One telling criticism it reports remains a tendency toward unscientific approaches to case data that are geared in favor of prosecutors' interests in simply obtaining convictions. See *Strengthening Forensic Science in the United States: A Path Forward*, National Academies Press, 2009. The PDF version of this report is available online through a link at www.finalanalysisforensics.com.

the rational point of doing natural science: the point is to identify the rational steps taken to reach and to explain the answer. In cases of guesses that do not involve simply making predictions, the question to ask remains "But how do we know that the answer is correct?" The answer is to use a reliable scientific method. To use a reliable method entails showing your work: clearly showing the steps required to support the conclusion wherever possible.*

In science, reaching an answer—even a correct answer—is not sufficient. Answers must be reached through reliable rational steps, and through rigorously identified, documented, and collected data. This data must be observable to anyone despite his or her lack of training. The old adage written on many a chemistry and physics examination applies: show your work—show how you derived the answer, do not just provide the answer. Show that you have not simply copied the answer from someone else or simply guessed.

Through applying this method, even though we may initially reach innocently incorrect or honestly incomplete answers, progress and knowledge eventually develop. Thus, many of the conclusions developed from the scientific analysis of the given data made available in any crime scene must remain inherently modest in scope. We may sometime develop new data that will force us to reconsider those conclusions. This in fact has most recently occurred with the development of more sophisticated DNA technologies applied to investigations of both old and new evidence. The additional data developed have resulted in strong scientific support either for convicting or for exonerating the accused. Science continues to develop. Forensic scientists must also continually develop. We must test the applications of these scientific and technological developments in an effort to more reliably explain the data mined from any type of physical effects. We must resist the oversimplifications of the uninformed and still remain clear and coherent. More than that would be foolish to request. More than that should be received suspiciously.

> You cannot choose your battlefield, God does that for you; but you can plant a standard where a standard never flew.
>
> **Nathalia Clara Ruth Crane (1913–1998), "The Colors"**

Attributions and Limitations of Accuracy

This volume contains cases I investigated over many years. Obviously, I did not work these cases alone. Nor do I mean that I worked entire cases. A scientist's part in a homicide case remains a small part of a larger whole. Sometimes that part is very significant, other times that part is completely insignificant. Much material in any case owes a debt to the hard

* This becomes an issue for the crime laboratories that require their members to issue one-line reports, just providing the conclusion but excluding the supporting materials used to develop that conclusion. One must wonder: what do they fear? The approach is often justified by a fear of providing too much information that can be twisted by the attorneys into any monstrous form suitable to support their diabolical purposes. However, the practice is in fact unscientific and misleading. It violates the most basic principle that has advanced the sciences for centuries: the testing and falsification/verification of the reasoning that helps develop new insights and accounts for much of what we call "scientific progress." If no reasons are presented, another forensic scientist has no basis for knowing how the conclusion was reached or what work was done to develop the data involved. Unfortunately this reduces reputable science to the same low level as the storyteller who appeals simply to experience as his authority.

Nathalia Clara Ruth Crane.

work of others. The lawyers must deal with the scientific testimony. It is wise to remember that without lawyers, there would be no such thing as forensic science.

In these chapters, I obscured the nonscientific elements to leave the people, places, and jurisdictions anonymous. The changes do not alter anything scientifically relevant or change any features of a case integral to discussions of the scientific foundations for event reconstructions. This book presents a picture of scientific crime scene reconstruction and a discussion of the role that the natural sciences play in any such cognitive* endeavor.

If the reader's goal is merely noncognitive (not rational) empty entertainment, devoid of scientific fact or principle, then this book has nothing to offer that reader. If any of the names of people in these chapters happen to correspond with actual individuals, then they do so simply by accident and not by the author's design. Of course, any errors are the responsibility of the author alone.

* "Cognitive" refers to any enterprise or undertaking that makes statements that must be either true or false, even though we might not know their actual truth values. "Noncognitive" refers to conjectures feeling or displays of emotion that lack statement-making structures, such as music or the visual arts.

Shootings

I

Three Strikes and Out 1

> I have seldom known anyone who deserted truth in trifles that could be trusted in matters of importance.
>
> **Attributed to William Paley**

Shooting reconstructions apply many well-known elements from physics, mathematics, biology, and chemistry. Describing the chemistry required to recognize and to measure the effluents produced when a firearm discharges could supply material for an entire chapter. Another chapter could cover the design of various types of ammunition and their various physical properties. With the advent of lead-free bullets, primers, and powders, both detection methods and discharge analyses have developed considerably from the early days of black powder firearms.

The designs of various handguns (revolvers, pistols, and fully automatic machine pistols), different rifles (single shot, semiautomatic, or fully automatic) and different machine guns (light or heavy) also play a significant role in determining the type of analyses required to understand a particular shooting event. The azimuth and incident angles of a bullet's path, as well as other features influencing a bullet's trajectory, such as deflection angles, the number of targets, and wind and range, often must be determined.

The order of multiple targets also must be considered if a bullet perforates one object (the primary target) before striking another object (the secondary target). The physical properties of the targets themselves also may become a focus for testing and experiments. The behavior of different bullet designs when shot through glass, plastics, metals, and fabrics, or various combinations of these may also become a relevant consideration.

Shotguns with cylinder bore, full choke, or modified choke barrel designs can shoot either lead or steel shot of various sizes and descriptions, as well as sabots or "slugs." Some shotguns have rifled barrels and some do not. Some firearms involved in death cases may have been modified illegally by their users, and some, such as the old-fashioned zip gun (an unauthorized product of industrial arts classes), may be homemade.

Even if several shootings involve the same type of gun, not all shootings are subject to the same scientific treatment. Even a list of "required steps" may simply serve to sidetrack a shooting investigation. Basic intellectual skills must be applied to recognize and to collect data, as well as to organize and to apply relevant sciences to the case at hand. Obviously the required steps differ by case.

A scientist must approach each case with an eye clearly focused on what we do know, what we can know, what we can learn, and how we can learn it through the data we have before us. This ability is part of what it is to approach reconstructions by being a certain kind of person—by *being* a scientist. When I say being a scientist, I really mean being scientific.

Being a scientist has its advantages. Circumstances such as unique "one-off" events, coincidences, or just plain lucky happenstance do not sidetrack one's ability to develop a scientifically reliable shooting reconstruction. Given such complex events, being a scientist

Figure 1.1 Some of Deal's inventory, which was recovered after he went out of business.

by being scientific, with its attendant intellectual stance, is preferable to knowing a set of generic approaches, with its attendant abstractions, listing what to think and how to think it.

This first case begins to illustrate what I mean by being a scientist through being scientific. The remaining chapters have the same intent. Among their pages invariably lurks some telling of course, but I offer more showing* than telling. I do so with the hope that these modest efforts exhibit the nature of scientific methods applied to casework. I attempt to do so in a manner that reveals the basic aspects of being scientific.

This case involves the shooting death of a wife-beating drug dealer named Rick Deal. While his wife worked on base as an Army cook, Deal spent his days watching cable TV, drinking, selling marijuana, and generally avoiding any duties loosely associated with being a househusband (Figure 1.1). Like any respectable drug dealer, Deal usually carried a pistol and at least one knife.

The Criminal Investigation Command (CID) learned of Deal through his incautious efforts to sell "product" on base to an MP. The MP played along, but immediately informed his superiors. As a result, Deal had been under surveillance for the last 2 months.

One afternoon, the MPs were called to the Deal home on base by a neighbor who reported some sort of domestic disturbance. Officers arrived to find Deal's wife sitting dazed on their porch nursing a swollen black eye and a bloody lip the size of a Wisconsin brat. Deal exercised his calling as a troublemaker a bit too enthusiastically, and his story is a short one. In fact, it ended that same afternoon.

Introduction to a Shooting Reconstruction

Base officials knew Deal's criminal history, and that he usually carried a gun and at least one knife. They knew his business and knew it was time to end Deal's entrepreneurial

* In kindergarten, I liked the showing part of show-and-tell more than the telling part. I was severely disciplined one day when I forgot to bring something to show. My teacher was not happy when I showed her my tongue. She showed me the vice principal's office.

efforts on base. When the MPs and CID arrived, negotiations began. Deal adopted the strategy of ignoring any words spoken in his general or specific direction.

This "negotiation" continued for over an hour. The MPs ordered Deal to take his hands out of the pocket of his hoodie, but he did not respond. One MP stood at the base of the concrete stairs while another circled around a planter on the right side of the steps. His refusal to comply earned him a chance to feel the effects of a Taser. He stiffened, pulled the wires from his clothing and continued his refusal to respond.

When Deal saw the officer to his left, he turned in that direction and appeared to be drawing his hand out of the hoodie. At that point, three shots were fired. Deal was dead.

The inevitable investigations followed. The MP who fired the fatal shots was, officially, cleared of any wrongdoing. The press, however, had a field day. The officer was vilified as the cold killer of an innocent and unarmed civilian, a citizen with great promise and a rosy future who was just out on his front steps enjoying a day in the sun. A team of lawyers jumped into the drama. They developed an account of the shooting that was published in installments with the Sunday supplement. The gist was that the unfortunate citizen Rick Deal was first Tasered, and thereby rendered helpless. As he lay "twitching" on the planter box beside the stairs, the merciless MP shooter walked up the stairs, leaned over the victim and shot him three times. He was not just shot, the media decried. He was murdered!

Of course, the team of lawyers was able to find multiple people who witnessed the entire episode. Each swore that the account serialized in the Sunday paper was exactly correct. They each reported the twitching, the falling, the climbing, and the shooting with strangely similar detail. The Department of Defense, Office of Law Enforcement Policy and Support, contacted me and asked me to develop a scientifically reliable shooting reconstruction to help sort out the mess. They wanted to know what had happened to Deal, according to the physical evidence. The project was clear, but how to produce the reconstruction was not. How is always dictated by the available information—the data—not by a list of procedures and protocols.

After my work was complete, I provided a report to those who hired me. I always begin any report that I write for the government or for the military with an executive summary. I am always afraid that this summary will be all that they read. In this case, the results were easy to abstract from all the work it took to derive them.

> At approximately 1 P.M., Rick Deal was shot three times in the upper torso while standing and rotating to his left on the steps of an on-base apartment complex at 612 East Patterson. This overall conclusion results from an independent scientific shooting event reconstruction based upon the physical evidence available in this case. This shooting reconstruction was made possible given a perforating gunshot wound through the victim's neck area that resulted in a bullet impact on the brick wall of 610 East Patterson. This reconstruction focuses only on Mr. Deal's position when he was struck by each of the three shots fired by PFC Henry Metcalf of the Military Police, Fort Lewis, Washington. The reconstruction describes only about 1 to 2 seconds of the shooting events.

That couple of seconds was enough time to answer the question. Following the summary, my report listed some conclusions, giving each a number. (I have come to believe that this makes it easier for busy people to skim report highlights without wading through all of the details.)

The intermediate results of this shooting reconstruction that support this overall conclusion follow:

1. The three shots fired by PFC Henry Metcalf were fired in rapid succession. Testing conducted at the *Final Analysis Forensics* laboratory determined that the time required to fire three shots (a triple tap) would have been approximately 1 second. The locations of the wounds on Mr. Deal's body confirm this time frame.
2. The three shots were fired while PFC Henry Metcalf was standing with both feet firmly planted on the ground and while aiming directly at Mr. Deal. Therefore, the three shots were fired along the same general trajectory. The location of the wounds on Mr. Deal's body, the location of the bullet chip on the brick wall, and the locations of the spent cartridge cases help confirm this trajectory.
3. Two bullets were recovered by the medical examiner. The lowest bullet, furthest from Mr. Deal's head, struck the lower left chest and was excised from the posterior midline. The next highest bullet entered near his right nipple, exited on his back near his left armpit, and was recovered in Mr. Deal's clothing. The highest bullet striking Mr. Deal, an intermediate range perforating gunshot wound to Mr. Deal's upper right torso, entered near his neck and exited to the left of the posterior midline. This third bullet, recovered on the porch of 610 East Patterson, left a chip on the adjacent street-facing brick wall. These facts correlate Mr. Deal's injuries with his intermediate position(s) at the scene and establish his location on the steps of 612 East Patterson.

It was therefore possible to determine accurately (to a reasonable degree of scientific certainty) Mr. Deal's position for each of the three shots fired by PFC Henry Metcalf.

The results of the reconstruction are summarized in the 3D rendering of Mr. Deal's body position for each of the three shots (as shown in Figure 1.2). The blue and yellow line shows the straight line trajectory (i.e., approximate entrance [blue] and exit [yellow]) followed by each of the three shots with relative body positions for each entrance wound on the decedent's body as his position changes over the course of approximately 1 second. The yellow line shows both the actual exit for

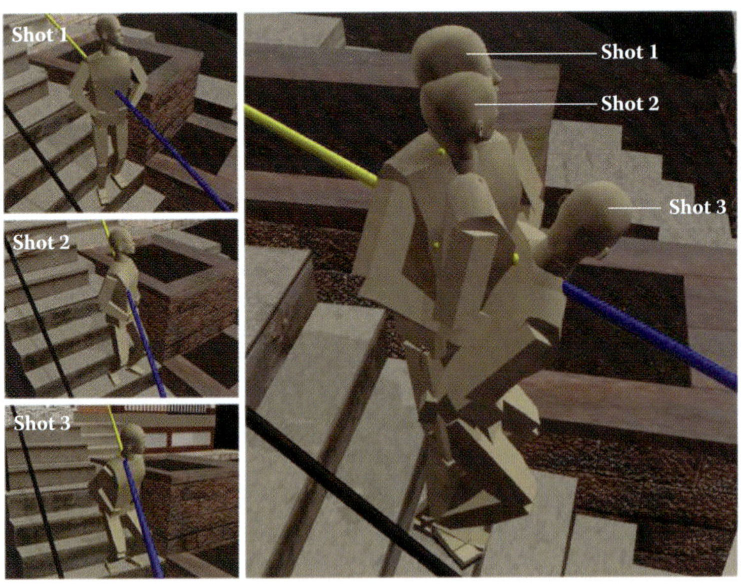

Figure 1.2 Three-dimensional rendering of Deal's body position for each of the three shots.

the perforating shot (last) and the trajectory of the other two shots *if* they would have exited both his body and his clothing and if they would have struck the wall.

Mr. Deal was in a standing position with his body facing in the general direction of PFC Henry Metcalf when the first shot was fired. Mr. Deal's body was rotating to his left and it would have taken him approximately 0.3 seconds to rotate from Position 1 to Position 2. Mr. Deal's body began to drop after the first shot struck him and he continued to drop while shots 2 and 3 struck him.

Mr. Deal's head would have been turned to his left when he fell, as evidenced by the injury on his right temple and right eye with the associated damage to his glasses. This further supports the fact that Mr. Deal was rotating to his left during the course of the 3-shot sequence.

In the illustrations (in Figure 1.2), the highest figure (along the vertical Z axis) represents Mr. Deal's position for the first shot; the second-highest figure represents his position for the second shot; the lowest figure represents his position for the final shot, which perforated his neck and struck the brick wall on the porch of 610 East Patterson.

Ballistics and Scientific Method

The data developed in a reconstruction influence the specific steps in a scientific reconstruction. I followed the steps and let the applications of method develop as we went along. First, physics.

Ballistics is a branch of classical physics covering the flight of projectiles. The famous American forensic scientist Calvin Goddard (circa 1920s), said to be the father of modern ballistics in the forensic setting, applied the term much more broadly to include the study of firearms as tools that may leave distinct marks on the projectiles, or bullets, that they deliver. Ballistics in the forensic setting, as evolved from classical physics, has at least three branches: internal ballistics (the study of a projectile's movement within a firearm), external ballistics (the study of a projectile's movement from muzzle to target), and terminal ballistics (the study of a projectile's movement through a target, usually, but not always, restricted to damage inflicted upon human tissues).

Scientists suitably trained in ballistics also may incorporate research conducted on the ejection patterns of semiautomatic or fully automatic weapons to help determine the possible positions of shooters using this type of armament while leaving ejected cartridge cases, or research the many principles and properties involved in the areas of internal, external, and terminal ballistics.

The Scope of This Reconstruction

Given the bullet impact on the brick wall of the neighboring porch, given the single perforating gunshot wound to the victim resulting in this impact on the brick, and given external photographs of both the anterior and posterior showing bullet wounds to his body, I determined that a scientific shooting reconstruction was possible to produce. I therefore consulted with Eugene Liscio of AI^2 Forensic Animation, Imaging & Illustration to assist with the various methods of measurement applied in this effort. To that end, we used a total station, together with three-dimensional (3D) software, to produce a forensic digital model of the scientifically relevant shooting events (to my scientific specification) and using my independently derived measurements. The Sokkin total station used in this case was validated and calibrated before our use. Before each set of our measurements, we performed our own validations using alternate measurement methods.

This model has the advantage of displaying accurate scene measurements of known facts and allowing the physical display of these scientifically relevant facts, which must together combine to help account for the shooting sequence. This involves the application of physics, ballistics, and crime scene measurement techniques. The resulting work is a limited shooting reconstruction—a form of scene reconstruction as described earlier. The work is limited to the facts that are established through the sciences of physics and ballistics together with the basic assumptions described next.

Summary of Shooting Reconstruction Preliminaries

The goal of this shooting reconstruction was simply to determine the position of the decedent when the three shots struck his body. For this purpose, the best evidence in this case becomes Mr. Deal's body itself as documented by external photographs. To this end, independent measurements were made by using the photographs, as described later. These independently derived measurements were compared with those taken during the autopsy by the medical examiner. It is important to note that both sets of measurements represent the same relative position of the body when Mr. Deal was struck by the three shots to his torso.

The most important physical evidence used in reconstructing both the trajectory of the shots and the position of Mr. Deal when he was struck by the three shots involves the perforating gunshot wound to the neck. This bullet perforated his body, did not strike bone, and continued on its path, striking the brick wall on the porch of 610 East Patterson. Additional significant facts developed during my scientific work include the following elements:

a. The evidence shows that PFC Henry Metcalf fired three shots, a triple tap, in quick succession. (Our testing showed that the three shots took approximately 1 second to fire using the 40 cal Smith & Wesson pistol involved in the shooting. Additional evidence supporting this observation is detailed below.)
b. That PFC Henry Metcalf was within about 8 inches of his stated position (as demonstrated on January 8, 2009) when he fired these three shots. PFC Henry Metcalf had his feet planted firmly on the ground with both hands on the gun. He was not moving at the time he fired the shots.
c. That the bullet perforating Mr. Deal's neck, and striking the brick wall leaving a measurable defect on the brick, was the third and final shot fired.
d. That for practical purposes, the three bullets fired by PFC Henry Metcalf left his pistol's muzzle in approximately the same straight line.
e. That the line described by the third shot perforating the decedent, starting from the muzzle of PFC Henry Metcalf's pistol and ending at the impact point on the brick, represents approximately the same straight line followed by each of the three shots striking Mr. Deal.

This perforating wound and the impact on the brick, together with PFC Henry Metcalf's position, allows us to reconstruct Mr. Deal's position when he was struck by each of the three shots. This reconstruction is developed in what follows later. Note that many of the traditional methods for developing a shooting reconstruction are applied to accomplish this goal. Many methods are dictated by recognizing the helpful circumstances that remain a part of the shooting event.

Evidence Examination and Analysis

1. The three shots in a 1-second time frame—Three facts support the 1-second duration of the three shots striking Mr. Deal. The first involves the fact that we know the position for the shot that perforates Mr. Deal's neck and strikes the brick on the neighboring porch. This perforating shot is the highest on his body. After this shot perforates his body, he already is leaning forward, turning to his left, and falling downward. The two points through his body and the impact on the brick combine to describe the trajectory of this shot. This fixes his body position in relation to the shots fired at the scene. We must therefore reason backward from this shot to establish what must have been the two prior shots striking his body.

   In order for the other two shots to strike his body while it occupies this space, the next highest shot has to strike him before the perforating neck shot: that is, before he completes his fall due to gravity. This shot strikes him in his right nipple, and must be a prior shot since he is falling downward. The next shot, the lowest shot, has to strike him before both these other higher shots since he is falling downward. The first shot in time, then, is the lowest shot on his body. This shot destroyed the apex of his heart, and he fell downward due to gravity. He continued his rotational movement until this force was overcome by or conjoined with gravitational, physical, or muscular influences. Each of these three shots strike Mr. Deal's torso moving from left-center to upper right; that is, moving right and upward. The space described by the distance between the lowest and highest shot is the distance he fell over an approximately 1-second time frame. Any more time, and at least the third shot would not have struck Mr. Deal. It would have missed his body and have simply struck the brick.

   The second fact involves the actual time it takes to fire the Smith & Wesson 40 cal pistol, which PFC Henry Metcalf fired three times at Mr. Deal. The following table summarizes our testing using an Xnote Stopwatch:

Trial Number	Time
1	0.96 seconds
2	1.04 seconds
3	0.93 seconds
4	1.10 seconds
5	0.78 seconds
Average Time	0.96 seconds

   The third fact involves the actual training that MP soldiers receive: they are trained to fire multiple rounds in rapid succession and then to reassess the threat level.* This 1-second duration, therefore, is dictated by the physical evidence, supported by testing, and supported by the training that MP soldiers receive.

2. Collected Evidence Items: Including Deal's Clothing, Pistol, Bullets, and Cartridge Cases—In reports of this type I describe the details by providing my laboratory notes, including relevant analytical results. I also attach photo documentations of the evidence as received, opened, and processed in the laboratory along with photomicrographs of relevant items. That way, a reader can refer to the information, or not, as the reader's need dictates. Of course, I always hope that they read it all.
3. Summary of Examined Evidence Items and Their Roles in the Reconstruction—Clothing examined for any shooting reconstruction presents specific challenges for determining entrances and exits, and thereby determining trajectories. These challenges involve measurement issues given the fact that clothing can move over a human body: for example, the entrance defect on an item of clothing may not accurately reflect the entrance location on the human body wearing the clothing during the shooting since the clothing can move over the body. Since items of clothing can "bunch up," wrinkle, and move, the anatomical location of any entrance wound to the body may not correspond with the anatomical location of the corresponding entrance on an item of clothing when it is repositioned for examination. This challenge is exacerbated when several layers of clothing, as in this case, are perforated by a bullet or bullets.

   The trajectories through the clothing items perforated (by what we have determined to be the last of three shots) supply the best evidence for further analysis. The clothing was installed, in order, on a mannequin:

   Layer 1—The orange T-shirt
   Layer 2—The Champion sweatshirt
   Layer 3—The blue jean jacket
   Layer 4—The black "hoodie"

   Photogrammetry targets were installed on the "hoodie" and a 3D virtual model of the sweatshirt, as worn by the decedent, was constructed (see Figure 1.3). A trajectory was then established with a line through the entrance to the exit on the model. This approximates the trajectory through the model of what I have called the third shot.†
4. Use of the autopsy photos in the shooting reconstruction: The external photographs of Deal's body showing the three entrance and the two exit wounds were used as described earlier. (The results of this photoanalysis in developing independent measurements are summarized in a chart included later.)

Serendipity

In forensic science, we must play the hand we are dealt. I had a bit of luck here: The bullet caused a perforating gunshot wound, missed bone, pursued a relatively straight trajectory,

* All MP soldiers who undergo mandatory firearms training are taught to fire multiple rounds, typically two to four rounds in rapid succession, when facing situations calling for the use of deadly force.

† When shot, Deal dropped down, therefore, the first shot is the lowest shot, the second shot is the next lowest, and the third and final shot is the highest. The fact that Deal dropped from the stairs down to the sidewalk after Metcalf fired three shots in approximately 1 second helps establish the order of the shots, in this case, from the lowest shot to the highest shot on his body.

Figure 1.3 Upper-body clothing.

and struck a brick wall. And that strike left a clear mark. That is one way that reconstructions are data driven. It also is one way that we are forced to think on the fly, making adjustments among our scientific tools along the way, but always with an eye firmly fixed on sound methods. No single set of approaches applies to every case. Only sound scientific method applies to every case. That means, sometimes, inventing a way to travel from A to B, applying a tool from another area in a novel way, or borrowing expertise and learning a new skill. All of that is a part of critical intelligence. It is part of what I mean by being scientific.

The highest shot striking the decedent (i.e., the perforating intermediate range gunshot wound to the neck*) perforated the hoodie before the black Champion sweatshirt and

* Entrance (quoted from Dr. Grunewald's report; see later section for additional measurements): "On the lateral lower aspect of the neck (right side) centered 9½ inches from the top of the head and 3 inches right of the anterior midline, there is a round ½-inch gunshot wound of entrance with abraded margins. … There is no soot, stippling or searing. The bullet creates a hemorrhagic wound path which goes through the soft tissue of the neck." Exit (quoted from Dr. Grunewald's report; see later section for additional measurements): "On the upper back (toward the left side), centered 10½ inches from the top of the head and 1¾ inches left of the posterior midline there is a somewhat round ½-inch gunshot wound of exit … with abraded margins. The abrasion is more prominent from 7 o'clock to 10 o'clock positions. No bullet is recovered."

the orange T-shirt. It also exited the back of the cutoff jean jacket worn under the hoodie. The bullet missed entering the blue jean jacket given its shape and its position as worn on the body.

Using the "hoodie" entrance and exit holes to investigate Mr. Deal's position when struck by this bullet, the trajectory through the model (entrance to exit) was then placed on the straight line from Metcalf's muzzle position to the impact on the brick. As discussed later, this closely approximates the positions established by using the external autopsy photographs of the entrance and exit wounds left by this third shot. Since the best evidence of trajectory and position is the body itself, no further trajectory work was done with the clothing items.

The methods and the results of the photogrammetry work with the hoodie clothing and the perforating shot to the neck follow.

Hoodie Analysis: Photogrammetry Method

Photogrammetry is the science of taking reliable measurements from photographs. ("Photogram" is a photograph and "metry" refers to the science of measuring.) Since a photograph takes objects from the 3D world and projects them on a flat 2D image plane, we lose the depth information. Photogrammetry, however, can be viewed as the process in the reverse. By knowing the exact position and orientation of the camera in 3D space, along with the focal length of the lens, we can regain some 3D information.

The focal length of a camera is the distance between the imaging plane (which is the digital sensor on a digital camera, or the negative in a film camera) and the point where all the light rays intersect (i.e., the optical center, the point at which all the light rays entering the camera cross). The focal length of a camera can also be considered, in simple terms, as the amount of zoom/magnification that was used to take a particular photograph. The focal length changes when someone zooms in and out to capture an image using a zoom or a macro lens.

There are several techniques used to develop a photogrammetry project. The choice of technique often depends upon the type of information desired. Sometimes an existing photograph permits the extraction of an important piece of 3D measurement information. In other photographs, one can use a camera as an accurate measuring instrument by first taking various photos of an object and then processing these photographs to obtain the desired coordinates or measurements.

In this particular case, it was important to accurately document and to compare the entrances and exits on the black hooded sweatshirt, or hoodie, as worn by Deal over the rest of his clothing when Metcalf shot him. As is often the case, practitioners will take measurements of a piece of clothing at the locations of the bullet entrances and exits. This is normally done by placing the clothing on a flat table or bench, then documenting the linear measurements to the edges or other reference features in the clothing. (My lab notes follow.) However, for 3D photogrammetric purposes, taking these measurements with the sweatshirt over Deal's clothing on a mannequin provides more accurate 3D spatial information.

Therefore, the process began by mounting Deal's T-shirt, Champion sweatshirt, jean vest, and the hoodie sweatshirt on a representative mannequin. This permitted the sweatshirt to occupy a more natural position. Photogrammetry targets were then placed around

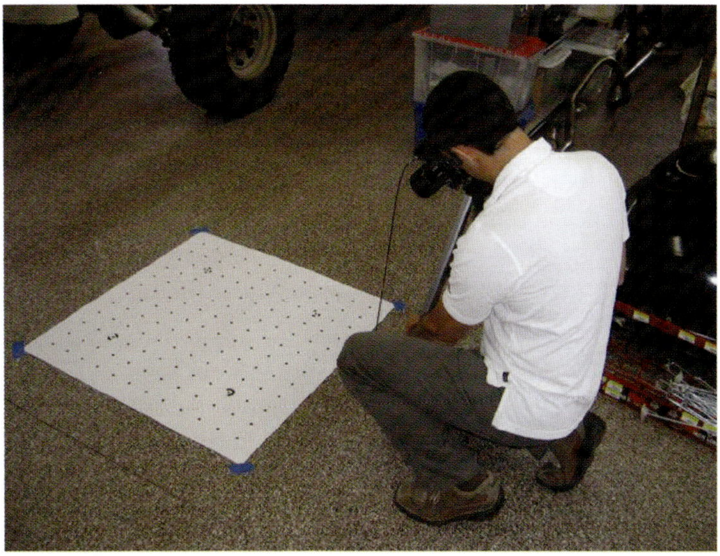

Figure 1.4 Eugene Liscio, president of AI[2] calibrates a camera in the Final Analysis Forensics laboratory.

the sweatshirt. This included targets on each of the bullet entrances and the one exit near the back of the neck. Each target becomes a calculated point in 3D space and measurements can be taken between any of the solved targets.

The camera is calibrated to calculate the intrinsic properties of the lens and camera assembly (Figure 1.4). Camera calibration allows for greater accuracy by accounting for small lens distortions. The calibration process often uses a grid of points, a checkered template, or known coordinates to be used as a "control."

A number of photos were then taken around the sweatshirt to permit various, overlapping photos. Each of these photos was then used as part of the photogrammetry solution.

Targets placed on the garment are photographed and measured to develop the 3D data (Figure 1.5). The results of the analysis produce a set of points in 3D space. Each point reflects the center location of a target placement. Measurements were then taken of the entrances and exits of the bullets on the hoodie (Figure 1.6).

This was done strictly for comparative analysis. This data was not used in the final 3D reconstruction of the hoodie, since it is common for clothing to move relative to the body and it is also common, therefore, for the bullet entrance and exit locations to be significantly different. All of this depends upon the actual orientation of the body. Yet that is the object of reconstructive study. However, it is important to note the significant alignment of the bullet entrances and the exit on the black hooded sweatshirt with the wounds on Deal's body in this case.

Software

The software used for this photogrammetry work was PhotoModeler Scanner made by EOS Systems Inc. PhotoModeler is a well-known 3D photogrammetry software package allowing 3D measurements from photos. PhotoModeler also allows for "inverse camera" projects that solve for photos where the camera focal length and orientation are unknown.

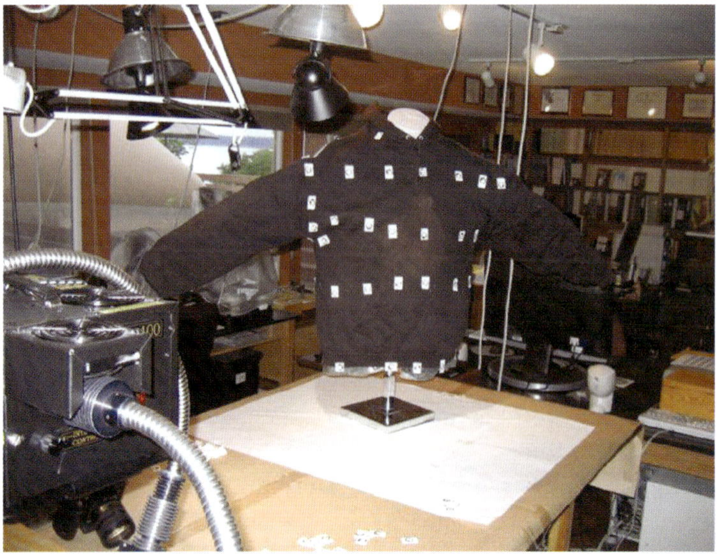

Figure 1.5 Targets are placed on the garment, then are photographed and measured to develop the 3D data.

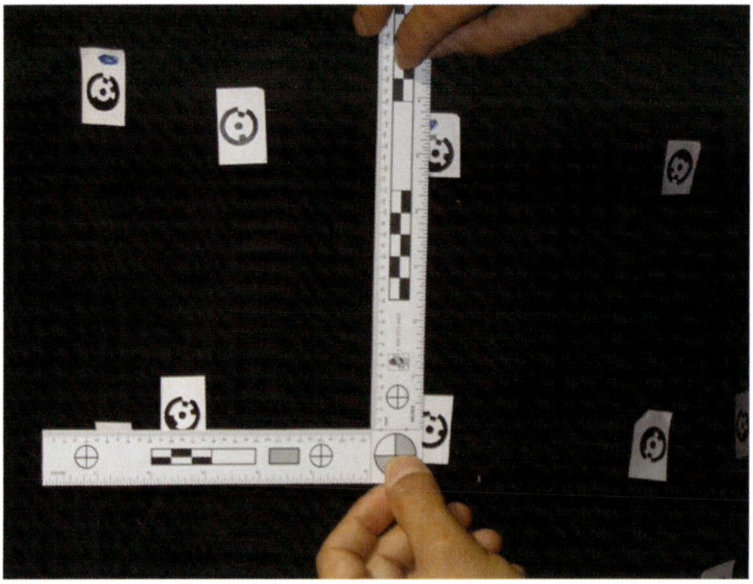

Figure 1.6 Measurements being taken of the entrances and exits of the bullets on the hoodie.

Accuracy and Error

The accuracy of a measured point in a photogrammetry project is defined within PhotoModeler as a maximum "residual" value given in pixels. A pixel (short for "picture element") is the smallest unit in a computer image or display. The residual value is defined as the difference between where a point is located and where the software calculates the point to be.

The maximum residual value for the measurements taken on the black hooded sweatshirt is 1.41 pixels. According to the PhotoModeler Help Files, "All projects should have the largest point marking residual of under (sic) 10 pixels. For projects with known and calibrated cameras the largest residual should be less than 3 pixels and for projects done with all sub-pixel targets, the largest residual should be less than 1 pixel and smaller if possible."

Therefore, a maximum residual error of 1.41 pixels is an acceptable variance for this type of photogrammetry project. A conservative estimate of the error for measurements based on the strength of the photogrammetric solution is approximately ±1 mm (±0.040 inches) and this was verified by checking known distances between some of the photogrammetry targets placed on the sweatshirt.

The fact that a mannequin was used as a representative model of how the sweatshirt might have appeared on Deal would in itself introduce some error, since there could be some differences between Deal's relative shape, size, position, and orientation, and similar features of the mannequin. For example, Deal's chest and abdomen size could have been larger or smaller in some areas, which would deform the sweatshirt in slightly different ways. Also, testimony places Mr. Deal's hands in the front pocket of the black hooded sweatshirt. This could have caused some tension on the front of the sweatshirt, thereby lowering the front portion of the sweatshirt and consequently placing the entrance defects slightly higher on the sweatshirt than if his hands had been by his side.

Results

The photos documenting the development of the point measurements appear under the hoodie in the clothing section described later. The result is a 3D movable image that can be rotated in any possible direction with the software in order to see the object from indefinitely many orientations. The still images capture only a few of these orientations. The photogrammetric images showing these still images are in Figures 1.7, 1.8, and 1.9.

Reconstruction Methods

The 3D Total Station Scene Measurement Method

One increasingly common method for recording and documenting measurements at a crime scene involves using what is called a total station. A total station is a commonly used laser-based measuring device. A Sokkia SRX3 total station was used to take one set of measurements in this case.

This device uses a "modulated laser, phase comparison method" with red laser diode for calculating distances. The phase comparison method involves using a transmitted laser beam composed of the basic (laser) carrier signal with a modulation signal superimposed

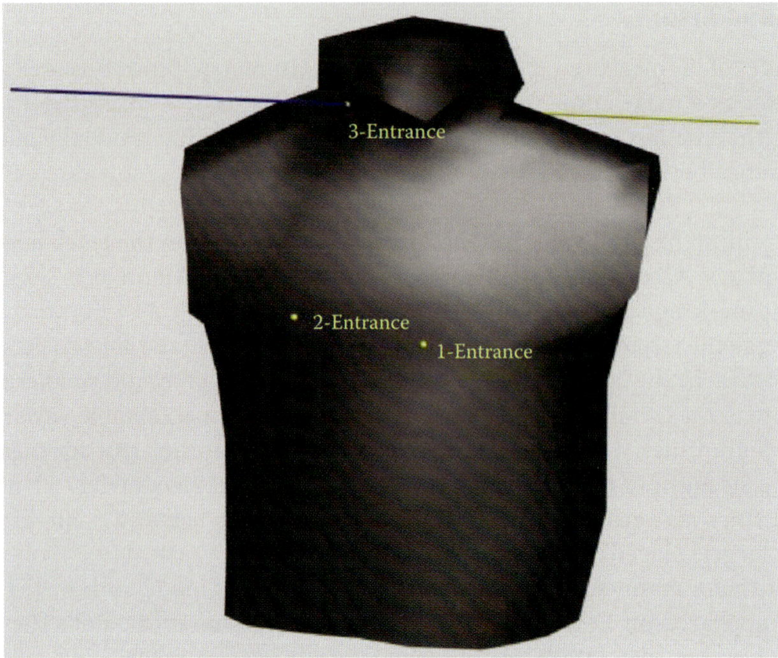

Figure 1.7 The entrances on the front of the hoodie. The blue line represents the entrance trajectory and the yellow line represents the exit trajectory as the clothing is viewed in anatomical position. Note that this is not representative of Deal's position at the scene.

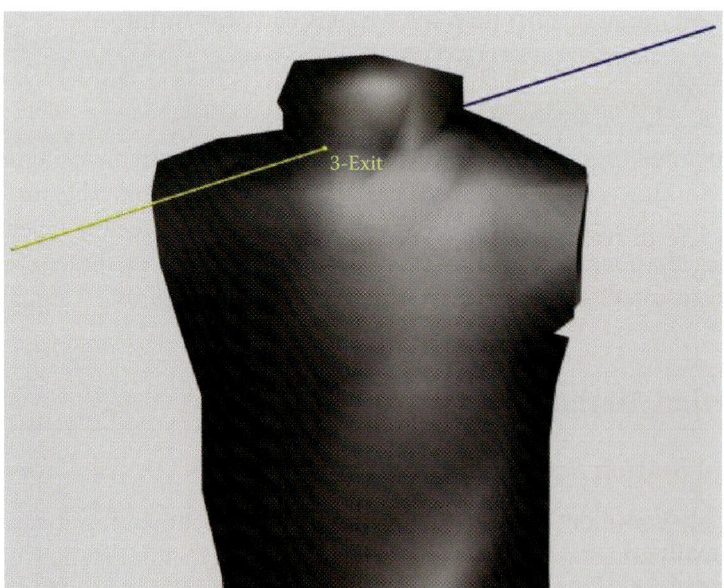

Figure 1.8 Back view: This shows the hoodie from the back with the yellow line again representing the exit trajectory and the blue line representing the entrance trajectory. This hoodie is viewed in anatomical position and is not meant to represent Deal's position when the shot struck only the trajectory through the hoodie.

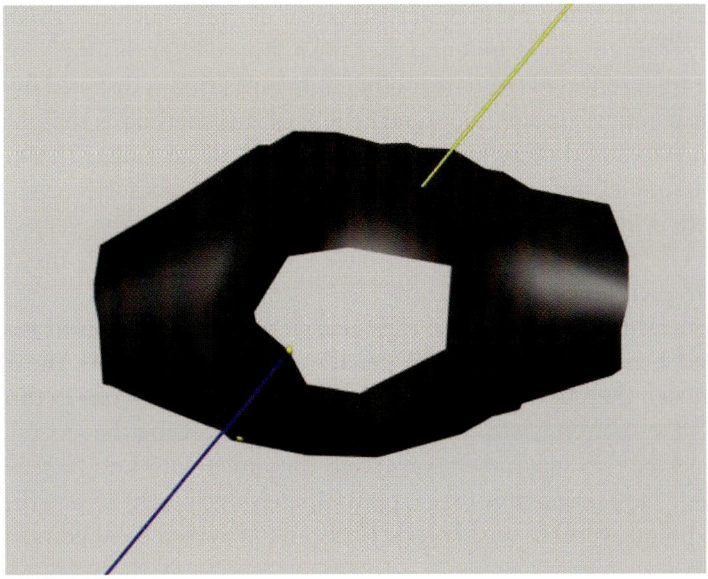

Figure 1.9 Top view: The hoodie from the top. The blue line represents the entrance trajectory on the front and the yellow line represents the exit trajectory on the back.

for the purpose of measurement (a specific, known frequency wave). This modulation signal, or measuring wave, is derived from and held at a constant value using a stable frequency generated by this equipment.

The beam sent out by the unit is continuous: It then strikes the object being measured and a small part of the signal is reflected back along the identical path to the instrument, where it is detected using a receiver. The weakened reflected signal is then processed such that the measuring wave is compared with the signal of the original transmitted beam (or reference signal).

The resulting computation produces a measurement by calculating the difference in phase (or phase angle) between the two signals. These phase differences are a direct derivative of the distance the laser light has traveled.

Although the distance measurement process is completed within 2 seconds, each measurement is actually an averaged value of approximately 3000 individual measurements. The standard deviation for these individual measurements is calculated and must fall within an acceptable value. If the standard deviation is not acceptable, the instrument will not determine the distance measured and will indicate a distance measurement error. (Additionally, the distance meter is equipped with various other redundancy checking processes to assure the accuracy of the distance measured.)

Two Methods of Measurement with a Total Station

There are normally two methods of measurement with modern total stations of this nature. One is with the use of a "prism and pole" for which the operator walks to various positions in the scene and places the pole on a specific point for measurement. The height of the pole (i.e., from the point being measured to the prism resting on the pole) is known and is used to correct for the prism's offset to the point being measured. The benefit of the prism-and-pole method is that measurements can be made over very long distances with high levels of accuracy and minimal error.

The second method of measurement is known as "reflectorless." This method is carried out directly from the main unit and uses the onboard laser to simply target any visible point. The laser light reflected from the point being measured is detected by the unit, again using the phase comparison method. One benefit of this method is that measurements of both vertical surfaces and objects not easily reached can be made. These measurements normally have a smaller range of error for perpendicular surfaces. However, there are some surfaces that are difficult to measure (e.g., very dark surfaces and very oblique angles). Should the surface become too oblique or have too many errors, the instrument will not return a measurement.

All of the measured points can be exported in various common formats accepted by standard computer-aided drafting (CAD) and 3D animation programs. The imported points of measurement can then be used as a basis for reconstructing a scene in three dimensions.

The points of measurement for this case were captured using the second method over a 2-day period and the two separate surveys were combined with known reference points to produce the scene measurements of the (approximately one block) area of the shooting. The same reference points were surveyed on each day and provided fixed points for alignment.

Reference Measurements and Accuracy

The SRX3 total station has a manufacturer's stated accuracy of approximately 3 mm up to 200 m (0.118 inches up to 656 feet). Normally there are other factors that contribute to errors in measurement, such as the surface properties of the object being measured and how steady the prism is being held while measurements are being taken. Therefore, a good method of checking errors is by taking reference measurements with a steel measuring tape.

Reference measurements made at the scene across the top of the garage at 612 East Patterson, show that the errors in measurement are approximately ±0.25 inch (¼ inch), however, the estimated accuracy of any given point is approximately ±1 inch. This allows for any errors that may generate from holding the prism slightly off center and for some of the larger distances measured. It should be noted that tape measures also generate such error. However, it is best to think of these "errors" as establishing a *range of acceptable variance* since they are not actual errors in the sense of scientific mistakes, they are simply variations that fall within the scope of the relevant scientific technology applied, in this case, to measurement. That feature is common to all types of measurement devices, including tape measures and rulers, as well as total stations.

Creation of the Virtual Environment Model Using Total Station Data

The data collected from the total station survey can be exported as a digital file that can be interpreted by other software packages. The digital models were created based on the measurements that were imported into 3D Studio Max (a modeling and animation software package often used in forensic reconstructions). These measured points were used as the basis of the reconstruction in 3DS Max and objects in the virtual environment (e.g., sidewalks, steps, and parts of structures) were modeled to scale and based on the available data.

Virtual Anatomical 3D Model of Rick Deal

In order to depict the bullet trajectories through the body accurately, a virtual 3D anatomical model was constructed based on the known dimensions of Deal's body. Since Deal's body was not available for direct inspection and measurement, some dimensions were obtained through the available autopsy photographs, while others were obtained based on his known clothing and shoe sizes.

The digital anatomical model is used to represent Deal's body and can be articulated in ways that reflect how the human body can and does actually move.

The known and measured dimensions of Deal follow:

- Known Measurements
  - Overall height—67 inches (170.18 cm)
  - Deal waist size—32/30 (from pants)
  - Deal was wearing shoes with 1¼-inch heels and ¾-inch soles (which add approximately 1.5 inches overall to his height)
- Deal Body Measurements from Photographs
  - Vertex of head to chin—approximately 30 cm (11.81 inches)
  - Vertex of head to nipple—approximately 52 cm (20.47 inches)
  - Upper arm length—approximately 29 cm (11.42 inches)
  - Forearm length (to wrist)—approximately 24 cm (9.45 inches)
  - Base of foot to knee—approximately 43 cm (16.98 inches)
  - Chest width—approximately 35 cm (13.78 inches)
- Measurements for Autopsy Table
  - Horizontal (spacing across table width for a total of 5 holes)—4⅞ inches (12.3 cm)
  - Vertical (spacing along length of table for a total of 10 holes)—5⅝ inches (14.3 cm)
  - The holes in the autopsy table are 3/8 inch in diameter.
- Estimated Torso Thickness
  - Deal's chest depth, or the thickness of his torso, is estimated to be approximately 7¾ inches to 8 inches deep, anterior to posterior. This estimate is based upon photos, clothing sizes, and other related measurements, such as his waist size. These measurements point to an average torso depth of about 8 inches based upon anthropological data.*
- Accuracy and Error
  - Since Deal's body was not available for direct measurement and inspection, it is impossible to say that the anatomical model perfectly matches Deal in every way. However, the available data and measurements taken from the medical examiner's report provide sufficient information to allow for a reconstruction of reasonable accuracy and trajectory error within approximately ±4 inches .

Method: Establish a Trajectory Envelope

Constructing what I call a trajectory envelope is a method developed here for eliminating impossible or improbable bullet trajectories based on key pieces of physical evidence, such as the location of the shooter, objects obstructing the path of a bullet track, and the impact location (mark) of an actual bullet striking a target: either a primary target or a

* American Society for Testing and Materials (renamed ASTM International); ASTM D6240—98(2006), "Standard Tables of Body Measurements for Men Sizes Thirty-Four to Sixty (34 to 60) Regular," ASTM International, West Conshohocken, PA, 2006, www.astm.org. See also NIST government standards with references to ASTM committee work.

Figure 1.10 The lateral trajectory of the bullet was limited to an area between the wall on the left at the entrance to the porch and the pillar on the right, also at the entrance to the porch.

secondary target.* Constructing a trajectory envelope provides some insight into the range of possible positions a shooter and a victim could occupy by documenting relevant boundary conditions for the shots. This step helps limit erroneous and unproductive avenues of research and helps rule out physically impossible shooting scenarios. It also helps to demonstrate visually the meaning of "acceptable variance" or "error rate" as applied to trajectory development.

In this case, a bullet chip defect, as a secondary target, was observed on the front porch brick wall of 610 East Patterson. Based on the available evidence locating all involved persons at the time, the trajectory envelope was created as follows:

1. There was no evidence that the bullet interfered or struck the pillars or floor at the entrance of the porch to 610 East Patterson. Therefore, the lateral trajectory of the bullet was limited to an area between the wall on the left at the entrance to the porch and the pillar on the right, also at the entrance to the porch. The envelope occupies the area represented in the set of scale images, as seen from various perspectives, reproduced in Figure 1.10.
2. No evidence suggests that the bullet interfered or struck the porch floor prior to its impact with the brick wall. Therefore, the lower portion of the trajectory envelope was adjusted to accommodate the lowest possible path of the bullet into the porch avoiding the porch floor (see Figure 1.11). This also defines Metcalf's lowest possible muzzle position when the shots were fired. The trajectory envelope also allows

* A primary target is simply the first object or person struck by a bullet; a secondary target is simply the second object struck by that same bullet (a bullet that first hits something else and then hits that target, etc.). The trajectory envelope is not the "shooting cone" commonly associated with a single shot; rather, it represents the boundary conditions for the entire shooting event. These boundary conditions contain the variance allowable by the data and as such depict the error rate, or acceptable variance, visually.

Figure 1.11 As seen, the lower portion of the trajectory envelope was adjusted to accommodate the lowest possible path of the bullet into the porch avoiding the porch floor.

for any slight deviations in flight path that may have influenced the bullet's flight path after first perforating the soft tissues of Deal's neck and upper back area. The exit wound photographed during the autopsy appears smooth and symmetrical, further supporting a straight trajectory through the body with little evidence of yaw, pitch, wobble, or deflection.

3. The upper limit of the trajectory could be no higher than Metcalf's highest reach from his stated standing position close to the west of the stairs, on the sidewalk at 612 East Patterson. Metcalf's position was demonstrated, documented, and measured on January 8, 2010.

This tapered rectangular envelope indicates the area of physical space within which the shots could have been fired (i.e., given the possible positions of Metcalf relative to the chip in the wall, both of which are boundary conditions for the trajectory values). This also shows the area Deal would have had to occupy when he suffered the perforating gunshot wound to the neck. Given the 1-second time frame for the three shots, the trajectory envelope also defines the space Deal occupied when he was struck by the two prior shots.

The bullet chip in the wall, as a secondary target, provides the central piece of physical evidence that ties the path of the bullet to the muzzle of Metcalf's gun, dictates Deal's position when he received this perforating shot to his neck area, and provides boundary conditions for each of the positions Deal must have occupied for all three shots during a 1-second interval. The trajectory envelope is widest at the point from which the gun was fired, allowing for all the possible locations of the muzzle that still articulate to the point established by the chip in the wall. This envelope eventually converges to a single point at the bullet chip. Therefore, the possible positions of Deal are confined to the area within this trajectory envelope (see Figure 1.12).

This view of the trajectory envelope shows the possible areas within which both Metcalf's pistol muzzle and Deal must fit in order for the perforating gunshot wound

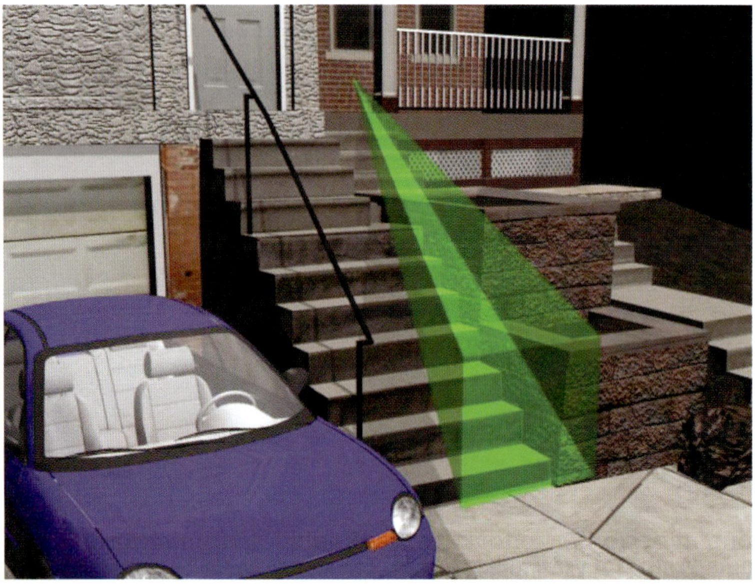

Figure 1.12 This view of the trajectory envelope shows the possible areas within which both Metcalf's pistol muzzle and Deal must fit.

through Deal's neck to strike the brick, leaving the chip. The specific measurements are given in what follows next. For practical purposes, the bullet perforating the soft tissues of Deal's neck approximates a straight line over the distance at issue. Gravity plays no practical physical role in these trajectory considerations; the velocity is too high and the distance is too short.

Trajectory and Muzzle Movement during Triple-Tap Firing

Testing was performed to investigate any influence of "muzzle climb" on the trajectory of each of the three bullets fired in rapid sequence. For purposes of this testing, the behavior of the Smith & Wesson 40 cal pistol when firing various 40 cal ammunition, including the Federal 165-grain 40 cal rounds involved in this shooting, was investigated. Specifically, this question of muzzle climb was both documented and measured. The testing was documented with a Casio High Speed camera shooting up to 1200 frames per second. Measurements of relative muzzle positions were taken and error rates were calculated. The results show that three rapidly fired shots, triple tapped in about 1 second, can pursue very similar relatively straight line trajectories independent of muzzle climb or other mechanical issues resulting from simply firing the pistol. This work is documented in what follows. The measurements of the envelope at the muzzle appear in Figure 1.13.

The three spent cartridge cases ejected from the Smith & Wesson 40 cal pistol may also help locate the muzzle's relative location. Spent casings eject to the right and slightly backward by design when this model of gun is fired. The eventual resting position of ejected cartridge cases can be influenced by many variables, such as wind, the angle and nature of the surface they strike, and the movement of either the shooter or the surface. When shots are fired in rapid succession, the cartridge cases may also strike each other in flight before they strike or come to rest on any surface. In addition, they could be kicked, stepped on, or otherwise moved after their deposition. In this case, the cartridge cases were recovered on

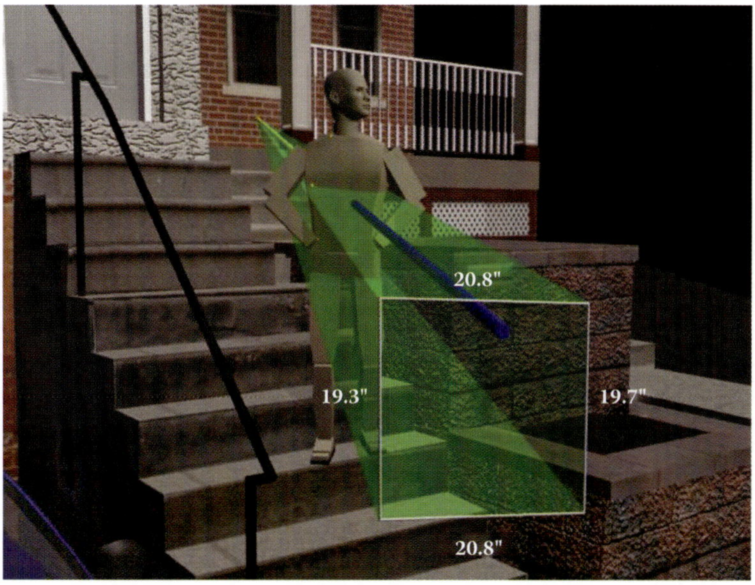

Figure 1.13

the asphalt roadway of East Patterson. The gun's position when the shots were fired was on the sidewalk level above the roadway.

The locations of the cartridge cases at the scene were measured from the approximate muzzle position, using the trajectory heading as 0 degrees for convenience. The first ejected cartridge case landed 13.5 feet from the pistol at 117.3 degrees. The second ejected cartridge case landed 16.6 feet from the pistol at 125.5 degrees. The third ejected cartridge case landed 14.9 feet from the pistol at 157.7 degrees. The arithmetical mean for these three shots is 15 feet at an average (mean) 133 degrees. The angles range fall within about a 40-degree range to the right and behind the pistol. Their locations at the scene are pictured as shown in Figure 1.14.

These or any cartridge case locations can seldom be experimentally reproduced or duplicated even when exact conditions tests are conducted.* Despite this limitation, testing can inform us about the significance of their relative positions. The test distances for three shots fired in less than 1 second are summarized in the following table:

Trial Number	Time	Shot 1 (ft @ degrees)	Shot 2 (ft @ degrees)	Shot 3 (ft @ degrees)	Range (degrees)
1	0.96 seconds	16 @ 100	15 @ 140	12 @ 155	55
2	1.04 seconds	23 @ 110	15 @ 145	8 @ 120	35
3	0.93 seconds	18 @ 135	13 @ 138	20 @ 122	16
4	1.10 seconds	18 @ 120	22 @ 125	19 @ 158	38
5	0.78 seconds	11 @ 110	14 @ 130	9 @ 135	25
Average (mean)	0.96 seconds	17 @ 115	16 @ 136	14 @ 138	34

* "Exact conditions" means the same location, same ammunition, same pistol, same direction, same targets, same lighting, same temperatures, same surfaces, and so on. These tests are not exact conditions tests. They are laboratory test shots fired with a muzzle at 67.2 inches (same height) from a concrete floor in a shooting range.

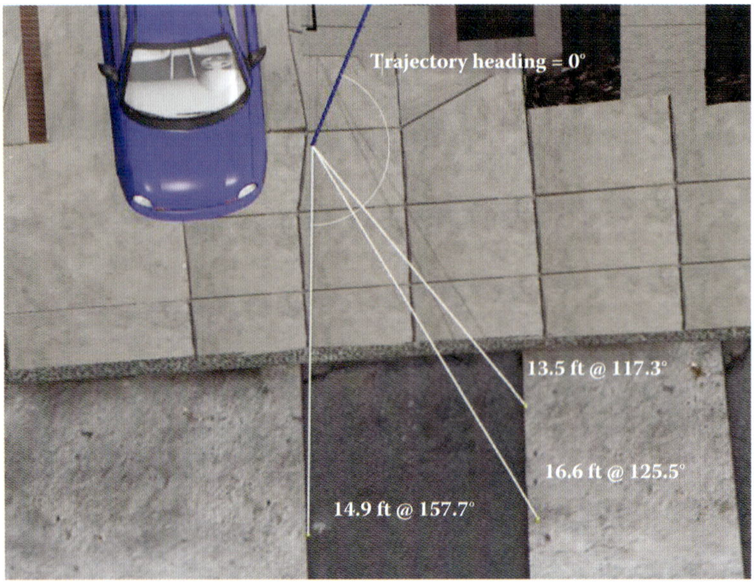

Figure 1.14 The arithmetical mean locations for these three shots.

The ejection pattern testing reveals simply that the location of the cartridge cases at the scene of the shooting falls within the boundary conditions typical for cartridge cases ejected from that model 40 cal Smith & Wesson pistol. The cases do not appear to have been kicked, moved, or otherwise disturbed at the scene before their locations were documented. Their positions, therefore, do not indicate that the position of Metcalf's pistol differed from that defined by the trajectory envelope's boundary conditions.

The measurements of the envelope at the victim's body are shown in Figure 1.15. This figure shows the area that Deal's body must occupy in order to be struck by each of the

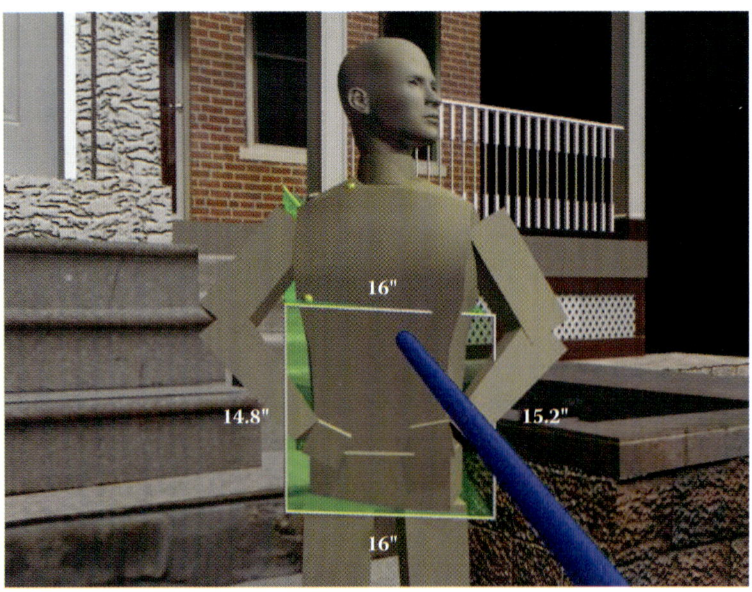

Figure 1.15 Measurements of the envelope at the victim's body.

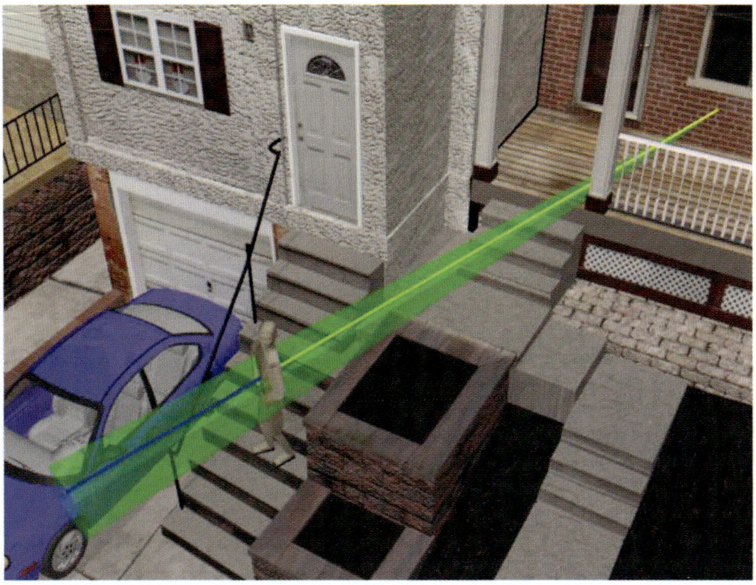

Figure 1.16 This rendering shows the trajectory envelope, seen in green, from an oblique top view with Deal's position represented in the space required for each of the three bullets to strike him.

three shots fired from Metcalf's pistol. As his body continues its rotation to the left and drops downward as a result of the first shot, the entrance points, seen as yellow dots, move both to occupy this space and to meet the blue trajectory line.

Figure 1.16 shows the trajectory envelope, seen in green, from an oblique top view with Deal's position represented in the space required for each of the three bullets to strike him, inflicting the injuries documented in the external autopsy photographs. The perforating neck shot leaving the chip in the wall provides the data used to orient Deal's position and tie the shooting events to the physical scene.

For diagrammatic clarity, the blue line represents a bullet's entrance trajectory while the yellow line represents the straight-line path which each of the bullets would have traversed if each had exited Deal's body without significant deflection from striking the primary target.

Figure 1.17 shows the trajectory envelope again in green as seen from Metcalf's approximate perspective (from the front). Figure 1.18 shows the trajectory envelope, again in green, from the backside facing Metcalf with the brick wall, secondary target, off the picture to the right.

Distances

1. Metcalf's position was demonstrated and recorded during a scene visit. However, during the time of the incident, there was a blue Dodge Neon in the driveway at 612 East Patterson. Based on his demonstrated location, a lateral adjustment to Metcalf's position was necessary since his lower body would have interfered with the front driver's side bumper/fender of the blue Dodge Neon by approximately 7 inches. The adjustment was well within the trajectory envelope.

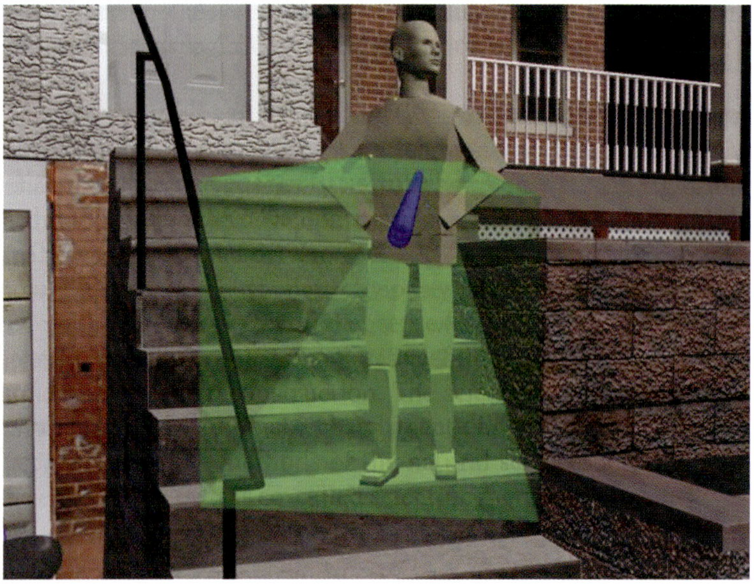

Figure 1.17 The trajectory envelope in green as seen from Metcalf's approximate perspective (from the front).

Figure 1.18 The trajectory envelope in green, from the backside facing Metcalf with the brick wall, secondary target, off the picture to the right.

2. Muzzle to each target point—The distance from the muzzle to Deal for each shot:
   Shot 1: 87 inches (7.25 feet)
   Shot 2: 76 inches (6.3 feet)
   Shot 3: 69 inches (5.75 feet)
3. Metcalf's distance to Deal—Shot 1 is 87 inches from the muzzle of the gun. Considering the gun's length with the length of Metcalf's outreached arms we add about an additional 30 inches. Therefore, the body-to-body distance becomes approximately 117 inches (9.75 feet).
4. Deal's distance from the planter (approximate midline to wall)—19.8 inches (1.65 feet)
5. Trajectory distance from exit of muzzle to chip in wall—333 inches (27.75 feet)

Angles

1. Angles from muzzle to chip at 0, 0, 0 using the brick wall as the X axis baseline:
   Top view: 26.52 degrees (116.52 degrees from the top view with the wall as zero); Figure 1.19
   Side view: 8.64 degrees; Figure 1.20
2. Deal's rotation angles as he moves toward the left.
   Shot 1: Deal standing approximately 41.93 degrees to the right (Figure 1.21). (The zero-degree position is established if Mr. Deal were standing facing forward on the step (toward the street) with his shoulders parallel with the brick porch wall at 610 East Patterson [with the bullet chip]). The left-facing green line/arrow is drawn parallel to the brick porch wall. These measurements use the bullet chip on the porch wall as the origin (0, 0, 0) with the X axis parallel to the porch wall, and the Y axis perpendicular to the porch wall. (The Z axis represents the vertical dimension.)
   Shot 2: Deal is now standing approximately –30.15 degrees toward the left (Figure 1.22).
   Shot 3: Deal standing approximately –10.71 degrees to the left (Figure 1.23).

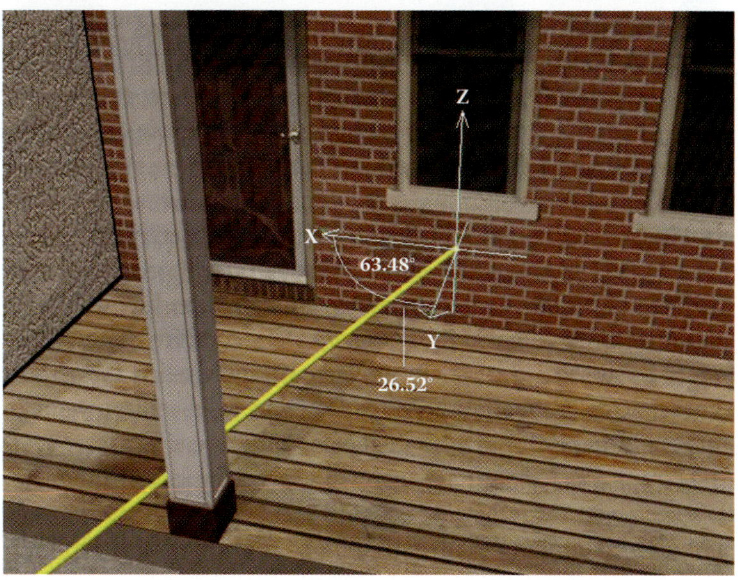

Figure 1.19 A top view of the measured impact angle to the brick wall.

Figure 1.20 Looking at the chip in the brick from the east looking toward the west; the lower line is the center line of the trajectory representing the impact angle. It is 8.64 degrees below the horizontal, which is represented by the top line that is drawn parallel with the horizon (level line).

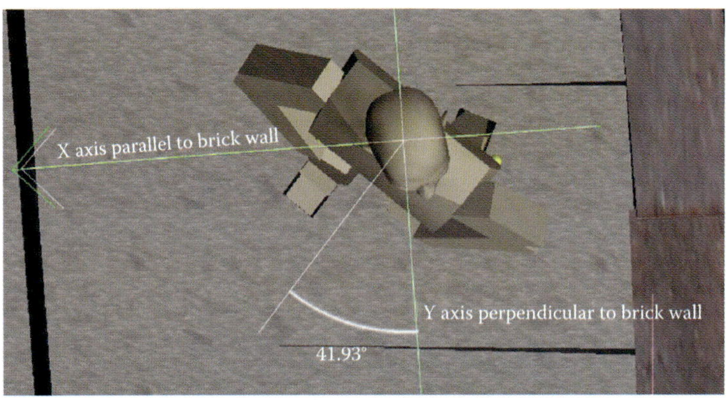

Figure 1.21 Shot 1: The zero-degree position is established if Mr. Deal were standing facing forward on the step (toward the street) with his shoulders parallel with the brick porch wall.

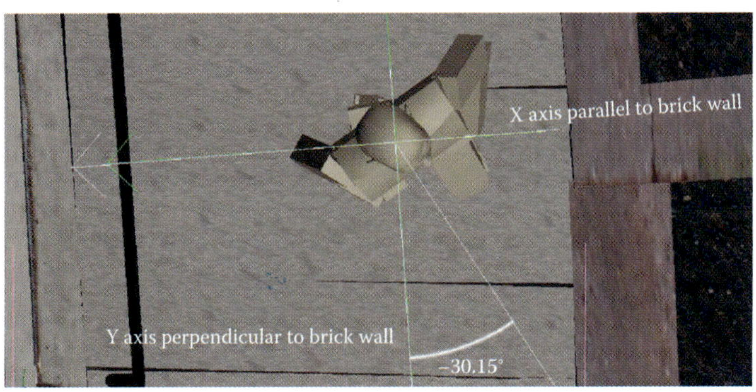

Figure 1.22 Shot 2: Deal now standing approximately –30.15 degrees toward the left.

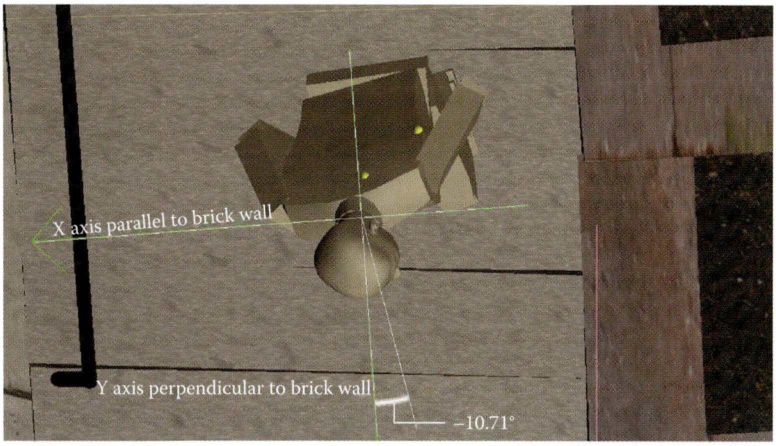

Figure 1.23 Shot 3: Deal standing approximately –10.71 degrees to the left.

3. Mr. Deal's vertical drop over the 1-second duration for the three shots. These measurements along the Z axis are based on the differences in vertical position of entrance wound shot 3 compared to shot 1 as measured on the body. The vertical drop over 1 second is approximately 13.2 inches. Shot 1 vertical coordinate: Deal's height off the top of the step level with lower planter to sidewalk: 100 inches (8.33 feet).

External Autopsy Photographs and Developed Measurements of Entrances and Exits

Measurement checks are described as follows and refer to analytical work done on the photographs of the decedent taken during the external examination, specifically documentation of the entrance and exit wound(s).

Measurement from the Vertex (Top of the Head) to the Entrance Wound of the Neck

Figure 1.24 reveals the effort required to correct the image to eliminate the lens distortion adversely affecting the measurements. The sides of the photo are no longer parallel but exhibit a curve that results from correcting the image by squaring the autopsy table so that the sides of the table are in fact parallel in the photograph. This, plus the presence of the ruler or scale in the photograph, allows us to develop accurate measurements using standard photogrammetric techniques. These measurements serve as a check on the measurements developed at the autopsy. It is useful to remember that the medical examiner's role does not necessarily involve making exact measurements. His work, in this case, is simply to orient the bullet wounds, one relative to the other, distinguishing entrance wounds from exit wounds.

Again, note the curved sides of the image (Figure 1.24) and the squared lines along the sides of the autopsy table. Rectifying the image is a step required to develop more accurate measurements. Errors result if one assumes that the presence in the picture of the scale or ruler alone allows one to develop measurements directly from the photograph.

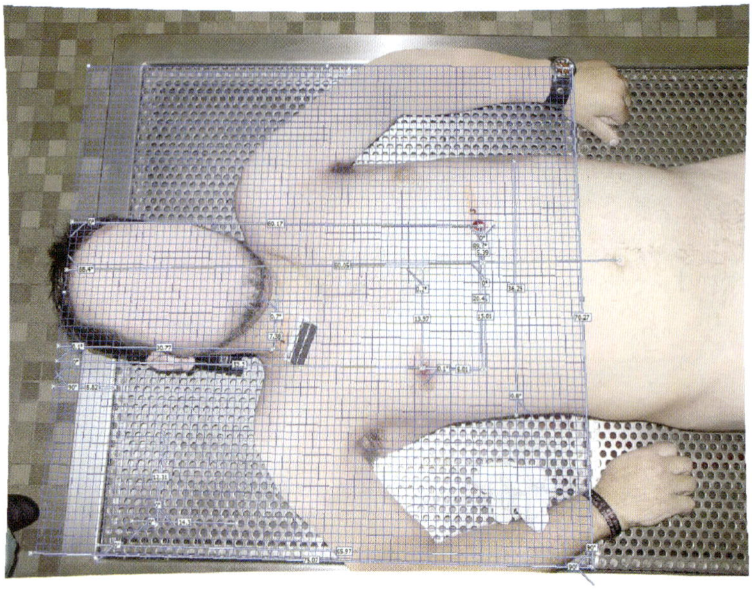

Figure 1.24 This rendering reveals the effort required to correct the image to eliminate the lens distortion, which is adversely affecting the measurements. This is called *warping*.

Measurement from the vertex to the neck is 30.77 cm (12.11 inches).
Measurement from the vertex to the neck is 30.17 cm (11.88 inches).
The average of both measurements is 12 inches (rounded to the nearest ½ inch).
(The X dimension measurement was taken by measuring the width of the neck and then dividing by 2.)
This distance is 13.54 cm (5.33 inches). (See Figure 1.25.)

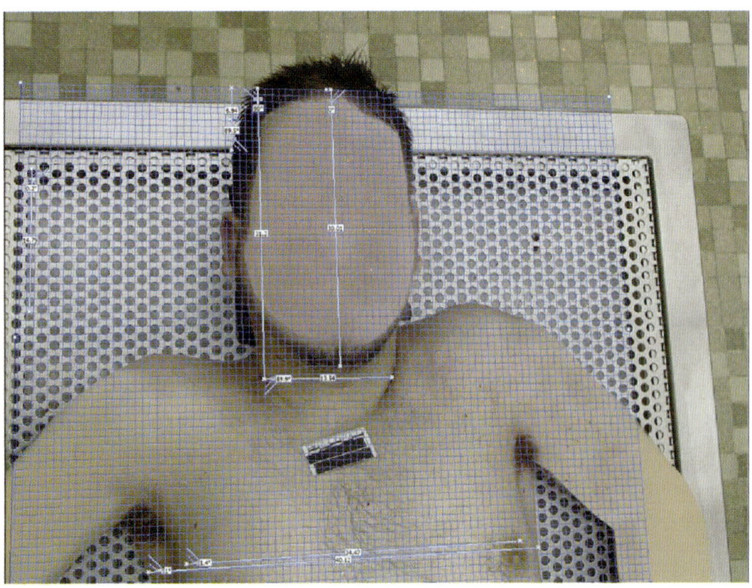

Figure 1.25 The X dimension measurement was taken by measuring the width of the neck and then dividing by 2.

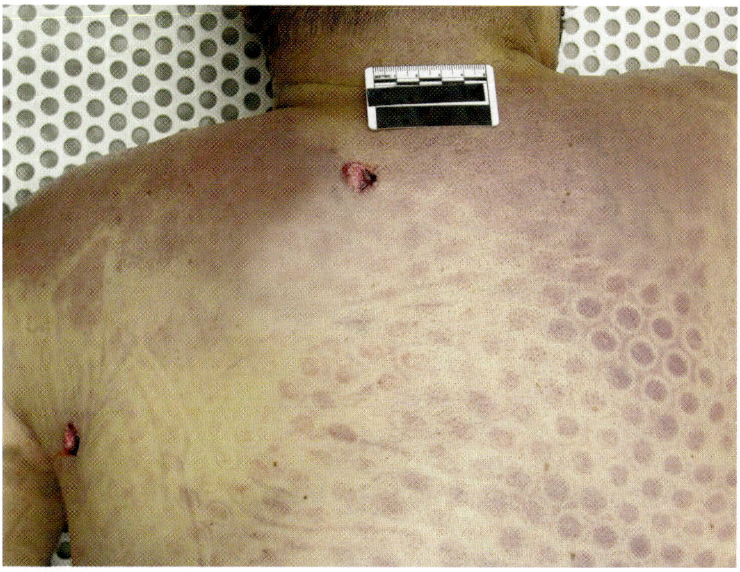

Figure 1.26 Measurements are taken from the apex of the shoulders since the vertex of the head is not visible.

5.33/2 = 2.66 inches
Rounded to the nearest ½ inch is 2.5 inches.
The coordinates for this entrance are therefore (–2.5 inches, –12 inches).

Measurement from the Vertex to the Entrance Wound of the Right Nipple

The measurement to the right nipple is 52.20 cm (20.55 inches) = 20.5 inches rounded to the nearest ½ inch.
The X dimension measurement was taken by measuring to the center of the chest.
The distance was 13.97 cm (5.5 inches).
The resulting coordinates are therefore (–5.5 inches, –20.5 inches).

Measurement from the Vertex to the Entrance Wound of the Left Side Front

The measurement to the left entrance is 60.17 cm (23.68 inches) = 23.5 rounded to the nearest ½ inch.
The X dimension measurement was taken by measuring to the center of the chest.
The distance is 5.39 cm (2.11 inches) = 2 inches rounded to the nearest ½ inch.
The resulting coordinates are therefore (2 inches, –23.5 inches).
The right neck entrance wound appears to be in line with the apex of the shoulders. Therefore, measurements taken from Figure 1.26 are taken from the apex of the shoulders since the vertex of the head is not visible.

Measurement from the Apex of the Shoulders to the Exit Wound of the Neck

The measurement to the neck exit wound is 5.31 cm (2.09 inches) = 2 inches rounded to the nearest ½ inch (below the apex of the shoulders, and therefore 14 inches below the vertex).

The X dimension measurement was taken by using a scale placed in the photograph.
The distance is measured as approximately 4 cm (1.57 inches) = 1.5 inches rounded to the nearest ½ inch.
The resulting coordinates are therefore (1.5 inches, –14 inches).

Measurement from the Apex of the Shoulders to the Chest Exit Wound (Just below the Left Armpit)

The measurement to the neck/chest exit is 17.78 cm (7 inches) = 7 inches rounded to the nearest ½ inch (below the apex of the shoulders, and therefore 19 inches below the vertex).
The X dimension measurement was taken by using the overall chest width estimate.
The maximum distance is measured as approximately 40.12 cm (15.79 inches) while the minimum distance is measured as approximately 34.47 cm (13.57 inches). The average of these two measurements is 14.68 inches, which when rounded to the nearest ½ inch = 14.5 inches.
The exit wound is almost at the edge of the chest and therefore has been estimated as 7 inches from the midline.
The resulting coordinates are therefore (7 inches, –19 inches).

Estimated Position of the Bullet Recovered at the Midline

This bullet is at about the same position as the left exit just below armpit, which is also at the midline. Its coordinates are therefore (0 inches, –19 inches).

Projected Path of the Recovered Bullet

This was an estimated exit based upon the fact that the bullet had not been deflected by striking any internal structures, such as bone, and continued along a relatively straight path. Dr. Grunewald's complete autopsy establishes this fact for the shot perforating Deal's neck and striking the brick. This shot remains the most critical piece of evidence for this shooting reconstruction when paired with the duration of Metcalf's triple tap. A complete discussion of the autopsy, Grunewald's report, and an assessment of its conclusions must be completed to correct some potentially misleading inferences made by the medical examiner as developed from the data gathered through the autopsy. These corrections affect the accuracy of the shooting reconstruction, but they do not affect the conclusions concerning the cause and manner of death developed by the medical examiner. The purposes of each activity do not necessarily converge. What may be good enough for an autopsy measurement may not be good enough for a shooting reconstruction.*

These measurements, as constructed through the analysis of these photographs, are summarized in the autopsy drawing in Figure 1.27.

* Measurements serve various purposes. The "exactness" of a measurement depends upon the reason for measuring. Suppose that I ask, "How far is it from Spokane to Seattle?" because I want to decide how to get from one to the other. Can I get there by walking, or taxi, or do I need a bus or an airplane? If the reply is "about 300 miles," that measurement is accurate enough to aid my decision. If the reply is 268.37 miles, the accuracy offered does not fit the question asked: "How does one derive that figure? From where, exactly, to where, exactly are you measuring?" We must not present the illusion of accuracy that is not supported by either the data or the context of measurement.

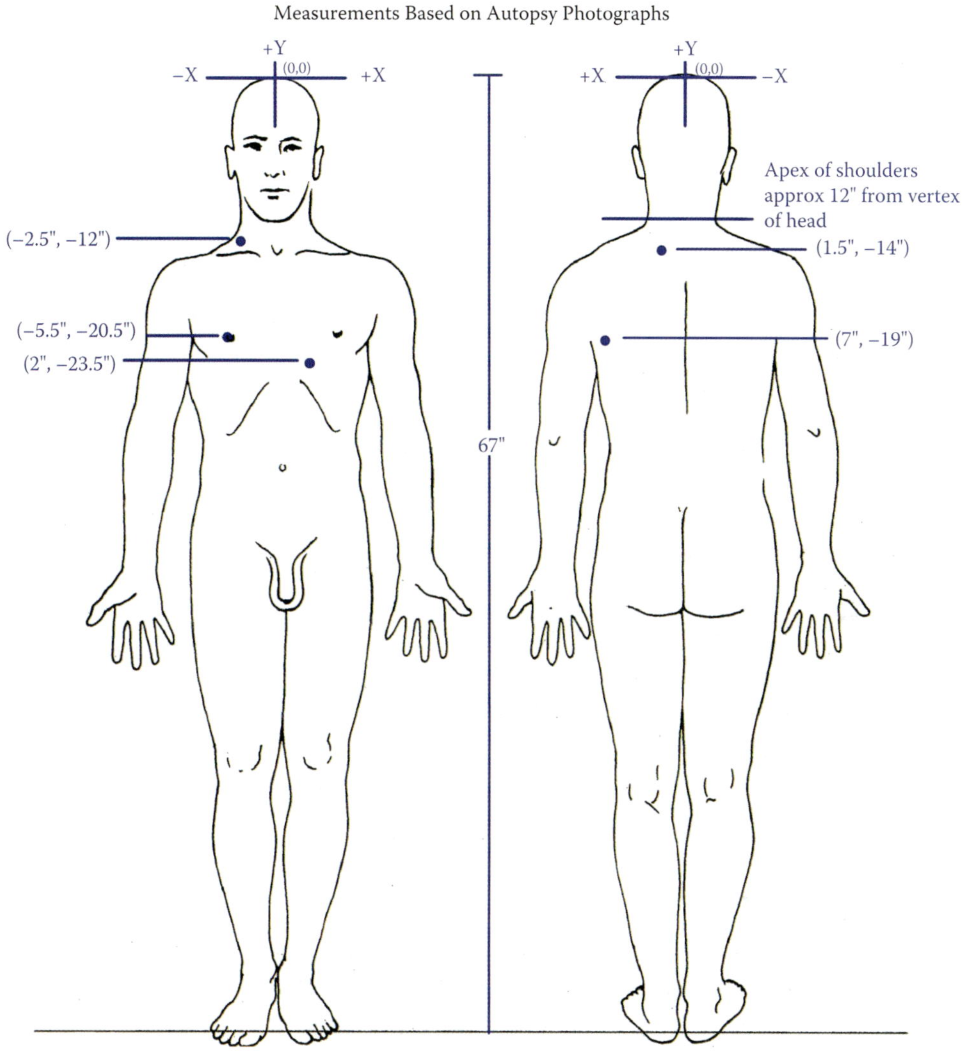

Figure 1.27 These measurements, as constructed through the analysis of the previous photographs, are summarized in the autopsy drawing.

Shooting Reconstruction of Deal's Position for the 1-Second, Three-Shot Sequence Only

The following renderings show three positions for each shot from a different perspective. (Figure 1.29 shows Deal's position for each of the three shots from a different perspective.)

The 3D digital model in Figure 1.28a shows Deal's position for the second of the three shots, again as seen from Metcalf's point of view.

The 3D digital model in Figure 1.28b shows Deal's position for the third of the three shots from Metcalf's point of view. This is the shot that perforated Deal's body and struck the brick on the neighbor's porch.

The 3D digital model in Figure 1.29 shows the combined view of Deal's positions for each of the three shots. This combined view shows a straight-line trajectory with relative positions for each entrance wound on the decedent's body as his position changes over the

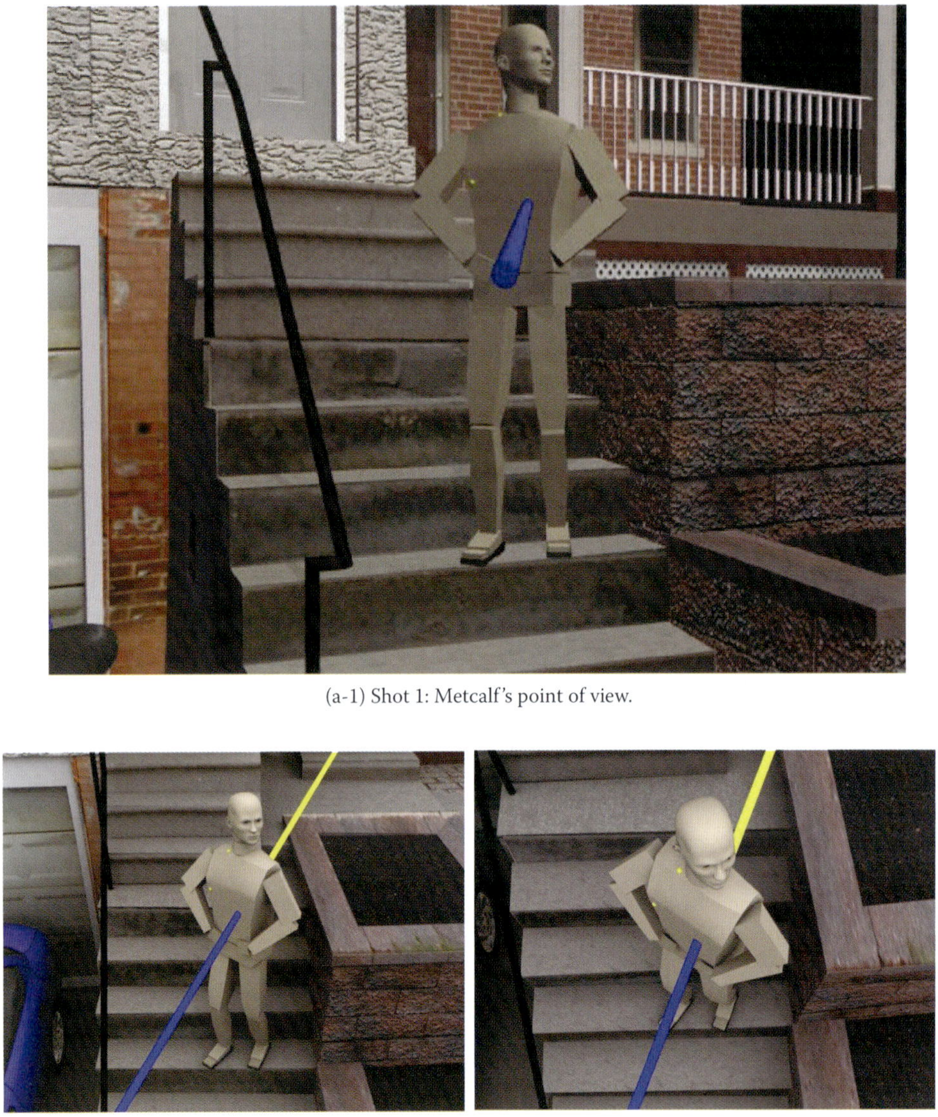

(a-1) Shot 1: Metcalf's point of view.

(a-2) Shot 1: Front upper-angle point of view.

(a-3) Shot 1: Vertical point of view.

Figure 1.28 These 3D digital models show Deal's positions for the three shots from various points of view.

Continued

(b-1) Shot 2: Metcalf's point of view.

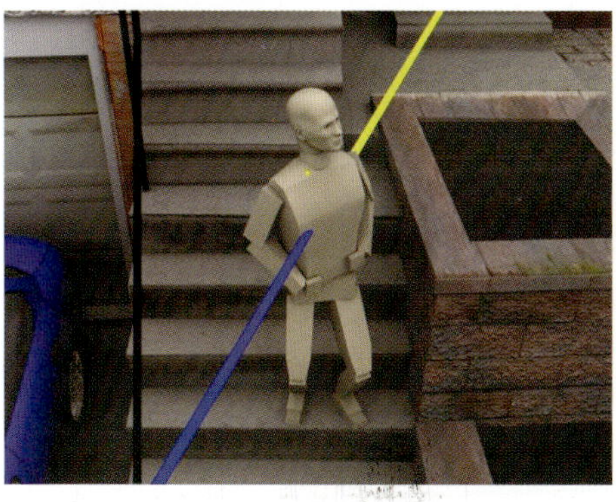

(b-2) Shot 2: Front upper-angle point of view.

(b-3) Shot 2: Vertical point of view.

Figure 1.28 (*Continued*) These 3D digital models show Deal's positions for the three shots from various points of view.

Continued

(c-1) Shot 3: Metcalf's point of view.

(c-2) Shot 3: Front upper-angle point of view.

(c-3) Shot 3: Vertical point of view.

Figure 1.28 (*Continued*) These 3D digital models show Deal's positions for the three shots from various points of view. *Continued*

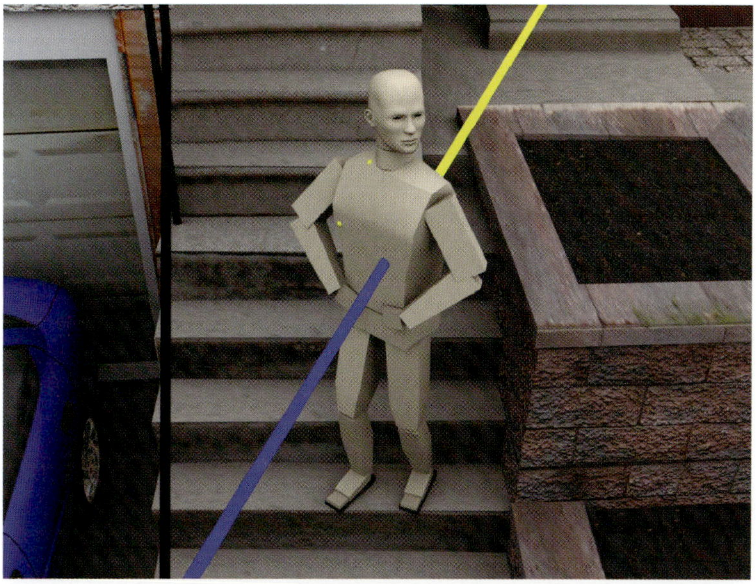

(d) Shot 1: Front vertical view of first shot with yellow dots showing each of the two successive shots.

(e) Shot 1: Upper view of first shot with yellow dots showing shots 2 and 3.

Figure 1.28 (*Continued*) These 3D digital models show Deal's positions for the three shots from various points of view. *Continued*

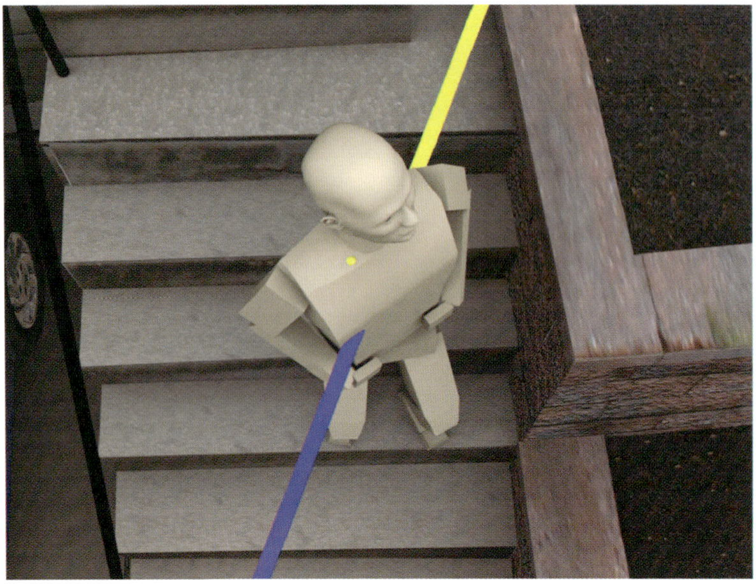

(f) Shot 2: Upper view of second shot with yellow dot showing shot 3.

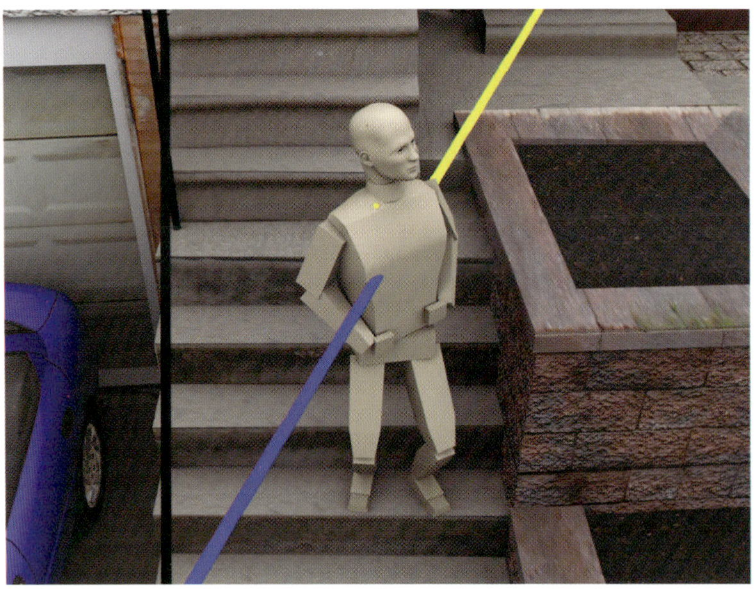

(g) Shot 2: Upper frontal point of view of second shot.

Figure 1.28 (*Continued*) These 3D digital models show Deal's positions for the three shots from various points of view. *Continued*

(h) Shot 3: Vertical perspective of shot 3.

(i) Shot 3: This shows the third shot through Mr. Deal as he falls downward and forward.

Figure 1.28 (*Continued*) These 3D digital models show Deal's positions for the three shots from various points of view.

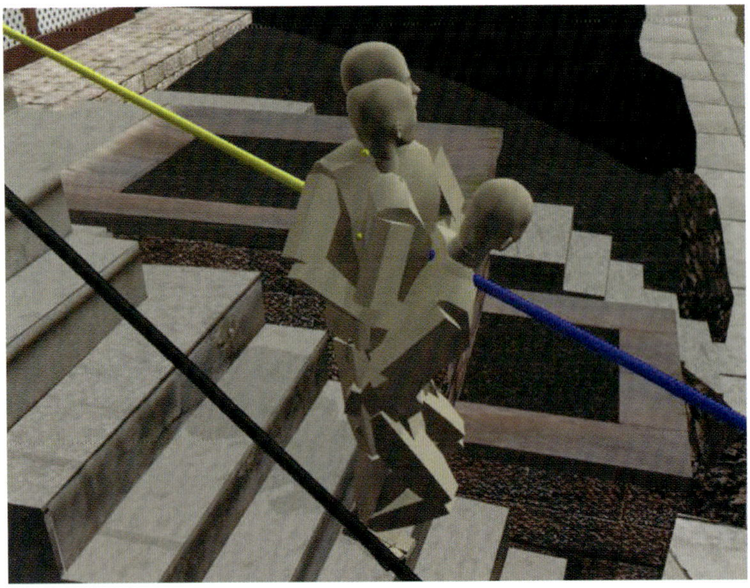

Figure 1.29 This 3D digital model shows the combined view of Deal's positions for each of the three shots. Note that the trajectory for each shot remains relatively constant. As the shots hit, Mr. Deal moves, not the gun's muzzle.

course of approximately 1 second. (The measured changes in his position are explained and documented earlier.) The model appears as seen from above and west to illustrate Deal's body movements involved during these three shots.

The figure highest in Figure 1.29 (along the vertical Z axis) represents Deal's position for the first shot; the second-highest figure represents his position for the second shot; the lowest figure represents his position for the final shot, which perforated his neck and struck the brick wall on the neighboring porch.

The 3D digital model showing these three shots (Figure 1.29a) shows a straight-line trajectory with relative positions for each entrance wound on the decedent's body as his position changes over the course of approximately 1 second along the Z axis (vertical) as well as along the Y axis (moving toward the street) and the X axis (moving toward the planter box). The figure highest (along the vertical Z axis) and further toward the porch brick wall to the right (along the Y axis) represents Deal's position for the first shot; the second-highest figure represents his position for the second shot; the lowest figure represents his position for the final shot, which perforated his neck and struck the brick wall on the porch. Deal moves along the X, Y, and Z axes as follows: he moves toward the planter box (X), he moves toward the road (Y), and he moves downward (Z).

The 3D digital model showing three shots (Figure 1.29b) again shows the straight-line trajectory with relative positions for each entrance wound on the decedent's body as his position changes over the course of approximately 1 second. The figure highest (along the vertical Z axis) and furthest to the left (along the perpendicular Y axis) represents Deal's position for the first shot; the second-highest figure represents his position for the second shot; the lowest figure represents his position for the final shot, which perforated his neck and struck the brick wall on the porch (toward the left side of the illustration). Recall that

Figure 1.29a This 3D digital model shows these three shots in a straight-line trajectory with relative positions for each entrance wound on the decedent's body as his position changes over the course of approximately 1 second. The yellow dots represent the entrances and exits of the first (lowest) and the second (next lowest) shots.

Figure 1.29b This 3D digital model again shows the straight-line trajectory with relative positions for each entrance wound on the decedent's body as his position changes over the course of approximately 1 second.

the chip represents the origin (0, 0, 0) of the scene coordinate system defined by the X, Y, and Z axes.

Deal's Movements and the Bullet Impacts

Popular images from movies and television greatly distort the realities of a bullet's physical effects when striking a human body. On both the big screen and the small screen, actors' bodies fly backward great distances when they are shot by handguns, rifles, or shotguns, often to great dramatic effect. But that movement is sheer fantasy, presented merely for dramatic effect. It has no basis in scientific fact. Despite the laws disobeyed in the storyline, including the laws of physics, real human bodies struck by real bullets must behave consistently with the laws of physics. This provides a sound scientific basis for assessing the actual effects of real bullets striking real human bodies.

Ballistics, as we have seen, is an area of classical physics obviously relevant to this study.* The science of physics helps us both understand and explain the interactions among the various physical forces involved as they interact when a human body is shot. Specifically in this case, physics will help us understand Deal's movements when struck by the three bullets fired by Metcalf.

Physics dictates that bodies in motion tend to remain in motion until other forces overcome the force both producing and sustaining that motion. Bodies at rest tend to remain at rest until acted upon by forces sufficient to initiate movement. Here, Deal was turning toward his left prior to being struck by the first bullet. We know this because his body continues rotating to the left after being struck by the first bullet, until the force of gravity overcomes the force supplied by this voluntary movement. Evidence for this is supplied by the location of the second two entry wounds on his body, which are to the right and upward.

The mass of the bullet was not sufficient to "move" Deal's body backward or to his left. This is an instance of classical physics represented by the basic formula $F = MA$, where A is a vector of acceleration, M is mass, and F is force. If the first bullet was replaced by a moving automobile, for example, then both the mass and the size (surface area) of the automobile striking his body would be far greater than the mass and the size of the bullet, while the velocity of the automobile would be far less than the velocity of the bullet. His body would move along with the automobile, but his body does not move along with the bullet.

To underscore this point, Deal's body continues its rotation toward the left, initiated as a voluntary movement, despite being struck near the midline of the chest and continues that rotation between the first and the second shot, with the second shot striking him on the right side at the right nipple. His body continues its rotation toward the left until the force of gravity overcomes both this rotational force and the muscular force supporting his vertical position. As the force of gravity overcomes these other forces, he falls downward as his leftward rotation ceases, as seen by the location of the third bullet entrance which appears higher and to the right side of his body at the base of his neck toward the end of the 1-second shooting interval. The three bullet impacts, therefore, serve to end his voluntary and involuntary movements and enable the force of gravity to influence his fall down the concrete steps.

Shooting Reconstruction Summary

The following summarizes Deal's movements during and immediately after the 1-second shooting sequence. The aforementioned reconstruction remains exclusively based upon physical evidence and only addresses Deal's position(s) during this approximately 1-second, three-shot sequence. For this purpose, no assumptions were made about Deal's head position or his hand/arm positions and nothing in the aforementioned work depends upon his arms, hands, or head being in any specific position (as long as their positions remain consistent with the known position of his torso as scientifically established earlier).

* Equally obviously, Toolmark examination is not an area relevant to this study of trajectory.

The positions of the torso as depicted in the 3D models are established scientifically through the analyses described earlier. The legs are positioned as required to support the torso. The positions of the head, arms, and hands illustrate both witness testimony and derive from an analysis of the injuries to his right eye and the right side of the head, and the damage to his glasses.*

The summary includes the physical evidence of Deal's fall toward the sidewalk. Assessments of his fall are based upon examining the photos of his head injuries as well as examining the damage to his glasses, including photomicrographs of the abrasions on the right lens and the right bow.

Deal was standing on the steps as depicted in the shot sequences in Figure 1.28 and Figure 1.29; turning his torso from right to left as shown when he was struck by three bullets. While being struck, his rotation continued from right to left. When struck he dropped downward (gravity). He initially struck the right side of his face at the right eye and temple on the concrete, damaging his glasses as described.

Laboratory Notes and Drawings

Figures 1.30 to 1.45 are my lab notes and drawings. As an undergraduate, I studied art to learn how to make careful observations. To draw something is to see that thing in detail. Even if I don't draw it well, I have at least learned to observe it well. I find that art and science have many skills in common. Each enterprise can learn much from the other.

* Witness testimony describes Deal's hands in his hoodie pocket, with his head turning leftward.

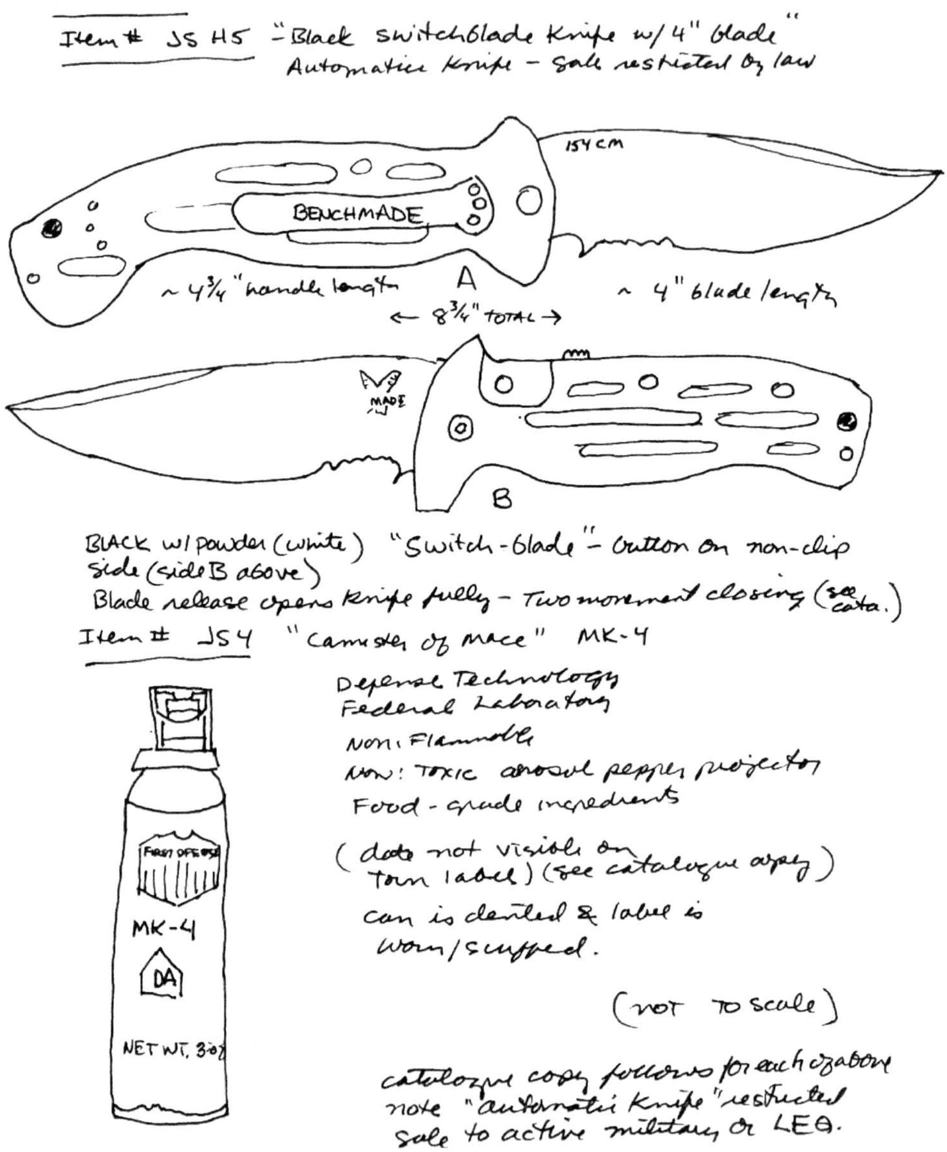

Figure 1.30 I make it a practice to draw items of evidence to improve my observations of relevant details.

BENCHMADE

BLACK CLASS KNIVES

Automatic Knives

Auto Presidio

- Patented Bi-Directional Grip Pattern for Superior Grip
- Non-Reflective Components for Tactical Applications
- Automatic AXIS® Allows for Ambidextrous Actuation
- NSN # 4220-01-546-0478

- Blade Length: 3.42"
- Blade Thickness: 0.130"
- Blade Material: 154 CM Stainless Steel
- Blade Hardness: 58–60
- Blade Style: Drop-Point or Trainer
- Clip: Black, Reversible, Tip-Up • Weight; 5.62 oz.
- Overall Length: 8.25" • Closed Length: 4.87"

Part #	Description	Price
5300SBK	Auto Presidio: Combo Edge, BK Coating	$187.95

Auto Presidio: Tanto

- Patented Auto-AX1S™ Mechanism
- Integrated Safety
- Machined Aluminum Handles
- All Components are Non-Reflective
- Blade Length: 3.44"
- Blade Thickness: 0.115"
- Blade Material: 154 CM Stainless Steel
- Blade Hardness: 58–60 HRC
- Blade Style: Tanto
- Clip: Black, Movable, Tip-Up. • Weight: 5.80 oz.
- Overall Length: 8.30" • Closed Length: 4.86"

Part #	Description	Price
5300SBK	Auto Presidio: Tanto, Combo Edge, BK Coating	$187.95

Auto Stryker/Mini Auto Stryker

- BK 1 Blade Coating Provides Unrivaled Blade Protection
- Non-Reflective Components for Tactical Applications
- Tanto Blade Shape has Piercing Power
- Thick Blade Spine Absorbs Shock, Enhanced Tip Strength
- Blade Length: 3.70" • Mini: 2.90"
- Blade Thickness: 0.1.20" • Mini: 0.100"
- Blade Material: 154 CM Stainless Steel
- Blade Hardness: 58–60
- Blade Style: 9100/9500 Tanto: 9130/9530 Spear-Point
- Weight 4.1 oz. • Mini: 2.90 oz.
- Overall Length: 8.25" • Mini: 7.10"
- Closed Length: 4.68" • Mini: 4.20"

Federal regulations restrict the sale of automatic knives to anyone other than active military or law-enforcement with proof of photo identification. If you are currently active military or law-enforcement and wish to purchase Benchmade automatic knifes from LEED, please contact us. Please note that the models presented here represent our most commonly stocked knives. We stock more than what is pictured and can special order any knife that Benchmade manufactures, please contact us for pricing.

Part #	Description	Price
9100SBK	Auto Stryker: Tanto, Combo Edge, BK Coating	$183.95
9130SBK	Auto Stryker: Spear-Point, Combo Edge, BK Coating	$183.95
9500SBK	Mini Auto Stryker: Tanto, Combo Edge, BK Coating	$153.95
9530SBK	Mini Auto Stryker: Spear-Point, Combo Edge, BK Coating	$153.95

Figure 1.31 Part of any examination involves researching items of interest—this knife, for example, is not legal for anyone outside the police force or the military to own or carry.

DEFENSE TECHNOLOGY
FEDERAL LABORATORIESTM

OC AEROSOL/PEPPER SPRAY

First Defense® Duty Aerosols are the world's most widely used pepper spray products in law-enforcement and corrections. Tested and proven on the streets, in the lab and the courtroom. Defense Technology has never lost or settled a legal claim against First Defense®. Ever.

Every First Defense® product is completely non-toxic, non-flammable and environmentally friendly.

Part#	Description	Price	Case of 25
5039	MK-3 Stream: 1.47 oz.	$12.95 ea.	$281.95
5049	MK-4 Stream: 3.0 oz.	$14.95 ea.	$316.95
5069	MK-6 Stream: 0.68 oz.	$12.95 ea.	$281.95
5139	MK-3 Inert Stream: 1.47 oz.	$9.95 ea.	$238.95
5149	MK-4 Inert Stream: 3.0 oz.	$11.95 ea.	$279.95
5169	MK-6 Inert Stream; 0.68 oz.	$9.95 ea.	$238.95

The original, patented, fully tested formulation shown in MK 3, MK-4 and MK-6. (Left of Right)

First Defense® 360°

Part#	Description	Price	Case of 25
5439	360 MK-3 Stream: 1.47 oz.	$12.95 ea.	$281.95
5449	360 MK-4 Stream: 3.0 oz.	$14.95 ea.	$316.95
5239A	X2 360 MK-3 Stream: 1.47 oz.	$12.95 ea.	$281.95
5249A	X2 360 MK-4 Stream: 3.0 oz.	$14.95 ea.	$316.95
5439I	360 MK-3 Inert Stream: 1.47 oz.	$1.95 ea.	$238.95
5449I	360 MK-4 Inert Stream: 3.0 oz.	$11.95 ea.	5279.95

A true breakthrough in aerosol performance, First Defense' 360° delivers a consistent stream pattern at any angle and any orientation. Using our trusted OC formula, it is propelled by compressed air with barrier separation from the liquid formulation. No additional training of education is required for use..Available in MK-3 and MK-4 sizes.

First Defense® X2

First Defense" X2 has all the industry-leading attributes of the original First Defense" but with twice the capsaicin concentration for extra pungency. It also uses an environmentally safe propellant, nitrogen, to project a stream 12 feet.

Part#	Description	Price	Case of 2S
5239	X2 MK-3 Stream: 1.47 oz.	$14.95 ea.	$310.95
5249	X2 MK-4 Stream: 3.0 oz.	$15.95 ea.	$365.95
5239A	X2 360 MK-3 Stream: 1.47 oz.	$14.95 ea.	$310.95
5249A	X2 360 MK-4 Stream: 3.0 oz.	$15.95 ea.	$365.95
5139	MK-3 Inert Stream: 1.47 oz.	$9.95 ea.	$238.95
5149	MK-4 Inert Stream: 3.0 oz.	$11.95 ea.	$279.95

Formulation Percentages

Model	OC Solution	Major Capsaicinoids
First Defense	10%	0.20%
X2	20%	0.40%
360°	10%	0.20%
PapperFoam	10%	0.20%

Figure 1.32 Information covering OC Aerosol/Pepper Spray.

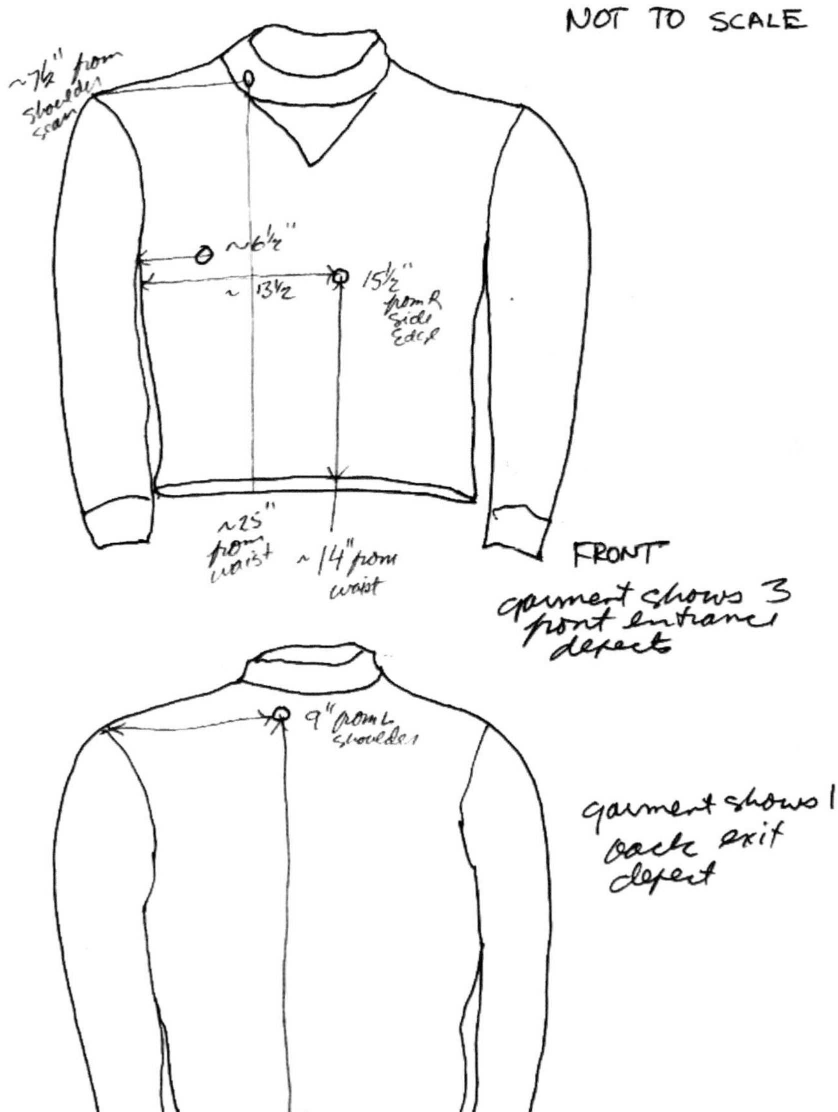

Figure 1.33 Lab notes also help identify and record relevant measurements, and the locations of other evidence on garments.

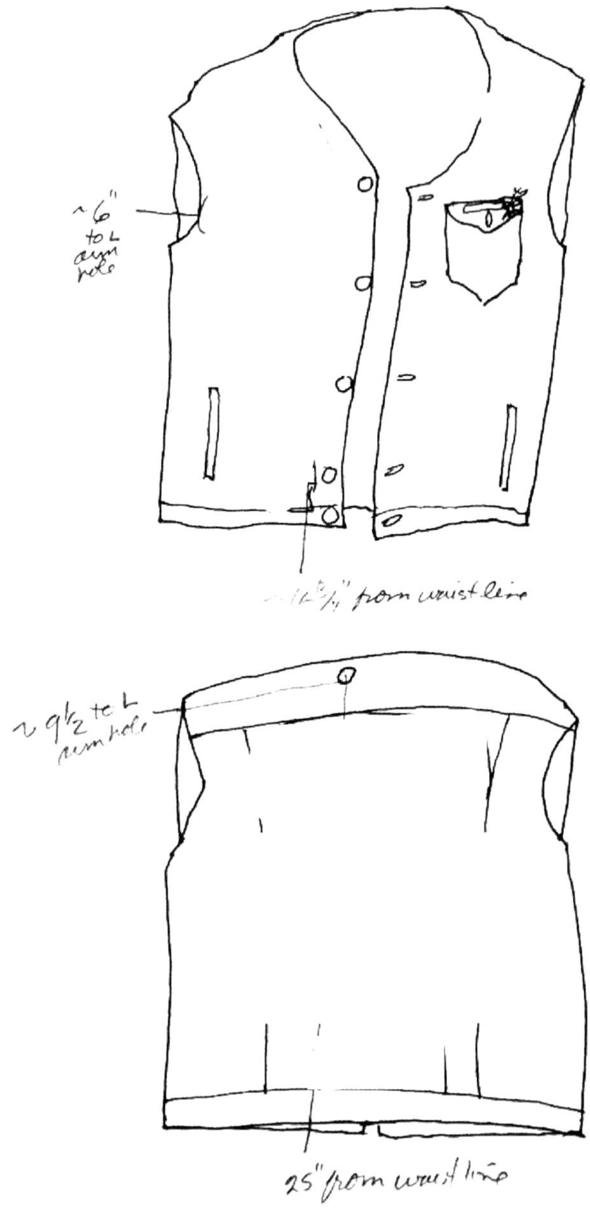

Figure 1.34 Each layer of clothing has some story to tell.

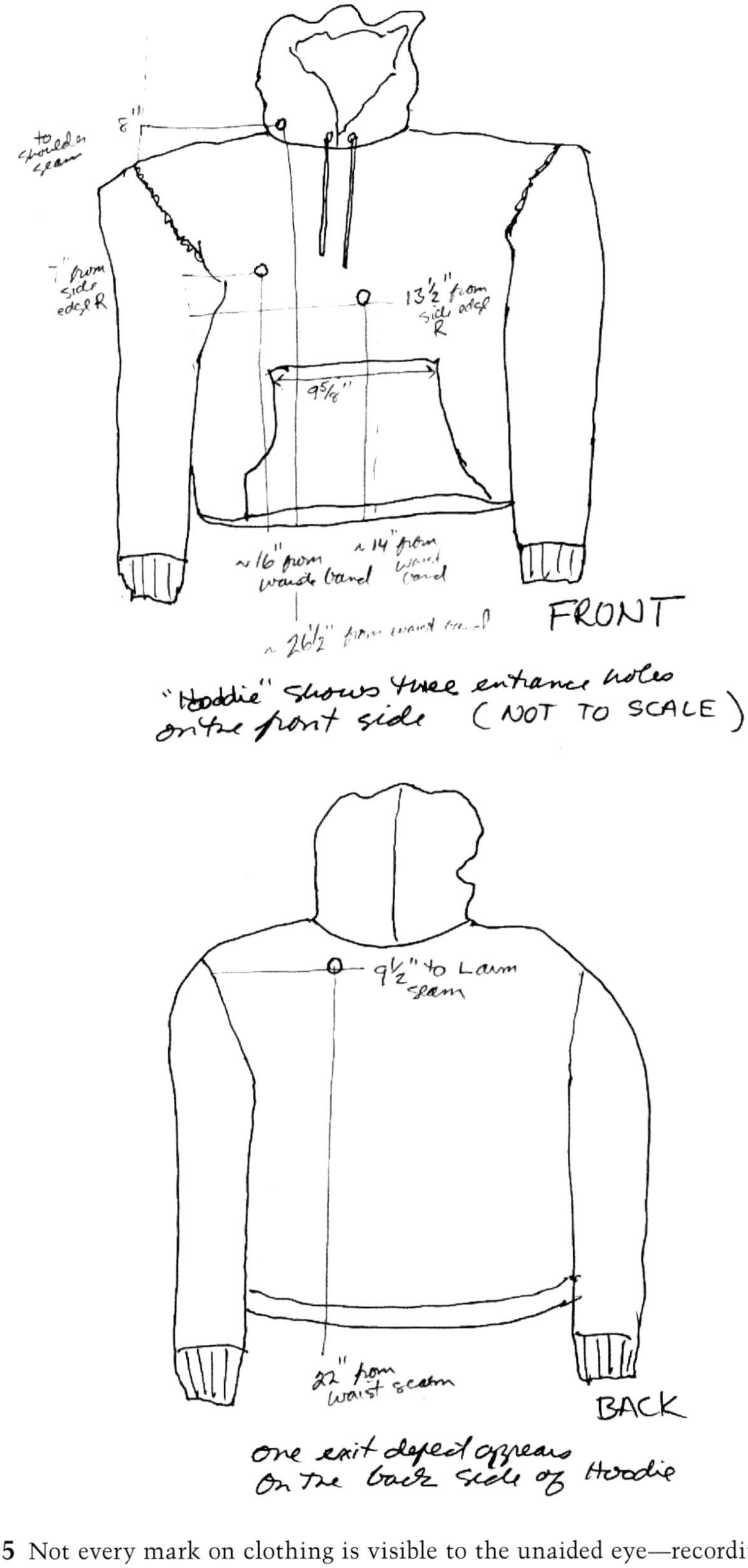

Figure 1.35 Not every mark on clothing is visible to the unaided eye—recording their positions can assist in finding their locations later.

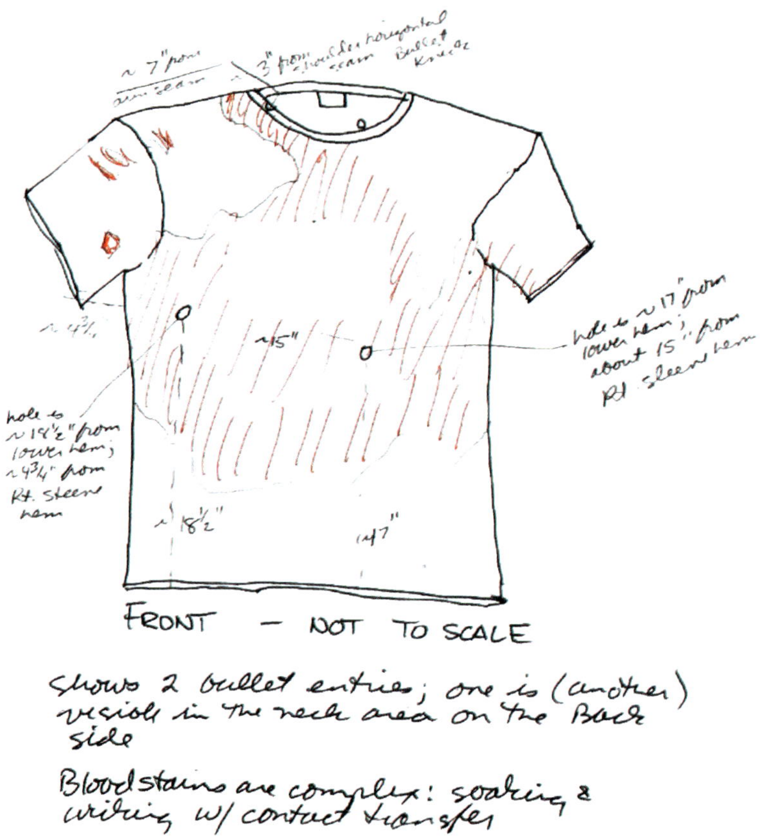

Figure 1.36 Lab notes and drawings help locate and describe bloodstains and other pattern evidence.

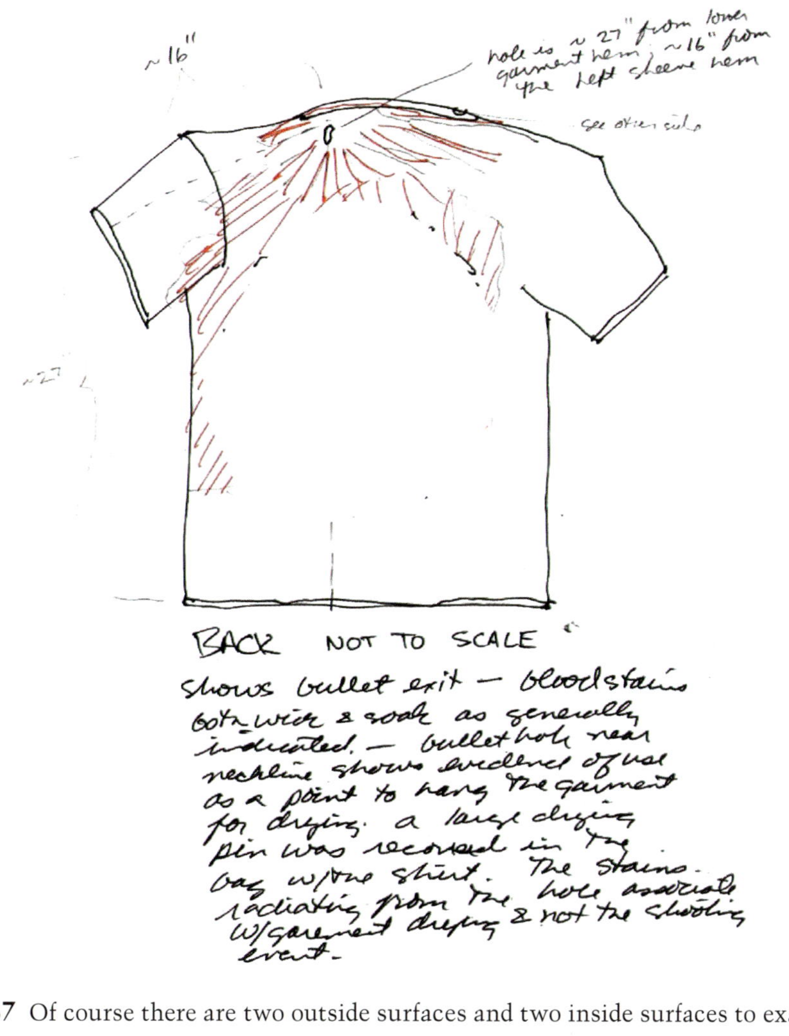

Figure 1.37 Of course there are two outside surfaces and two inside surfaces to examine.

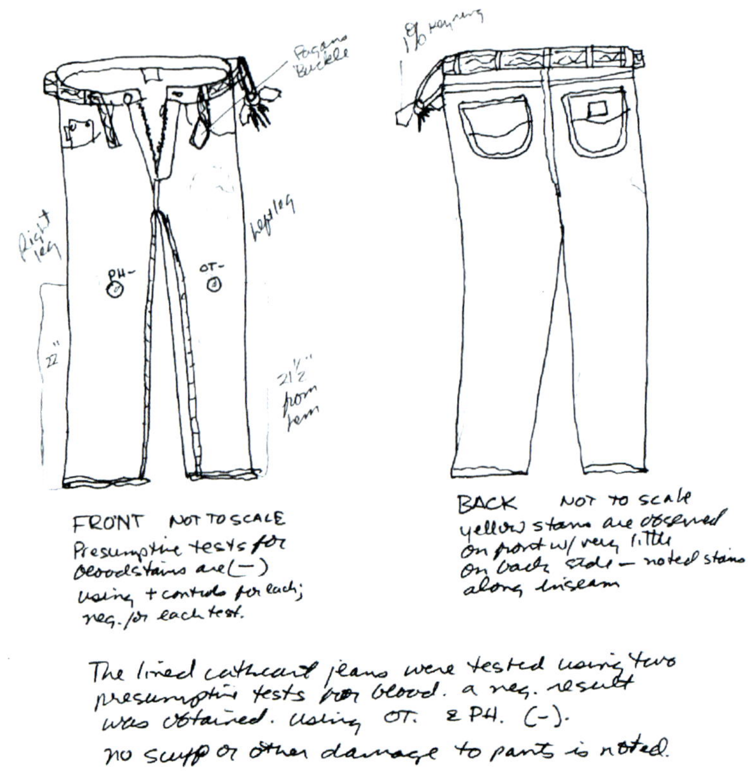

Figure 1.38 Stains as well as any tears or abrasions on the fabric must be identified and explained—are they relevant to this event or some prior event?

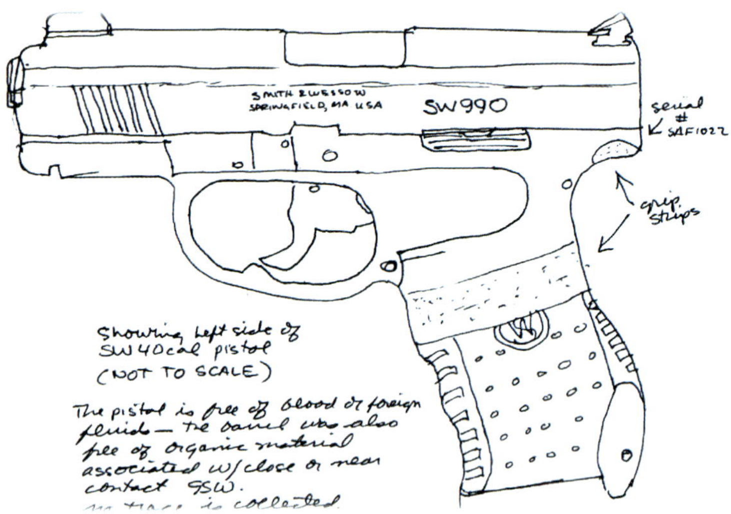

Figure 1.39 The pistol was received without a magazine and no unfired bullets were provided. Class characteristics are only reliable for analysis. The pistol is in good operating condition with no operational defects. Trigger pull, over 10 tests = 8 lb 7.5 oz (double-action only). The pistol is well oiled and well maintained. Modifications are limited to a grip strip and a web strip grip. A magazine from our inventory, fitting the Smith & Wesson Walther P99 design, was used for testing purposes. Testing also used Federal 40 cal JHP rounds.

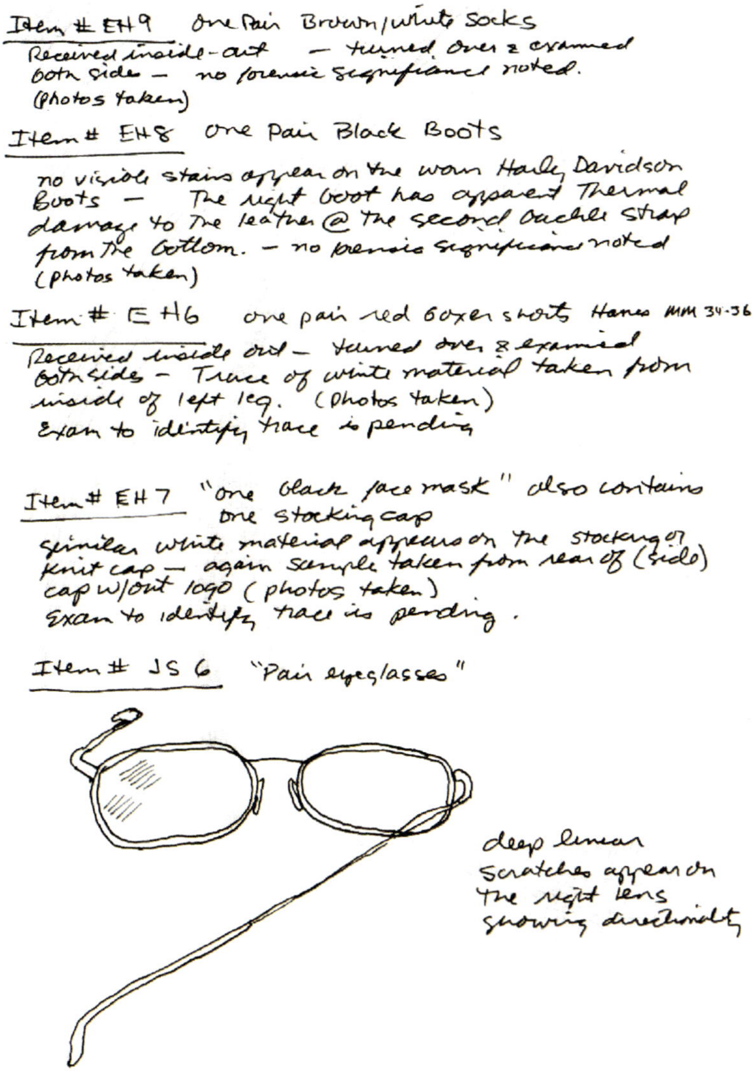

Item # EH9 One Pair Brown/white socks
Received inside-out — turned over & examined
both sides — no forensic significance noted.
(Photos taken)

Item # EH8 One Pair Black Boots
no visible stains appear on the worn Harley Davidson
Boots — The right boot has apparent Thermal
damage to the leather @ the second buckle strap
from the bottom. — no forensic significance noted
(photos taken)

Item # EH6 one pair red boxer shorts Hanes MM 34-36
Received inside out — turned over & examined
both sides — Trace of white material taken from
inside of left leg. (Photos taken)
Exam to identify trace is pending

Item # EH7 "one black face mask" also contains
one stocking cap
similar white material appears on the stocking or
knit cap — again sample taken from rear of (side)
cap w/out logo (photos taken)
Exam to identify trace is pending.

Item # JS 6 "Pair eyeglasses"

Figure 1.40 Lab notes on other items.

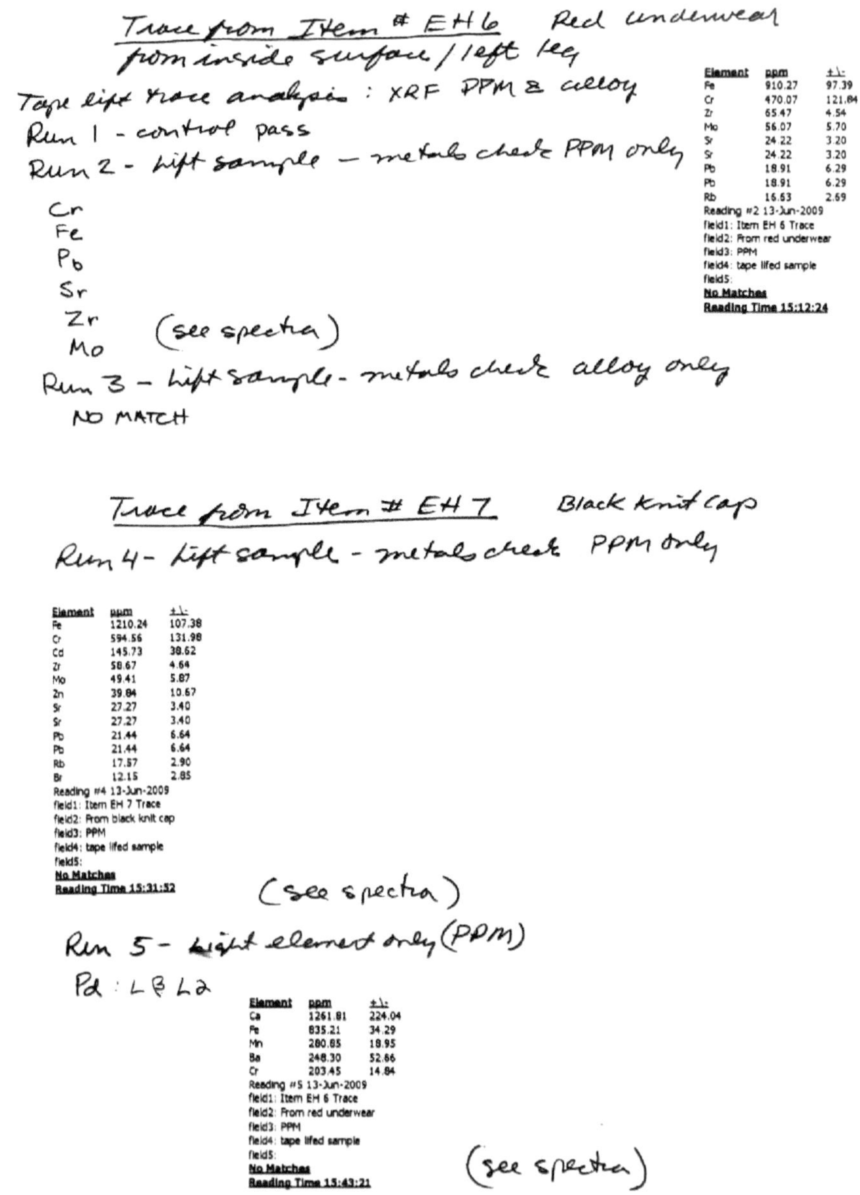

Trace from Item # EH6 Red underwear
from inside surface / left leg

Tape lift trace analysis : XRF PPM & alloy
Run 1 - control pass
Run 2 - lift sample – metals check PPM only

Element	ppm	+\-
Fe	910.27	97.39
Cr	470.07	121.84
Zr	65.47	4.54
Mo	56.07	5.70
Sr	24.22	3.20
Sr	24.22	3.20
Pb	18.91	6.29
Pb	18.91	6.29
Rb	16.63	2.69

Reading #2 13-Jun-2009
field1: Item EH 6 Trace
field2: From red underwear
field3: PPM
field4: tape lifed sample
field5:
No Matches
Reading Time 15:12:24

Cr
Fe
Pb
Sr
Zr
Mo (see spectra)

Run 3 – lift sample - metals check alloy only
NO MATCH

Trace from Item # EH 7 Black knit cap
Run 4 - lift sample - metals check PPM only

Element	ppm	+\-
Fe	1210.24	107.38
Cr	594.56	131.98
Cd	145.73	38.62
Zr	58.67	4.64
Mo	49.41	5.87
Zn	39.84	10.67
Sr	27.27	3.40
Sr	27.27	3.40
Pb	21.44	6.64
Pb	21.44	6.64
Rb	17.57	2.90
Br	12.15	2.85

Reading #4 13-Jun-2009
field1: Item EH 7 Trace
field2: From black knit cap
field3: PPM
field4: tape lifed sample
field5:
No Matches
Reading Time 15:31:52

(see spectra)

Run 5 - light element only (PPM)
Pd : Lβ Lα

Element	ppm	+\-
Ca	1261.81	224.04
Fe	835.21	34.29
Mn	280.85	18.95
Ba	248.30	52.66
Cr	203.45	14.84

Reading #5 13-Jun-2009
field1: Item EH 6 Trace
field2: From red underwear
field3: PPM
field4: tape lifed sample
field5:
No Matches
Reading Time 15:43:21

(see spectra)

Figure 1.41 X-ray florescence spectrophotometry provides nondestructive analyses of items, supplying information about the presence of any metallic elements.

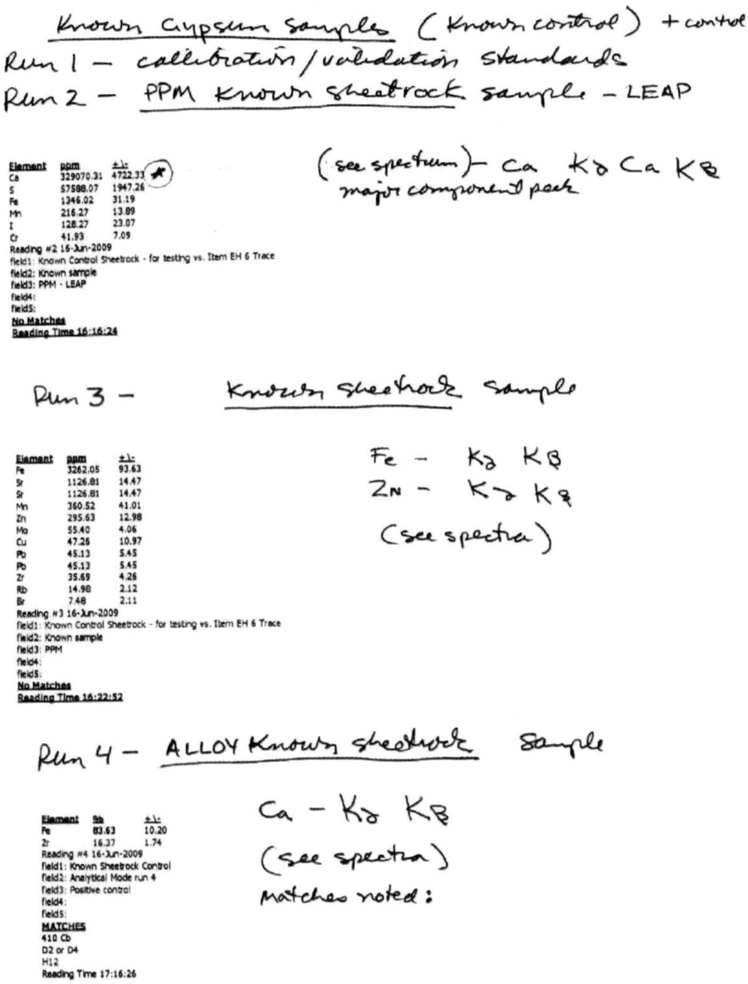

Known Gypsum sample (Known control) + control
Run 1 — calibration/validation standards
Run 2 — PPM Known sheetrock sample – LEAP

Element	ppm	+\-
Ca	329070.31	4722.33
S	57588.07	1947.26
Fe	1346.02	31.19
Mn	216.27	13.89
I	128.27	23.07
Cr	41.93	7.09

Reading #2 16-Jun-2009
field1: Known Control Sheetrock - for testing vs. Item EH 6 Trace
field2: Known sample
field3: PPM - LEAP
field4:
field5:
No Matches
Reading Time 16:16:24

(see spectrum) – Ca Kα Ca Kβ
major component peak

Run 3 — Known sheetrock sample

Element	ppm	+\-
Fe	3262.05	93.63
Sr	1126.81	14.47
Sr	1126.81	14.47
Mn	360.52	41.01
Zn	295.63	12.98
Mo	55.40	4.06
Cu	47.26	10.97
Pb	45.13	5.45
Pb	45.13	5.45
Zr	35.69	4.26
Rb	14.98	2.12
Br	7.48	2.11

Reading #3 16-Jun-2009
field1: Known Control Sheetrock - for testing vs. Item EH 6 Trace
field2: Known sample
field3: PPM
field4:
field5:
No Matches
Reading Time 16:22:52

Fe – Kα Kβ
Zn – Kα Kβ
(see spectra)

Run 4 — ALLOY Known sheetrock sample

Element	%	+\-
Fe	83.63	10.20
Zr	16.37	1.74

Reading #4 16-Jun-2009
field1: Known Sheetrock Control
field2: Analytical Mode run 4
field3: Positive control
field4:
field5:
MATCHES
410 Cb
D2 or D4
H12
Reading Time 17:16:26

Ca – Kα Kβ
(see spectra)
Matches noted:

Figure 1.42 Analytical results are printed out by the XRF—spectra are graphed for further study.

Material Notes: Alabaster is a fine-grained, translucent gypsum. Three cleavage directions. Occurrence: Formed by evaporation; found in sediments and desert floor. Common in efflorescence. Crystallization Pressure for $C/C_S = 2$ is 282 atm at 0°C. Crystallization Pressure for $C/C_S = 2$ is 334 atm at 50°C. Crystallization Pressure for $C/C_S = 10$ is 938 atm at 0°C. Crystallization Pressure for $C/C_S = 10$ is 1110 atm at 50°C. Crystallization Pressure for $C/C_S = 50$ is 1595 atm at 0°C. Crystallization Pressure for $C/C_S = 50$ is 1900 atm at 50°C.

Property values are typical of naturally occurring gypsum ($CaSO_4 - 2H_2O$). Actual property values will vary from sample to sample.

Key Words: Gypsum ($CaSO_4 - 2H_2O$)

Vendors: No vendors are listed for this material. Please click here if you are a supplier and would like information on how to add your listing to this material.

Physical Properties	Metric	English	Comments
Density	2.32 g/cc	0.0838 lb/in^3	
Solubility	1750–2150 ppm	1750–2150 ppm	Freshwater
Molecular Weight	127 g/mol	127 g/mol	
Chemical Properties	**Metric**	**English**	**Comments**
Molar Volume	55.0 cc/mol	55.0 cc/mol	
Mechanical Properties	**Metric**	**English**	**Comments**
Hardness, Knoop	46.0–54 .0	46.0–54.0	
Hardness, Vickers	60	60	
Hardness, Mohs	2.00	2.00	
Abrasive Hardness	1.25	1.25	
Drilling Hardness	8.30	8.30	
Descriptive Properties			
Color	White		

Figure 1.43 The analytical results often require research into the items found on garments—it is useful to present that work along with the lab notes.

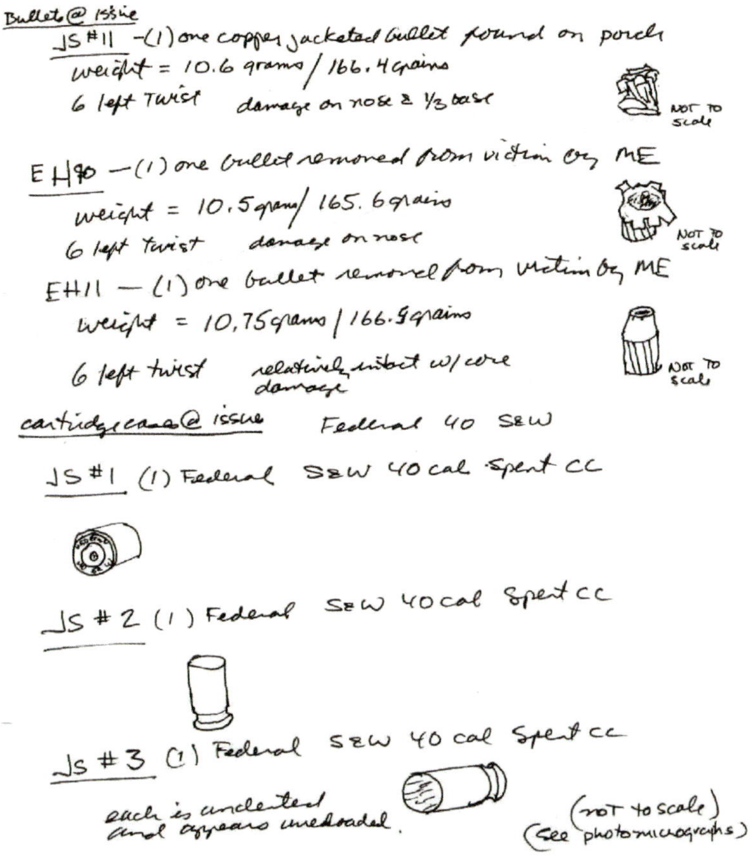

Figure 1.44 Sketches of the relevant bullets and cartridge cases usually precede any microscopic examinations. Often such examinations are performed to investigate the possibility of more than one shooter.

Figure 1.45 The knife described earlier—photographs with a scale are useful data—a known measurement grid can be helpful if exact 90-degree photos are not taken.

Laboratory Photographs: Clothing and Other Items Submitted to the Laboratory

Knife and pepper spray: Figures 1.46 to 1.49
Orange T-shirt: Figures 1.50 to 1.53
Jeans jacket: Figures 1.54 to 1.56
Black hoodie sweatshirt: Figures 1.57 to 1.61
Photogrammetry preparations with black hoodie: Figure 1.62
Blue jean pants: Figures 1.63 to 1.69
Socks: Figures 1.70 to 1.73
Red underwear: Figures 1.74 to 1.78
Black knit cap: Figure 1.79
Glasses: Figure 1.80
Smith & Wesson 40 cal pistol: Figures 1.81 to 1.84

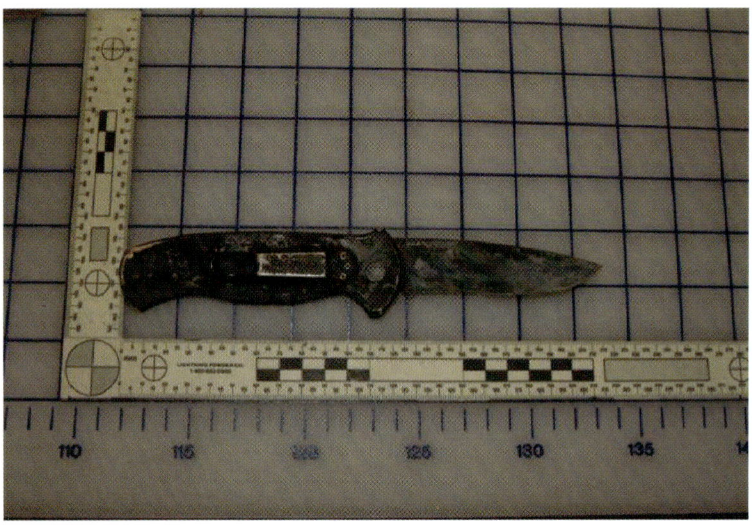

Figure 1.46 Knives also have more than one side—all sides should be examined and documented to address any future questions which may not be anticipated now.

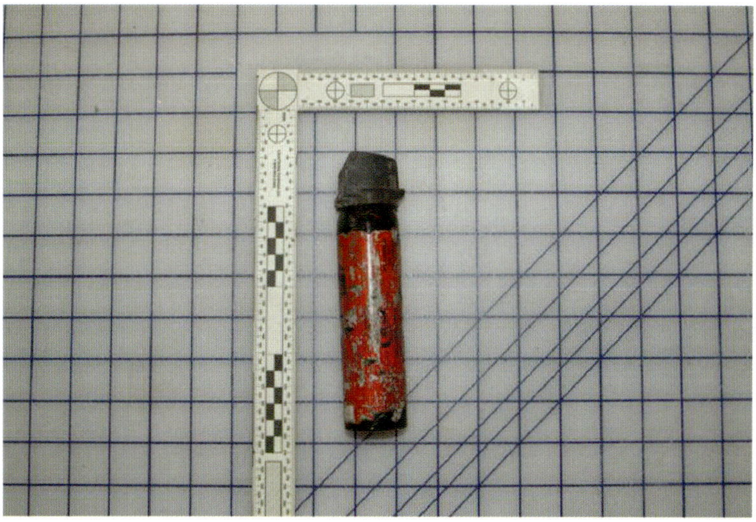

Figure 1.47 The pepper spray showing signs of wear (again photographed on a grid).

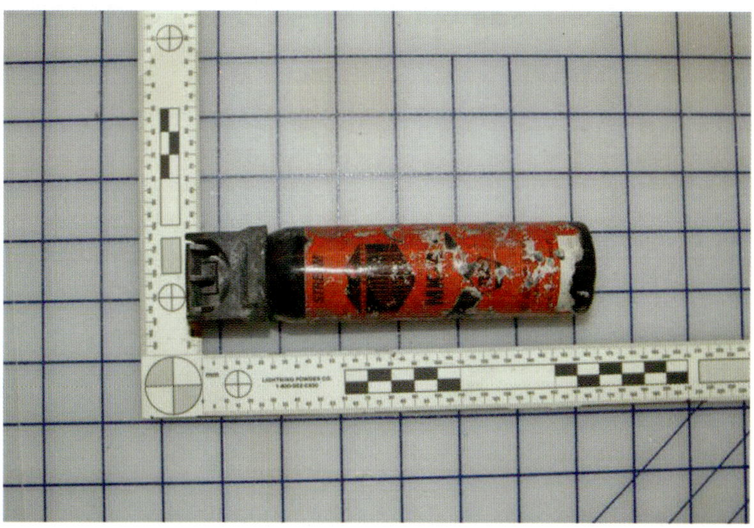

Figure 1.48 Pepper spray from another angle.

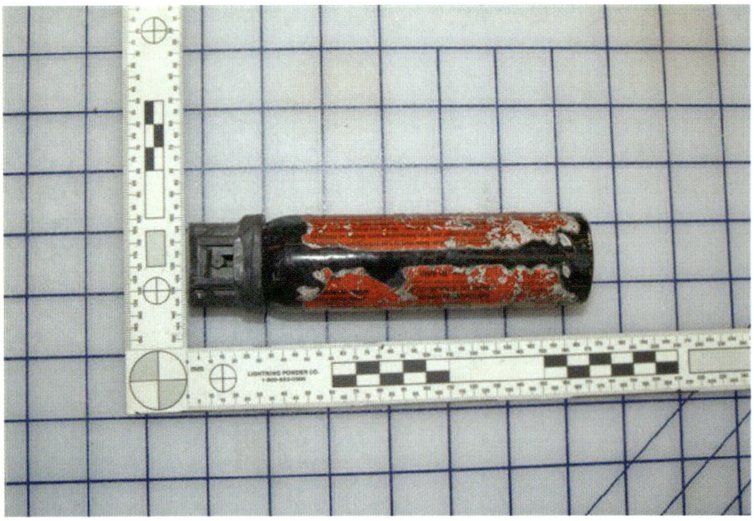

Figure 1.49 Pepper spray from yet another angle.

Figure 1.50 This shows the T-shirt along with the soaking and wicking bloodstains.

Figure 1.51 Further detailed and close-up photos may be required to document each item.

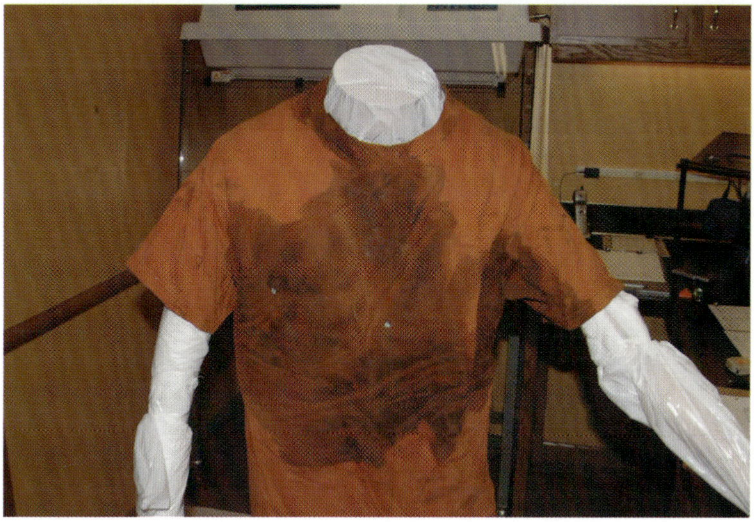

Figure 1.52 Full front view of the orange T-shirt on a mannequin.

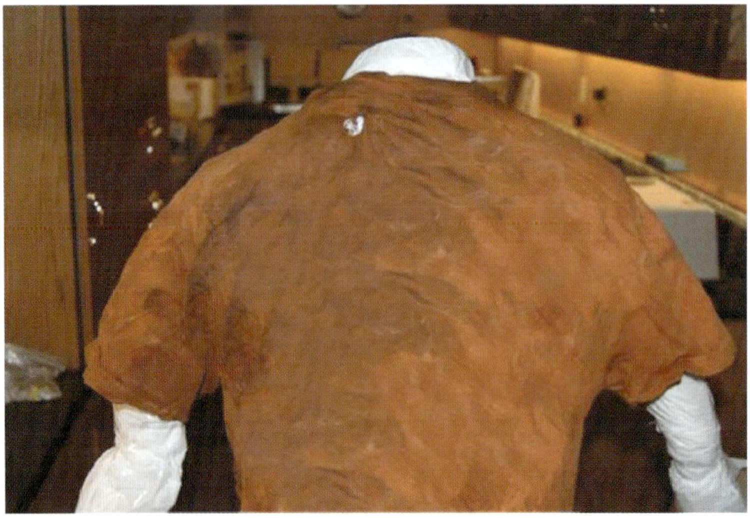

Figure 1.53 The back of the T-shirt showing the bullet exit from shot 3.

Figure 1.54 Full front view of the jean jacket.

Figure 1.55 Close-up of the back neck of the jean jacket showing the exit defect from shot 3, which perforated each layer of clothing.

Figure 1.56 Full front view of the jean jacket on a mannequin.

Figure 1.57 Full front view of the black hoodie sweatshirt.

Figure 1.58 Back view of the hoodie—I initially examine all biohazards under a fume hood with the appropriate filters for both biohazards and the chemical contaminants, which are likely to be encountered in casework.

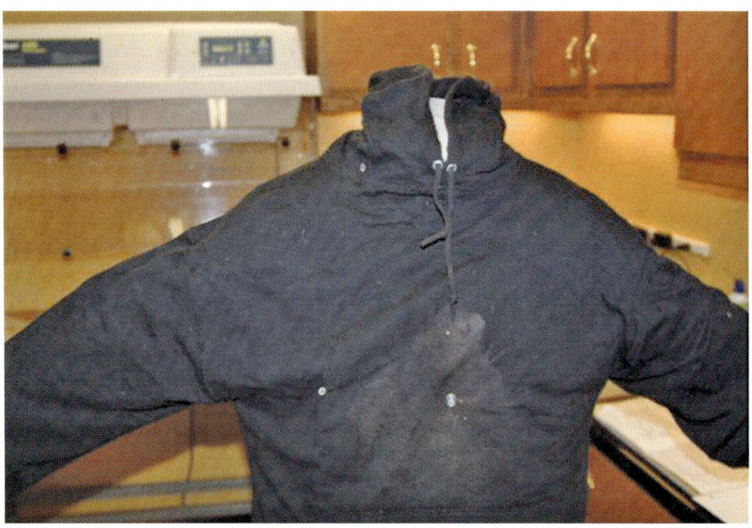

Figure 1.59 Front view of upper torso and neck of the hoodie on a mannequin with pushpins showing the bullet entrance locations on the garment.

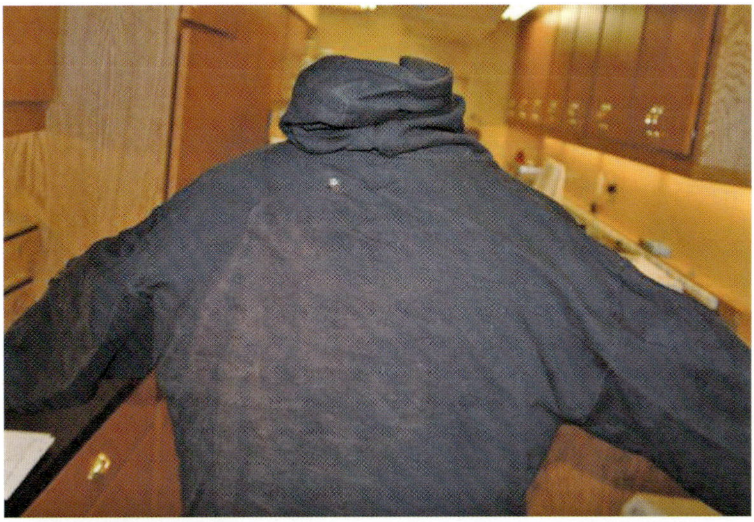

Figure 1.60 Front view of upper torso and neck of the hoodie on a mannequin.

Figure 1.61 Long shot, full front view of the hoodie on a mannequin. Above the exam table, a filtered duct helps remove any potential airborne hazards from the clothing items..

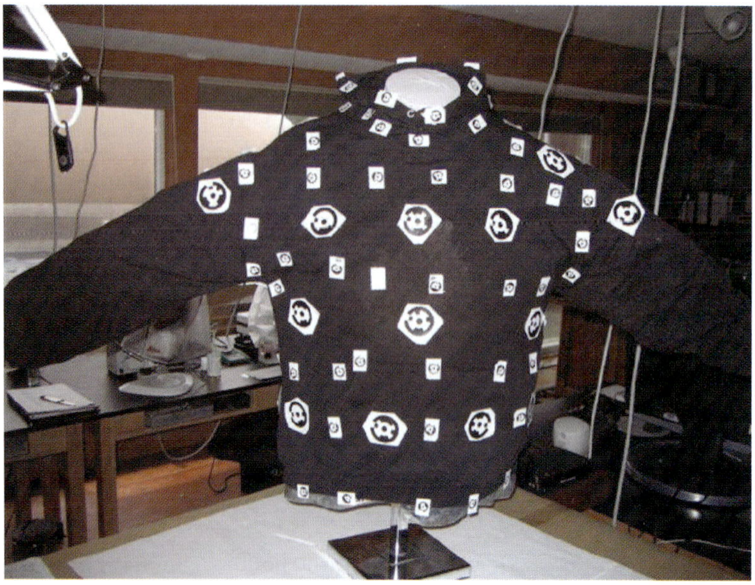

Figure 1.62 Photogrammetry targets documented specific points allowing a calibrated lens to work with the 3D software to produce scaled images. Measured points provide some distance confirmations.

Figure 1.63 Front view of the blue jean pants with belt.

Figure 1.64 View of only the front legs of the blue jean pants.

Figure 1.65 Back view of the blue jean pants with belt.

Figure 1.66 View of only the back legs of the blue jean pants.

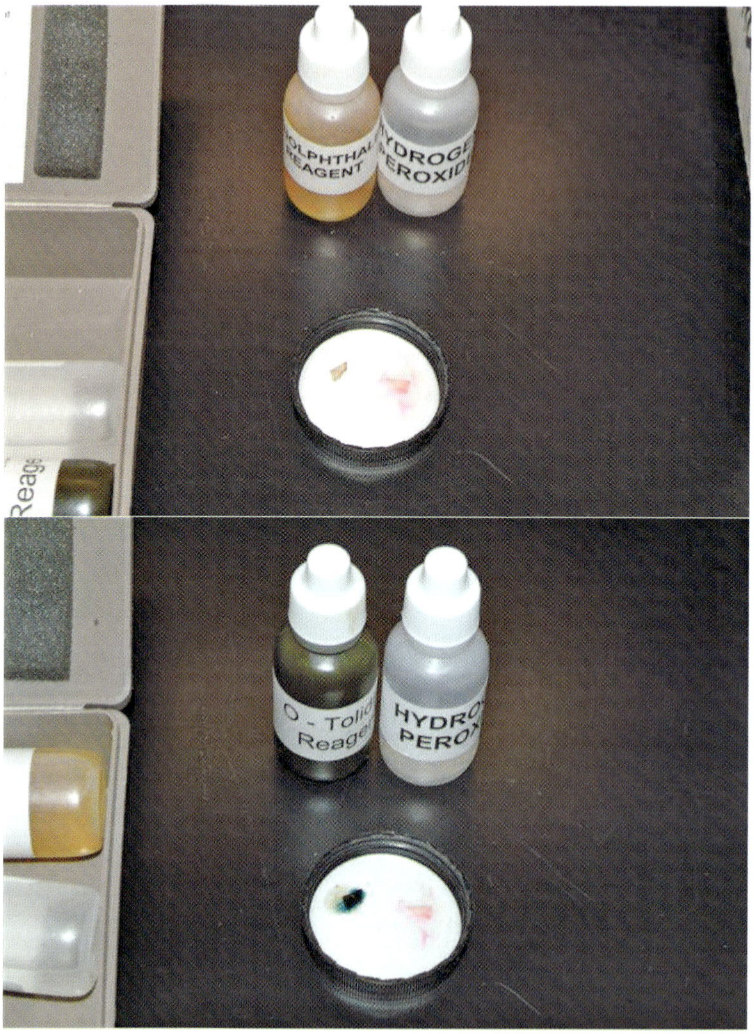

Figure 1.67 Presumptive testing: The control samples test the efficacy of the reagents. Selected stains on the pants were tested and some were found to be negative for blood. The results are usually noted with OT– or Ph– if negative, or OT+ or Ph+ if positive.

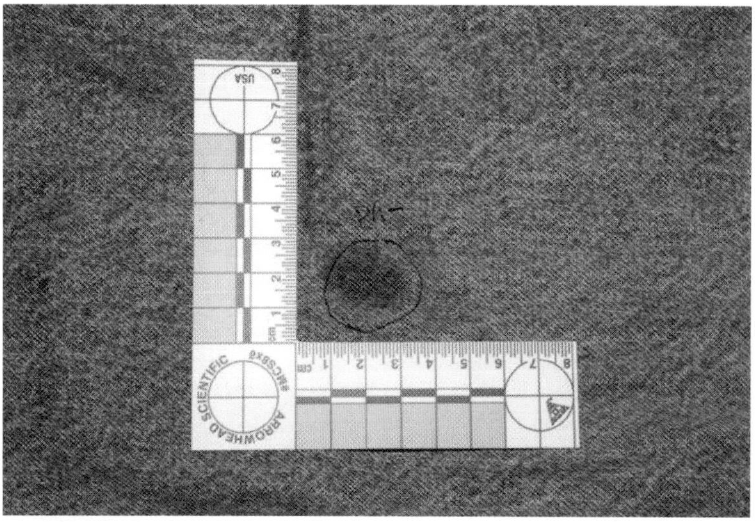

Figure 1.68 An important step for any bloodstain pattern analysis is to locate and document areas of suspected bloodstains and perform a presumptive test for blood (human or nonhuman)—in this case, the stain was phenolphthalein negative, so the stain is not blood. FTIR (Fourier transform infrared) spectroscopy could help identify the substance if necessary.

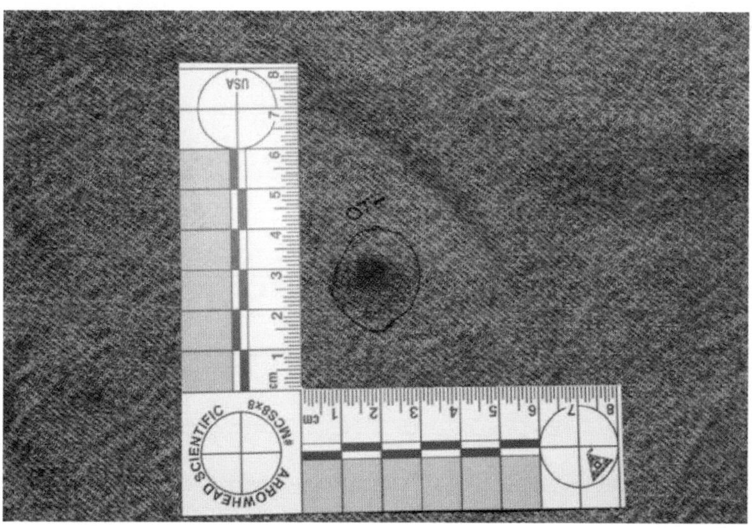

Figure 1.69 Another presumptive test for blood is negative for a sample from the questioned spot—the spot is not blood.

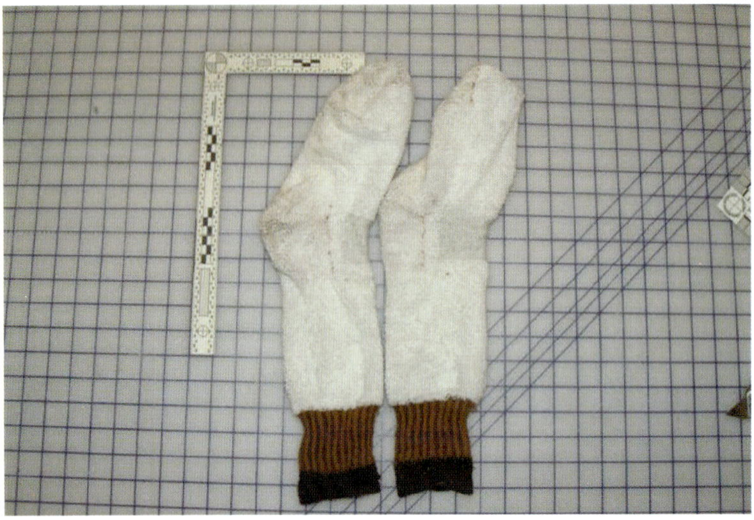

Figure 1.70 Various angles of measurement of the socks. Socks can be overlooked, but are often sources for trace evidence visible with alternate light sources or IR (infrared) photography.

Figure 1.71 Photos taken at higher resolution and saved as TIFF files permit future examinations of evidence by zooming in on the images. We may not yet know what is significant when the photos are taken.

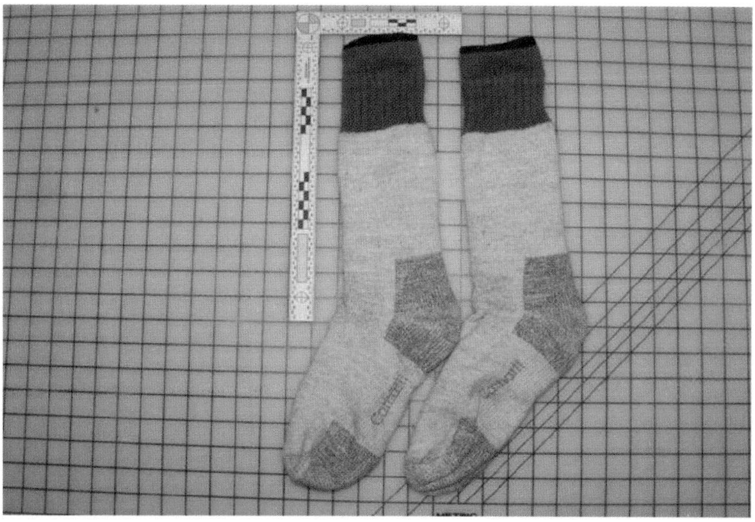

Figure 1.72 Various angles of measurement of the socks. Socks also have an inside and an outside—the analysis of both can often become significant—if not, the time spent taking the extra images is still time well invested.

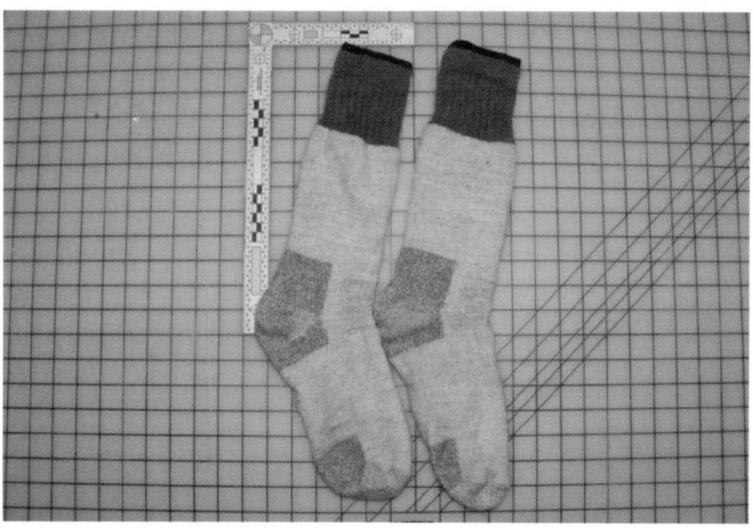

Figure 1.73 Various angles of measurement of the socks.

Figure 1.74 Underwear often supplies a hiding place for items of interest—traces often appear to help identify those items.

Figure 1.75 Tape lifts can aid in testing after microscopy and photomicrographs are taken—tape lifts can be tested in many nondestructive ways.

Figure 1.76 Various angles of measurement of the red underwear. Underwear also has two sides.

Figure 1.77 Various angles of measurement of the red underwear. The grid helps for any necessary warping should that become necessary and if the photos are not all taken at 90-degree angles.

Figure 1.78 Various angles of measurement of the red underwear.

Figure 1.79 Back and front views with measurements of the black knit hat. Tape lifts provide trace evidence collection for any observations requiring further analysis and examinations with test equipment.

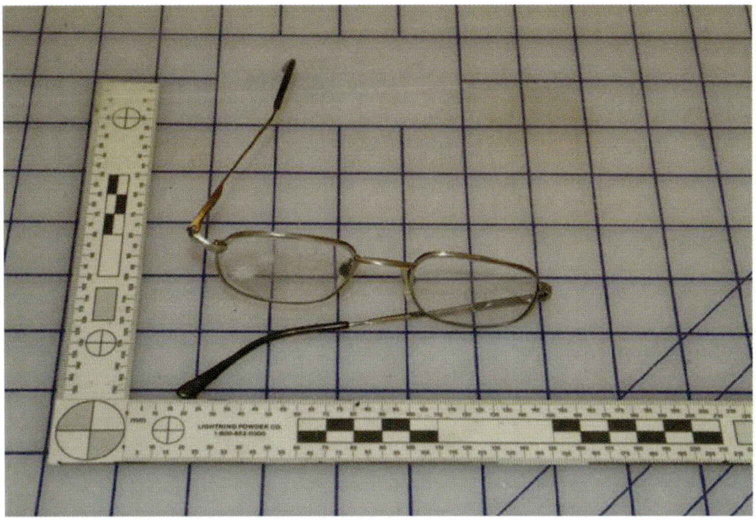

Figure 1.80 Measurement of the glasses. The abrasion patterns and damage to the glasses can help us interpret any fall or impact to the head with the glasses in place.

Figure 1.81 Various abraded material on the surface of the glasses.

Figure 1.82 S&W 40 cal pistol. Pistols are not examined or transported as evidence until they are made safe: for this purpose, cable ties are used to secure the slide and breach, as well as the receiver.

Figure 1.83 Photo of the S&W 40 caliber pistol with measurement.

Figure 1.84 Photo of another angle of the S&W 40 caliber pistol.

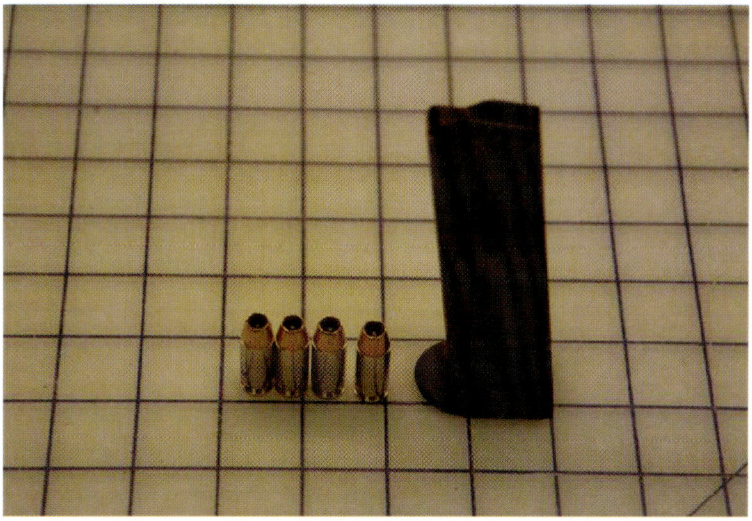

Figure 1.85 Bullets and magazine from the S&W 40 cal pistol used in the shooting.

Figure 1.86 The bullets striking Deal exhibit class characteristics identical with those from bullets test fired for these analyses (Figures 1.85 to 1.95). The cartridge cases recovered on the street also exhibit firing pin impressions, with ejection and extraction marks having the same class characteristics as those exhibited by cartridges test fired from this pistol.

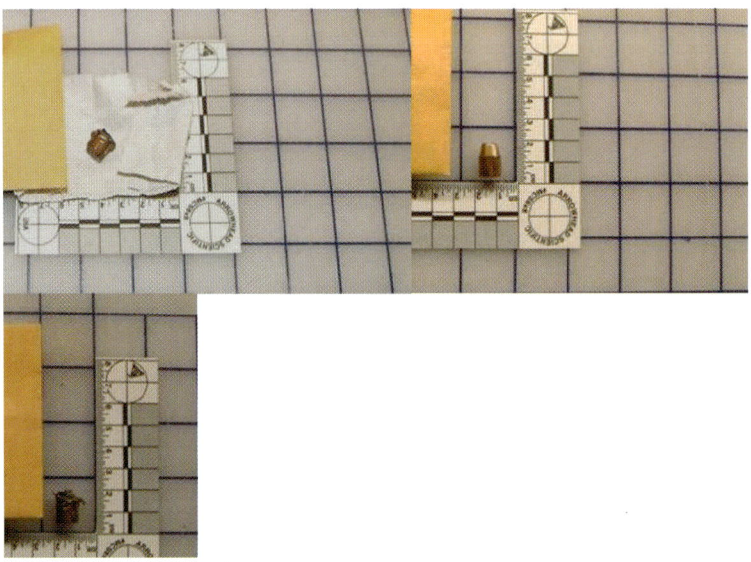

Figure 1.87 Each of the three shots striking the decedent appear on the grid.

Figure 1.88 Close-up of the pistol.

Figure 1.89 The nose of the bullet that hit the brick wall after perforating the decedent's neck—further examination will show the trace evidence from the brick in the nose of the JHP (jacketed hollow point) 40 cal bullet.

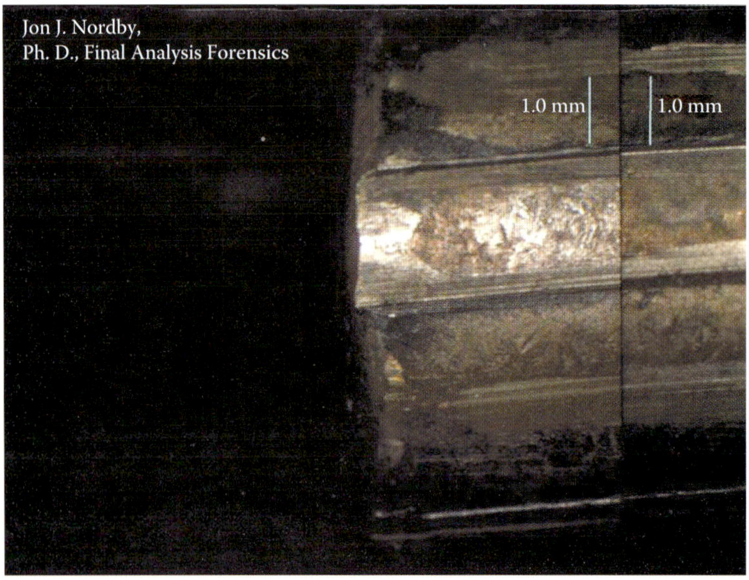

Figure 1.90 This photomicrograph shows a comparison of two bullets: one recovered from the decedent's posterior midline on the right and one recovered near the decedent's armpit in his T-shirt on the left.

Figure 1.91 The bullet striking the brick appears on the right and the bullet from the armpit of the T-shirt appears on the left.

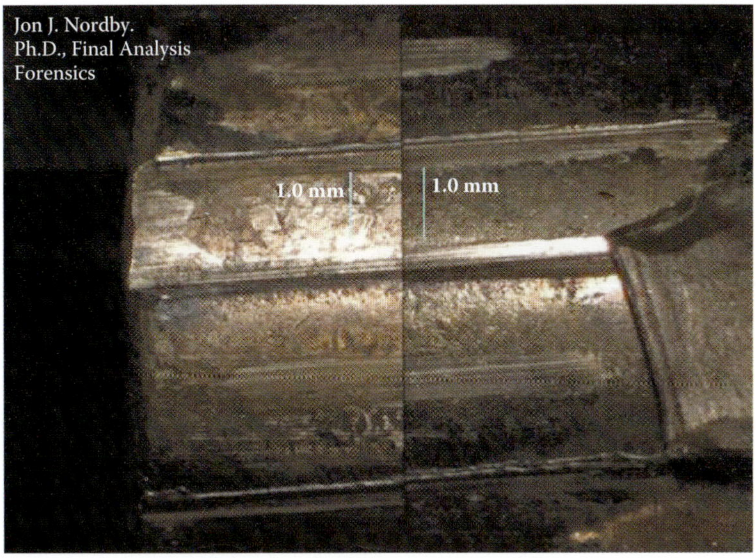

Figure 1.92 A test-fired bullet from the pistol in our shooting range appears on the left and the bullet striking the brick appears on the right.

Figure 1.93 This photomicrograph compares the two cartridge cases recovered at the shooting scene—the firing pin is compared and later the head stamp, the extractor, and the ejector marks on the case are compared and documented. In an optical microscope, depth of field becomes an issue: focus is only possible for one depth at a time.

Figure 1.94 Similar comparisons help document all the remaining cartridge cases found at the scene. Test fires are also compared to help confirm that the same pistol fired each of the shots.

Figure 1.95 Close-up, analysis, and measurements of the nose of the bullet. Comparison of test-fired cartridge cases with cases found at the scene is documented with photomicrographs.

Figure 1.96 Comparison of firing pin impressions from combinations of cartridge cases found at the shooting scene.

Figure 1.97 Close-up of the tail of the bullet. This shows the deformation produced by the bullet striking the brick wall. Traces from the brick also appear on the tail.

Figure 1.98 Microscopy showing the brick trace on the base of the bullet that struck the brick wall. The bullet recovered on the porch displays evidence of its impact with the brick wall: the brown material on the bullet nose and the bullet base is material from the brick (Figure 1.89 and Figure 1.97). On the brick, testing by both XRF and sodium rhodizonate also found traces of lead left by the bullet nose on the brick. This transfer of materials is common and is often referred to as Locard's principle of transfer.

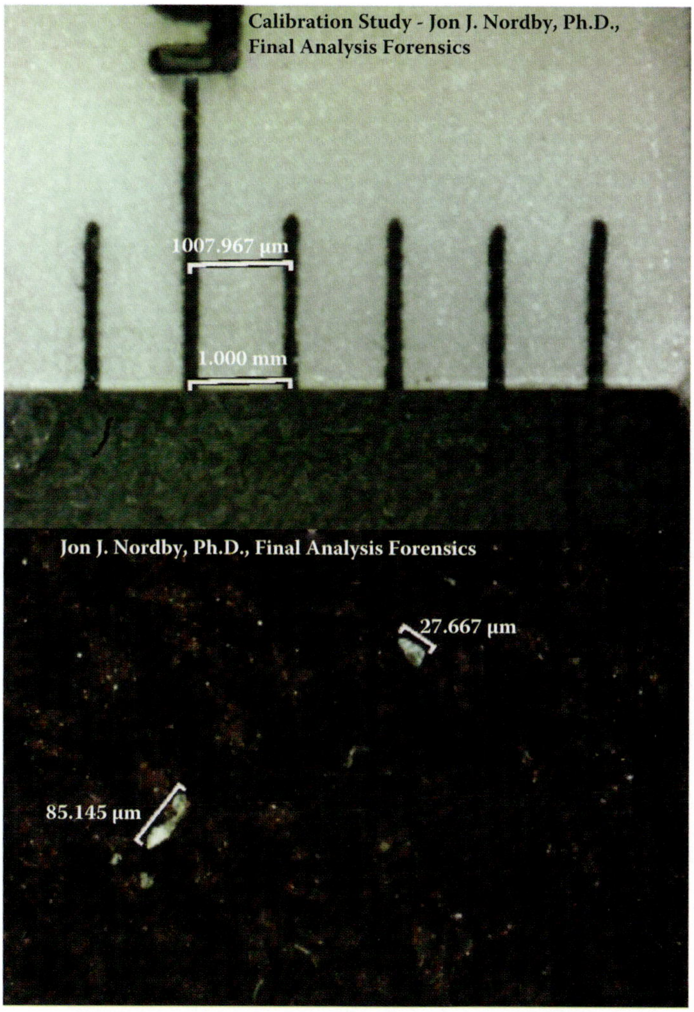

Figure 1.99 Calibration study. Calibration of measurement tools remains an essential first step. Usually, we try to use at least two or three measurement methods for each item—this helps understand the relevant or acceptable variations among measurement methods.

Microscopy

Analysis showed that the trace particles found both on the black knit cap (Figure 1.100) and on the underpants (Figure 1.101) were a mixture of sheetrock dust and cocaine.

XRF testing results appear in the laboratory notes. The evidence has been preserved so that it can be further analyzed by other independent agencies if the need for such analysis ever arises. Further testing by this laboratory identified a mixture of sheetrock dust and cocaine.

Figure 1.100 Trace particles—a mixure of sheetrock dust and cocaine—found on the black knit cap.

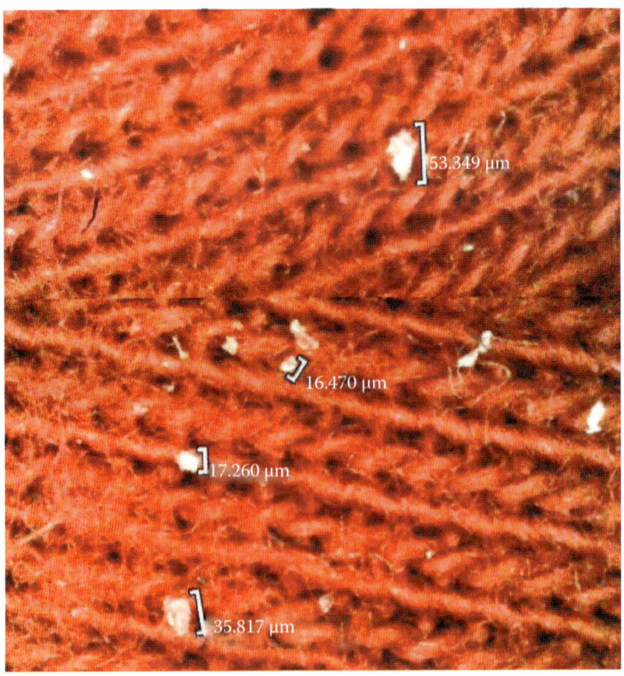

Figure 1.101 Trace particles—a mixure of sheetrock dust and cocaine—found on the red underwear.

Entrance Wounds and Scene Measurements, Photogrammetry, and Total Station Measurements for Limited Shooting Reconstruction

The size and geometry of the autopsy table, together with the distances between the vents, were used to develop and to confirm the basis for more accurate measurements (Figure 1.102).

Blue lines represent vanishing point image orientation, necessary to scale the photo given the fact that the photo was not taken at a measured 90-degree angle from lens to subject (Figures 1.103 to 1.109).

Photogrammetry targets were used to develop 3D modeling of the hoodie as discussed in the report and referenced earlier (Figure 1.110).

Test items: bullets and cartridge cases; Microsil casts of barrel and bullet chip (Figure 1.111)

Trace evidence tape lifts were taken from black knit cap and red underwear for microscopy and analytical testing (Figure 1.112). Nondestructive analyses are always preferred and remain a central standard operating procedure (SOP) in the Final Analysis Forensics laboratory.

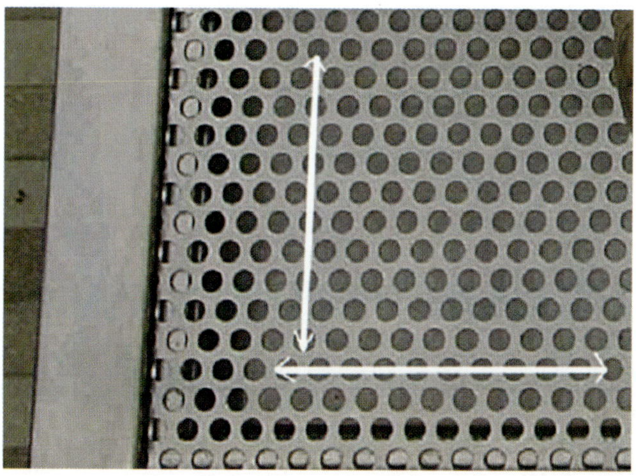

Figure 1.102 The size and geometry of the autopsy table, together with the distances between the vents, were used to develop and to confirm the basis for more accurate measurements.

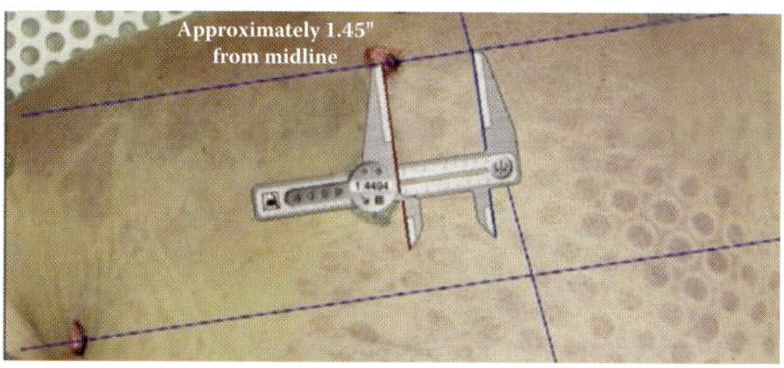

Figure 1.103 Photo and measurement of entrance wound (purple lines represent vanishing point image orientation).

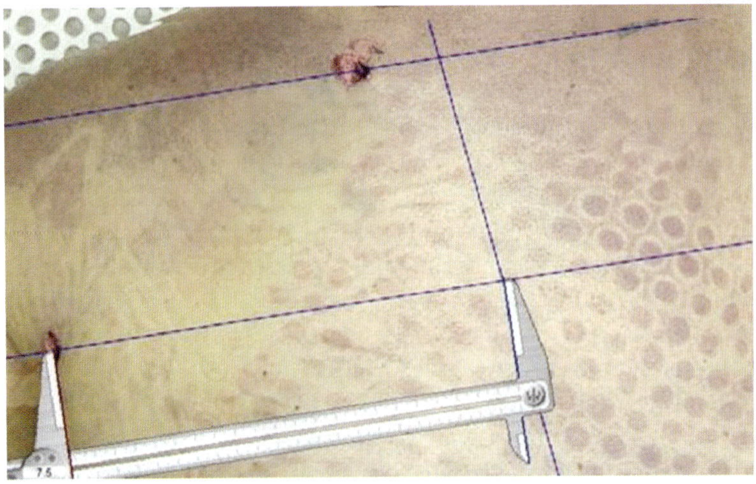

Figure 1.104 Another measurement of the exit wound using the vanishing point correction.

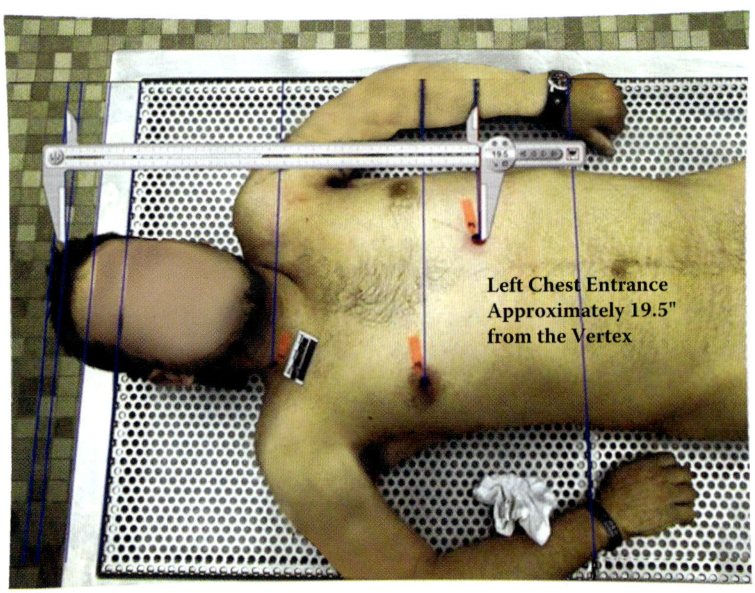

Figure 1.105 Photo and measurement of an entrance wound, again applying the vanishing point image orientation.

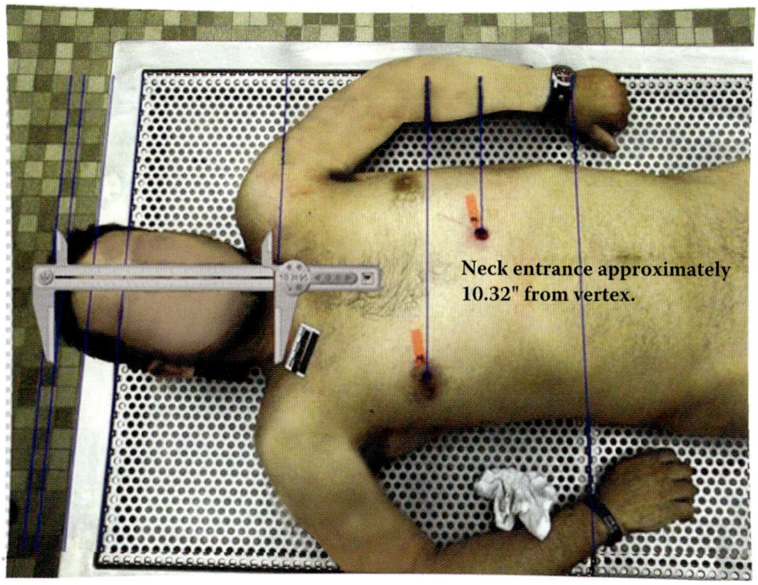

Figure 1.106 Entrance wounds oriented using the vanishing point method. Note that this must be done for each photograph that is used.

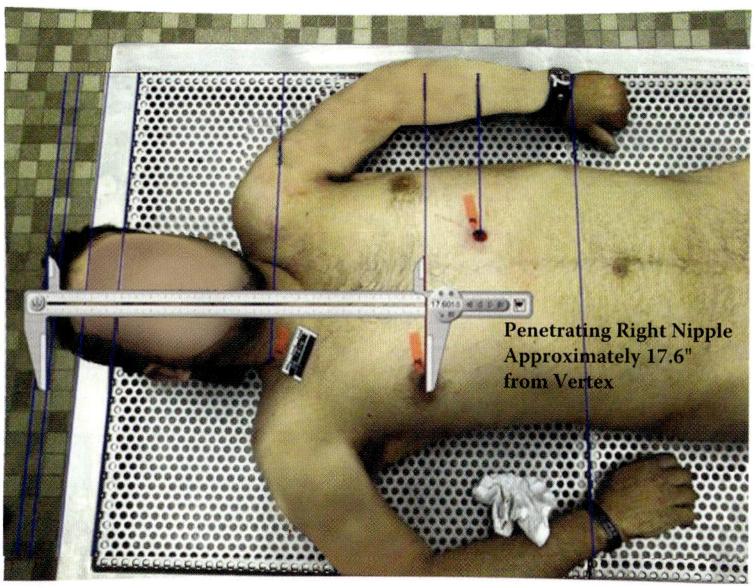

Figure 1.107 The entrance wounds can be oriented from the perpendicular perspective using the vanishing point orientation.

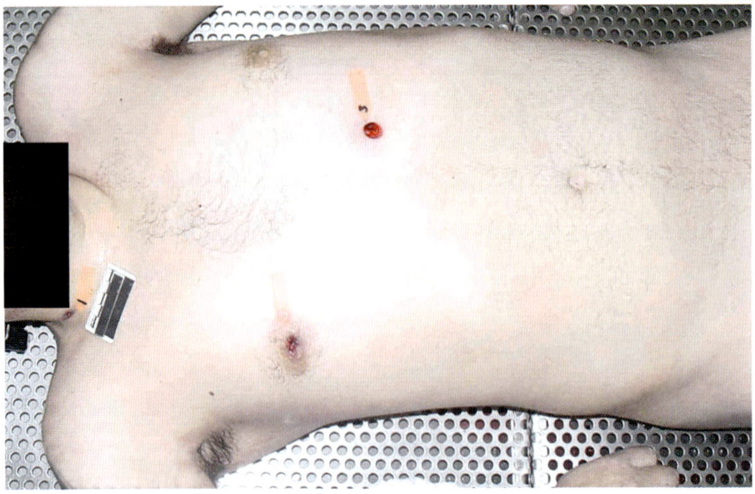

Figure 1.108 The entrance wounds are oriented from the perpendicular perspective using the vanishing point orientation.

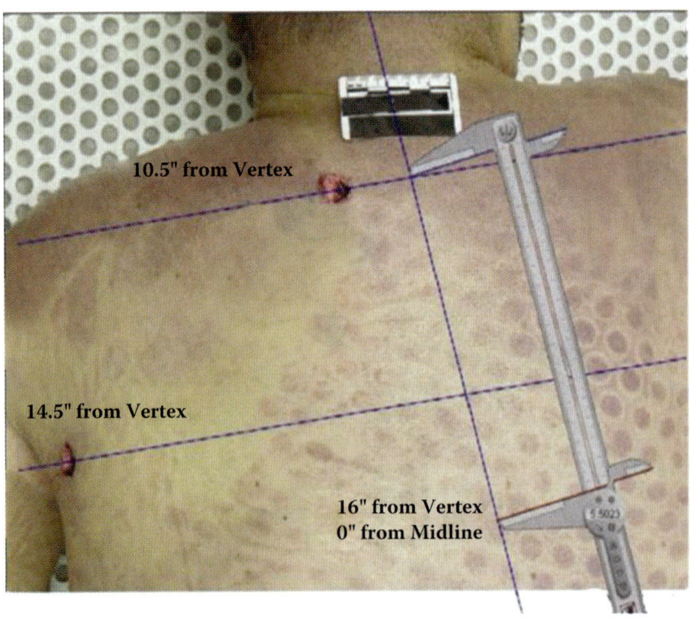

Figure 1.109 Photo and measurement of the entrance wound (purple lines represent the vanishing point image orientation).

Figure 1.110 Photogrammetry targets were used to develop 3D modeling of the hoodie as discussed in the report and referenced earlier.

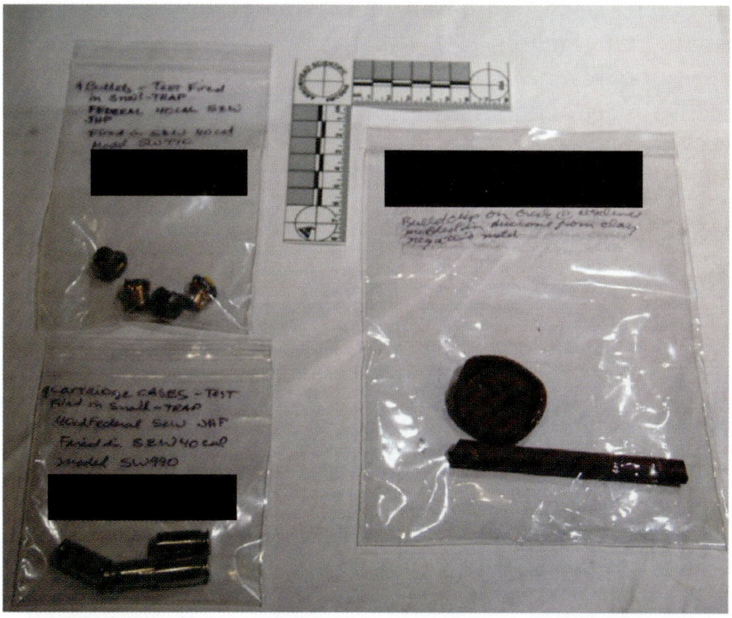

Figure 1.111 Microsil casts of the barrel and bullet chip.

Figure 1.112 Trace evidence tape lifts were taken from the black knit cap and red underwear for microscopy and analytical testing.

The 3D Modeling and Work on the Bullet Chip in the Brick

A cast was taken of the chip using inert modeling clay and the impression was cast with Microsil. The brick was swabbed and tested for lead residues using sodium rhodizonate, the accepted chemical test for lead transfers. The test was positive even after the elapsed period of time between the bullet impact and the testing. (Note that the photomicrograph of the bullet recovered on the porch also exhibits traces of the brick materials composing the wall.) Figures 1.113 to 1.115 show the clay impression with trace present and the Microsil cast from this impression as well as the 3D view of this bullet impact developed from the data.

The chip in the brick exhibits characteristics of a secondary target struck by a bullet that is not tumbling or yawing. The trace evidence testing mentioned earlier confirms what is visible in Figure 1.113. Lead residue is present on the left side (gray discoloration) of the chipped area.

A clay impression was made from the bullet impression on brick in order to make a mirror image of the chip (Figure 1.114).

A Microsil casting was made of the bullet impression made from clay mold. The clay mold is the negative image of the chip; the Microsil cast is the positive image mirroring the chip on the brick.

Point cloud data was developed from photogrammetry targets pasted on the brick wall (Figure 1.116) and photos of the chip were taken in order to display the brick as a secondary target.

Figure 1.113 The chip on the brick wall provides the necessary connection between the perforating shot, the victim, and the scene.

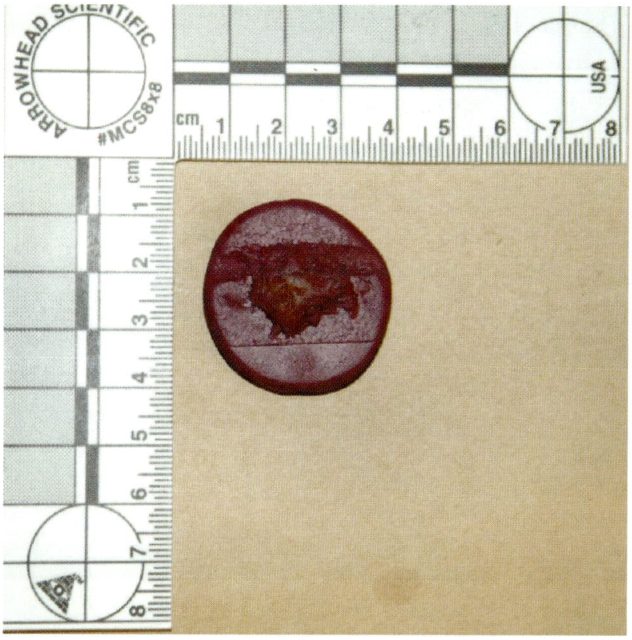

Figure 1.114 A clay impression was made from the bullet impression on the brick in order to make a mirror image of the chip.

Figure 1.115 This clay impression was made from the bullet impression on the chip creating a mirror image.

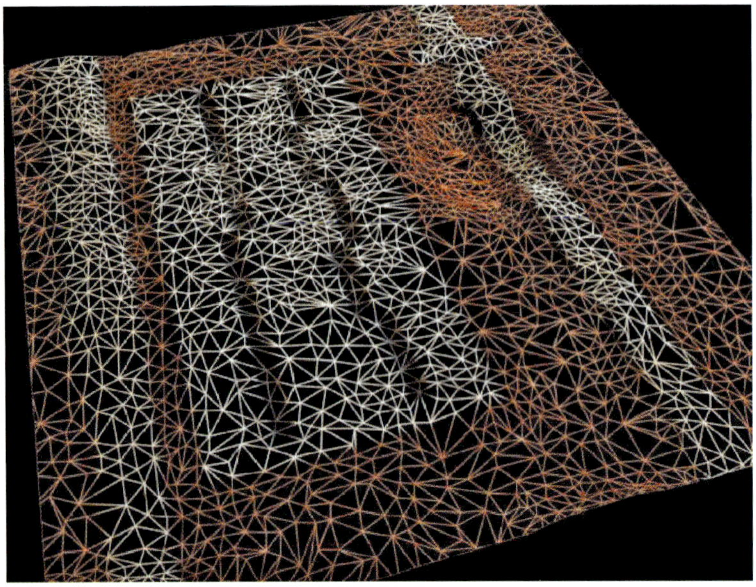

Figure 1.116 Point cloud data was developed from photogrammetry targets pasted on the brick wall and photos of the chip were taken in order to display the brick as a secondary target.

Point Cloud Data Representing the Chip

The 3D image representing the chip in the brick made by the bullet results from these efforts (Figure 1.117). The 3D imagery allows for more precise and accurate metrology with both impact patterns and any cartridge case or bullet comparisons required by the reconstruction (Figure 1.118).

The wall struck by the bullet that first perforated Deal's upper torso and neck area before hitting the brick is shown in Figure 1.119. The design of the porch, including the support pillars, deck, and furniture, helped support and develop trajectory data (Figure 1.120). It is important to remember to take exhaustive measurements when the scene is undisturbed in order to document the positions of items that may be moved, such as furniture. The deck and support pillar, for example, are less likely to be adversely affected simply by the passage of time.

Photographic documentation of the scene remains critical in order to develop any scientific crime scene reconstruction. Some basic skills allow one to turn a digital camera into a sophisticated measurement tool and to transform what might appear to be ordinary photographs into documentations not dissimilar to total station data—for a lot less cost (Figures 1.120 and 1.121).

Figure 1.117 The three-dimensional image representing the chip in the brick made by the bullet.

Figure 1.118 Another angle of the three-dimensional image representing the chip in the brick made by the bullet.

Figure 1.119 The wall struck by the bullet that first perforated Deal's upper torso and neck area before hitting the brick.

Figure 1.120 The porch, the wall, the deck, and the pillars showing their relationships.

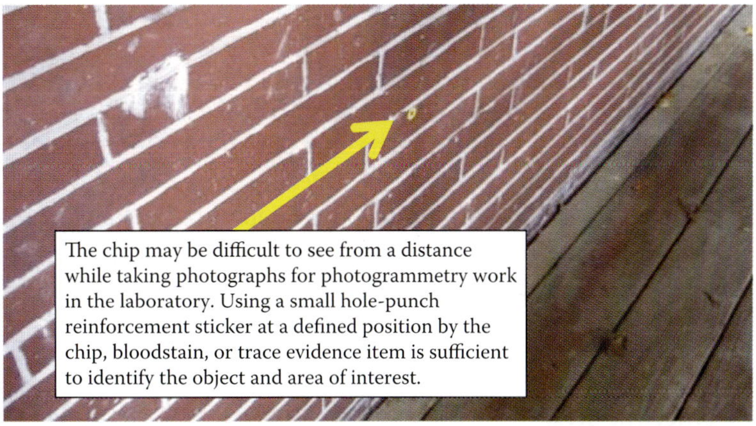

Figure 1.121 Long shot of the wall struck by the bullet that first perforated Deal's upper torso and neck area before hitting the brick.

The Medical Examiner's Autopsy Report

Introduction

When developing a shooting reconstruction, no data should be merely accepted at face value. All data must be reviewed, regardless of source or science, and in some cases this data must either be developed or corrected. Often the data may have been developed for purposes quite distinct from the purposes involved in developing a shooting reconstruction. The autopsy specifically requires careful scrutiny.

An autopsy report serves to document the cause and manner of a human being's death. It describes any injuries sustained that either caused or contributed to the death whenever such descriptions are possible. In this case, the cause of Deal's death is unequivocally attributed to "multiple gunshot wounds" and the manner of Deal's death is unequivocally classified as "homicide"—a death caused by the actions of another human being. In medical examiner's terminology, of course, *homicide* does not have the same meaning as *murder*—the type of homicide, such as justified, negligent, and aggravated, that falls into legal and moral rather than medical and scientific categories. The scientific goal of an autopsy remains limited to determining both the cause and the manner of death. It does not include establishing the exact medical mechanism(s) of death, nor does it include a precise or accurate reconstruction of the production of any fatal injuries.

A scientific shooting reconstruction, however, has a decidedly distinct purpose. As discussed earlier, the science of ballistics, including terminal ballistics, considers data sufficient to establish the trajectories of bullets into, through, and out a target, such as a human body. The purpose is to establish, if scientifically possible, additional facts not readily available through a relatively quick examination of a decedent through an autopsy. The purposes remain distinct, and although related, the sciences applied remain equally distinct. (See comments about accuracy and precision as footnoted earlier in the discussion of scientific method.) This type of analysis requires relevantly accurate or precise measurements and must account for all the relevant physical evidence whenever possible.

Autopsy Report Summary

The cause and manner of Deal's death—the scientific and medical goal of the autopsy—is reported to be homicide by multiple gunshots. The medical examiner carefully states that he ascribes no order or firing sequence to the shots by the order in which he describes them. He proceeds to describe the injuries to Deal. The major injuries involve three injuries on the anterior surface of Deal's body identified as bullet entrance wounds, and two injuries on the posterior surface of Deal's body identified as bullet exit wounds. He opens the tissues around the midline of the back and recovers the third bullet, which did not exit.

Deal Injuries Documented

The multiple gunshot wounds to the chest and neck are numbered for convenience only (the medical examiner states that "Photos must be viewed for clarity") and locates the injuries on the body as follows.

Indeterminate Range Penetrating Gunshot Wound to the Left Chest

Dr. Grunewald (the medical examiner) finds an entrance on the left chest wall measuring 19½ inches from the top of the head and 1½ inches left of the anterior midline. He says that it is a ½-inch round gunshot wound with abraded margins at the 5 and 10 o'clock positions.

Associated with this entrance, he describes an exit near the left axilla (armpit area) on the back 14 inches from the top of his head and 7½ inches left of the posterior midline. A bullet is recovered in Deal's clothing (it did not perforate all layers of his clothing).

Indeterminate Range Perforating Gunshot Wound to the Chest

Dr. Grunewald documents an entrance into the right nipple measuring 16½ inches from the top of the head and 4 inches right of the anterior midline with abraded margins. He finds no soot, stippling or searing, which means the muzzle was at an intermediate or long range from the injury.

Indeterminate Range Perforating Gunshot Wound to the Neck

Dr. Grunewald finds an entrance on the lower lateral aspect of the neck on the right side centered 9½ inches from the top of the head and 3 inches right of the anterior midline. He describes a round ½-inch entrance with abraded margins with no soot, stippling, or searing with a hemorrhagic wound path perforating the soft tissues of the neck.

Associated with this entrance is an exit on the upper back (toward the left side) centered 10½ inches from the top of the head and 1¾ inches left of the posterior midline with abraded margins at 7 and 10 o'clock. He adds that no bullet is recovered.

Autopsy Wound Measurement Summary

The following chart summarizes Grunewald's measurements of the three entrances and two exits.

Medical Examiner's Description—No Sequence Implied	Medical Examiner's Label Numbers in Report	ME Report	
		X Coordinate	Z Coordinate
Perforating gunshot wound to the left chest	3	1.5	−19.5
Penetrating gunshot wound to the right chest (at nipple)	2	−4	−16.5
Perforating gunshot wound to the neck	1	−3	−9.5
Exit left of the posterior midline	4	7.5	−14
No exit, recovered 16″ at the midline on the back	None assigned	No exit	No exit
Exit left of the posterior midline	5	1.75	−10.5

Note: All measurements in inches. Measurements taken from vertex of head. Negative Z direction is toward feet. Negative X direction is toward right side of body starting at midline.

Deal's Autopsy X-Rays

The X-rays of Deal provide some additional information about his injuries. The films were taken of Deal before his clothing was removed. Items such as the large "safety pin," as well as the buttons and pin on his jeans, jacket, and necklace, appear in the images (Figure 1.122). The bullet recovered in his clothing can also be seen just below the left shoulder in Figure 1.122. Figure 1.123 shows Deal's head and Figure 1.124 shows his torso; with items that appear in his pockets.

Figure 1.122, the right side image, was produced by combining the X-rays of the left and the right anterior chest. Figure 1.122 also shows a fracture of the condyle of the second left rib (the condyle is the rounded surface at the end of the bone). This evidence of a fracture helps establish the trajectory of the shot into the right nipple area, correlating it with

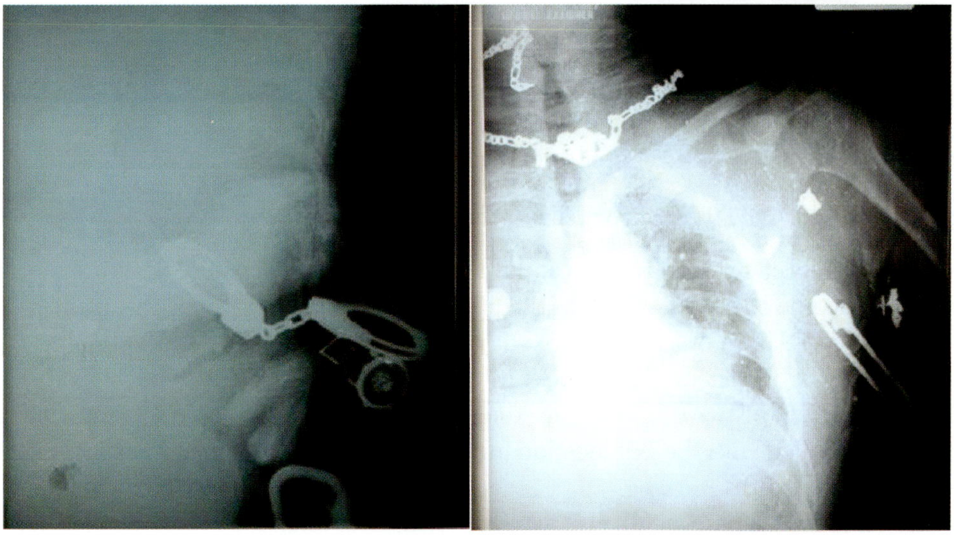

Figure 1.122 Autopsy X-rays show items such as the large "safety pin," as well as the buttons and pin on Deal's jeans, jacket, and necklace. The bullet recovered in his clothing can also be seen just below the left shoulder.

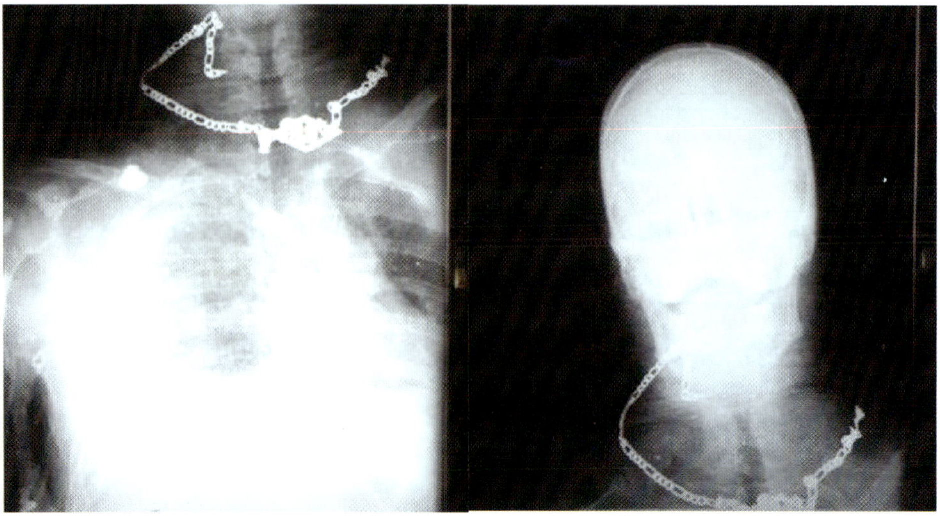

Figure 1.123 X-rays of Deal's neck and head showing the necklace.

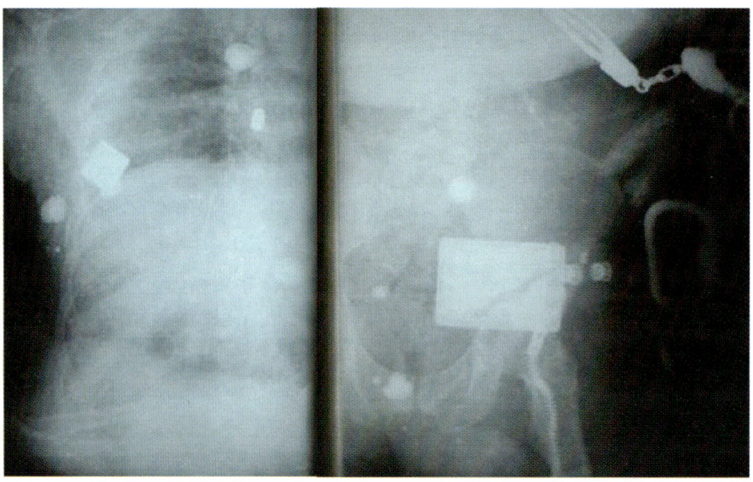

Figure 1.124 X-rays of Deal's torso. Items appear in his pockets

the exit above the left armpit area (left axilla). This leaves the lowest shot (striking the apex of the heart) as the bullet recovered in the paraspinal space. This physical evidence helps establish the correlation of the entrance with the exit of the first two shots striking Deal. The discussion is further developed next in the consideration of the medical examiner's confusion between the two exit locations, the exit location for shot 1 (to the lower left chest, a penetrating bullet) with the exit location for shot 2 (to the right nipple, a perforating bullet with an actual exit).

The X-rays are not high-quality images filled with clear anatomical details. In fact, most X-rays taken postmortem often leave a lot to be desired. Often the equipment is old, well worn, and, in some cases, even antiquated by comparison with newer machines. Budgets remain inadequate in many facilities. Poor images can sometimes be improved by using software such as Photoshop. The software work must be documented and all the steps taken must be repeatable, with identical results, when followed by anyone and applied to

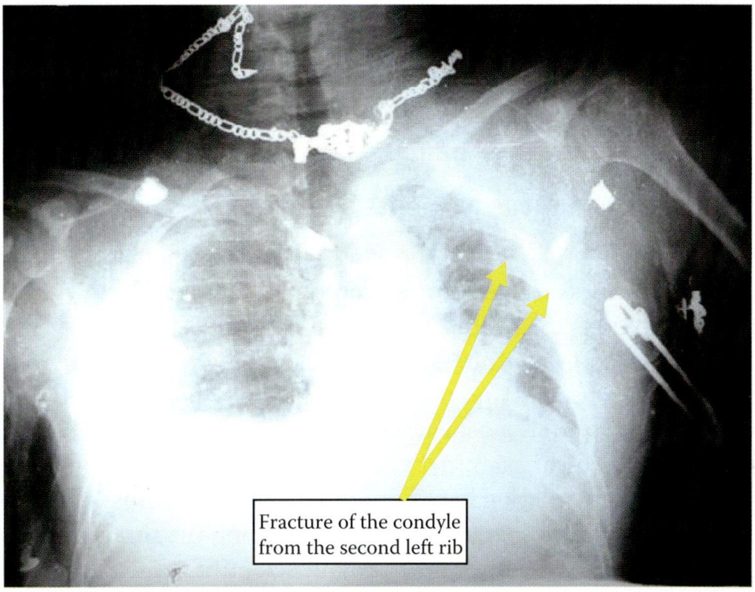

Figure 1.125 This X-ray was produced by combining the X-rays of the left and anterior chest, showing a fracture of the condyle.

the same original data. However, the results are often disappointing. Nothing is a substitute for taking readable postmortem X-rays (Figure 1.125), whatever that may require in specific circumstances. Such clear images can help avoid or resolve future confusion.

Scientific Shooting Reconstruction Independent Measurements

The following summarizes the independently derived entrance- and exit-wound measurements developed for this reconstruction from the medical examiner's photographs using the photogrammetry techniques discussed earlier.

Shot, Entrance/Exit Reconstruction Sequence	X Coordinate	Z Coordinate
Shot 1 Entrance	2	−23.5
Shot 2 Entrance	−5.5	−20.5
Shot 3 Entrance	−2.5	−12
Shot 1 Exit	No exit	No exit
Shot 2 Exit	7	−19
Shot 3 Exit	1.5	14

Note: All measurements in inches. Measurements taken from vertex of head. Negative Z direction is toward feet. Negative X direction is toward right side of body starting at midline.

The following two sections demonstrate the similarities and the differences involved, first by using the *measurements from the autopsy* in place of the independently derived measurements from the ballistics analysis to reconstruct the shooting and, second, the differences involved by using the *trajectories from the autopsy* in place of the independently derived trajectories from the ballistics analysis to reconstruct the shooting. The similarities and differences are discussed next. Often such a visualization of differences is sufficient to initiate appropriate revisions.

The Medical Examiner's Measurements

Figure 1.126 to Figure 1.28 show the shooting reconstruction model developed using the medical examiner's measurements of entrance and exit locations with reconstructed shooting sequence and trajectories. These images show that Deal would occupy essentially the same position for each of the three shots whether using the medical examiner's measurements locating the entrances and the exits on his body or using those measurements developed independently from the data alone.

The purpose of the medical examiner's measurements was not to produce an accurate or a precise shooting reconstruction based upon the physics of ballistics. Rather they were made to document the relative spatial relationships among the three shots striking Deal. Such relative measurements are sufficient to provide scientific medical support for his conclusion regarding Deal's manner of death.

The purpose of the autopsy was not to produce a traditional ballistics analysis in the field of physics. Indeed, if we were to accept the medical examiner's trajectories through Deal's body, the third shot would have missed him completely and the second shot would either have missed him or would have struck him at a completely different angle and location. These facts are developed and explained next.

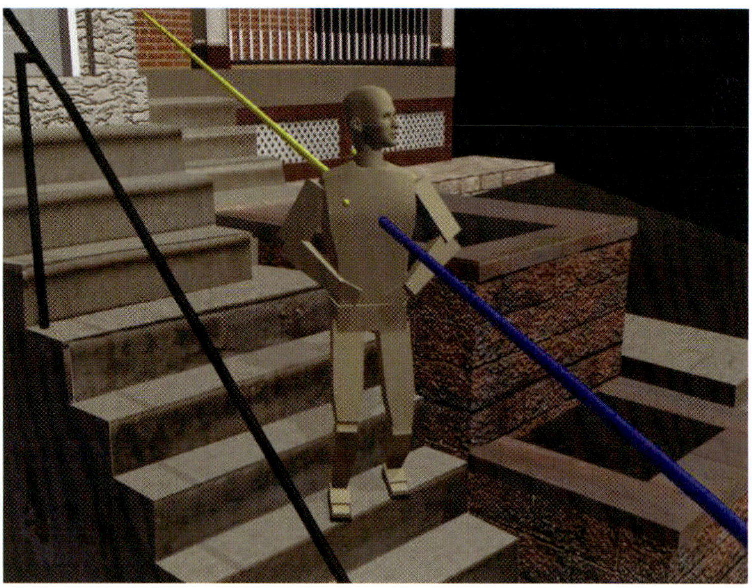

Figure 1.126 Shot 1 using the medical examiner's measurements for the location of the bullet's entrance and exit, with reconstructed shooting sequence and trajectory. This orients the trajectories of each shot (using the medical examiner's measurements) approximately within the trajectory envelope developed in the scientific shooting reconstruction based upon independently derived measurements.

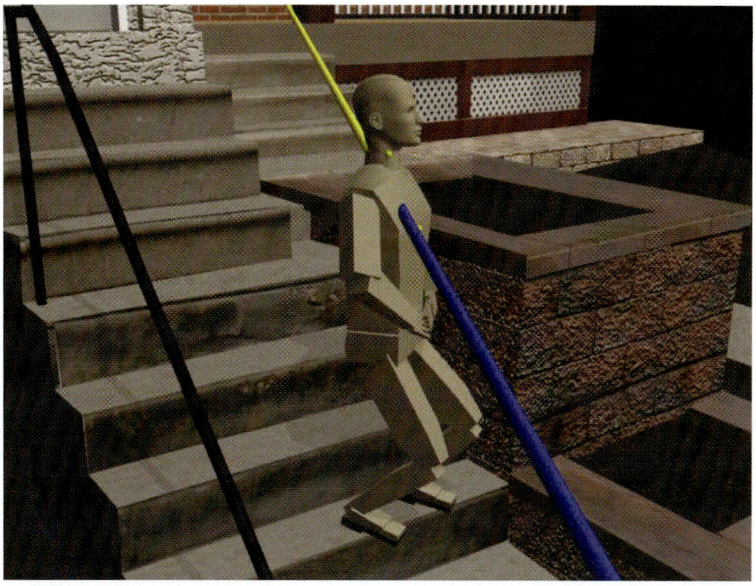

Figure 1.127 Shot 2 using the medical examiner's measurements for the location of the bullet's entrance and the exit, with reconstructed shooting sequence and trajectory. Again this orientation (using the medical examiner's measurements) falls within the trajectory envelope and places Deal in the same relative position as the ballistics work.

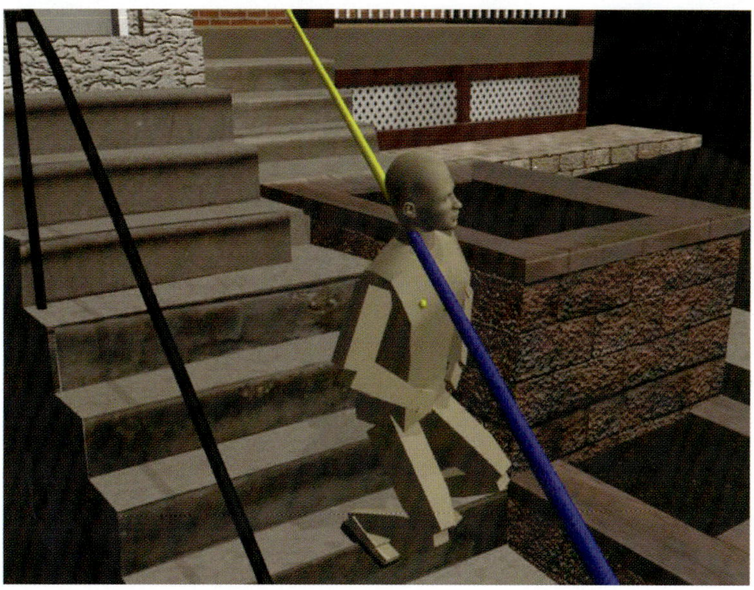

Figure 1.128 Shot 3 using the medical examiner's measurements for this bullet's entrance and exit, with reconstructed shooting sequence and trajectory.

The Medical Examiner's Trajectories

Figure 1.129 to Figure 1.141 show the shooting reconstruction model developed by using the medical examiner's measurements of entrance and exit locations together with the medical examiner's trajectories of the two lower shots striking Deal.

Given the trajectory proposed by the medical examiner, the first shot would place Deal in a backward-leaning position (Figure 1.129). According to his autopsy, this shot essentially destroyed the lower portion of his heart muscle* (the apex of his heart). If he were in the position necessary to accommodate the proposed trajectory when he was struck,† Deal would then begin to fall backward immediately after suffering such an injury. The fact that he did not fall backward at all but that he in fact fell forward shows us that this cannot be the correct trajectory for this shot.

Additional facts conflict with this proposed trajectory associating this entrance and exit. If the bullet entering Deal's lower left chest damaged the heart as described and if that same bullet exited from his left axilla, then the entrance wound would appear differently than it did. The medical examiner describes the entrance as having abraded margins. These abrasions around the circumference of a bullet entry wound, referred to as "abrasion rings," help to distinguish entrance wounds (which have such marks) from exit wounds (which lack such marks). They can also help supply information about the angle of entrance.

If one section of the wound's circumference exhibits pronounced abrasions and another section of the circumference lacks such abrasions, then the bullet's angle of entry is more acute with (leaning more toward) the abraded region and more obtuse toward (leaning more away from) the clear region. If the angle is acute enough, then the abrasion ring, as

Figure 1.129 Shot 1 using the medical examiner's trajectory from the entrance in the lower left chest to the exit he associates with it near the upper left axilla.

* He describes damage to the heart muscle as "fragmentation." This would disrupt the heart's capacity to pump physically as well as to function electrically.

† Accommodating this trajectory essentially involves connecting the point of entrance with the proposed point of exit through his body. These two points determine the proposed trajectory.

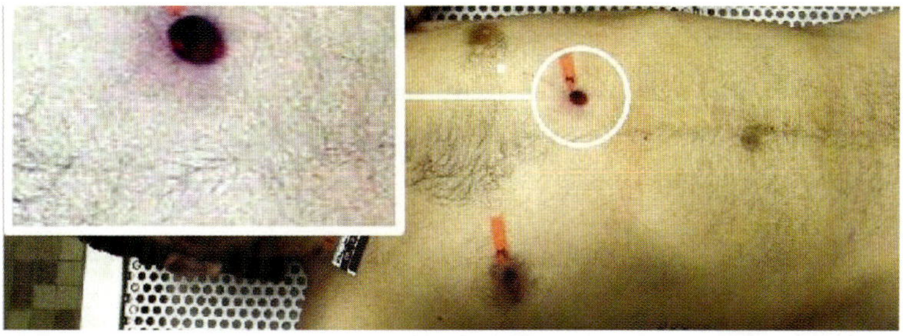

Figure 1.130 The photo shows the lower entrance wound with a close-up inset to the left isolating the lowest entrance shot.

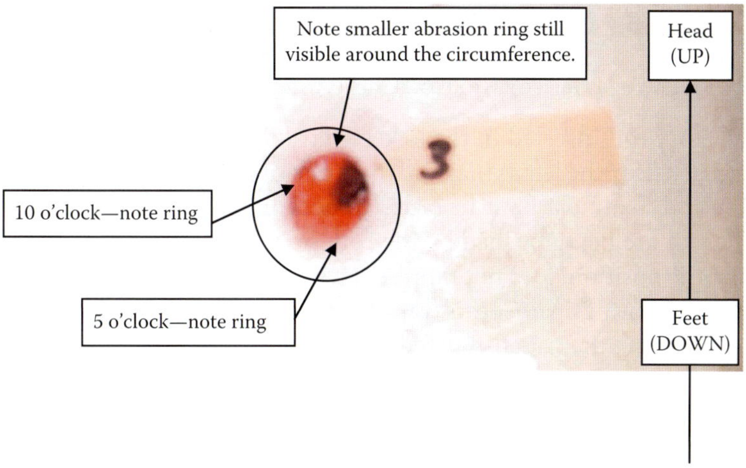

Figure 1.131 A close-up view shows the lowest entrance. Note that the abrasion ring is visible all around the circumference of #3 with more pronounced thickness in the locations indicated by the medical examiner.

well as the entry defect, appears elliptical rather than spherical. In general, the more round the entrance, the closer the trajectory approximates 90 degrees.*

The medical examiner describes the abrasion ring from this shot as follows: "the abrasion is more prominent from 5 o'clock to 10 o'clock." This observation is documented in the details taken from the autopsy photo reproduced in Figures 1.130 and 1.131. The location and the relative size of this ring indicate that the angle of entrance is closer to 90 degrees than to any acute angle (as noted earlier) required by the proposed trajectory. If the proposed trajectory were accomplished by this shot, then one would expect a much more pronounced abrasion appearing at the 6 o'clock position with little or no abrasion appearing at the 12 o'clock position. The relatively spherical appearance of the actual entrance, then, as well as the actual abrasion ring noted, also count against this proposed trajectory.

Another fact conflicting with this proposed trajectory can be seen in a photo taken of a trajectory rod (Figure 1.132). Such trajectory rods can be inserted through the body to explore connections between any two perforating gunshot wounds. In this case, the photo

* Factors influencing this rule of thumb include the thickness of the tissues involved, their proximity to bone, and the thickness and density of any body fat or muscle in the area. It does not mean that the trajectory is identical with 90 degrees.

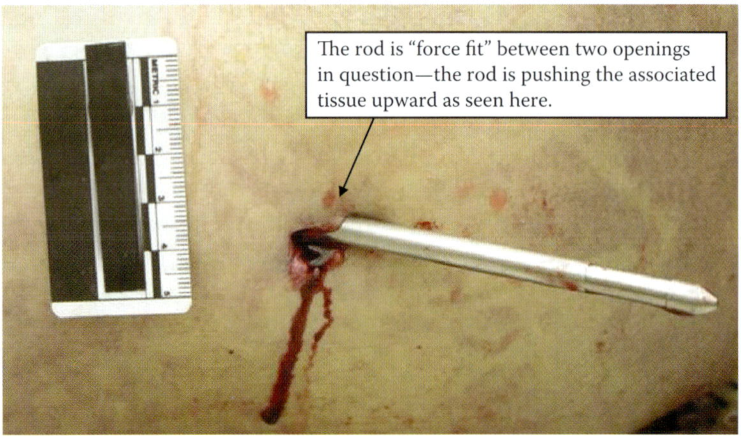

Figure 1.132 A photo of a trajectory rod force fit to connect an entrance with the incorrect exit.

pictures a trajectory rod inserted in the lowest entrance, discussed earlier, and the exit on the decedent's back near the left axilla. The rod's position as pictured does not approximate a good fit. The rod is almost bent in order to accommodate the association. The upper margins of the exit are being pushed by the rod, severely altering the exit wound's shape, suggesting that the position does not represent a best-fit condition. However, the most significant evidence counting against this trajectory further clarifies and resolves any possible ambiguities concerning the data discussed earlier.

The most conclusive fact counting against this proposed trajectory involves not just the locations of the other two shots: the most conclusive fact remains that Deal was, in fact, struck by the two additional shots at all. If the trajectory rod inserted through the lowest entrance and the left axilla exit were the correct trajectory for the first shot, then it would be physically impossible for the remaining two shots to strike Deal. He would have initially fallen on his back, dead from a single fatal gunshot wound to the lower left chest.

The locations of all three gunshot wounds are documented in the autopsy photographs. Given the single perforating gunshot wound, which also struck the brick, we know Deal's position for what had to be the last shot. This has been established through the boundary conditions documented by the trajectory envelope discussed in this report.

Deal's position, required for the first shot trajectory proposed by the medical examiner, is illustrated by Figure 1.129. We know that this would be his position first because of the information provided by the perforating shot that struck the brick and, second, because this is the angle required when linking the lowest entrance and the axilla exit wounds in question under the medical examiner's hypothesis.

Given this trajectory, when struck Deal would have fallen both backward and downward. He would fall backward because of the leaning position necessary to accommodate the proposed trajectory within the trajectory envelope. (His body would have to be within the trajectory envelope for this and for every shot given the unequivocal data provided by the perforating shot that struck the brick as well as the approximate 1-second time frame given for the three-shot sequence.) He would fall downward when the force maintaining this leaning position was overcome by the force of gravity. The damage to his heart (Figure 1.133) was sufficient to initiate this fall.

Under ideal conditions, classical physics tells us that the distance of a free-falling object in space can be derived by applying the basic fact that a free-falling object will

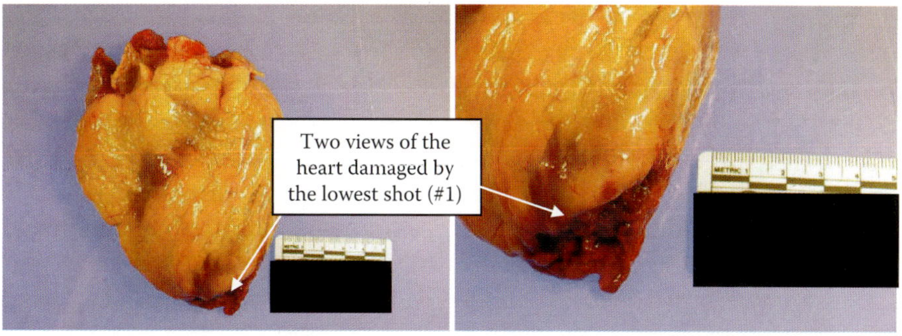

Figure 1.133 Two views of Deal's heart, showing the damage by shot 1.

accelerate at 32 feet per second per second (sec^2) during free fall. Deal fell downward 13.2 inches between the first shot's entrance and the third shot's entrance (see measurements in report). Classical physics tells us that a free-falling object, under assumed ideal conditions, would fall these 13.2 inches in .261 seconds or, rounded up, in .3 seconds.* We know, however, that such idealized conditions seldom exactly apply in nature.† Indeed, competing forces do relevantly influence Deal's rate of fall. In effect, these forces slow his fall. While the exact nature of these forces may be unknown, their effect is known: they slow his fall over this distance from .3 seconds to about 1 second. These data, then, can provide reliable information only about the time it took Deal to fall 13.2 inches between the first and the third hypothetical shot. The time it took Deal to fall the distance between the first and the second hypothetical shot remains obscured by incomplete data.

For example, this second shot could have hit Deal, but if it hit him in his right nipple, then we would expect to find a completely different trajectory than we actually find, given this backward leaning and falling position necessary to accommodate the first hypothetical shot. Thus we would expect to see a different characteristic abrasion ring than the one actually observed in the nipple entrance. But it is also possible that this second hypothetical shot could be a miss. We simply do not have the data to determine the actual rate of Deal's fall between the first hypothetical and the second hypothetical shots. The data we do have can be supplied by simple arithmetic, but this data is not sufficient to provide us with any facts that serve to limit the many possibilities governing various entrance locations for the second shot under these hypothetical conditions.

However, given this mathematical work, the third shot strikes Deal after his body falls 13.2 inches in 1 second. This puts Deal's body beyond the known trajectory established for the third shot; so under the hypothetical first shot, the third shot would miss Deal entirely. This outcome is sufficient to falsify the trajectory associating the proposed trajectory of the first shot linking the lowest entrance with the upper left axilla exit. This in turn renders speculations about the entrance of the hypothetical second shot moot.

* "Ideal conditions" means that we do not consider any other factors such as atmospheric density, the aerodynamic properties of the object; or its color, cost, and so forth. In fact, the classical theory tells us explicitly that such conditions are not relevant for the determination of an object's ideal free fall.

† See a discussion of the laws of physics as "idealizations" in *How the Laws of Physics Lie*, by Nancy Cartwright, with Jon Nordby coauthoring Essay #6, "For Phenomenological Laws," pp. 100–127; further acknowledgment of my relevant work is found in the book's Introduction, pp. 1–20; Clarendon Press, Oxford University Press, New York, 1983. See also "How Approximations Take Us Away from Theory and Toward the Truth," Nancy D. Cartwright and Jon Nordby, *Pacific Philosophical Quarterly*, July 1983.

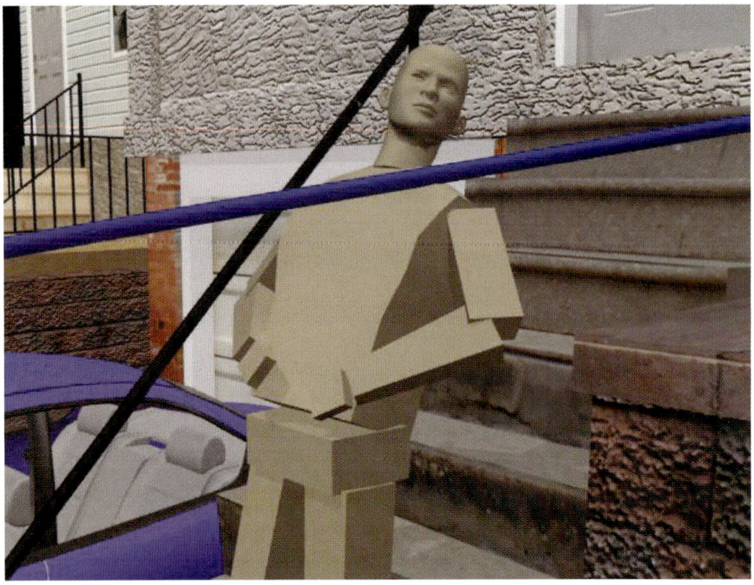

Figure 1.134 This rendering shows that a minimum miss distance for the second shot is about 4 inches.

This 1-second body-fall time interval represents the maximum time for Deal to fall a distance of 13.2 inches. Following with simple math, in one-third second, he would fall about 4.4 inches, but note that simple math also tells us that he would fall 6.6 inches after the first shot and 6.6 inches after the second shot. (This is, of course, because the first shot represents the start of the fall and the last shot represents the stop of the measured distance.) The relevant distances, then, are the first and the second 6.6 inches, since the distance he falls after the third and final shot takes him all the way to the sidewalk below.

In this hypothetical situation, then, after being struck by the proposed first shot, simple math tells us that he will fall 6.6 inches. But as stated earlier, simple math does not allow us to account for the unknown influences on Deal's movements between the first and the third shot. We do not know in fact how far he could have fallen before the second shot struck him because we do not know the relevant facts influencing his movement. We lack the necessary data to help us sort out this hypothetical situation.*

Figure 1.134 shows that a minimum miss-distance for the second shot is about 4 inches. This distance takes Deal outside the given trajectory established by the chip in the brick. However, as mentioned earlier, this hypothetical would leave us with great uncertainty about what his hypothetical movements could have been. This depends only upon the number and type of nonfactual assumptions one decides to entertain. Since by definition there is no data to support this hypothetical or to select from among the various possibilities, it is therefore impossible meaningfully to say whether the trajectory of this second shot would or would not strike Deal in some fashion.

* These two subsequent shots—shot 2 and shot 3—are illustrated in the following figures. The second hypothetical shot appears in Figure 1.134 and Figure 1.139 (Figure 1.139 differs by not accounting for the backward-leaning positional requirements imposed by the proposed trajectory of the first shot). The third hypothetical shot, retaining the backward-leaning positional requirements imposed by the proposed trajectory of shot 1, appears in Figure 1.135 to Figure 1.138.

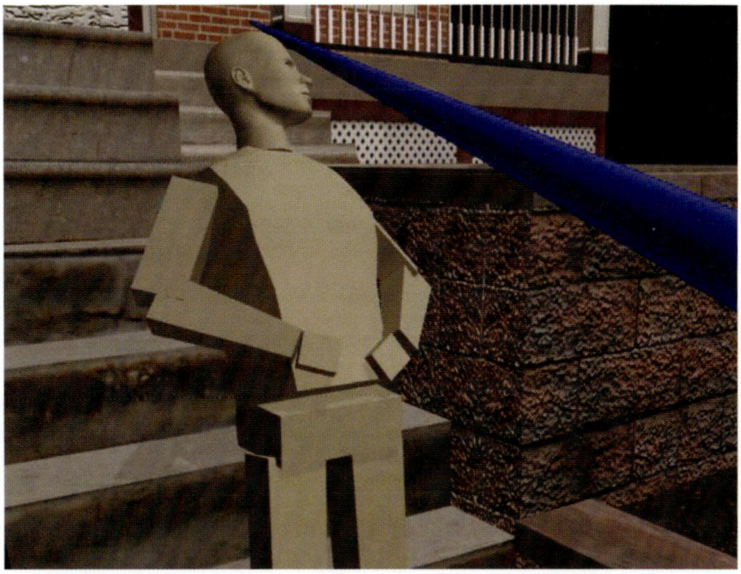

Figure 1.135 Deal's body position represents the distance fallen between the first shot and the third shot—the maximum distance of 13.2 inches.

Since his body could fall about 6.6 inches after the first shot under the hypothesis that the first shot's trajectory links the lowest entrance with the upper left axilla exit, and since about 4 inches (4.4 inches by simple math) is sufficient for a miss, then the second shot could still miss his body entirely. However, the impossibility of constructing speculative movements of hypothetical positions places this conclusion outside the realm of physical evidence. Fortunately, matters are much clearer for the third shot.*

In Figure 1.135, Deal's body position represents the distance fallen between the first shot and the third shot—the maximum distance of 13.2 inches. It is clear that this body position places Deal outside the relative straight-line trajectory established for each of the three shots fired by Metcalf.† Given this fact, the reasoning that follows is fairly straightforward: If the medical examiner is correct in his assumption that the first shot's trajectory links the lowest entrance with the upper left axilla exit, then the third shot would have missed Deal entirely.

Therefore, the medical examiner's proposed trajectory leads to a contradiction: the third shot both hit and did not hit Deal. Therefore, the proposed entrances and exits establishing the trajectory leading to that contradiction must be incorrect. If we assume that this proposed trajectory is correct, then certainly the third bullet fired would have missed Deal. We know that it did not. *Modus tollens* provides support for the inference from premises to the logical conclusion.

Figure 1.136 to Figure 1.140 show the hypothetical third shot, the known trajectory of the shot striking the brick figure, and the trajectory envelope establishing the boundary conditions for after reconstruction.

* This third shot can be seen in Figure 1.135. Figures following Figure 1.135, that is, Figure 1.136 to Figure 1.138, show a top view and two side views of this third hypothetical shot with the trajectory envelope in place.

† This in fact is established for shot 3 since the physical evidence provided by this perforating gunshot wound resulted in a bullet chip in the brick.

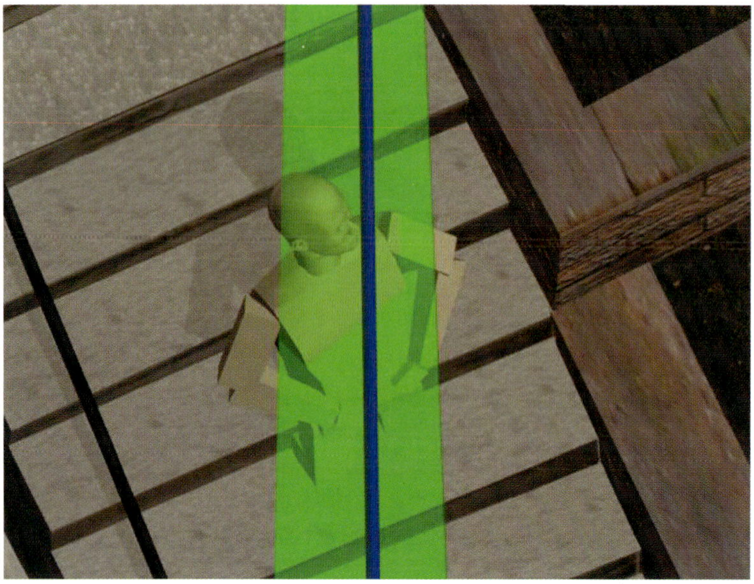

Figure 1.136 This rendering shows the third shot and Deal's body position from a top view given the proposed trajectory of the first shot linking the lowest entrance with the upper left axilla exit.

Figure 1.136 shows the third shot and Deal's body position from a top view given the proposed trajectory of the first shot linking the lowest entrance with the upper left axilla exit. Although parts of his body remain within the trajectory envelope, his torso falls well outside the trajectory, which is, in fact, established by the third shot. (This third shot struck the brick providing evidence of this fact.)

Figure 1.137 shows the third shot and Deal's body position from a side view given the proposed trajectory of the first shot linking the lowest entrance with the upper left axilla exit. Again, although parts of his body remain within the trajectory envelope, his torso falls well outside the established trajectory of the third shot.

Figure 1.138 again shows the third shot's actual trajectory and Deal's body position from a side view given the proposed trajectory of the first shot linking the lowest entrance with the upper left axilla exit. Parts of his body remain within the trajectory envelope, but his body position falls well outside the established trajectory of the third shot and the established location of this shot that struck his body.

Figure 1.139 shows the position of Deal's body if we accept the medical examiner's proposed trajectory for the first shot without its consequences for his position (refuted earlier) and if we accept the medical examiner's proposed trajectory for this second shot (switch of the exit with the first shot's exit).

To accomplish this pictured trajectory, given the trajectory proposed by the medical examiner for the first shot, Deal would have to move upward rather than downward after being struck by the first shot in the position required and discussed above. Moving upward after receiving such a first shot remains physically impossible given the body position required by its proposed trajectory. He would have to be leaning backward, have the apex of his heart destroyed by the first shot, and then he would have to move upright again in time to turn left and fall forward (less than fractions of a second). This also could not happen under the proposed circumstances.

Figure 1.137 This rendering shows the third shot and Deal's body position from a side view given the proposed trajectory of the first shot linking the lowest entrance with the upper left axilla exit.

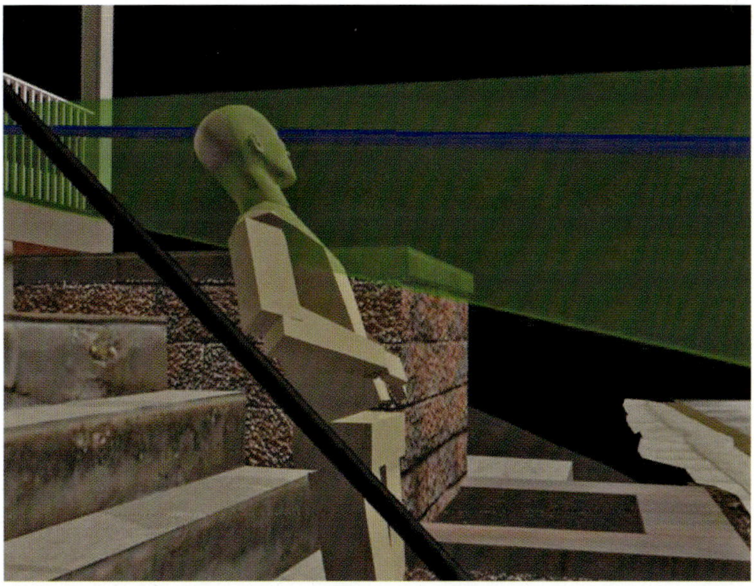

Figure 1.138 Again, this rendering shows the third shot's actual trajectory and Deal's body position from a side view given the proposed trajectory of the first shot linking the lowest entrance with the upper left axilla exit.

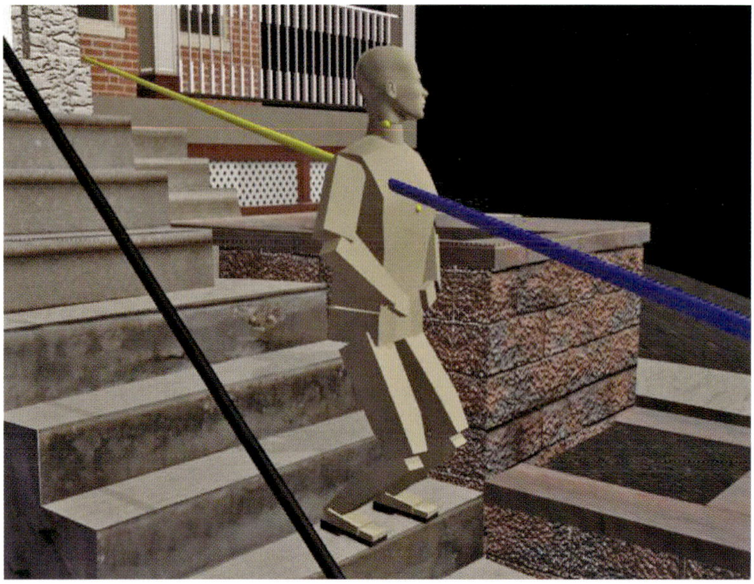

Figure 1.139 Shot 2 using the medical examiner's trajectory from the entrance at the right nipple to the recovery point of the then penetrating bullet at the midline of his back.

Note that using his measurements places shot 3 a bit too high on his body, lowering his body to accommodate the trajectory while raising the shot's location on his body (Figure 1.140). Despite these differences, the entrance and exit trajectory approximates the trajectory developed in the ballistics analysis for this report. The entrance and the exit that he identifies for this shot are in the correct relative positions placing Deal's body in a position that can account for his forward fall toward the sidewalk.

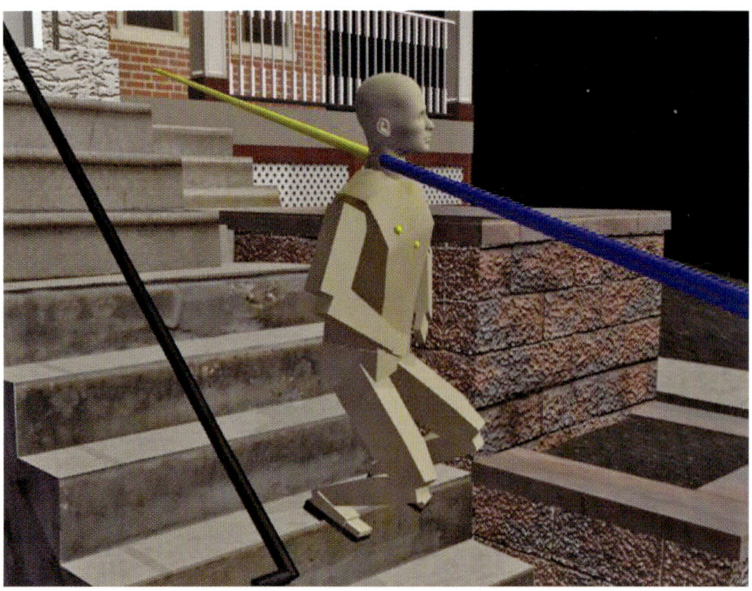

Figure 1.140 Shot 3 using the medical examiner's measurements and trajectory for the perforating shot to Deal's neck.

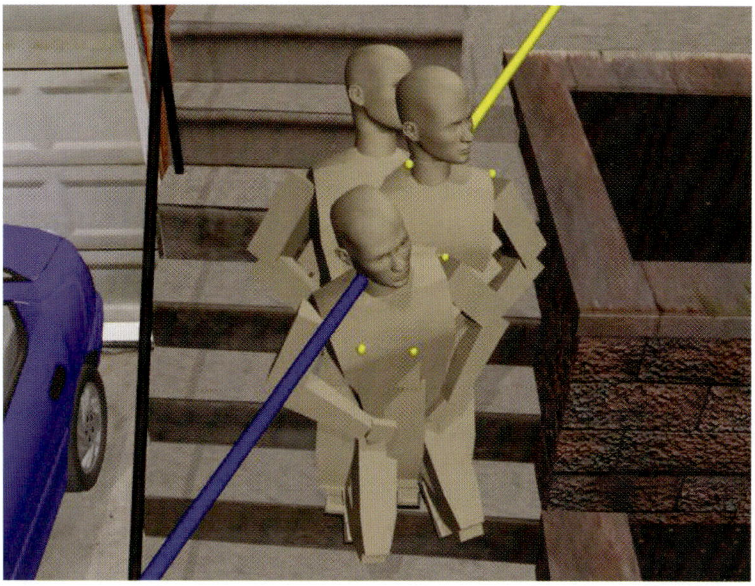

Figure 1.141 This model shows the impossible relationships among the three shots.

However, confusing the exit locations of the first two shots blocks the development of any coherent shooting reconstruction. Despite this misidentification as discussed here, the autopsy result remains sufficient for determining both the cause and the manner of Deal's death—the purpose of the autopsy report. In fact, the medical examiner's identification of the trajectory of this (third) shot contradicts the trajectory proposed for the (first) shot. Since we know that the account of the third shot is approximately correct, it follows that the account of the first and second shots cannot be correct at the same time.

Figure 1.141 shows the impossible relationships among the three shots if we accept the claim that the shot to the lower left chest exits by the left axilla and the claim that the shot to the right nipple penetrates the tissues of the back at the midline. The shot pictured in Figure 1.141 (the figure closest to the foreground) shows the perforating neck wound that struck the brick on the porch. The other shots, as the medical examiner measured them and as are represented by the yellow dots on each of the figures, become physically impossible, both to locate on his body and to link with the path followed by the bullets.

The association of the first shot's entrance with the second shot's exit in the autopsy report leads to these problems. The first two entrances mix up their exits: Deal cannot simultaneously fall both backward and forward in the relevant 1-second time interval.

Trajectory Conclusion Summary

The fact that the medical examiner misidentified the two lower exits is neither surprising nor unusual. First and foremost, it was not the medical examiner's job to produce a ballistics reconstruction. His job was to establish the cause and manner of Deal's death. Further, he did not have sufficient information to reach scientifically reliable conclusions involving bullet trajectories. He did not have information about the bullet chip on the brick wall and therefore could not develop the trajectory envelope (the boundary conditions limiting possible trajectories) in this case. The association of these two entrances with these two exits does not account for the fracture to the condyle of the second left rib, which shows directionality moving from the right to the left. It also does not account for Deal's movement after the apex of his heart was destroyed by the bullet (if the trajectories are correct, then Deal would have fallen backward after the first shot), nor does it adequately account for the damage to the heart itself (if the trajectories are correct, then the damage to the heart would be from right to left, not from front to back). The obvious conclusion is that absent sufficient information provided by the trajectory of the third shot striking the brick, the lowest two shots have simply had their exit/recovery location switched. With this correction made, the physical evidence, including the injuries and Deal's body movements, makes sense and the shot trajectories can then consistently account for all of this physical evidence both on the body and at the scene.

Trajectory and Muzzle Movement in Triple-Tap Firing

Introduction

Often questions arise in the course of work that must be both anticipated and addressed. In this case, one obvious question concerns the influence, if any, of muzzle climb on the working hypothesis that all three shots followed the same approximate trajectory. The best way to address this question involves designing and conducting tests. The best way to investigate muzzle climb is to examine and measure it. In that way, the question raised can and must be answered in order to present a scientifically reliable reconstruction.

Tests were performed to investigate the muzzle behavior of the Smith & Wesson 40 cal pistol when firing three shots in rapid succession. The relationship of shot discharge with the phenomenon known as muzzle climb provided the focus for this investigation. In addition, the general movements of the muzzle relative to a target during an approximately 1-second triple tap were investigated.

The purpose of this testing was to document muzzle movement during rapid fire of the Smith & Wesson 40 cal model SW99 pistol and to measure any effect this movement may have on maintaining an acquired target. Since three shots were fired in rapid succession in this case (triple tap), three shots were fired in rapid succession for this testing.

Summary of Overall Result

The major influence on maintaining a target for a three-shot 1-second shooting interval remains the skill of the shooter. This test shows that in the hands of a skilled shooter, it is possible to shoot three shots in about 1 second with each shot roughly following the same trajectory. No unusual mechanical influences affect the relationship of the muzzle orientation to the target when firing a SW99.

Testing Protocol

A target was established about 8 feet from the muzzle (see distance of muzzle to target in case) (Figure 1.142). (A grid with one-inch squares serves as background for the photos.) Photos were taken with a Casio EX-F1, using video capture at 300 frames per second and at 1200 frames per second. Best visual results were obtained using 300 frames per second capture. These images appear next.

Figure 1.142 A target that was established about 8 feet from the muzzle (see the distance of the muzzle to the target in case).

The Smith & Wesson 40 cal SW99 pistol was fired while aiming at the target using a typical two-hand shooting stance with center-of-mass targeting. The pistol was triple tapped (with each set of three shots fired in approximately 1 second). The three-shot sequences were photographed. The images that appear in Figure 1.143 to Figure 1.148 show the movement of the pistol during the discharge of three shots. The angles were measured and the target-acquisition results were associated with the discharge.

Error Rate (Acceptable Variance)

The error rates associated with this testing involve issues of measurement: statistical protocols are employed to help establish such error rates. The error rate relevant here (the range of acceptable variance) results from efforts to measure muzzle angles: the variance was found to be about ±4 cm at 8 feet. The variance was found to be about ±4 degrees. The overall significance of this variance remains negligible when considering measurements to assess the mechanical influences of muzzle-rise movement resulting from firing the pistol on the uniformity of a basically single trajectory for three bullets fired in rapid succession as discussed earlier.

Shooting Setup

Figure 1.142 shows the overall shooting setup: the target appears on the right and the grid and shooting area appears on the left. The pistol was held over the table, with lights and camera on the far left of the photo. Only the muzzle-to-target distance involved in Deal's shooting was approximated for purposes of this testing.

Caveats

If the question posed was "Can a Smith & Wesson Model SW99 semiautomatic pistol rapidly fire three shots, which pursue approximately the same trajectory?" a single positive instance of three such shots (within a suitable range of acceptable variance) would be sufficient to answer "yes." This testing examines the muzzle climb of a SW99 pistol in order to obtain this positive instance. More than this interpretation of the testing is neither warranted nor intended. The testing does not isolate distinct variables to distinguish, for example, the skill of a shooter from the physics of the firearm in question. The focus remains determining general muzzle rise and its potential influence on shooting three shots that pursue approximately the same trajectory.

Three-Shot Firing Sequence

Shot 1 of 3—Baseline target acquired

Figure 1.143 shows the typical orientation of the pistol to the target after the target is acquired before the first of three shots is fired (center-mass technique)—shot 1 baseline.

Shot 1 of 3—Bullet fired at target; first cartridge case (cc) ejected

Figure 1.144 shows the highest point reached by the muzzle after the first shot has been fired—shot 1 muzzle climb.

Shot 2 of 3—Baseline target reacquisition

Figure 1.145 captures the pistol after the gun has presumably been controlled by the shooter to reacquire the target in preparation for the second shot—shot 2's baseline.

Shot 2 of 3—Bullet fired at target after reacquisition; second cc ejected

Figure 1.146 shows the highest point reached by the muzzle after the second shot has been fired—shot 2 muzzle climb.

Shot 3 of 3—Baseline target acquired

Figure 1.147 captures the pistol after the gun has presumably been controlled by the shooter to reacquire the target in preparation for the third shot—shot 3's baseline.

Shot 3 of 3—Bullet fired at target after reacquisition; third cc ejected

Figure 1.148 shows the highest point reached by the muzzle after the third shot has been fired—shot 3 muzzle climb.

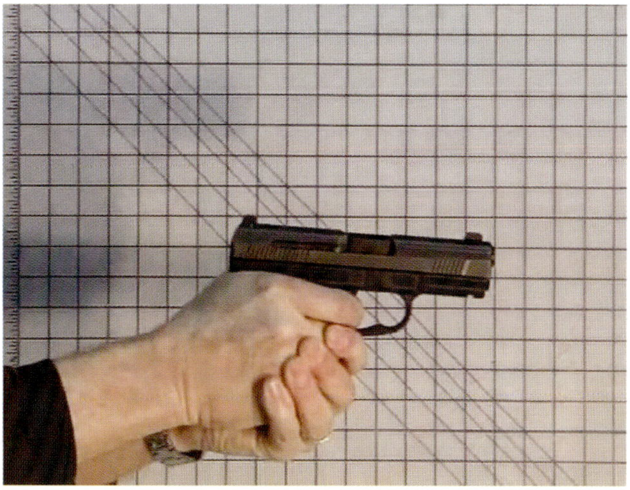

Figure 1.143 Typical orientation of the pistol to the target after the target is acquired before the first of three shots is fired (center-mass technique)—shot 1 baseline.

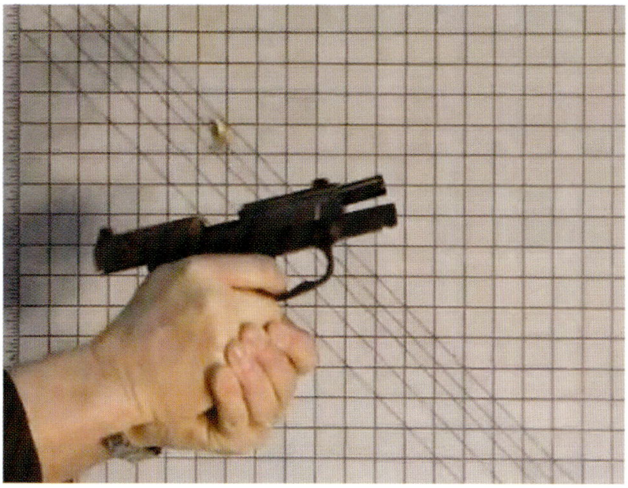

Figure 1.144 The highest point reached by the muzzle after the first shot has been fired—shot 1 muzzle climb.

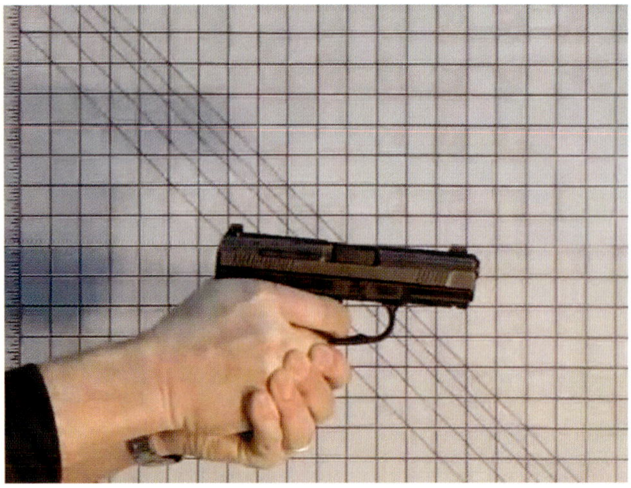

Figure 1.145 The pistol after the gun has presumably been controlled by the shooter to reacquire the target in preparation for the second shot—shot 2's baseline.

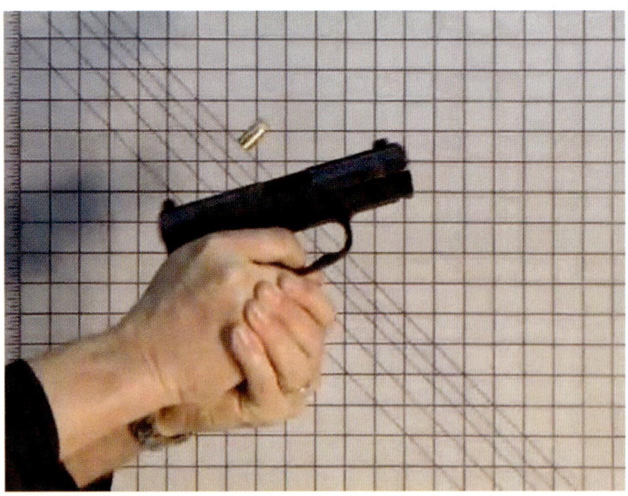

Figure 1.146 The highest point reached by the muzzle after the second shot has been fired—shot 2 muzzle climb.

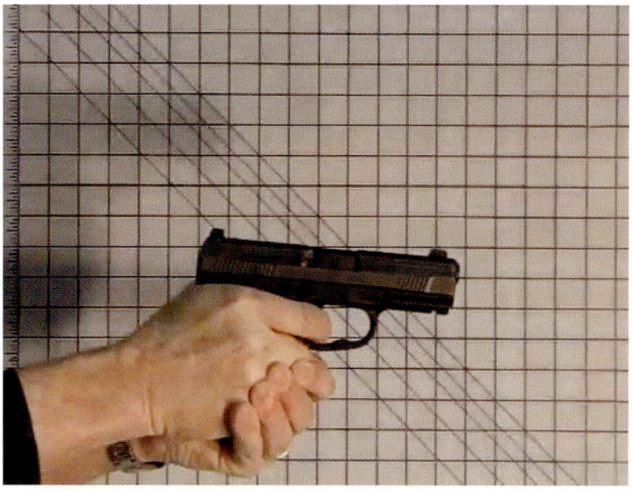

Figure 1.147 The pistol after the gun has presumably been controlled by the shooter to reacquire the target in preparation for the third shot—shot 3's baseline.

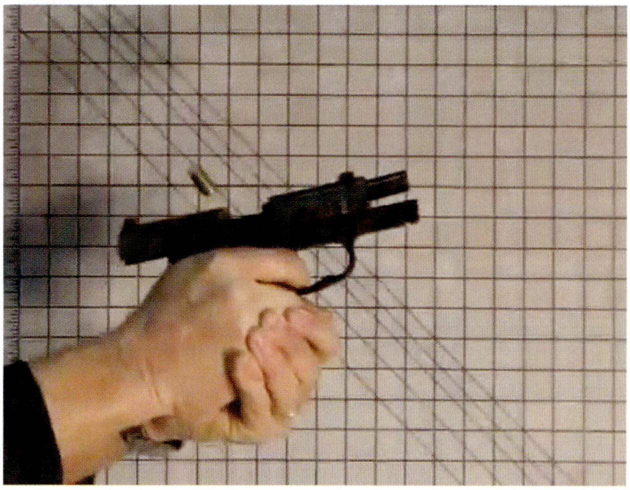

Figure 1.148 The highest point reached by the muzzle after the third shot has been fired—shot 3 muzzle climb.

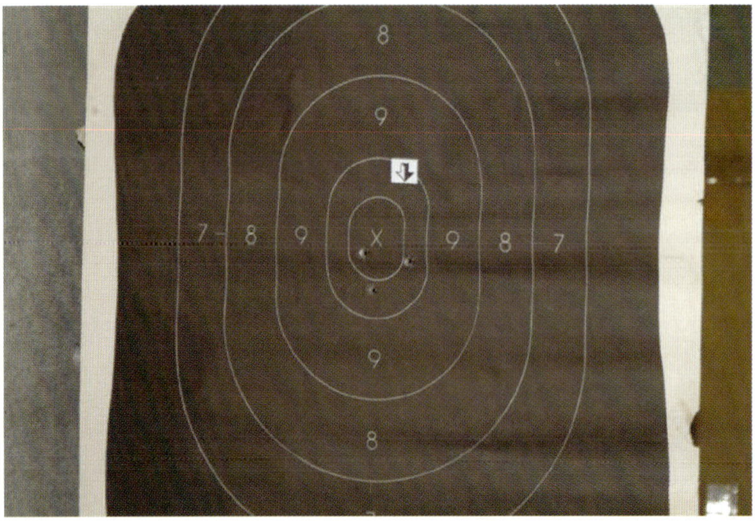

Figure 1.149 Results of the triple-tap fired in approximately 1 second with an 8-foot muzzle-to-target distance.

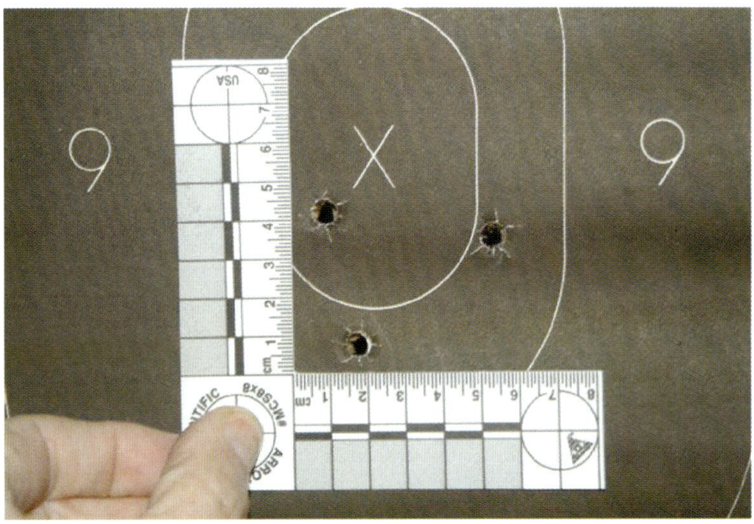

Figure 1.150 Results of the triple-tap fired in approximately 1 second with an 8-foot muzzle-to-target distance.

Three-Shot Firing Sequence Target

Figures 1.149 and 1.150 show the results of the triple-tap fired in approximately 1 second with an 8-foot muzzle-to-target distance.

Three-Shot Firing Sequence Angle Measurement

For visual clarity, only two shot images of the pistol are overlaid here to illustrate positions. Shot 3 is shown in the blue. The image is very similar to shot 2 (Figures 1.151 and 1.152).

The result—a relatively close grouping of shots on the target (precision)—shows only that a reasonably skilled shooter can fire three shots in rapid succession without each shot

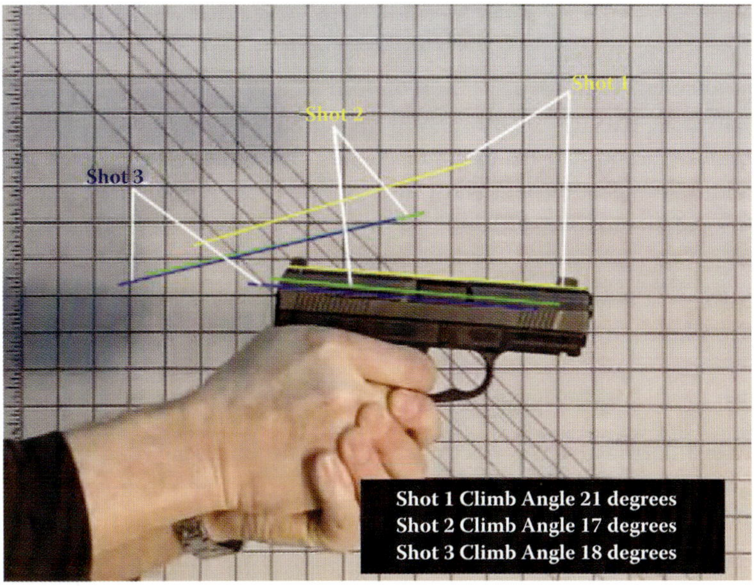

Figure 1.151 The positions for each of the three shots determined through photogrammetry using Figure 1.145.

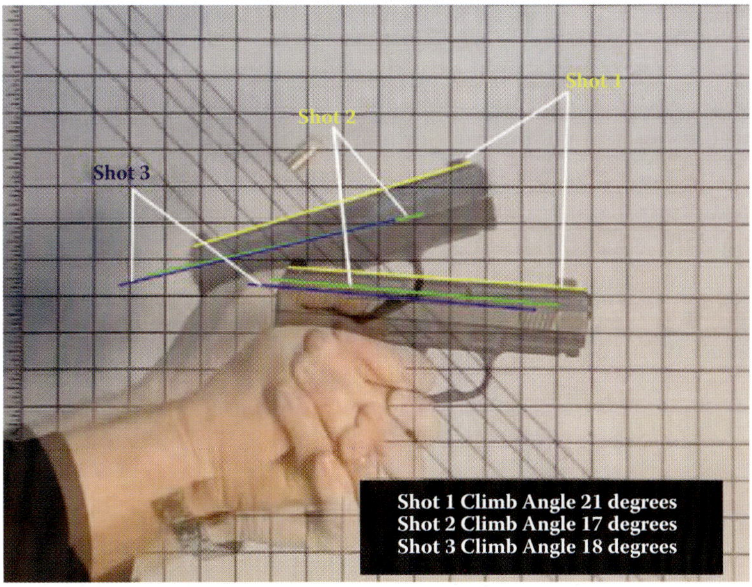

Figure 1.152 The overlay of Figures 1.147, 1.145, and 1.146 illustrate the results.

assuming a radically different trajectory due to the physical behavior of the pistol (muzzle rise). It does not show more, nor does it show less, than that.

In the hands of an unskilled shooter, of course, anything could happen, including the precision illustrated. This testing addressed the possibility of each of the three shots pursuing approximately the same trajectory under the conditions imposed. The range of acceptable variance due to measurement error also supports this possibility. The possibility that a Smith & Wesson 40 cal model SW99 can do so is negatively influenced by muzzle climb.

Instrument Calibration and Validation Data

Introduction

In addition to evaluating all the given data and the inferences based upon that data, another important component of a shooting reconstruction, designed to help ensure its scientific reliability, is the calibration and the validation of relevant scientific instruments to be used either to develop or to evaluate and analyze the data. Such accurate calibrations and validations, for example, not only help ensure reliability but also help establish any relevant error rates or measurements of acceptable variance among results. This is another element that should maintain scientific modesty when presenting the results of a reconstruction.

Simply put, in this context, a calibration is a comparison between two measurements: one of the measurements is the standard or known magnitude and the other is the measured result derived by correctly applying the instrument. The relevant features of the device are adjusted to bring the instrumental result in line with the given standard. So the known, correct, or assigned result is called the standard. The device being adjusted may have any of several names depending upon the device being calibrated. Validation is the process of determining the degree to which an instrument accurately derives reliable data given the intended uses of the instrument or tool. Calibration can be seen as properly setting up the device and validation can be seen as ensuring the correctness of the measured result.

In shooting reconstructions, for example, lasers, total stations, sonic measurement tools, laser measurement tools, light driven tools, and even protractors, ABFO scales, rulers, GPS devices, and tape measures along with microscopes, XRFs, FTIRs, and SEMs are subject to calibration and validation. Sometimes these steps for the specific devices are performed independently of the specific reconstruction and become a part of the laboratory's standard protocols, with results, along with the history of the device's performance, available should the issues be raised in court.

Calibration data from the Sokkia total station used in this reconstruction, for example, is generally added as part of the shooting reconstruction. This helps enable readers to assess the data developed in terms that in turn support an analysis of that particular data's reliability. Such calibration data for the Sokkia total station takes the following form (see Figures 1.153 and 1.154).

Conclusions

The theory of the shooting advanced by the group of attorneys was shown to be incorrect. The reconstruction, carefully stated, seemed to avoid political problems with the medical

EDM Baseline Data

Company: SOKKIA PURCHASE STOCK
Address: 16900 W. 118TH TERRACE
City, State, Zip: OLATHE, KS 66061

Jobno: QA
Technician: Gene Norris
Date: 09/25/2007 **Time:** 12:04:07

Instrument Type: SRX3
Serial Number: 100859

Temperature: 15.0 C **Pressure:** 733.8 mmHg **Humidity:** 100 % **PPM:** 10.3685

Prism Distance: 133.2217 **Meters**

133.221	133.221	133.221	133.221	133.221
133.221	133.221	133.221	133.221	133.221

Mean:	133.2210	**Corrected Distance:**	133.22238
PPM (Adj. for Elev.):	10.3685	**Standard Deviation:**	0.000
PPM Distance Adj.:	1.3813	**Distance Error:**	0.68

Prism Distance: 222.6123 **Meters**

222.610	222.610	222.610	222.610	222.610
222.610	222.610	222.610	222.610	222.610

Mean:	222.6100	**Corrected Distance:**	222.61231
PPM (Adj. for Elev.):	10.3685	**Standard Deviation:**	0.000
PPM Distance Adj.:	2.3082	**Distance Error:**	0.01

Prism Distance: 451.4364 **Meters**

451.434	451.434	451.434	451.434	451.434
451.434	451.434	451.434	451.434	451.434

Mean:	451.4340	**Corrected Distance:**	451.43868
PPM (Adj. for Elev.):	10.3685	**Standard Deviation:**	0.000
PPM Distance Adj.:	4.6807	**Distance Error:**	2.28

Figure 1.153 EDM Baseline Data.

examiner's office as well as produced a sound and reliable work product that withstood any robust scientific critique thrown its way. That does not mean that the work avoided all mistakes; it just means that doing the best science involves showing the work clearly so that any mistakes become clear for evaluators to find and hopefully easier for its authors to fix.

Certainly not all shooting reconstructions demand the same scientific or technological tools. Nor do they require identical approaches. What remains constant, however, is the critical thinking—critical intelligence—required for any such scientific endeavor. Critical intelligence involves the ability to generate novel approaches and to include in those generations a validation of the approach. That, in fact, is one characteristic of scientific methods most often overlooked. There is no set list to follow in all cases, and no unique set of rules to embrace. There is only scientific work to be done.

Linear Rail Calibration Report

Company: SOKKIA PURCHASE STOCK
Address: 16900 W. 118TH TERRACE
City, State, Zip: OLATHE, KS 66061

Jobno: QA
Technician: Gene Norris
Date: 09/25/2007 **Time:** 12:01:57

Instrument Type: SRX3 - Sheet Target
Serial Number: 100859
Temperature: 21.3 C **Pressure:** 733.8 mmHg **Humidity:** 0 % **PPM:** 15.4

Target Distance: 9.97733 Meters

9.976	9.976	9.976	9.976	9.976
9.976	9.976	9.976	9.976	9.976

Mean: 9.97600 Meters	PPM Adj: 0.1537 mm	Adj. Distance: 9.97615 Meters
Standard Deviation: 0.000 mm	Distance Error: -1.18 mm	

Target Distance: 11.97713 Meters

11.976	11.976	11.976	11.977	11.977
11.977	11.977	11.977	11.977	11.977

Mean: 11.97670 Meters	PPM Adj: 0.1845 mm	Adj. Distance: 11.97688 Meters
Standard Deviation: 0.483 mm	Distance Error: -0.25 mm	

Target Distance: 13.97729 Meters

13.976	13.976	13.976	13.976	13.976
13.976	13.976	13.976	13.976	13.976

Mean: 13.97600 Meters	PPM Adj: 0.2154 mm	Adj. Distance: 13.97622 Meters
Standard Deviation: 0.000 mm	Distance Error: -1.07 mm	

Target Distance: 15.97715 Meters

15.976	15.977	15.977	15.976	15.977
15.977	15.976	15.976	15.976	15.976

Mean: 15.97640 Meters	PPM Adj: 0.2462 mm	Adj. Distance: 15.97665 Meters
Standard Deviation: 0.516 mm	Distance Error: -0.50 mm	

Target Distance: 17.97712 Meters

17.976	17.976	17.976	17.976	17.976
17.976	17.976	17.976	17.976	17.976

Mean: 17.97600 Meters	PPM Adj: 0.2770 mm	Adj. Distance: 17.97628 Meters
Standard Deviation: 0.000 mm	Distance Error: -0.84 mm	

Target Distance: 19.97721 Meters

19.977	19.977	19.977	19.977	19.977
19.977	19.977	19.977	19.977	19.977

Mean: 19.97700 Meters	PPM Adj: 0.3078 mm	Adj. Distance: 19.97731 Meters
Standard Deviation: 0.000 mm	Distance Error: 0.10 mm	

Note: Mean distance is in meters. Standard deviation, distance error and cyclic error are in millimeters.

Cyclic Error: 1.28 **Maximum Error:** 0.10 **Minimum Error:** -1.18

Figure 1.154 Linear Rail Report.

Continued

Instrument Type: SRX3 **Serial Number:** 100859

Temperature: 21.2 C **Pressure:** 733.8 mmHg **Humidity:** 0 % **PPM:** 15.3

Target Distance:					
10.00002 Meters	10.001	10.001	10.001	10.001	10.001
	10.001	10.001	10.001	10.001	10.001

Mean: 10.00100 Meters	PPM Adj: 0.1532 mm	Adj. Distance: 10.00115 Meters
Standard Deviation: 0.000 mm	Distance Error: 1.13 mm	

Target Distance:					
11.99984 Meters	12.001	12.001	12.001	12.001	12.001
	12.001	12.001	12.001	12.001	12.001

Mean: 12.00100 Meters	PPM Adj: 0.1838 mm	Adj. Distance: 12.00118 Meters
Standard Deviation: 0.000 mm	Distance Error: 1.34 mm	

Target Distance:					
14.00003 Meters	14.001	14.001	14.001	14.000	14.001
	14.001	14.001	14.001	14.001	14.001

Mean: 14.00090 Meters	PPM Adj: 0.2145 mm	Adj. Distance: 14.00111 Meters
Standard Deviation: 0.316 mm	Distance Error: 1.08 mm	

Target Distance:					
15.99985 Meters	16.001	16.001	16.001	16.001	16.001
	16.001	16.001	16.001	16.001	16.001

Mean: 16.00100 Meters	PPM Adj: 0.2451 mm	Adj. Distance: 16.00125 Meters
Standard Deviation: 0.000 mm	Distance Error: 1.40 mm	

Target Distance:					
17.99980 Meters	18.000	18.000	18.000	18.000	18.000
	18.000	18.000	18.000	18.000	18.000

Mean: 18.00000 Meters	PPM Adj: 0.2757 mm	Adj. Distance: 18.00028 Meters
Standard Deviation: 0.000 mm	Distance Error: 0.48 mm	

Target Distance:					
19.99989 Meters	20.000	20.001	20.001	20.001	20.001
	20.001	20.001	20.001	20.001	20.001

Mean: 20.00090 Meters	PPM Adj: 0.3064 mm	Adj. Distance: 20.00121 Meters
Standard Deviation: 0.316 mm	Distance Error: 1.32 mm	

Note: Mean distance is in meters. Standard deviation, distance error and cyclic error are in millimeters.

Cyclic Error: 0.92 **Maximum Error:** 1.40 **Minimum Error:** 0.48

Figure 1.154 (*Continued*) Linear Rail Report.

Misplaced Method

2

The greatest obstacle to discovering the shape of the earth, the continents and the oceans was not ignorance, but the illusion of knowledge.

Daniel J. Boorstin, Librarian of the U.S. Congress, Emeritus

Introduction

All too often when facing a new shooting reconstruction, one faces not only a distinctly nonscientific audience in the courtroom but also a nonscientific "expert" who has produced, for one side or the other, a "reconstruction" that is lacking any semblance of a testable or a data-driven work product. Such efforts are often encapsulated in very short one- or two-paragraph reports. Sometimes there will be no report at all to confirm the elaborate storytelling delivered from the stand.

In the courtroom, the attorney working with such a witness appeals to an extensive list of the many years this "expert" spent working with crime scenes, the work often overlapping decades of scientific advances about which this expert remains blissfully unaware. If an opposing attorney questions this "reconstructionist" on his lack of knowledge, the reply remains an appeal to the authority of years spent doing the same work, over and over, always with painfully similar results. Members of the jury get a good story, which they seem to prefer over the more methodologically modest detail of dry scientific discourse.

Often all one can do is to present the facts and the results of the scientific work to the best of one's ability, letting the metaphorical chips fall where they may. One must simply heed the advice offered by Dr. Paul Brouardel, the French pioneer in forensic medicine: "If the law has made you a witness, remain a man of science: you have no victim to avenge, no guilty or innocent person to ruin or save. You must bear testimony within the limits of science." It helps to have faith in the critical thinking skills of the jurors too; most people have an innate ability to distinguish intellectual wheat from self-serving chaff most of the time.

The Scene

The state's renowned police expert had experience gleaned from years spent attending to the aftermath of murder, and he had seen it all before. An expert in crime scenes, bloodstains, firearms, and all things homicide, he made his determinations at the scene quickly and concisely after walking through the small single-story house, looking at the ransacked living quarters, seeing the dead woman lying face up on the kitchen floor, and examining the bloodstains he found on the husband's face. He was, after all, an experienced "reconstructionist." His conclusions were firm and unyielding.

The crime scene, he said, was staged merely to look like the work of an intruder. But the blood on the husband's glasses told him the story of domestic murder. The husband was guilty. He shot his unsuspecting wife in the back with a shotgun sabot slug sending her into the next world with a single blast from his 12-gauge while she worked at the kitchen sink. The expert gleaned facts sufficient to craft the end of another violent story. He told the photographer to document the scene, as he packed up his notebook and left through the front door.

When he left for the city to write his report, he had been at the rural mountain house about 2 hours, give or take an hour, including travel time. Only the photographer and three patrol officers, including the first officer to respond to the original 911 call, remained awaiting the deputy coroner. The deputy coroner's chief duty involved collecting and transporting the body for autopsy. The lead detective had taken the suspect into custody and transported him to jail pending his formal arraignment.

The photographer looked around for something other than the body and the blood to document, like a shotgun, for example. He did not see one. Maybe someone had removed it, along with the suspect. He got the patrol officer's notes and began a hasty review to catch up on the case while he loaded his camera.

The husband had made the 911 call at 7:12 P.M. He called from a neighbor's house, which was about a 5-minute walk or a 3-minute dash from his home. The kitchen phone dangled from one remaining sheetrock screw still stuck into the wall; its severed wire barely touching the floor. The husband told the 911 operator that his wife had been shot. He said he knelt in front of her, felt for a heartbeat, saw she was dead, tried to use the dysfunctional phone, ran to use the neighbor's phone, and called for help. The operator kept him on the line until the first patrol officer arrived at 7:35 P.M. The officer's narrative had not yet been copied onto the appropriate report form. The photographer looked up from the officer's notebook toward the kitchen, into the adjoining dining area and through to the living room/family room. He moved slowly through the small house taking photos as he went.

The dining room table was pushed against the wall with two of its four chairs tipped precariously on spindly legs. The end tables just inside the living room, drawers open, spilled miscellaneous contents onto the beige carpet. The buffet doors were open, as were the desk drawers. Papers littered the proximate surrounding flat surfaces. The home entertainment center displayed empty slots and protruding tangles of wire once feeding now-absent components. An empty holster and a box of 357 cartridges lay beside the open coat-closet door. His flash helped illuminate the details for his camera. He changed rolls of film and moved on through the residence.

The bedroom revealed a neatly made bed that had become a dumping ground for drawers pulled from an oak highboy. Their various contents, including male and female clothing, appeared randomly strewn about the room. The open closet door barely concealed a pile of possessions once suspended from plastic hangers.

The husband, in his statement to police, had listed various missing property including audio components, a revolver, a hunting rifle, some jewelry, and some cash from the bedroom. He had said he did not own a shotgun.

The arrival of the deputy coroner provided the photographer with his opportunity to document matters in the kitchen. Along with the husband's hands, clothing, and glasses, the body and the kitchen area had already occupied most of the official attention allotted the case. The sabot had entered the decedent's back and exited along the midline of her chest. Three holes appeared through the back of her sweater when the body was moved by the deputy coroner—the sabot had separated as designed, leaving one hole made by the slug, and two holes made by the separating sabot.* The front of her chest was covered with blood. Bloodstains of varying sizes appeared on the arms and the lower waistband of the once-white fuzzy sweater. The slug and the two halves of the sabot appeared on the counter beside the sink. They would need to be collected and taken into evidence. And a thorough search would have to be made for the murder weapon.

Shotguns, Shells, and Sabots

One half of the sabot was found in the kitchen on the floor. The slug fits between two of these plastic pieces; in this design, the plastic separates and the slug continues (Figure 2.1). However, in practice, the plastic piece(s) also do considerable damage if a target is close to the shotgun muzzle.

Shotgun shells come in many designs for different purposes. Figure 2.2 illustrates some of the different shotgun shells in 410-, 16-, and 12-gauge sizes. A loaded 12-gauge sabot is pictured on the upper right. In Figure 2.3, a sabot has been cut open to show the powder and the plastic that holds the slug as it leaves the shotgun's muzzle.

Some shotgun barrels are smooth bore and some are rifled to accommodate sabots of various designs. Shotguns are evaluated according to their choke (the amount of constriction at the muzzle designed to keep the pellets in a tight pattern for a longer distance). Shotgun chokes include the following:

- Full choke—Produces a dense pattern with about 70% of a shell's total pellets striking within a 30-inch circle at 40 yards (120 feet)

Figure 2.1 Part of the recovered sabot with soot and dried blood.

* A sabot (pronounced "sab bow") is designed to fire a single large slug through a shotgun barrel rather than multiple numbers of shot contained in a shot cup. The plastic surrounding the slug is designed to breakaway in flight after exiting the muzzle. This becomes a secondary missile, which may also strike the target depending upon the muzzle-to-target distance.

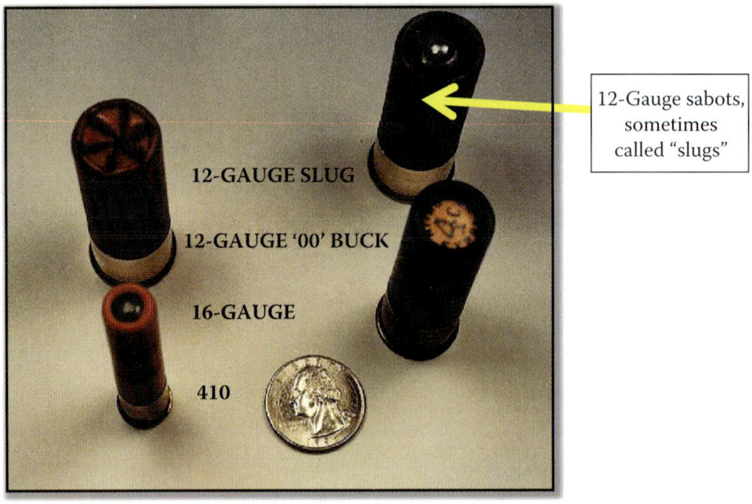

Figure 2.2 Shotgun shells.

Figure 2.3 Components of a 12-gauge shotgun shell, showing the shell, powder, sabot, and lead slug.

- Modified choke—Less dense pattern with about 60% of a shell's total pellets striking within a 30-inch circle at 40 yards (120 feet)
- Improved cylinder choke—Even less dense pattern with about 50% of a shell's total pellets striking within a 30-inch circle at 40 yards (120 feet)
- Cylinder bore choke—No constriction (the evidence shotgun) with about 40% of a shell's total pellets striking within a 30-inch circle at 40 yards (120 feet); most useful at shorter ranges

The other half of the plastic was found on the kitchen countertop. Often, if we have the shotgun that was suspected of firing the fatal shot, we can compare test-fired sabots with those recovered at the scene. The success of this endeavor depends upon the number of unique scratches, marks, and defects in the barrel of the gun (Figure 2.4). Some of these marks, when viewed under the microscope, may be useful if a suspect shotgun can be recovered for testing and comparison. However, in this case, no shotgun has been recovered.

Figure 2.4 Soot on the sabot.

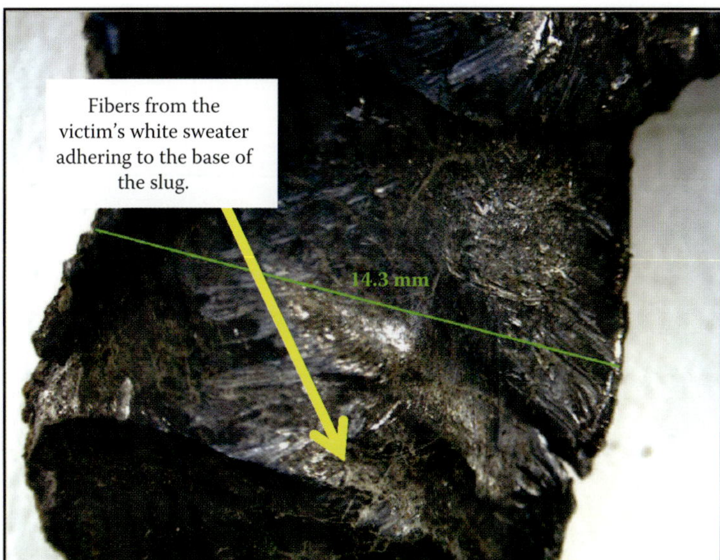

Figure 2.5 Trace on the base of the slug.

The lead component was also recovered and examined for possible trace or other evidence. This item revealed little information useful for pursuit. This component of the sabot can be seen in Figures 2.5 and 2.6. The lead slug perforated the victim and retains some fibers from her sweater.

In another case, a woman was shot with a 12-gauge shotgun firing shot shells as she ran from her home. The blast struck her hair and yanked a tuft right out of her scalp, embedding it into the wall (Figure 2.7).

Figure 2.6 A single large shotgun slug causes damage from its lead and sabot components.

Figure 2.7 From an unrelated case, showing how a multiple-pellet shotgun shell spreads its pellets. This hair from a female victim was ripped from her scalp and pushed through a sheet-rock wall. The shot cup and the pellets both did the damage.

Other Shotgun Injuries to Document

With the sabot, depending upon the range, there can be two or three entrance wounds: one from the lead slug and one from each side of the plastic sabot. In this case, we find three defects, or entrance holes in the decedent's sweater. Different loads have various sizes and numbers of pellets instead of one lead slug. We focus on the injuries to the victim when a shotgun is fired, but sometimes it is useful to examine the shotgun (if recovered) for data that may have injured the shooter. Those injuries, if the shotgun is recovered, can help identify the shooter.

In a different shotgun homicide, the shotgun was a Remington 870 pump shotgun that was "sawed off" to an overall length of about 25 inches (both the barrel and stock had been modified). This shotgun, along with a section of cutoff barrel, was reportedly found in a pond sometime after the two victims were recovered from the same pond (Figures 2.8 and 2.9).

Figure 2.8 Packages for a typical shotgun case, as delivered to the Final Analysis Forensics laboratory.

Figure 2.9 The shotgun as received, was illegally modified.

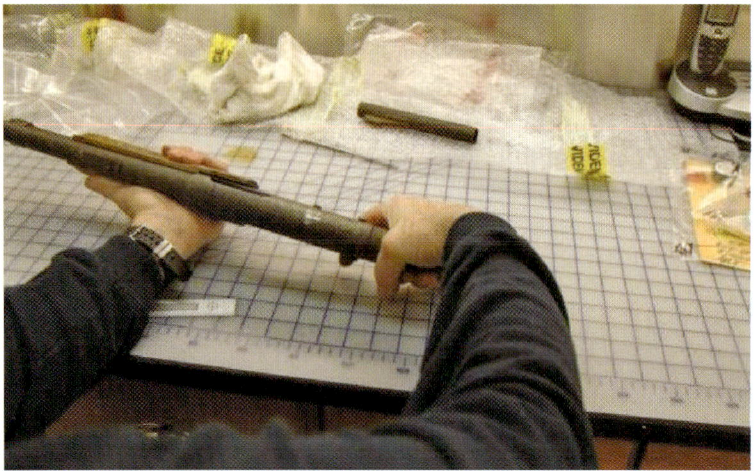

Figure 2.10 The pins above the trigger were replaced by common bolts with damaged heads. Both the gun and the sawed-off length of the barrel were found in the same pond as the two bodies.

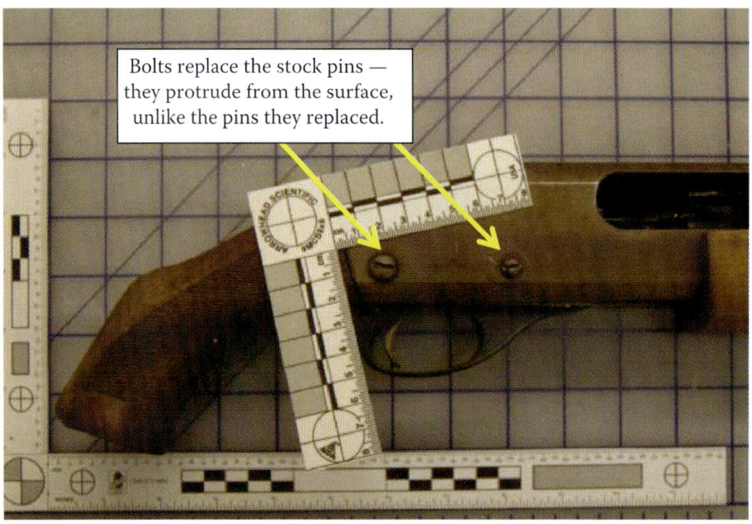

Figure 2.11 The two bolts replacing the stock pins. The rearmost bolt may cause abrasions on the shooter's hand.

The shotgun was illegally modified (Figure 2.10 to Figure 2.14). As modified, it is about 25 inches long. The cut barrel removes any constriction, leaving what is called a cylinder bore. No mechanism holds the shot pattern together, providing the widest spread of pellets when the shotgun is fired. The spread remains indistinguishable from a full-length cylinder bore shotgun.

After three shots with this shotgun, I noticed that my right hand's first finger had sustained an abrasion toward the proximal end facing the thumb. The more shots discharged, the more the aggravation, abrading the skin and resulting in bleeding. If gloves were worn by a shooter of this gun, and if this gun were used to fire at least 10 or 11 shots without interruption, even the glove would show signs of wear or damage at this location. If no gloves were worn, the shooter's trigger finger would be abraded (Figure 2.15 to Figure 2.18).

Most shotguns that are well used or modified have similar potential to mark the shooter. It always helps to look carefully for this data when examining the shotgun. Unfortunately, this shotgun was recovered in a pond. Any useful data, if it appeared, had washed away.

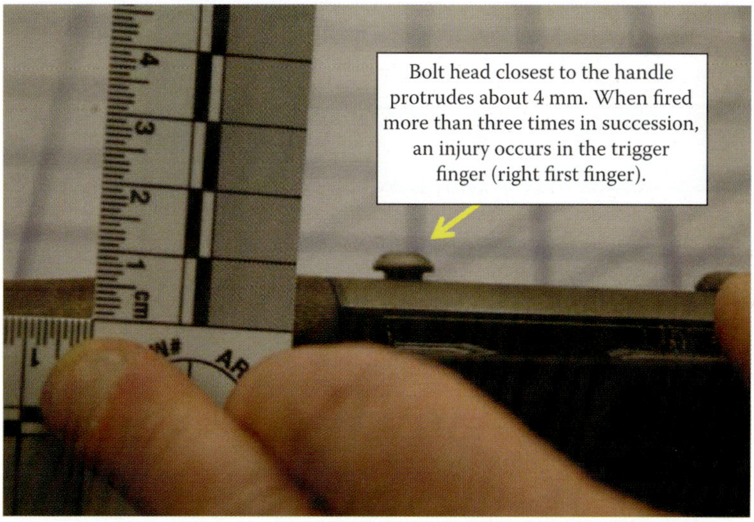

Figure 2.12 Potential for injury to the shooter.

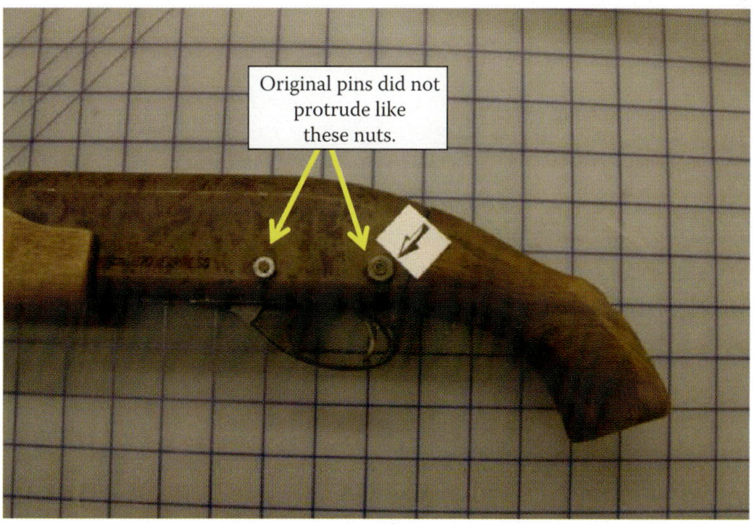

Figure 2.13 The nut on the bolt does not remain tightly secured and moves in and out about 4 mm.

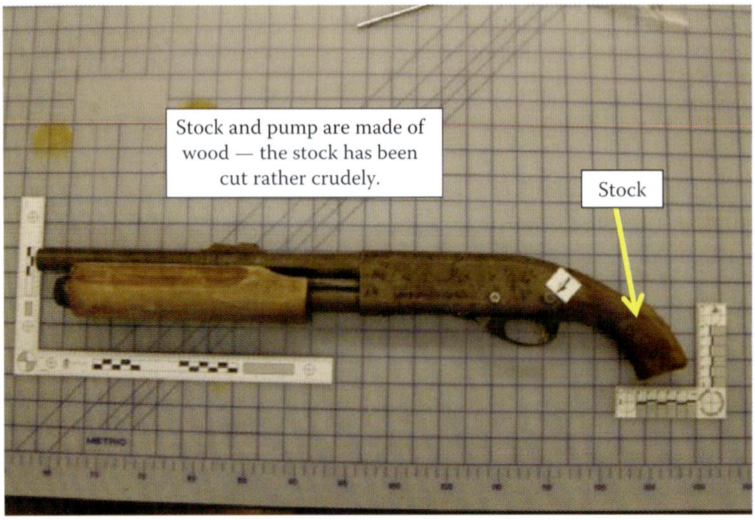

Figure 2.14 The barrel and stock have been shortened.

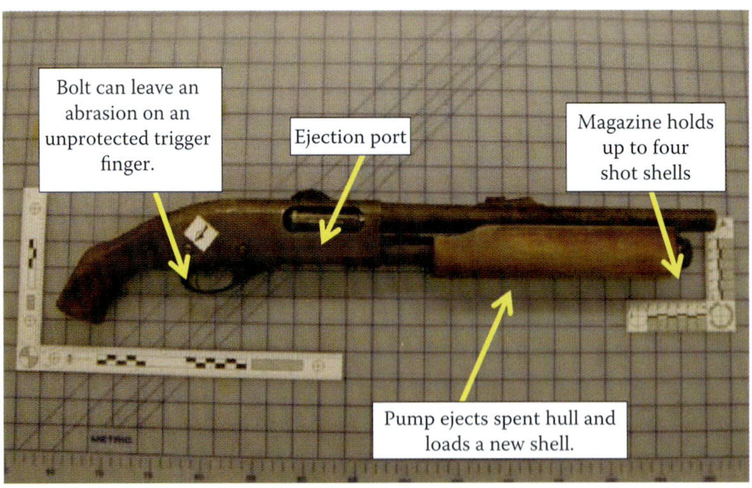

Figure 2.15 If gloves were worn when firing this shotgun, they would show wear at the point of contact between the bolt head and the trigger finger.

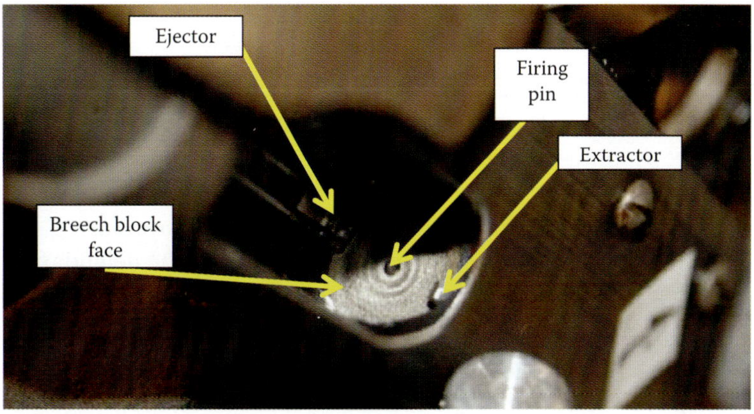

Figure 2.16 Each component noted above leaves predictable markings on any shot shell fired through the shotgun's action.

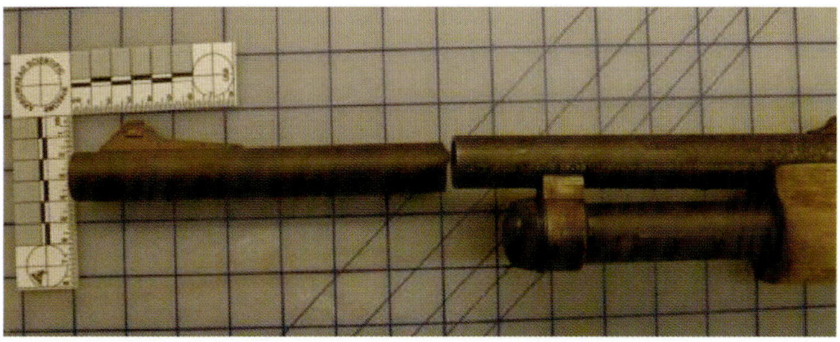

Figure 2.17 The barrel in this case was also recovered in the pond. It appears to have been cut with a band saw, the teeth move in one direction to make the cut.

Figure 2.18 This protruding bolt head caused a predictable pattern injury on the shooter after three shots. With no shotgun recovered in this case, no such evidence could be developed. The defendant had no marks of any kind on either hand. Without the shotgun, that data remained to be of limited value.

Search of the Premises

Searching the property took crime scene techs the better part of three full days. The family truck was found parked near the garage with the passenger door open, and the contents of the glove box strewn onto the bench seat and floor. No shotgun was found anywhere on the property, and none of the items reported to be missing from the house by the husband were found either. None would ever be either discovered or recovered.

Officers also confirmed that the car pool had dropped off the defendant in front of his house at about 6:55 P.M. the night of the murder.

The State's Expert "Reconstructionist" Reconstructs the Shooting Murder from the Clues

Like many official reports, the one produced by the state's expert was short in length and short on data, and long on assumptions and long on unsupported assertions. The conclusions forming the heart of a crime scene or shooting reconstruction remain only as good as the supporting data, together with its sound logical and scientific analysis. Both the data and the analysis offered in this report, absent logic, were found wanting.

The report stated unequivocally that the disarray in the small house resulted from the actions of the husband. These actions, the report claimed, were designed to be interpreted as the actions of some unknown intruder who had killed the decedent (Figure 2.19), robbed the residence, and departed undiscovered. Instead, the report indicated, they merely exposed the husband's clumsy plan to cloak his own murderous actions. According to this expert, the husband shot his wife as she worked in the kitchen, then he proceeded to conceal his nefarious act by pulling out drawers, opening doors, and strewing the household contents around each room. Support produced for this conjecture centered on some inadequately documented broken glass.

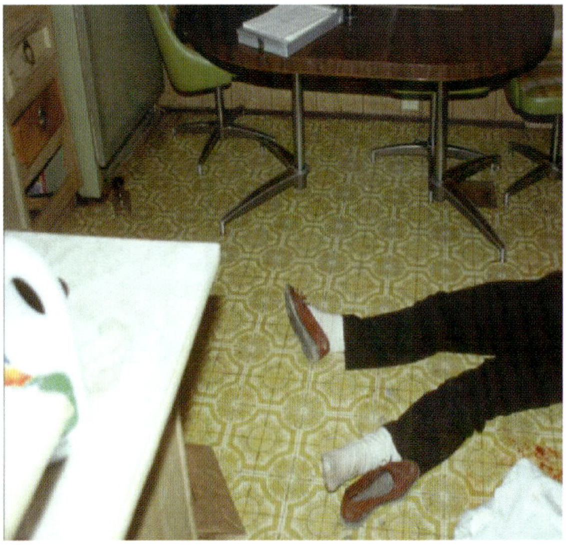

Figure 2.19 Kitchen scene showing the decedent's position on her back.

The ballistics of the sabot and the bloodstain analysis dovetailed, according to the report, to prove the husband's guilt. The report included a diagram of what was purported to be the trajectory of the fatal shot. The expert's assemblage of physical evidence, including the bloodstains on the defendant's glasses, proved that the defendant shot the victim in the back.

When taken into custody, the husband had small bloodstains on his glasses as well as bloodstains on the palms of both hands. Wipe stains appeared over the pockets of his Levi's. No other bloodstains of any significance were noted and none were described in the reconstructionist's report. The stains on the defendant's glasses were referred to as "high-velocity back spatter." (These "high-velocity stains" result when a bullet strikes a blood source and the resulting impact has sufficient energy to produce small droplets that fly backward toward the muzzle of the weapon firing the shot.) The report described the stains on the glasses as being "1 mm or less" in diameter. It further explained that such small stains "can *only* be produced through gunshot impact." The only mechanism with sufficient energy to produce such high-velocity spatter, according to some readers and some writers of the bloodstain pattern analysis literature at least, is a gunshot.

Certainly, back spatter from bullet impact with a blood source is a documented phenomenon. My own experiments have documented this fact, as seen in Figure 2.20.

Even if we considered the hypothesis that these stains on the glasses were back spatter from the shotgun, it is surprising that the expert seemed to ignore the fact that the suspect's white T-shirt showed no bloodstains at all (Figure 2.21). Presumably high-velocity blood is too fast to bother about cotton and heads as fast as possible for clear glass or plastic.

The fact that these stains appear on the defendant's glasses offers the only support for the expert's ballistics analysis: He concluded that the shot was fired at a close range, close enough for these tiny droplets to leave the target and strike the defendant's glasses at 90 degrees. The claim often advanced is that this high-velocity back spatter travels about 18 to 36 inches from the target (blood source) toward the muzzle. The round nature of the stains, the report claimed, supports this angle of impact at 90 degrees—when blood drops strike a surface, their angle can be determined by the shape of the stain (Figure 2.22). If the drop is round, the angle is 90 degrees. If the drop is oblong, then the length and width of the stain can be used to compute the exact angle of impact according to a well-established formula: Angle of impact = $\sin^{-1}(W/L)$, where W is the width of the stain in millimeters and L is

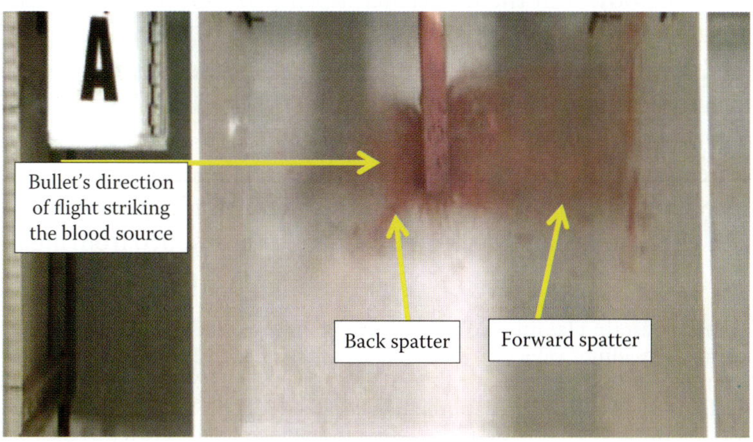

Figure 2.20 Basic physics at work.

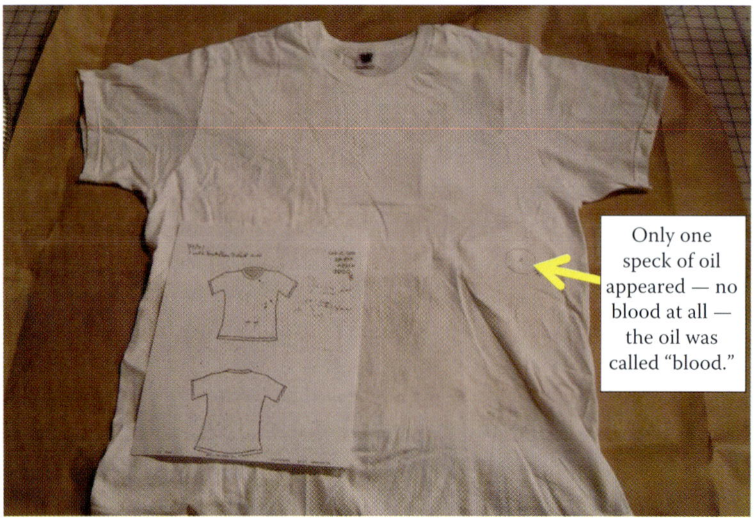

Figure 2.21 The suspect's shirt.

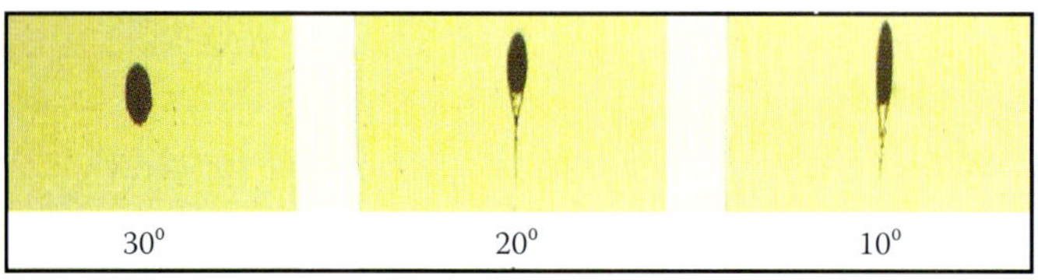

Figure 2.22 Blood drops striking at calculated angles.

the length of the stain in millimeters. The round stains, each 1 mm or less in diameter, absolutely prove that the defendant fired the fatal shot, the report concluded.

In addition, the report emphasizes the size of these stains appearing on the glasses; each stain on the glasses is reported to be less than 1 mm in diameter. The scientific claim underlying these conclusions remains unstated: The report's author assumes that only high-velocity bloodstains* created by gunshot can produce such small bloodstains. This assumption has been repeated frequently in courts and has gone mostly unchallenged. This assumption joins the others in the report that remained untested and undocumented by the expert. Nevertheless, the report concludes by offering a "crime scene shooting reconstruction" based upon the data selected for this analysis while notably ignoring most other data available for examination and consideration.

The reconstruction provided in the reconstructionist's story to the jury offered an explanation for each of the "significant" facts of the case: the disarray in the house resulted from the husband's efforts to stage the scene to make the death appear to be the result of a robbery gone bad; the bloodstains on the husband's glasses resulted from high-velocity back

* According to one bloodstain classification taxonomy, stains are distinguished as being "low velocity" (mostly large stains), "medium velocity" (mostly stains greater than 1 mm in diameter), or "high velocity" (stains 1 mm or less in diameter) in nature. This terminology, while still used, has lost favor with most scientifically trained bloodstain pattern analysts since it is inaccurate, imprecise, and not clearly descriptive. The energy applied to the blood source to produce blood spatter, rather than some vague notion of a droplet's velocity (a vector), is not adequately accounted for by distinguishing low-, medium-, and high-velocity stains.

spatter that could only come from him firing the shotgun sabot that struck his wife's back; and the husband's presence at the scene simply underscores his involvement in the crime.

When initially questioned by officers and detectives about the inability to find the murder weapon at the scene, the reply was that "absence of evidence is not evidence of absence," a quip that some attribute to the patriarch of bloodstain pattern analysis Herb McDonell. It states that our simple inability to discover or develop specific evidence in a case does not somehow *mean by itself or prove independently* that the evidence was never there to be discovered in the first place. However, inadequate science cannot be covered over or hidden by clever quips, no matter how soundly they were originally intended. Indeed, the problem with the scientific work in this case remained to be the lack of any scientific work at all, beginning with the data.

Revisiting the Scene Science: The Problem of Data

Viewing and approaching the scientific project merely in an adversarial light—like two teams clashing on a sporting field with an eye only toward winning the battle of "our results versus their results"—destroys any potential for the impartial examination of evidence. Of course, forensic science remains scientific work to be presented in court. This automatically puts the presentation of scientific results and their assessments into an adversarial context—a context in which interested parties become very partial. But the practice of science itself, the science applied to the data in order to reach those results, must never become guided by an eye fixed solely toward some desired or favored outcome based upon extrascientific interests. In science, the data should prevail, and the results should fall within the proper limits of the science applied to this data. This demand must constantly and consistently overcome all extrascientific, merely adversarial interests. Follow the data.

One potential adversarial problem involves the close associations among interested police investigators, enthusiastic detectives, and ambitious prosecutors with "police scientists" working in "police crime laboratories." This police influence can, at the very least, appear to be a conflict of interest with respect to the practice of independent data assessments and disinterested applications of natural science.* If it does not become an actual conflict, this fact is due in no small measure to the individual integrity and scientific professionalism of the specific crime laboratory analyst involved in the case and the majority of her colleagues.

To avoid any impression of bias, any application of science to crime scene data must remain independent. This can be a problem if one party does not want to share data or refuses to explain the scientific protocols, techniques, and reasoning applied to derive a particular result. These omissions become evident in any report that presents only conclusions with none of the work involved in their derivation. When this occurs, one important element of the scientific method, independent assessment generating either mutual

* Some crime laboratory analysts even openly state "I won't talk with the defense" or they accuse those who do of "going over to the dark side." This hardly reflects a "disinterested" scientific approach to data, which lets the science applied to that data stand openly, allowing other scientists the opportunity to check, test, and either to concur or to supply a chain of reasoning supporting any differences or disagreements. This notion of potential independent assessment remains an essential part of the scientific method, robbed of significance by any such proprietary, parochial attitude. This attitude is by no means limited to crime laboratory scientists. Too many scientists regardless of affiliation seem to fear being wrong and try to mold the data to minimize the appearance of error in their work. This does little to advance the cause of science seeking truth, or the best that we can do from the data.

support or raising additional doubts, remains stillborn. Scientists hired by the defense can be hampered by a prosecutor who retains ultimate control of the evidence.* Obviously, maintaining an interest only in the reliability of the sciences applied to the data in a case becomes increasingly difficult in everyday practice. Saying that one must remain, somehow, true to science is easy. It is straight out of the textbook. But actually doing so, while working among warring attorneys, is not easy for any scientist whether affiliated with a crime laboratory or affiliated with some private agency or lab.

This adversarial problem is not unique to those experts working with the prosecution. Of course, a similar problem can arise in the close association between defense attorneys and the experts they hire. Unfortunately, rather than supplying reliable scientific analyses of independently developed data, some experts are more interested in selling their testimony by making it favorable to defense or to prosecution interests. Most of these pseudoscientific showmen continue in the tradition of the Sophists so disparaged by Socrates in ancient Athens. These spokesmen-for-hire in ancient Athens possessed rhetorical skills more than sufficient to sell sea water to sharks. Socrates pointed out through Plato's dialogues that they functioned to make the true seem false and the false seem true. According to Plato, they lacked a harmonious soul. Even now their intellectual descendants, as reconstructionists unencumbered by fact or science, offer their stories to juries.

These ethical issues and the resulting problems for scientific method are by no means novel and are by no means exclusive to either the prosecution or the defense. Suffice it to say that ethical lapses, wherever they occur, denigrate the application of scientific method as well as the reputations of those who take ethical shortcuts either for personal gain or for the sake of some independently perceived morally right outcome trotted out to justify such scientific sin.

The intellectual baggage revealed by the state expert in this case provides additional insights into how scientific method, or at least what may superficially pass for it, can go badly wrong. The data remain to be the focus of the problem.

Asking questions is always preferred to answering questions, at least when it comes to the initial generation of data. When applying science to the investigation at a crime scene, an unfortunate cookie-cutter approach to the sciences in the expert's toolbox can take hold. In this approach, the data are recognized and collected according to nothing less than specific pseudoscientific prejudices seemingly based upon the statistical frequency of some observation rather than upon the observation of actual data discovered and documented at the scene. Such pseudoscientific generalizations can be confused with covering scientific laws, laws supposedly allowing the deduction of particular instances as logical consequences of these scientific laws. The data, therefore, become recognized as an instance of the familiar law, and the law then provides the scientific explanation of the data. The potentially inappropriate circular nature of this reasoning remains obvious. One example involves claims about high-velocity bloodstains resulting only from a gunshot (Figure 2.23).

The 1-mm bloodstains on the defendant's glasses were both recognized as data through the application of this inferential process (less than 1 mm is high velocity; high velocity is only from a gunshot, therefore this blood on the glasses is high velocity from a gunshot)

* In one case, when the prosecution denied me access to the physical evidence I was to examine, I simply asked the attorney who hired me to ask the prosecution why they bothered to collect, document, and save the evidence if it was to remain locked away, never to see the light of day. Sometimes this jogging of the legal memory is sufficient to stimulate the release of materials for scientific analysis.

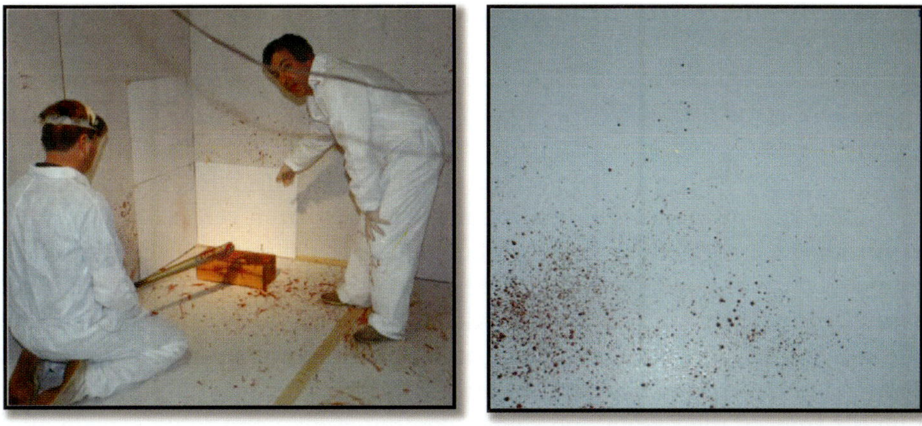

Figure 2.23 Spatter produced by striking a baseball bat into a blood source also produces stains indistinguishable from those produced by a bullet's impact with a blood source.

and explained as the result of a gunshot through this deductive process. Scientific methods, however, only begin with the recognition of some data as puzzling: data that require an explanation. It cannot end with the imposition of an explanation simply imposed at the point of recognition. Of the scientific issues apparent in this case report and expert reconstruction story, the high-velocity bloodstains reported on the suspect's glasses, remains to be one of the most informative about the applications of scientific method.

Testing to Develop Scientific Inferences from Data

The blood evidence available in this case included some scene photos, the decedent's clothing in paper bags, the suspect's clothing (also in paper bags), and the suspect's glasses. Bloodstains can tell us a great deal about any blood-shedding event. Bloodstain pattern analysis, as a discipline, involves the scientific study of the static consequences resulting from dynamic blood-shedding events. As observations, bloodstains must be described and documented. If equivocal mechanisms could produce any given stain pattern, then testing those mechanisms becomes essential. The tests are designed to help determine, if possible, the mechanism that produced the given stains at the scene in this given instance. Such questions posed about the stains on the suspect's glasses and on the decedent's sweater revealed essential information permitting scientifically supported inferences about the shooting event. These inferences replace the apparently circular inferences resulting from the cookie-cutter approach to applied "science" described earlier.*

* This method, if one could call the approach a method, precludes essential elements of what we refer to today as the scientific method. In what I have called the cookie-cutter approach to crime scene data, the science appears in the form of wise sayings, which function scientifically to inform us about the data. (This can even be taught as a form of deductive reasoning.) Sometimes this approach results in the correct answer, but, if one recalls one's early scientific training, it makes none of the correct scientific methods evident. That is why, presumably, we were asked to show our work when solving chemistry, physics, or mathematics problems. The right answer, without supporting reasons, is scientifically meaningless. As I recall my own scientific education, some students even got the right answer by copying from other students who coincidentally had derived the correct answer. The method of copying, however, is not to be confused with any scientific method of computation being measured through the work.

Scientific Inferences from the Decedent's Sweater

The first task in a scientific approach to the case involves observation, followed closely by description. I observed the decedent's sweater, front and back. I noted that the stains on the front, where the slug exited, appeared to be of several different types: they were the typical soaking and wicking stains from exposure to a plentiful blood source, but there were also drip stains and cast-off stains, stains with satellite spatter from vertical drops and stains with tails showing an arc of movement throughout their deposition. This meant that the various stains observed on the front of the sweater resulted from different mechanisms. These mechanisms included soaking and wicking, dripping from a blood source higher than the sweater's surface, and droplets cast off from some blood source moving in specific directions above the sweater's surface. So multiple events had occurred to deposit bloodstains on the front of the sweater; no single shooting event alone could explain all these different stains (Figures 2.24 and 2.25).

Examining the back of the sweater also revealed compelling bloodstain data. Recording overall macro observations is critical, but simply looking at bloodstains with the unaided eye seldom reveals sufficient detail to help classify and distinguish different bloodstain patterns. Unfortunately, the state expert's work did not include microscopic examinations and consequently included no photomicrographs of any relevant bloodstains. The failure

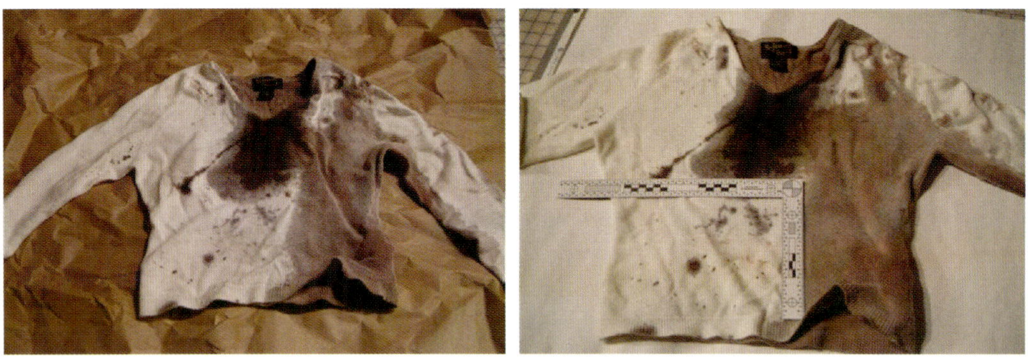

Figure 2.24 Stains on the lower-right quadrant of the garment show drip stains; the midline and the lower-left quadrant show soaking and wicking stains.

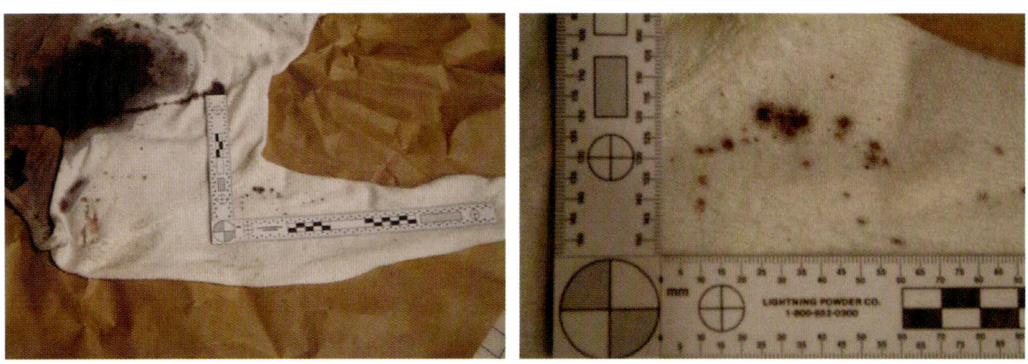

Figure 2.25 Stains on the right arm exhibit directionality, as well as a smaller size characteristic of stains cast off from some moving blood source. The photo on the right shows the mixture of these stains, the larger drips with some smaller cast-off stains.

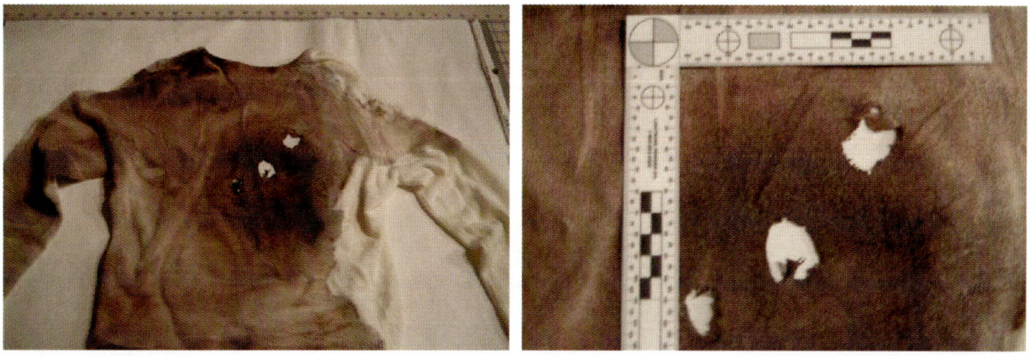

Figure 2.26 The overall views seen here on the left and the right show evidence of the soaking and wicking stains, as well as the holes in the sweater made by the slug and sabot, which guided the slug through the 12-gauge shotgun's barrel. Recall that the shotgun itself was never recovered.

to make adequate or detailed observations becomes a familiar flaw with the so-called cookie-cutter method: there is no point raising any further questions when the answers have already been supplied by the generalization applied.

The back of the sweater presented key pieces of information about the shooting including the three entrance holes: one from the slug, and one from each of the two sides of the sabot (Figure 2.26). But it also revealed another bloodstain pattern visible only microscopically, a pattern that raised further questions about the ballistics issues involved in the shooting.

The sweater was designed with multiple fuzzy fibers covering the entire white surface as part of its weave pattern: a shaggy sweater. Under the microscope, near each of the entrance holes, small droplets of blood appear stuck to these fibers. Recall that when a projectile strikes a blood source, the impact produces small blood drops that travel back toward the source of the projectile. This phenomenon is called "back spatter." It is commonly observed around gunshot wounds of entrance.

If the gun recovered at a shooting scene has small blood droplets present in and around the barrel, this data can be used to help estimate the muzzle-to-target distance. Generally, by experiment, these blood droplets travel backward from the target no more than 20 to 36 inches.* If a shooter has blood droplets on his face, hands, arms or chest, and if these droplets originate from such back spatter, then we have reason to believe that the shooter's face, hands, arms or chest was 20 to 36 inches from the blood source when the bullet or projectile struck the target.

These fibers on the decedent's fuzzy sweater served to block this back-spattered blood, capturing most of the droplets as they traveled back toward the shotgun's muzzle (Figure 2.27). Thus, regardless of the actual distance of the shotgun's muzzle or the shooter's face from the victim's back, even at 20 to 36 inches, no back-spattered blood would be evident in this case. This demonstrates a common exception to the so-called rule, or

* There are two physical phenomena at work here: one is back spatter from impact with the target as discussed here; the other is a vacuum effect whereby the hot gases exiting from the gun barrel along with the bullet create a suction, drawing blood droplets or other materials into the gun barrel itself. Usually, small bloodstains inside the gun's barrel originate from this phenomenon rather than from back spatter alone. I use Microsil casting material applied inside the gun's barrel to document and recover such bloodstains, if present.

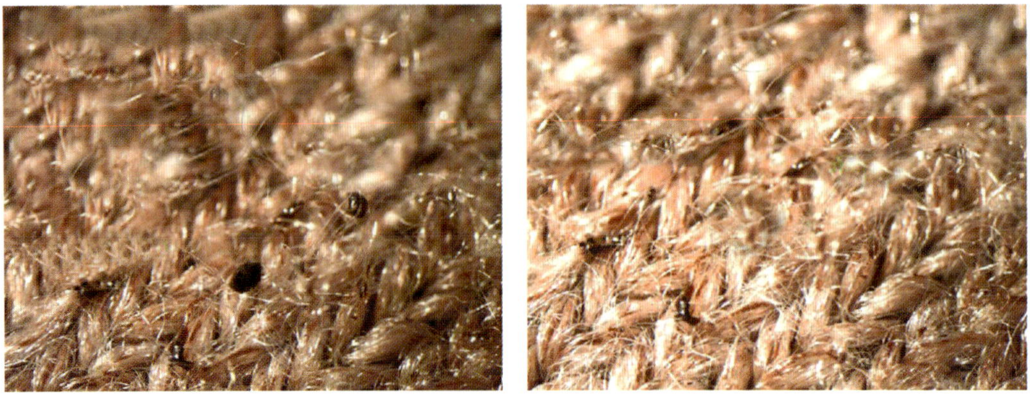

Figure 2.27 The photomicrographs show these back-spattered blood droplets captured by the sweater's fibers. Sometimes these small blood drops can be mistaken for burned and unburned gunpowder. In this case, a simple analytical test using X-ray fluorescence spectrophotometry will settle the issue: no nitrates and no other elements characteristic of gunpowder are present in the sample.

generalization that back-spattered blood will always provide information about muzzle-to-target distances involved in any shooting. The cookie-cutter method does not always cut it.

Interestingly, the data developed here help our additional scientific efforts to understand the shooting. We have no reason to use the back-spattered blood to determine muzzle-to-target distances. First, we know that the droplets were blocked by the sweater, and second, we do not have the shotgun to examine for the presence of bloodstains. But we do have some reasons to think that the muzzle-to-target distance is greater than 36 inches: we observed no muzzle effluents on the back of the sweater either macroscopically or microscopically, no gunshot residues, including no burned or unburned gunpowder and no shot shell talc is present on the sweater. The entrance holes from the slug and two halves of the sabot, and the absence of gaseous renting, tearing, and burning of the sweater, suggest an intermediate range between the muzzle and the target rather than a close-contact or near-contact gunshot wound of entrance. As the data develop, we can begin to see connections among their characters and the inferences that they can license through the careful application of scientific method.

Scientific Inferences from the Suspect's Glasses

The focus of the state's only remotely scientific argument remained to be the bloodstains observed on the suspect's glasses. The claim that only a gunshot entrance wound can produce this high-velocity back spatter has at least been put on notice by the examination of the sweater in this case. Clearly, the data from the sweater show that no back spatter from the entrance wounds could account for these stains. In addition, the data from the sweater show that the gunshot wound is at least an intermediate range injury rather than a close-range or contact gunshot entrance wound. This also precludes the stains on the glasses originating from high-velocity back spatter.

When faced with a question of the possible origin of bloodstains, a scientific approach suggests some sort of testing or experimentation. When hypotheses are developed, they must be tested. In the cookie-cutter approach to science, merely suggesting some hypothesis to explain some data appears to be sufficient to confirm that hypothesis: these cookie cutters seem to be self-confirming. The small stains on the suspect's glasses suggest high-velocity back spatter and the hypothesis of high-velocity back spatter are confirmed by the small stains appearing on the glasses.

Fortunately, the natural sciences do not work that way. Hypotheses must be tested independently of their derivations, and the results can be used in various ways depending upon what the test or experiment is designed to accomplish. Experimental results can suggest revisions of the original hypotheses as tested, as well as potentially providing confirmations or refutations of these original hypotheses. If our hypothesis that the small stains on the suspect's glasses originate from high-velocity back spatter does not work, then what does? The suspect himself provides another hypothesis suitable for testing. The scientific method intellectually demands that such tests be conducted.

The suspect's account of his actions at the death scene includes approaching the front of the body, bending over to touch her chest, finding that his hands became covered with blood, standing before the body for a second or two with his hands suspended in front of his own chest, and when bent over, shaking his hands in the air in an effort to remove the blood. He then stood upright and wiped his hands on his jeans before he tried to use the telephone. This suggests an alternative mechanism to account for the stains on the glasses. It also explains the lack of staining on his shirt and undershirt. The potential virtue of the hypothesis holding that the stains on his glasses originate from blood shaken off his hands is that this hypothesis can also account for the stains observed on the front of the victim's sweater. According to this hypothesis, the drip stains come from blood falling from his hands as they are held in front of his own chest and over her sweater while the cast-off stains come from shaking his hands. I developed a simple test using similar glasses and my own blood.

I drew 20 ml of my own blood and soaked it into a sponge surrounded by butcher paper on the lab floor. I put on the glasses. I squatted down, putting both hands on the blood-soaked sponge. I stood up with my hands suspended in front of my chest, letting the blood drip from my hands onto the butcher paper. I then simply looked down at the sponge on the floor and shook my hands. The observations successfully replicated the drip and cast-off bloodstains on both the front of the sweater and the bloodstains on the suspect's glasses (Figure 2.28 to Figure 2.31).

Figure 2.28 The suspect's glasses appear on the left while the test glasses appear on the right. Both were examined with the microscope to analyze the bloodstains present on the lenses.

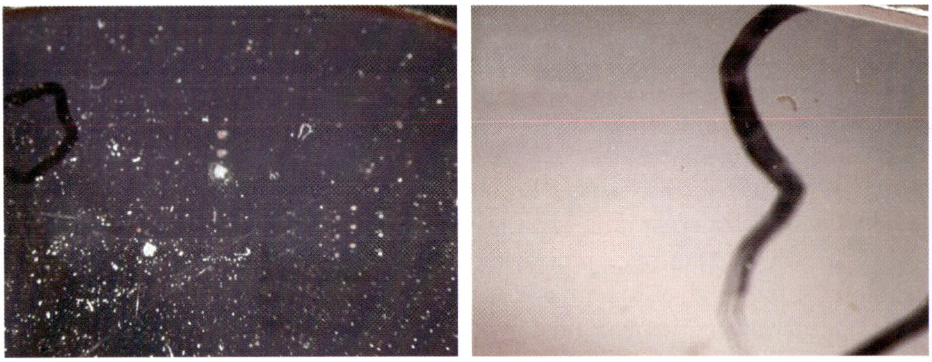

Figure 2.29 The suspect's glasses are seen under magnification. The black line here was introduced by the state's examiner. The glasses are dirty and scratched. The stain in the center of the picture is the blood spot. Note how small the stain is—smaller than the diameter of the lead in a modern mechanical pencil.

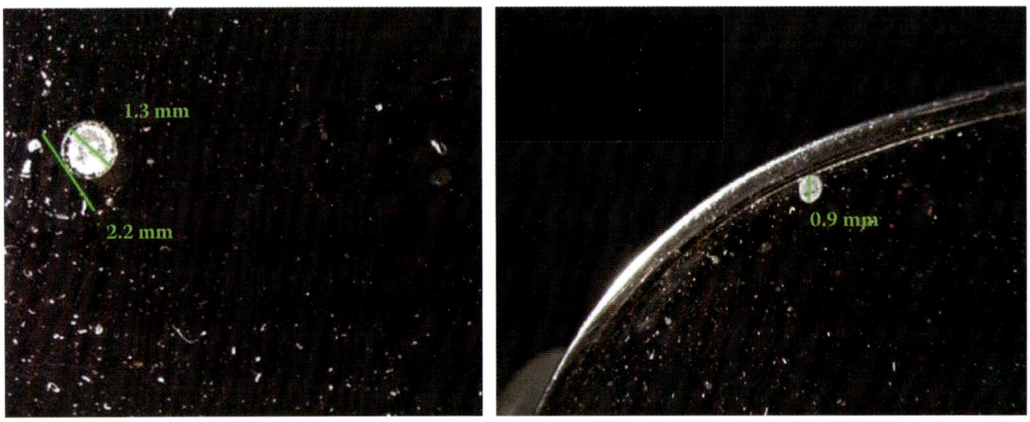

Figure 2.30 This slide shows an interesting characteristic of blood. Notice the small white spot of blood; it is 1.3 mm of blood. However, look closely behind it and you will see the original blood spot, which is 2.2 mm, almost twice as big. As the big spot dried, it shrank and left the smaller spot behind. The reason this happened is that the serum separates from the red cells and evaporates. The red cells then clump together and clot. The 1.3-mm spot is the clot of red cells. The photo on the right shows how small these stains are. (Remember the pencil lead example.) These are the smallest blood spots on the defendant's glasses.

I then wiped my palms on my jeans to conclude the experiment. The wipe stains on my Levi's also matched the stains that I observed on the suspect's jeans.

The more data that the hypothesis consistently accounts for, in general, the more reliable the hypothesis becomes. The foundation for this lies in "theories of truth," a province of the philosophical discipline known as epistemology, or theory of knowledge. The relevant point is that if the hypothesis is in fact true, then at the very least, all other facts in the world are logically consistent with that hypothesis.

To get an accurate result, I used test glasses that were also scratched and dirty (Figure 2.32 and Figure 2.34). The spots were produced by shaking both bloody hands while squatting on the floor (Figure 2.33). Remember the size of the spots on the defendant's glasses. They ranged from 2.2 mm to .04 mm in size.

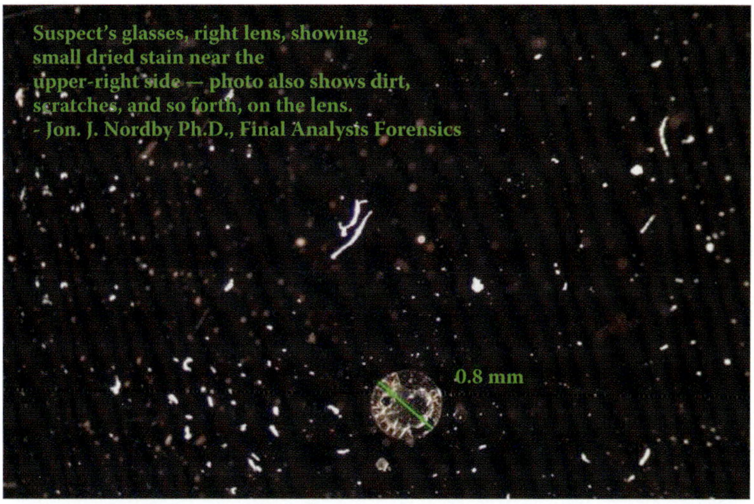

Figure 2.31 This is another example of the small size of the spots present on the glasses.

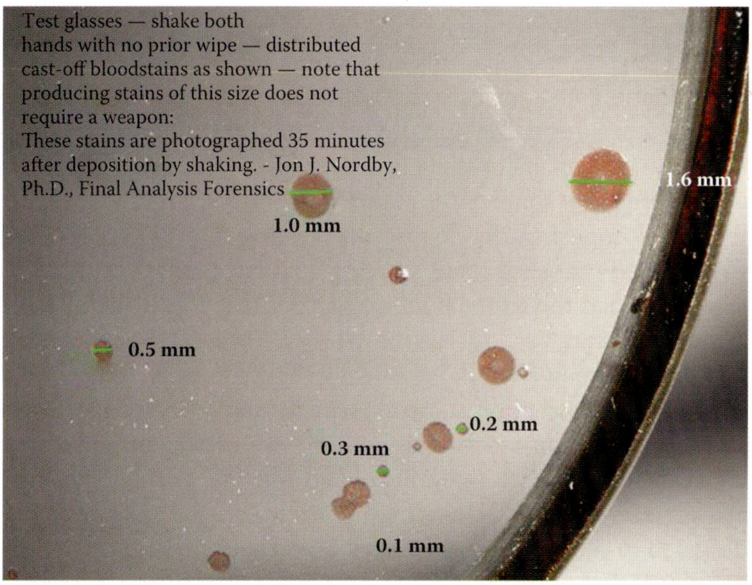

Figure 2.32 Here are blood spots produced by shaking hands that were not first wiped off. Notice the size. Some are bigger and some are smaller. These spots were photographed 35 minutes after the bloody hands were shaken. They have not yet started to clot and shrink. It takes approximately 4 hours for them to clot and shrink, depending upon the conditions.

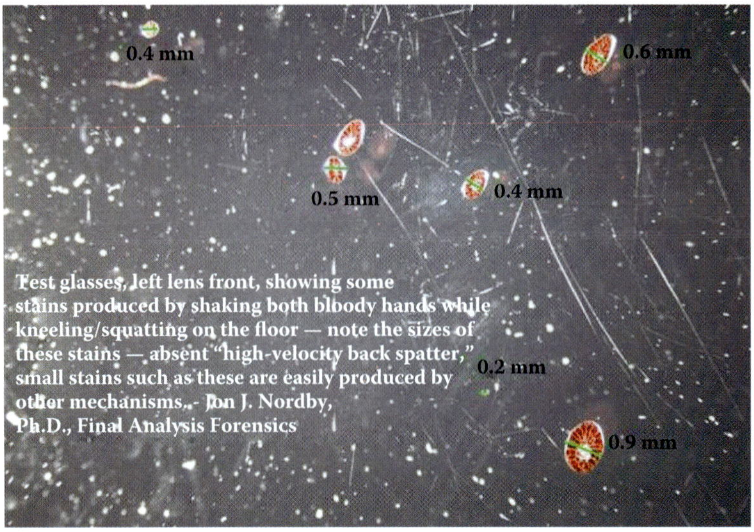

Figure 2.33 Test glasses lens.

Look at the spots in Figures 2.32, 2.33, and 2.34. They range from .9 to .02 mm in size. They are remarkably similar in size. In fact, there is no relevant difference between these spots and the spots on the defendant's glasses.

As you can see, the spots are identical to the spots on the suspect's glasses and my test glasses spots were not produced by high-velocity back spatter from a gunshot. They were produced only by shaking bloody hands. There are other ways to produce small spots than high-velocity back spatter, as my experiment has shown. The conclusion, then, is that the spots on the defendant's glasses could have been produced by hand shaking the way I produced these spots on the test glasses. This is exactly what the suspect reported doing.

The assertion that bloodstains as small as 1 mm must result from gunshot is simply proven false by this experiment. The steps involved in developing the hypothesis that hand shaking could explain the stains on the glasses involved microscopic examination of the glasses, in turn showing that some of the stains were indeed larger than 1 mm before they dried. The method of proposing, developing, and testing this hypothesis produced results that not only confirmed that such hand shaking could account for the stains on the suspect's glasses, but also had the added bonus of accounting for the stains observed on the front of the victim's sweater.

Of course, the experimental work is now only half complete. The original hypothesis offered by the state's reconstructionist also merits testing even though this had not been done as part of the state's reconstruction. If nothing else, testing provides an experimental context for the evaluation of both hypotheses and introduces modesty appropriate for a scientific, nonadversarial reconstruction. I found a similar fuzzy sweater, used a rifled barrel shotgun designed to shoot sabots, and put a blood source behind the back layer of the sweater. I shot a similar sabot from a 4-foot muzzle-to-target distance. I wore another pair of old and scratched glasses. The results were as I suspected.

The back spatter was collected onto the long fuzzy fibers forming the surface of the sweater. Some spatter appeared on the butcher paper underneath the sweater target. No spatter reached the glasses. I repeated the experiment several times from several distances

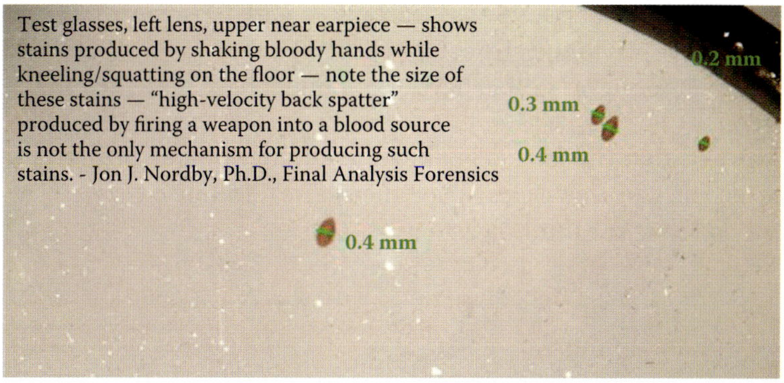

Figure 2.34 Test glasses lens near earpiece.

up to 2-feet muzzle-to-target distances. Relevantly similar results were produced: no stains reached the glasses given the fibrous sweater.

Scientific Conclusions about the Shooting Events

The dependence of good natural science and a sound shooting reconstruction upon data and its cogent development cannot be overemphasized. The state's reconstruction suffers from multiple instances of systematically ignoring significant data. For example, in addition to the problem of flawed bloodstain pattern analysis, if the suspect arrived home when witnesses indicated, and if he did the things that he stated, immediately phoning 911 when arriving at the neighbor's home, and if, as we do know, he stayed on the phone until police arrived, how then did he have time to stage the scene, hide the murder weapon, and dispose of the missing items from the home, let alone find time to shoot the victim? These questions must be raised. Once raised, they require at the very least some effort to provide synoptic answers. These features are certainly relevant to any reconstruction of the shooting events.

Neither the shotgun nor items reported missing from the home were ever found despite an extensive search. Such data cannot be systematically ignored just because a one-size-fits-all covering a law-like generalization suggests a solution to the case that is independent of these facts. Such pseudoscientific generalizations must never be taken at face value—doing so betrays the very essence of scientific method and flies in the face of its successful practice.

A crime scene reconstruction remains only as reliable as the sciences applied to the data. The sciences applied to the data are only as reliable as the completeness of the recognized, collected, documented, and preserved data. And the data can only be developed through the natural sciences, applied according to the scientific method, which distinguishes reliable scientific inferences from prejudices, lucky guesses, and mere random chance.

Conclusions

The jury listened patiently as I presented PowerPoint photos of my shotgun and bloodstain pattern testing, as well as documentations of my clothing analyses. On cross-examination,

I simply reiterated the facts of the case as clearly as I could in response to the questions. The judge interjected questions to the prosecutor such as "didn't you do that?" and "why didn't you?" I left the stand feeling, as I often have, that I might have presented the evidence more simply or clearly.

Later, I was told that the jury deliberated less than an hour before returning a not-guilty verdict. People can think clearly for themselves. In our criminal justice system, a great deal rides on that clear thinking.

The Room with the View 3

It is a glorious feeling to discover the unity of a set of phenomena that seem at first to be completely separate.

Albert Einstein in a Letter to Marcel Grossmann (1901)

Introduction

The three bodies lay in a 12 × 12 foot sitting room. All three women had been shot multiple times. The television broadcast provided the only artificial illumination in the room. The curtains over a small sliding aluminum-frame window were bunched on the left side, admitting the first rays of the morning sun.

One body occupied the daybed, extended from the convertible couch, lying over the neatly made bed, a pillow under her head, both arms at her sides, palms up. Her left hand appeared mangled. The first finger, beside the remaining stump of her left thumb, was missing. Her name was Lesella. Another body, that of Maxalline, covered what seemed to be at least half of the remaining floor space in the room. She lay face down, with the right arm extended over her head, in front of the TV. The third woman, later identified as Deneshia, lay on her side, wedged between the second woman and the TV. A rerun of the *Rockford Files* provided fictional gunfire to accompany its flickering multicolored light. The overall effect was, to say the very least, quite disturbing.

The four detectives working the scene quickly concluded that the three women had been moved into the small room after their deaths. A 44 caliber Magnum jacketed hollow-point (JHP) bullet was lodged in the subfloor, having passed through the carpet and the pad. The defect was surrounded by a soaking bloodstain. Another expanded 44 caliber JHP bullet lay resting on a side table between the wall and the daybed. Two other bullet fragments were recovered, one in front of the bed and the other in front of the television set. All were identified as 44 caliber Nosler Dual Partition JHP bullets.*

The spent bullets were thought to have been placed on the table and the floor to deceive any unwary, inexperienced investigator. At the scene, a detective retrieved one more spent bullet from a pillow, found at the end of the daybed, by cutting through the pillowcase, the pillow cover, and burrowing through the fibrous padding with a pocketknife. The other bullets awaited recovery during the autopsy.

All four investigators—three homicide detectives and one crime lab firearms examiner—agreed that if a 44 Magnum 250-grain Nosler Dual Partition JHP was fired through a standing victim in that small room, then the bullet would continue through the sheetrock wall and travel well into the surrounding neighborhood before either stopping on its

* Such ammunition is often used in 44 Magnum revolvers, which function as backup guns carried by bear hunters in Alaska. The Nosler Dual Partition bullets are both powerful and destructive, providing sufficient stopping power to kill a large bear.

own or striking its next target. The locations of the three victims and the fact that various bloodstains seemed to occupy areas well away from each of them seemed to support this hypothesis. So the search was on to find the real crime scene.

Case Review

The beginning of any shooting reconstruction based solely on science depends upon recognizing, documenting, and developing the available data. In some cases, the initial scene work may serve to block rather than help to develop sound scientific data.

When examining the physical evidence recovered in this triple homicide case, I began with a host of questions. To seek answers, I looked toward the original police reports, the scene drawings, and the scene and autopsy photographs and X-rays. I examined all recovered items of physical evidence including the bedding, two towels found in the room, the pillows, and the victims' clothing items. I examined a 44 Magnum Smith & Wesson Model 629-6 revolver found in shallow water off a nearby Florida beach, as well as bullets recovered from the room and from the victims. This included six jacketed Nosler 44 Magnum double-core bullets and some small lead and jacket fragments.

Everyone, including the crime lab's firearms examiner, claimed that shooting 44 cal Nosler Dual Partition bullets in such a small room without perforating the walls was impossible. The bullets would perforate the targets and exit the walls, flying through the neighborhood until they struck the next target or simply ran out of energy, finally overcome by gravity. I asked if anyone had ever fired such ammunition. The answer was no. Experience alone, they said, was sufficient to "know how such ammunition behaved."

Although I have over 25 years' experience, I am quick to confess the boundaries of my knowledge and the limits of that experience. So I built a room, 12 × 12 feet, got several boxes of 44 cal Nosler Dual Partition 250-grain bullets and set up my model of the crime scene. To investigate the behavior of the bullets in question, why not shoot a few hundred rounds to see what happens?

Summary

1. Scientific experiments were conducted with 44 cal semijacketed hollow-point Nosler ammunition in order to study the behavior and physical properties potentially unique to 44 cal Nosler Partition ammunition and its behavior with various targets.
2. Scientific experiments were conducted to analyze alternative mechanisms for the production of bloodstain patterns and gunshot wounds found in the investigation of this case.
3. The collection, documentation, and polarized light microscope analyses of fiber evidence was developed from Lesella's blue sweatshirt and from the Nosler Partition jacketed secondary core with single petal present recovered in the left chest of Maxalline.

My Conclusions Summarized

1. That the crime scene, as initially presented to investigators and as captured in official crime scene photos, has been altered or staged at some time after the shootings and before the official police documentation of the scene, as follows:

   a. That the pillow behind Lesella had been moved after the shootings
   b. That Deneshia had been positioned near Maxalline after both had already been shot
   c. That the bedding was rearranged after the shootings, including the blanket covering Maxalline and Deneshia, and the blanket put onto the bed
   d. That a receiving blanket and a yellow towel were used for at least two distinct purposes, perhaps at least once by the shooter and perhaps at least once by the stager
   e. That the nature of the staging suggests the actions of someone cleaning up the scene after the shootings; potentially someone other than the shooter or shooters; both person or persons unknown (from the physical evidence)
2. That the gunshot wound (GSW) injuries to Lesella and the evidence of such injuries on her clothing are best explained by three separate 44 cal Magnum Nosler bullets striking this victim, one in the hand, one in the back, and one in the head.
3. That the bullet holes found on the fabrics mentioned earlier—the bedding, blanket covering Maxalline and Deneshia, the blanket put onto the bed, the receiving blanket found near Lesella, and the yellow towel—support their origins as being from 25 cal ammunition unrelated to this case and 44 cal ammunition possibly related to the three homicides.
4. That the origin of the blood on the foot of the lower bedsheet, found under the pile of blankets, sheet, towel, and spread that were piled over the assembled bedding after the shootings, supplies further evidence of a staged crime scene (the blood was found to belong to Lesella via DNA analysis).
5. That Maxalline's injuries result from a single 44 cal Magnum Nosler jacketed hollow-point round that struck her head while tumbling, leaving a characteristic keyhole entrance after striking a primary target, then breaking apart and leaving four of its six jacket petals in her neck along the wound track from the top of her head to her left chest (with one more separated petal, the fifth of its six jacket petals, found in her left chest along with a lead core or part of a lead core). The primary lead core first left lead particles above the calvarium or skull bones (in her scalp), as well as along the wound track into her left chest, and either the primary core from this round separated at the skull, with part of this core entering the left upper axillary shoulder, or a separate primary lead core from a different Nosler round entering the left upper axillary shoulder.
6. That the bullet found on the desk is totally absent of any type of trace evidence, either organic or inorganic, and therefore does not scientifically relate to the injuries sustained by any of the three victims in this case.
7. That the bloodstains found on the scene cannot be explained if we assume that the three victims, as discovered and photographed in crime scene photos, remain in the exact positions in which they were originally shot, and if we assume that the blankets, victims, and thereby bloodstains have not been materially altered at some point after the homicides and before the police responded to the scene.
8. That a single 44 cal semijacketed hollow point struck Deneshia in the back, perforating her body, exiting her chest between her right arm and trunk.
9. That the injuries to Deneshia's leg resulted from a bullet fragment or fragments or from a bullet or bullets that may have first struck a primary target only to then strike Deneshia's leg as a secondary target.

10. That Lesella, the only victim with three separate gunshot wounds of entrance from intact, nonfragmentary bullets, appears to be the primary target of the murderer or murderers.
11. That Maxalline appears to have been a collateral target with entry wounds from one secondary missile, along with its fragments; or at most, from that missile with its fragments and another primary lead core from a different bullet.
12. That Deneshia may or may not have been a so-called collateral target given her injuries.

Autopsy Terminal Ballistics

In this case, we know the caliber and type of bullet used to murder the victims. In some cases, all we have are the entrances and maybe the exits left from some unknown bullet from an unknown gun. Sometimes, experiments can be developed to help determine the caliber of bullet producing the injuries. The data required begin with the autopsy.

If the bullet strikes the target at about 90 degrees, the work is somewhat easier. If we can determine the bullet's trajectory through the victim, then we can replicate that angle and fire into the test target. Measurements of the results can help determine the caliber of bullet used in the homicide. Of course great caution must attach to this work. Some calibers are very close to each other. In those cases, multiple shots with the various calibers at issue still may not resolve the issues.

Using white vinyl or leather over ¼-inch plywood works well to produce measureable entrance holes. It may seem counterintuitive, but the entrance hole is smaller than the bullet's caliber. The bullet wipe is the key data to observe. If the bullet wipe is not documented at the autopsy, or if the wound is produced by a tumbling bullet, then all bets are off. But experiments, as in this case, may be fruitful.* All that one can do is to consider the size of the entrance hole and the trajectory of the bullet penetrating or perforating the body.

If the hole in the body is close to 10 mm at 90 degrees, for example, one can rule out a 9-mm bullet; unless, of course, the bullet is tumbling or enters at an angle, or is striking a secondary target. Then the testing must attempt to replicate that angle. I find that reliably reproducing the characteristics of a secondary target does not work very well; the primary target may be unknown. Again, caution attaches to any conclusions. Error rates must be calculated based upon the measurement issues as well as on the varying angles and their effects on the diameter of the bullet wipe and entrance hole.

Figure 3.1 is one test out of 20, fired at about 90 degrees with a 9-mm Glock at a ¼-inch plywood target covered in white vinyl. If the muzzle-to-target distance is unknown or has not been determined, vary the distance from the target and record your results. That gives you more information about terminal ballistics.

Note that the trajectories of the bullets through each victim (as described next) are explained from the anatomical position. In this case, we know that the bullets are 44 cal Nosler Dual Partition—a fairly unique bullet, easily identified.

* In this case, the keyhole entrance into Maxalline's skull resulted from a tumbling bullet. Yet testing was able to reproduce the pattern of the entrance more than once. Bullets fired through a known primary target (ballistic gel in this experiment, representing Lesella's back) first perforated the target and then left a secondary pattern similar to the pattern left by the bullet striking Maxalline.

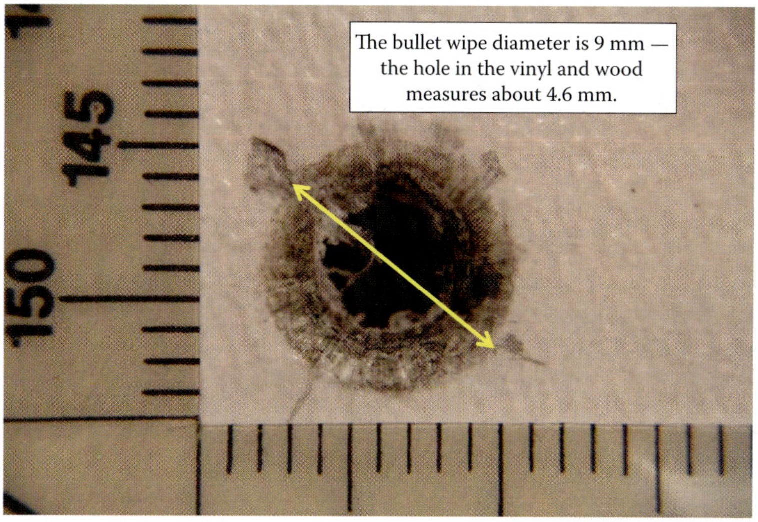

Figure 3.1 One test out of 20, fired at about 90 degrees with a 9-mm Glock at a ¼-inch plywood target covered in white vinyl.

Lesella

1. Enter dorsum of left hand; exit palmer surface, base of left thumb
2. Enter helix of left ear; exit behind right ear
3. Enter left posterior flank; exit right posterior lower back at T-12 L-1

The three bullets that perforated her hand, head, and flank (Figure 3.2) may have gone on to strike secondary targets (the physical evidence supporting the opinion that Maxalline's head wound provided at least one such secondary target will be discussed later). None of Lesella's injuries present evidence of stippling, tattooing, or soot deposition. Circular black marks around entry holes on her left hand, left ear, and on her blue sweatshirt resulted from abrasion rings left by the relevant 44 cal Nosler bullets, not from muzzle contact of any sort.

The bullet wounds suffered by this victim represent at least intermediate range gunshot wounds.

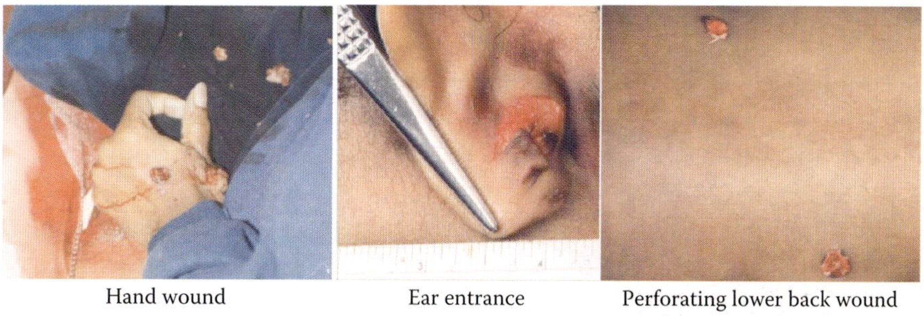

Figure 3.2 Hand wound, ear entrance, and perforating lower back wound on Lesella.

Maxalline

Three entrance wounds result from one bullet and its fragments that struck Maxalline after striking a primary target:

1. Enter occipital skull; exit left side of neck; reenter left chest
2. Enter apex of left shoulder; exit left side of left breast; reenter areola (fragment)
3. Enter posterior aspect left shoulder; tracks into left chest (fragment)

Recovered from Maxalline: one copper-jacketed lead bullet (secondary lead core), five separate copper jacket fragments (petals), and either one (split) or possibly two distinct primary lead core(s) listed as follows:

a. 44 caliber Nosler Dual Partition with secondary core, copper jacket and one of six petals remaining recovered from left chest
b. Four petals from Item #22 recovered from neck wound track mechanically articulate or fit onto the broken pieces of Item #22
c. One or part of one primary or upper lead core recovered with Item # 22 from left chest
d. One or part of one primary or upper lead core recovered with fibers from left breast
e. One petal from Item #22 recovered from victim

Maxalline apparently remained in a prone position when struck by a single Nosler round, which had previously struck Lesella. The round perforated Maxalline's back laterally from left to right and slightly upward. It appears that Maxalline was discovered approximately in the position that she occupied when she was struck by this tumbling bullet, which had first perforated Lesella's blue sweatshirt and back.

None of Maxalline's injuries present any stippling, tattooing, or soot deposition. All the wound tracks revealed via X-ray show the fragmentation order of the Nosler bullet, including its core, and its five petal fragments. This supports the view that the bullet produced a secondary keyhole entry at the top of her head, depositing lead fragments in her scalp tissues above the cranial vault:

a. Thereby fracturing the Nosler's upper core into two parts, one entering her left axillary shoulder and the other traveling with the jacketed secondary core with one remaining petal attached as they fracture the left clavicle, and are recovered in her left chest; or
b. If the Nosler's upper core was not fractured into two parts, then the lead core entering her left axillary shoulder must be the primary core, or a primary core part, from a different Nosler shot itself not directly striking Maxalline, resulting in two primary lead cores present and recovered in Maxalline's body.

The bullet wounds suffered by this victim represent intermediate range gunshot wounds since no soot, tattooing, or stippling appears to be associated with any of these injuries, and the size of the scene is approximately 12 × 12 feet with a single door leading into a hallway (see crime scene measurements).

Figure 3.3 depicts Maxalline's injuries by X-rays. Each has radio opaque objects, which are bullet fragments.

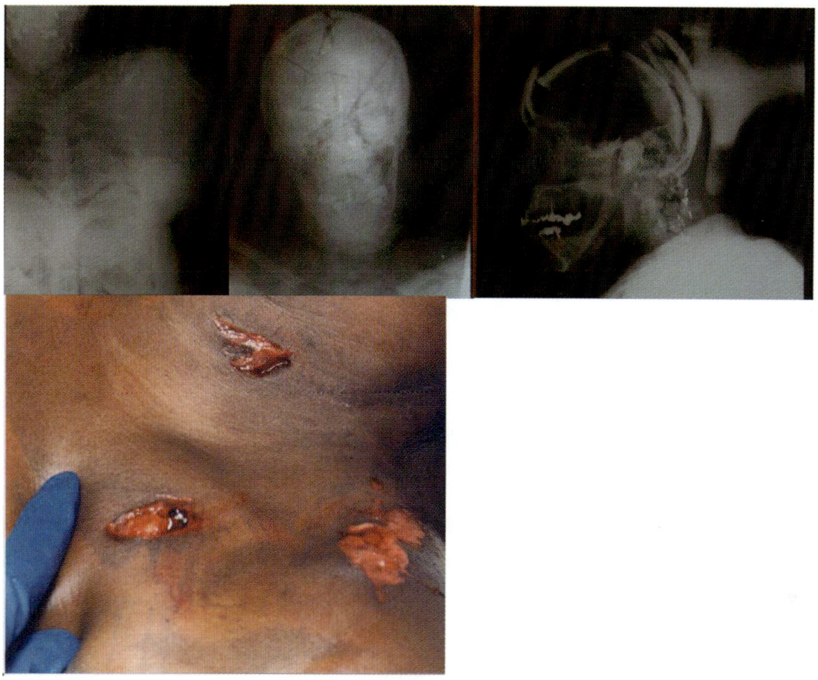

Figure 3.3 (Top) Films showing eight fragments; (bottom) the secondary and tertiary entrances.

Deneshia

Deneshia had two entrance wounds with exits and one apparent graze wound (Figure 3.4)—one entrance results from a fragment—either from a lead core(s) or jacket petal(s) (see ballistics testing for data on the behavior of 44 cal 250-grain Nosler Partition Winchester Gold ammunition fired through a 44 cal Smith & Wesson Model 629 revolver).

Her graze wound could result from a whole bullet traveling on and striking her (perhaps as a secondary target), or the wound could result from a fragment striking her as a secondary or even tertiary target. The clear single entrance to her back and its exit results from a single 44 Magnum Nosler round apparently striking her as its primary or first target:

1. Enter right superior back; exit lateral aspect right chest; reentry posterior aspect of upper right arm; exit anterior right biceps
2. Graze wound medial right lower thigh right to left and downward

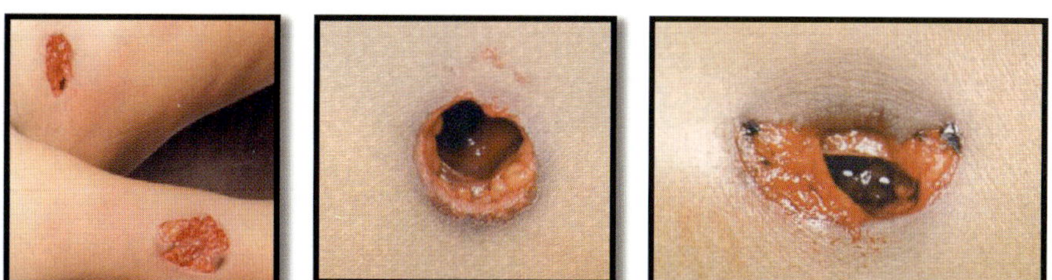

Arm and chest Back Shin

Figure 3.4 Arm and chest, back, and shin wounds on Deneshia.

3. Enter front of left lower thigh irregular shape; exit superior left lateral pretibial region also irregular

The perforating injuries sustained by Deneshia resulted in the recovery of no jackets, no jacket fragments, no lead cores or lead core fragments from her body.

None of her injuries present evidence of stippling, tattooing, or soot deposition. Given the small size of the room, the bullet wounds suffered by this victim represent at least intermediate range gunshot wounds.

Brief Summary of Each Victim's Injuries

GSW	In	Out	Found	Trace: Wound	Trace: Bullet	Trajectory
			Lesella			
A	Left ear	Behind right ear	Pillow bullet	Blood/GSR, no stippling, no soot	Fibers/tissues recovered in pillow?	L→R; F→B
B	Left lower back	Right lower back: T-12, L-1 spinal levels	Maxalline left chest	No stippling, no soot	From blue sweatshirt: fibers with tissues of polyester/cotton/flax blue in color	L→R; F→B
C	Left posterior hand	Left anterior palmer surface thumb	In floor after Deneshia injury to leg?	No stippling, no soot	Tissues/carpet fibers?	L→R; B→F
			Deneshia			
A	Left thigh by knee after striking Lesella's hand	Left shin	In floor carpet pad and wood?	None retained	Tissues and carpet fibers?	Horizontal with knee bent when struck: top to bottom, straight with slight leftward directionality to carpet
B	Right upper back	Right axillary area/right upper arm	In floor carpet pad and wood?	None retained	Tissues and carpet fibers?	Back to front, straight into carpet?
			Maxalline			
A	Top of skull/keyhole entrance with exit reentrance	Penetration into left chest	In left chest minus five petals and primary lead core	Lead fragments from primary core in scalp with skull as a secondary target of tumbling round	Tissues and blue cotton fibers best explained as originating from prior Lesella sweatshirt/back perforation (polyester, cotton, and flax composition)	Top to bottom, right to left and downward

Brief Summary of Each Victim's Injuries

GSW	In	Out	Found	Trace: Wound	Trace: Bullet	Trajectory
A 1	Fragments of five broken petals, some lead fragments perforate, penetrate the victim's left breast, shoulder, and left chest	Penetration into left chest and left axillary shoulder and left breast	Left chest and left breast	None recovered as noted	None recovered as noted	Top to bottom, right to left and downward

Comment on Hand Tests for Gunshot Residue

The presence or absence of gunshot residue (GSR) on the hands has long been considered of marginal value for the identification of shooters. GSR can persist on skin under certain conditions, but it is easily removed via washing, wiping, and everyday actions such as shaking hands, touching materials, or even holding objects in the hands. Of much more interest is the presence of any GSR on clothing—more specifically, the patterns of GSR, if identifiable, on the clothing. This data may at the very least indicate that the clothing was in the near vicinity of a firearm's discharge. The patterns may allow additional interpretations under certain specific, scientifically appropriate conditions.

Examination of the Smith & Wesson Revolver

The gun was reportedly recovered by beachcombers in shallow water off Bonita Beach in Scarsdale, Florida. The revolver, along with other ballistics evidence, was hand delivered to Final Analysis Forensics by Detective William Hartung of the Scarsdale Police Department. It arrived disassembled; the frame, crane, cylinder, and barrel were packed in a plastic zip-top bag, with the handle and internal mechanisms packed in other plastic bags.

The parts of any unserviceable revolver require careful examination in all cases. One issue might become the serviceability of the gun at the time it was fired at a victim. Figures 3.5

Figure 3.5 In another case, a Colt Trooper Mark III's fractured hammer sear is seen from the right side.

Figure 3.6 A Colt Trooper Mark III's fractured sear from the top. Note the location of what was later determined to be blood from a victim in that case. Also note its apparent absence along the fracture.

and 3.6 show the damage to a Colt Trooper III revolver's hammer sear, which became an issue in a different case. Did the damage occur before, during, or after the shooting?

In this case, however, the Atlantic Ocean seems to have eliminated any possibility of trace evidence from the Smith & Wesson 44 Magnum revolver. Its functionality, however, must be determined. It was later repaired sufficiently to test fire a bullet and examine known and unknown bullets to identify this revolver as the murder weapon.

Even a Smith & Wesson revolver, well made, safely designed, and easy to shoot (if a kick does not trouble the shooter) can be destroyed by careless or stupid handling. Figures 3.7 and 3.8 show a Smith & Wesson Model 629 (a revolver unrelated to this case) that literally blew up when the reloaded cartridges had been charged with too much gunpowder. Unfortunately, that did not happen in this case. If it had, perhaps the shooter would be dead and the three victims would be alive.

An assessment of a revolver should include at least the following information:

Item:	One 6-shot revolver
Description:	Smith & Wesson, Model 629, registered owner of record: unknown
Ammunition:	The revolver was reported to be inoperable as received. Essential parts were replaced and the weapon was cleaned before test firing. As received by Final Analysis Forensics, the weapon remains inoperable with no ammunition in the cylinder. The cylinder appears dirty with each chamber empty.
	Revolver Serviceability
The barrel:	Appears straight, free of bulges and tight in the frame.
	Pitting and rust are noted, but with no obstructions; the barrel is functional.
The sights:	The front sight is present, firmly attached, and the rear straight is missing.
The crane:	Appears to be free from major warping but presents rust, pitting, and other defects.
Ejector rod:	The ejector rod works normally, however, the locking rod inside the ejector rod sticks. As received, the cylinder fits and locks into frame. (However, repairs are required in order to fire this weapon safely.)

Figure 3.7 Although this gun is not related to this Florida case, these photos show the dangers of loading one's own cartridges. Meth cookers had reloaded these cartridges.

Figure 3.8 The damaged Smith & Wesson Model 629, as seen in Figure 3.7 above.

Cylinder:	Rotation is counterclockwise. Chambers are a bit rusty with no obstructions and are dirty. Cylinder exhibits rust, burrs, and some pitting. Dings and scratches are present on the cylinder.
Dry fire double action:	Space between the firing pin hole and the cylinder cannot be determined due to condition. The firing pin is deteriorated but attached to the hammer. The hammer shows signs of severe deterioration via oxidation and remains beyond repair.
Internal safety:	Internal safety functions cannot be determined in this weapon's present condition. Since this is determined for a Smith & Wesson by holding the cylinder thumb latch to the rear while pulling back the hammer, it cannot be done with such parts removed, and remains inoperable.
Cylinder rotation:	The cylinder rotates counterclockwise. That is all that can be determined in the weapon's present condition.
Cylinder lock and alignment:	A safe model would behave as follows: with the hammer down and finger off the trigger, the cylinder should not rotate. (N/A here.) The cylinder should lock without pronounced movement. (OK here.) The barrel should remain properly aligned with cylinder chamber. (N/A here.) A safe model would exhibit less than a 1/64-inch play in a properly functioning cylinder stop, with no apparent misalignment. (N/A here.)
Hammer:	When the hammer is cocked, pressing against the hammer spur should result in no movement. (N/A here.)
Trigger pull:	Double action not tested for reasons noted. No trigger pull measurements were conducted due to condition.
	Single action not tested for reasons noted.
Grips:	The handles appear to be wood, are caked with sand, algae, and salt but with suitable effort, it may be possible to restore them, at best, to a minimally serviceable condition.

Summary of the Condition as Examined

The revolver involved in this case remains inoperable as received (Figure 3.9). However, at the time of the shootings, we may infer that this weapon was in good working order. This inference derives from the fact that the PD crime lab firearms and tool mark section matched the individual characteristics of test bullets fired from this weapon (as repaired) with the individual characteristics of the bullets recovered both in Maxalline at autopsy and

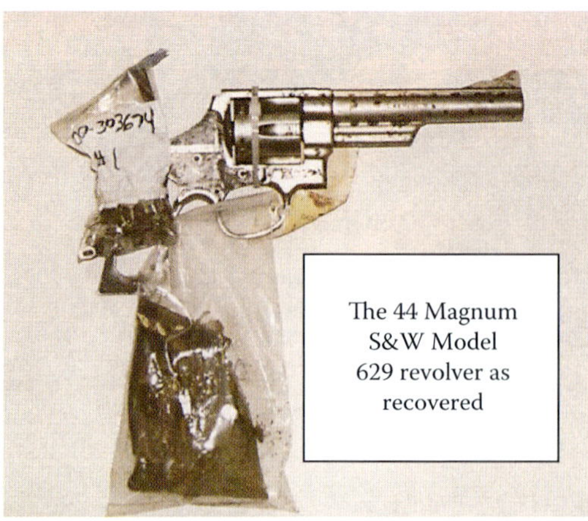

Figure 3.9 The revolver involved in this case.

Figure 3.10 The known bullet on the left and the unknown bullet on the right, recovered from Maxalline's autopsy for comparison.

Figure 3.11 The ammunition used in the Smith & Wesson revolver, shooting in a 12 × 12 foot room.

Figure 3.12 Replacement Smith & Wesson Model 629-6 revolver.

at the scene of the triple homicide (Figure 3.10). I confirmed this match independently by determining that all the identifiable bullets sent to me in evidence were fired from the same revolver. Therefore, these 44 Magnum Nosler rounds were fired through the barrel of this revolver (Figure 3.11). It is therefore the murder weapon used in these three homicides.

To test the performance characteristics of the ammunition striking the victim(s) at the scene, a relevantly similar Model 629-6 Smith & Wesson 44 Magnum revolver (Figure 3.12) was used with Nosler Partition 44 Magnum ammunition, the same ammunition recovered both from Maxalline and from the scene.

Summary of the Gunshot Residue and Weapon Testing

Method

Gunshot residue tests were performed as follows:

The 44 Magnum Smith & Wesson revolver was fired, using the same Nosler ammunition as noted, at targets placed at fixed distances from the muzzle. The spread of GSR was then measured to determine how the diameter of the cone-like projection of primer residues, unburned gunpowder, and copper/lead shavings (possibly scraped from the bullet by the barrel ring) exiting from the cylinder and the muzzle varied with distance from the muzzle. In addition, a study conducted with the cooperation of Keyence Corporation's Michael Graham documented the test Smith & Wesson Magnum's GSR and bullet discharge characteristics using a Keyence High Speed Camera capable of shooting 42,000 frames per second (Figure 3.13 to Figure 3.15).

The high-speed camera helps visualize the patterns and distributions of gunshot residues from a particular gun. In a revolver, the crane, or area in front of the cylinder, becomes one source of primer residues. But GSR is more than just primer residues. GSR includes what I have called "muzzle effluents," anything excluding the bullet that exits the muzzle.

Given the hypothesized proximity of the weapon to the three victims in the shootings under investigation, targets were placed as follows:

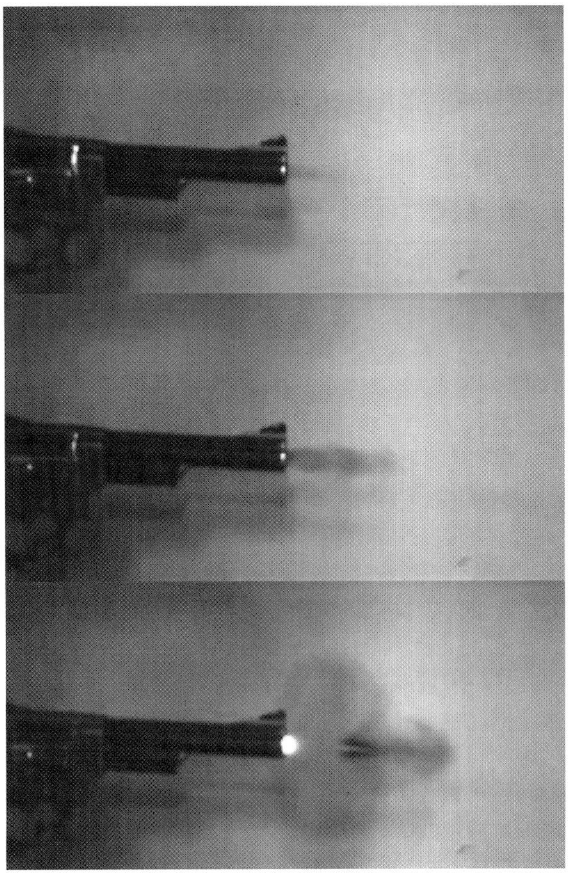

Figure 3.13 Showing images of a bullet fired from a 44 Magnum Smith & Wesson: The top image shows the shot initially fired, to the bottom image of the bullet leaving the muzzle.

Figure 3.14 Vapors flash as the bullet continues.

Figure 3.15 Additional residues and shock wave are visible.

1. Contact with the muzzle
2. 6 inches from the muzzle
3. At 18 inches from the muzzle
4. At 24 inches from the muzzle
5. At 36 inches from the muzzle
6. At 48 inches from the muzzle
7. At 72 inches from the muzzle

Targets consisted of white muslin fabric stretched across large hoops to increase the visibility of the entrance hole, any burned or unburned powder, and any attendant gunshot residues. The white target easily showed the distribution of residues at the various distances.

Measurements

The diameter of the residue spread at the following target distances resulted (Note: Any identifiable residues were used to extend the diameter):

1. Contact with the muzzle residues found on the target that was shredded by escaping gases
2. At 6 inches from the muzzle → 8-inch diameter residue pattern
3. At 18 inches from the muzzle → 20-inch diameter residue pattern
4. At 24 inches from the muzzle → 28-inch diameter scattered residue pattern
5. At 36 inches from the muzzle → 38-inch diameter widely scattered residue pattern
6. At 48 inches from the muzzle → no visible residue pattern appears
7. At 72 inches from the muzzle → no visible residue pattern appears

Conclusions

A. Residue patterns spread markedly at distances greater than 3 feet from the target. They remain apparent at distances much less than 3 feet from a target.
B. Shooting Nosler Partition ammunition in contact with a loose fabric target results in the destruction of the fabric target near the barrel as a result of the escaping gasses.
C. This testing supports the conclusion that Lesella was either struck three times by Noslers fired from a 3-foot or more than a 3-foot muzzle-to-target distance, or that all three shots were Noslers fired through some intervening target(s) that absorbed gunshot residues, such as towels, cloths, or other possible suppressive media. The latter remains a possibility even with Noslers fired from 3 feet or more.

Trace Evidence

Fiber evidence was recovered and analyzed from the following bullet and from the following fabric exhibits:

A blue fiber was recovered embedded among human tissues/fat and blood found adhering underneath the single remaining copper “petal” still attached to the jacketed Nosler round, which was recovered at autopsy from Maxalline’s left chest.

A blue thread composed of blue fibers recovered from Lesella's bra (which was bagged by police with the sweatshirt, shorts, panties, and hair tie), and a second blue thread composed of blue fibers recovered from Lesella's blue sweatshirt during my evidence examinations. The source of both blue fibers was Lesella's blue sweatshirt.

Blue fibers recovered from the bloody pillow associated with the 44 cal Nosler bullet that was recovered by police by cutting into the pillow at the scene, with multiple fiber sources present, resulting in potential contamination issues.

Blue fibers recovered during my evidence examinations from a receiving blanket found at the end of the bed. This item was bagged by police along with a yellow towel and a red bedspread. (This manner of packaging evidence also raises potential contamination issues.)

Conclusions from the Discoveries and Analyses

The blue fiber recovered on the bullet removed from Maxalline's chest remains the most interesting and the most important of the blue fibers identified and analyzed in this case: it is composed of a polyester strand, with smaller flax and cotton strands, identified through polarizing light microscopy.

The sweatshirt worn by Lesella is a navy blue Russell Athletic sweatshirt, size L, which is described on the label as being made from 100% cotton. However, when fibers from this 100% cotton sweatshirt were examined microscopically, it was found to be composed of blue fibers with a polyester strand, along with smaller flax and cotton strands. These fibers had the identical spectral properties of the fiber recovered from the bullet from Maxalline's chest, which adhered to tissues and blood underneath its one remaining copper jacket petal.

This determination strongly supports the conclusion that this bullet first perforated Lesella's navy blue Russell Athletic sweatshirt, Lesella's back, and then the other side of Lesella's sweatshirt before it struck Maxalline in her head. Further, the blue fiber was found to be embedded in the tissues on the Nosler bullet recovered at autopsy from Maxalline's left chest, and, therefore, the blue fiber could not be explained as a mere artifact of careless handling or as a result of contamination with other fiber evidence. This remains the problem with the blue fibers found among the pillow fibers. Such careless handling negates the scientific value of these blue fibers.

The entrance wound in Maxalline's head exhibits the characteristic keyhole configuration of a Nosler round that has previously struck a prior hard target (for example, see Figure 3.16 of keyhole entry on sheetrock resulting after perforation of a ballistics gelatin block covered with a similar blue Russell Athletic sweatshirt). The presence of lead fragments in Maxalline's scalp above the skull bones (visible on X-ray) also supports this scenario.

One important issue raised by detectives in any such triple-homicide investigation involves attempting to identify any specifically targeted individual as the primary victim, potentially separating that intended target from other victims who may simply have been in the wrong place at the wrong time. This effort may help investigators identify potential suspects by focusing resources upon that targeted victim's known associates, background, relationships, difficulties, and so forth.

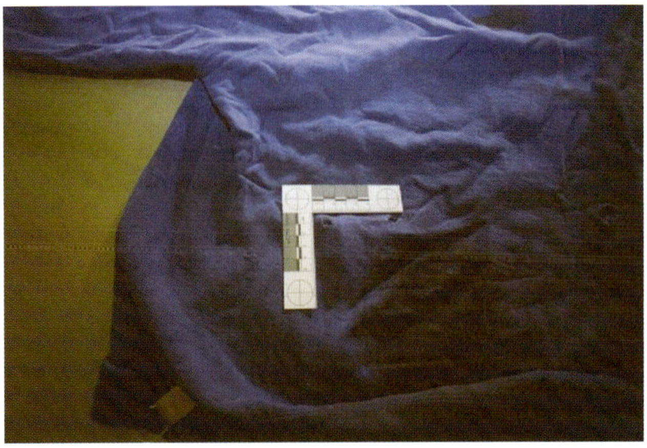

Figure 3.16 Back of victim Lesella's blue Russell Athletic not "100%-cotton" sweatshirt. In spite of claims made on the label, the fabric is actually a mix of cotton, flax, and poly fibers quite common in the garment industry.

Often scientific analyses are thought to provide little help with such investigative efforts. However, the analyses here can help reinforce the conclusion that the bullet striking Maxalline in the head first struck Lesella. This fact could cast doubt on any theory of the three murders (victimology) asserting that Maxalline was the primary intended victim or the specific target of the killer or killers.

In fact, Lesella has three separate first-target entrance wounds from three separate Nosler rounds, including one through her head. Deneshia has the only other first-target entrance wound, which appears on her back. Maxalline's only injuries resulted from a single Nosler bullet striking her head as a secondary target: the bullet had previously struck Lesella in the back (see fiber evidence, X-ray, autopsy description, and ballistics testing for confirmations). If Maxalline was struck by a second and distinct primary lead core from a second Nosler round that entered her axial left shoulder, then this second core's injury also resulted from a separate Nosler round itself not striking Lesella as a first target.

The presence of these other blue fibers, potentially originating from Lesella's sweatshirt, which were found on various items of fabric evidence collected from the crime scene, can best be explained through coincidental contact or by careless contamination. First, these fabrics could have contacted any blue fiber evidence when an unknown person "cleaned up" the crime scene for some unknown purpose. (For example, items were folded and repositioned on the bed.)

A further contamination of blue fibers with fabrics and other items of evidence, such as bullets recovered at the scene, may have resulted from incautious evidence collection and packaging, that is, removing the bullet from the pillow at the crime scene. This renders their significance questionable and unproductive at best in the scientific sense.

Microscopic Examinations of the Bullets and Trace Evidence

Of particular interest in this case is the trace evidence found on the bullet removed from Maxalline's left chest (Figure 3.17 to Figure 3.20). (The bullet found on the table beside the daybed surprisingly had no trace evidence at all.)

Figure 3.17 Trace evidence remains secured under the open petals of the Nosler round.

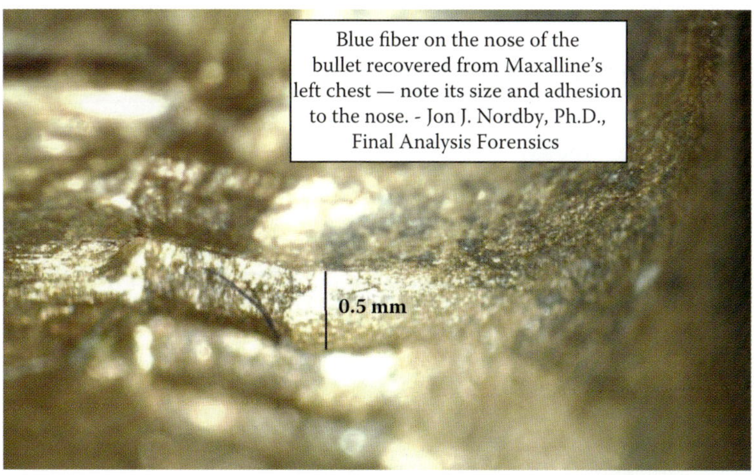

Figure 3.18 Blue fiber on the bullet did not originate from inside Maxalline's chest.

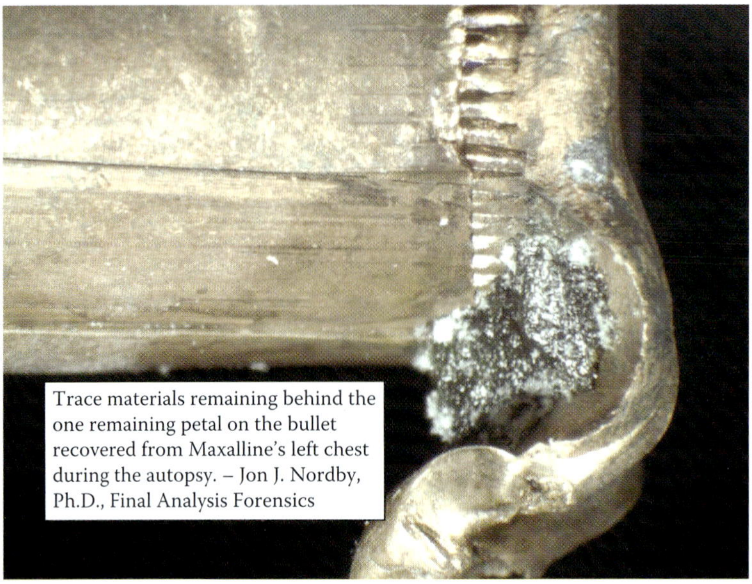

Figure 3.19 Trace evidence appears under the JHP petals.

Figure 3.20 Blue fiber evidence.

Bloodstain Pattern Summary: Clothing, Bedding, and Cloth Items

The scene revealed few significant bloodstains other than to show, by their locations, that the victims had bled in one location and had been moved to another location before they were discovered.

Most stains present on the various fabrics recovered from the crime scene are soaking stains, with many resulting from bleeding over an undetermined time period. The items and their stains are summarized and presented as rough drawings, notes, and photographs attached to the report I submitted.

Many stain patterns represent multiple bloodstaining events with differing time intervals. Their scientific interpretation remains hazardous at best. Some stains appear to be merely artifacts of incautious moving and inappropriate packaging during official crime scene processing activities. For example, items with wet soaking stains were packaged together in bags before they could be separately air-dried. After air drying, they also should have been packaged separately to avoid any further contamination of both the blood and trace evidence. This carelessness also affects some interpretations of the stains linked to specific victims by DNA analyses. The evidence is only as good as the collection and documentation practices applied to the scene.

Additionally, some fabric items appear to have been used to wipe or clean the crime scene after the blood-shedding event before their official discovery and before crime-scene processing. This fact accounts for many of the apparently random, angular, and geometric stains found on these items that appear unrelated to the actual blood-shedding events defining the homicides.

Reconstructionists of the nonscientific ilk develop elaborately entertaining stories to explain such complex angular stains. In doing so, they ignore the most obvious and the simplest explanation: the stains result from efforts to wipe up the blood. To test this explanation, an experiment can help. Spill some blood of a relevant volume and wipe it up with the same kind and size of fabric. The resulting stains on the fabric usually appear to be very similar to the unknown stains on the fabric found at the scene.

Especially note the absence of bloodstains where one might expect to find them: the room is 12 × 12 feet, a small area. A 44 cal Nosler fired at intermediate range into Lesella's head should result in forward and backward bloodstains, cerebral-spinal fluid stains, and brain tissue patterns. The walls and the other items in the room lack such stains. If the fabric items, repositioned by a person or persons unknown after the murders, contained such information, both the repositioning and the improper evidence collection may have contributed to our inability to address this issue scientifically. However, the loss of that evidence with the presence of fibers of unknown origin suggests that some fabric is now missing from the scene.

Ballistics

General Features of Nosler Ammunition

One bullet was recovered from Maxalline's left chest; five other bullets were recovered from the scene, along with both lead cores and jacket fragments. Therefore, six bullets were recovered from the bedroom where all three victims were discovered.

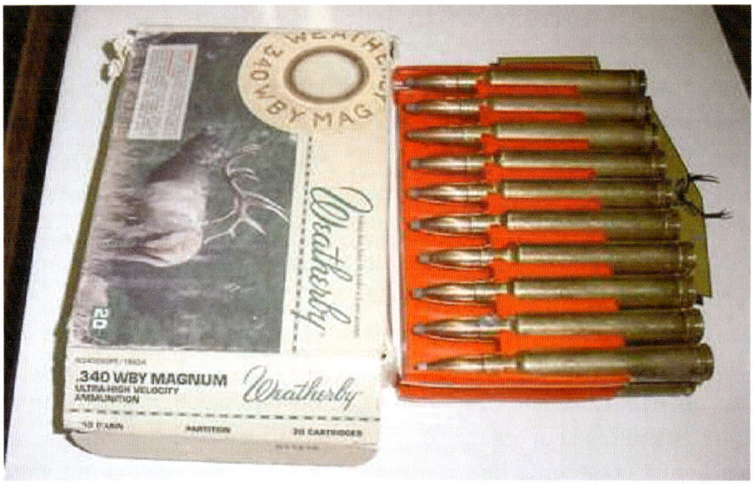

Figure 3.21 340 Weatherby 250GR Nosler Partition rifle bullets in a box of 20.

Each of the six recovered rounds was a Nosler Partition bullet. "Nosler Partition" refers to the bullets themselves, which may be reloaded or may come preloaded in various calibers of ammunition available in various calibers sold under several brand names such as Federal and Winchester. Like Federal and Winchester, other ammunition and weapon manufacturers also sell Nosler partition bullets (referring to the bullet itself, not to include the brand-name cartridge case, powder, and primer cap) loaded into their own shells (referring to the brass case holding the bullet, its primer cap, and the smokeless gun powder) under their own brand names (Figures 3.21 and 3.22). Ballistics terminology in casual use can be somewhat confusing given the relative informality with which words such as *bullet*, *cartridge*, *ammunition*, *shell*, and *fragment* are used.

In this triple homicide, the ammunition used involved Nosler Partition bullets in a 44 Magnum revolver round, presumably with 250 grains of powder, since this is the only load currently and readily available for a 44 cal revolver. The brand name of the ammunition (referring to the bullet along with its cartridge case) remains unknown since no cartridge cases were recovered at the scene and none were connected with the revolver's recovery. However, given the market availability of such 44 cal revolver ammunition, the best

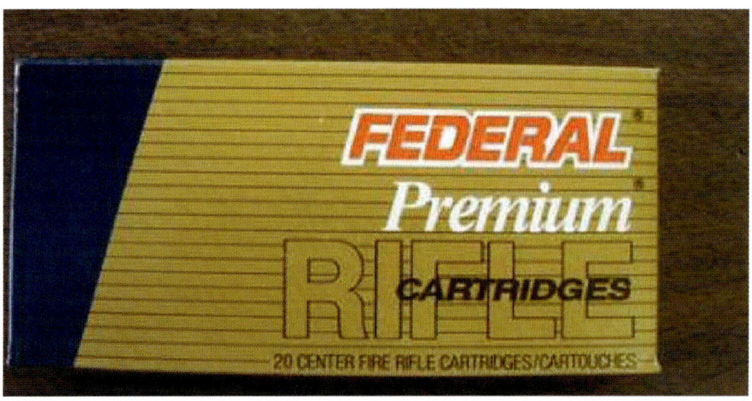

Figure 3.22 Federal 375 Holland & Holland Magnum ammunition with 300-grain Nosler Partition rifle bullets, again a box of 20.

Figure 3.23 Winchester Partition Gold Nosler bullets used in all testing with the Smith & Wesson 44 Magnum revolver Model 629-6 (the last 6 represents the model with a 6-inch barrel).

reasonable judgment remains that factory-loaded Nosler bullets on Winchester brand shells called Winchester 250-grain Partition Gold (see Figure 3.23) were used in the murders.

This Nosler bullet loaded in a 44 Magnum cartridge case has a powder load 10 grains greater than a standard 240-grain 44 Magnum load. The ammunition is designed for big game hunting, including bear hunting. Usually, a 44 Magnum revolver is used either as a backup big-game weapon or for killing big game at very close range.

Given the appearance of the recovered bullets under the comparison microscope, it is clear that these bullets were not reloaded ammunition. Therefore, they represent factory-preloaded ammunition. (For purposes of testing, Winchester Nosler Partition Gold 44 Magnum ammunition was used. These best approximate the ammunition used in this triple homicide.)

Nosler rounds are designed to inflict maximum damage to the target. Their design involves using two lead cores, one near the nose and the other at the base of the bullet. A metal jacket designed to open into six distinct petals separates the two cores. An excerpt from a description of the ammunition appears in Figure 3.24.

When Winchester 250-grain Nosler Partition Gold ammunition is fired in a 44 cal Smith & Wesson revolver, the following characteristics apply, both as researched and documented through my own measurements, and as predicted by theory and developed in the literature.

Range in Yards	Velocity in fps[a]	Energy in ft-lb[a]	Velocity in fps (chronograph)	Energy in ft-lb[b]
0	1230	840	1220	—
50	1132	711	1097	—
100	1057	620	988	—

[a] 2004 *Shooters Bible*, page 502.
[b] Not measured.

Since this Nosler 44 cal revolver ammunition is atypical in my own gunshot homicide case experience, testing became necessary to better understand the shooting events in this case. To this end, a series of tests was conducted to investigate its general properties when

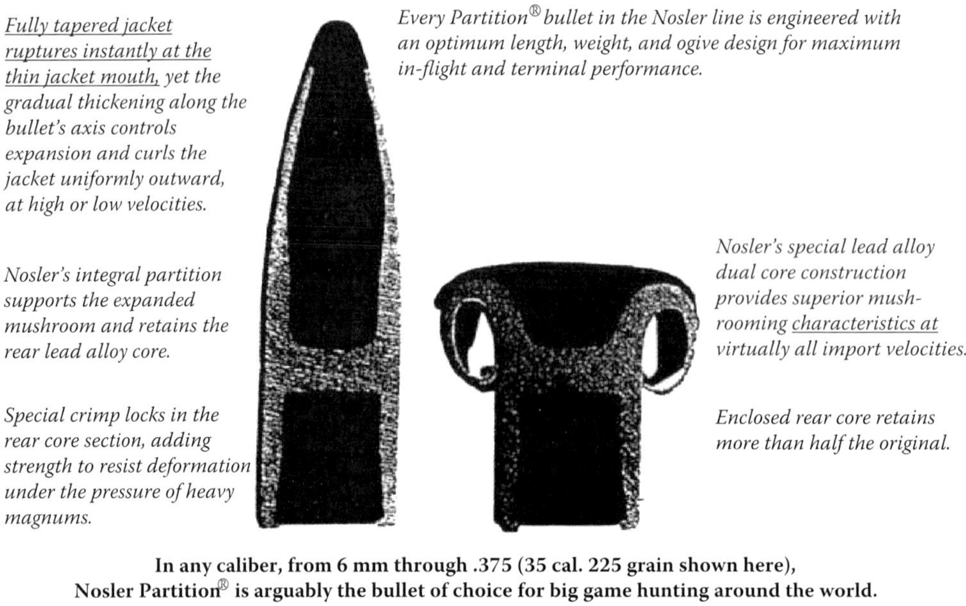

Figure 3.24 The dark areas represent the lead cores. Note the underlined phrase: "Fully tapered jacket ruptures instantly at the thin jacket mouth." The enclosed rear core retains more than half of the bullet's weight, which supplies added momentum to the round after expansion and fragmentation, pushing it further into the target than a normal jacketed hollow-point bullet would penetrate.

striking various targets, including its penetrating power through both primary and secondary targets, and the resulting condition(s) of the bullet or its fragments (Figure 3.25).

Issues with a Small Room and Powerful Ammunition

After this testing, a series of experiments was conducted to develop data to understand the mechanism of the victims' shootings and their injuries. The obvious question (answered by the detective's early untested assumption) is whether these injuries could be sustained within the relatively small confines of a 12 × 12 foot crime scene. Answering this question becomes more complex given the evidence that the crime scene has been altered or even staged after the shootings.

To test this hypothesis experimentally, a scale model shooting scene* was constructed in the shooting range. Walls were constructed by using materials similar to the walls at the actual crime scene. All experiments were conducted by shooting within the 12 × 12 foot confines of the model crime scene.

* The scene is not an exactly similar room. Rather, it abstracts (cuts off and preserves) the relevant similarities for the purposes at hand. The walls are made from the same materials: sheetrock over 2 × 4 frames at 16-inch centers filled with R12 paper-backed fiberglass insulation, sheeted with ½-inch plywood covered with 4-inch clear cedar siding. But they only represent the real wall; they are 4 × 8 feet and 4 feet tall. We built several. We located the items of furniture in the room and constructed a work surface to represent a bed of the same height as the daybed from the scene.

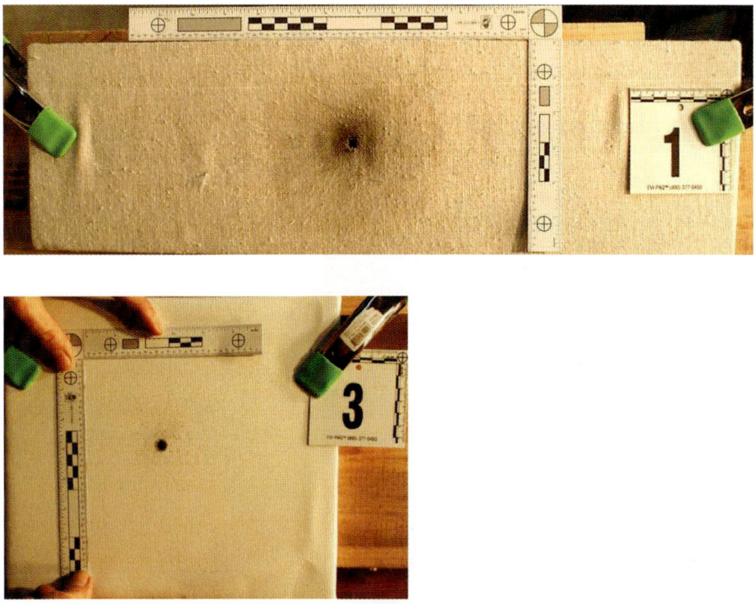

Figure 3.25 (Top) GSR as muzzle effluents appear around the entrance on a cloth-covered target. (Bottom) Here the residues appear as particles; the heaviest particles travel to the target at 2 feet.

Since the bullet wound of interest perforated Lesella's torso striking no bone, a ballistics gelatin block was used to represent this wound. Testing involved documenting the behavior of the bullets in the 12 × 12 foot room and documenting their penetration or perforation when striking their human targets. Ballistic gel is used since its density and elasticity closely approximate the density and elasticity of human tissues. If bones are involved in a bullet's trajectory through a body, then the relevant bones are added to the gel when it is molded. When a bullet strikes the gel, its effect on the target as well as its trajectory can be studied and compared with the autopsy and scene data. Figure 3.26 shows the initial effect of a bullet's entrance into a gel block.

Lesella wore a blue Russell Athletic sweatshirt with no undergarment when she was shot. So we bought several similar sweatshirts to model Lesella's back injury and to investigate the behavior of the 44 cal 250-grain Nosler ammunition after producing the injury at issue. Of course, the gel and the clothing are not going to tell us the entire story of the bullet's trajectory. A useful design feature of scientific experiments involves setting them up to answer as many related questions as possible. Since the objective here is to investigate the properties of this ammunition in a 12 × 12 foot room and since one hypothesis held that such a bullet fired in this room would not only perforate the victim but also perforate the walls, it became obviously necessary to reconstruct such a room. (This hypothesis concluded that since the room scene had been staged and since the bullets would perforate the walls, then the victims had been shot elsewhere and transported to the room after death.)

The room was modeled in our shooting range. The walls of the room were represented by tape on the concrete floor. Specific items found in the room were documented by tented numbers. The bed's outline appears in white tape (Figure 3.27 and Figure 3.32).

With the size of the room laid out, the walls were built using the same materials found in the murder house, including insulation (Figure 3.33). Several sections of wall

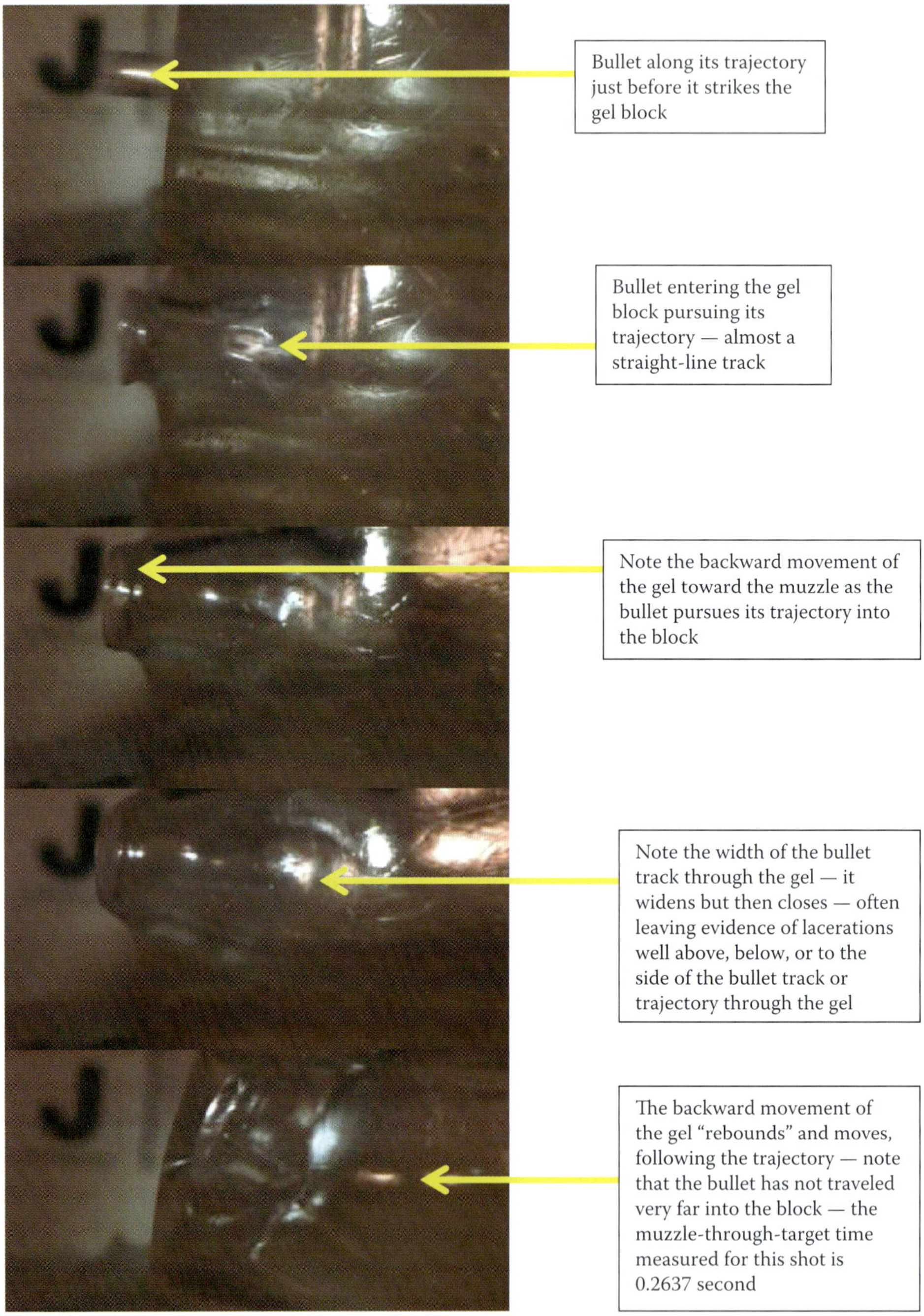

Figure 3.26 The initial effect of a bullet's entrance into a gel block.

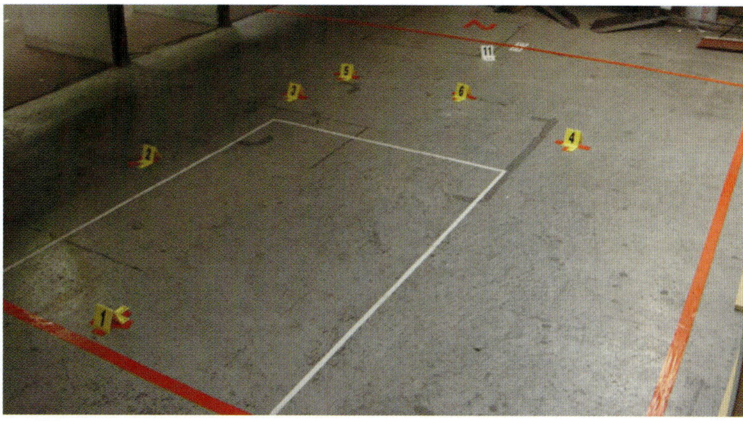

Figure 3.27 View of a model crime scene showing a bed outlined with white tape.

were produced in order to conduct a number of relevant tests. The exterior siding consisted of two different types: T-111 plywood on one section, and cedar siding on another (Figures 3.28 and 3.29). While this was not an exact conditions test, it was a reliable model of the original room.

The perforating wound to Lesella's torso provided the most interesting test possibilities. Since it did not strike bone and it perforated the body, its post-exit behavior could be further observed. This observation could be built into the testing. If the shooting is possible in this small room, then the bullet must either fail to penetrate the sheetrock or some other mechanism must stop it before it does. On the other hand, such a bullet would have to retain sufficient energy to strike and damage another human target. For example, Maxalline had a bullet wound injury characteristic of a tumbling bullet striking a secondary target. The best candidate for the source of that tumbling bullet was the bullet perforating Lesella's torso.

Figure 3.28 The exterior siding consisted of T-111 plywood on one section (left) and cedar siding on another (right).

Figure 3.29 Walls made of the same materials and with the same dimensions as the room.

The first step is a form of validation: The ballistic gel must reliably replicate the human tissues that it models and also allow for reproducible results. When I make our gel, I calibrate its relevant properties (powder-to-water ratios) by casting small blocks (Figure 3.30) and shooting an air-rifle BB or pellet into the block (Figure 3.31). I prefer to shoot at least two test pellets into the block. This calibration involves shooting the pellet or BB at 590±15 feet/second and penetrating the block 8.5±1.5 cm as measured from the side of the gel to the leading edge of the BB or pellet. An appropriate chronograph is the best way to measure velocity. Of course this is not anything more than an approximation. If the calibration data accompanies one's results, then reproducible results are at least, in principle, possible. If the power of the air rifle is greater than 600 ft/sec, then simply adjusting the muzzle-to-target

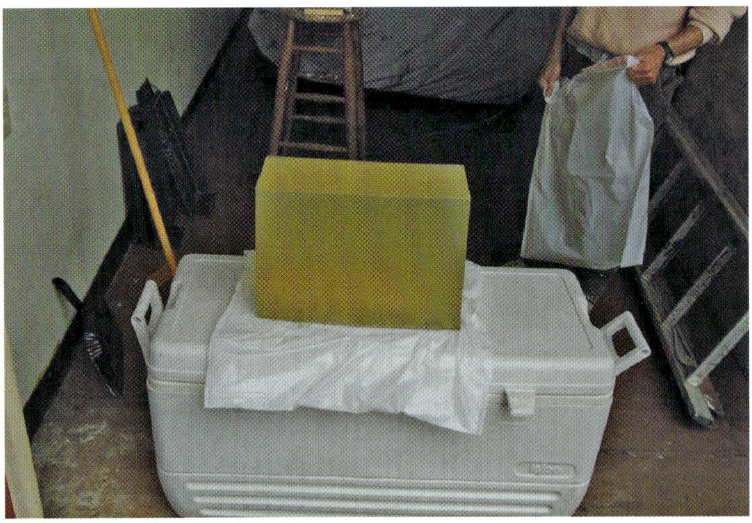

Figure 3.30 Ballistics gelatin blocks were prepared at Final Analysis Forensics and tested for their suitability as a stand-in for human tissues according to established protocols, in part developed and refined by the FBI and Dr. Martin Fackler, a trauma surgeon who studied wound ballistics (terminal ballistics). (Fackler correctly states that "ordnance gelatin is an approximation for living muscle, and most shots go through several tissue types." When we talk about calibrations, he also notes that this "can mislead in giving the impression that gelatin is a more exact technique than it really is.")

Figure 3.31 The air-rifle test penetration at both the specified muzzle velocity and penetration distance according to the protocol.

distance is a good way to achieve an acceptable result. If the results show that the gel is too rigid or too flaccid, then the mixture must be adjusted and the calibration test repeated until the test criteria are met.

With the 44 Nosler, the bullet perforated the gel block and still had sufficient energy to enter the sheetrock, the two components are characteristic of the Dual Partition Nosler bullet (Figure 3.34 to Figure 3.42). Since there was no defect in the sheetrock at the scene, it is unlikely that this victim was lying on the bed when shot. In this experiment, the bullet entered the sheetrock but did not exit the outer wall.

In another test, the bullet exited the victim (gel block) and simply struck the sheetrock without penetrating at all. Both components were recovered on the floor (Figure 3.43).

Figure 3.32 This photograph shows a scene set up with tent numbers, which are used to help represent the positions of the victims and evidence.

Figure 3.33 The walls were constructed of the same materials as the walls in the original murder scene.

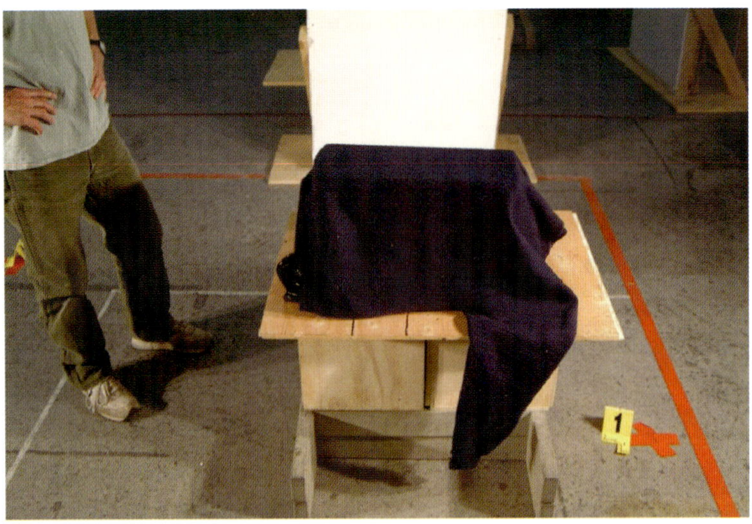

Figure 3.34 The gel block is inside the sweatshirt at the bed's height with the sheetrock as a secondary target to test the hypothesis that the Nosler bullet would go through the wall.

Figure 3.35 The gel block is inside the sweatshirt.

Figure 3.36 Damage to the sheetrock after the bullet perforates the sweatshirt and gel block.

Figure 3.37 Another view of the test result.

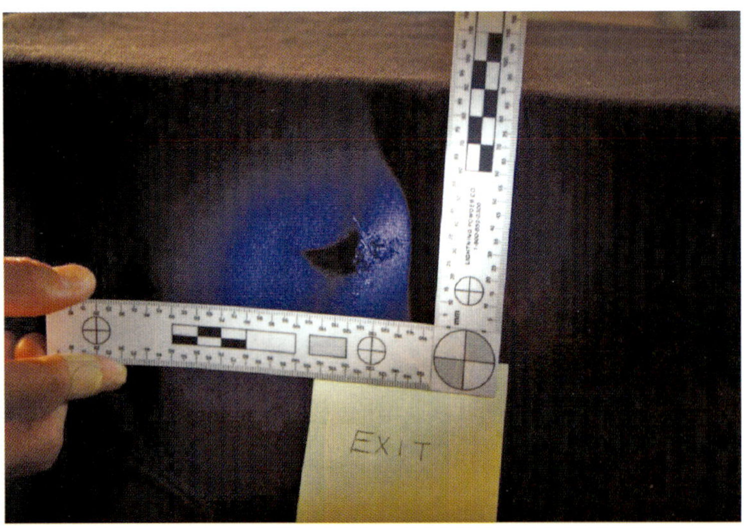

Figure 3.38 Showing the exit.

Figure 3.39 Energy dispersion is visible in the gelatin block as the Nosler bullet moves through the gel with its petals opened. The lacerations made in the gel by the bullet's temporary cavity can be seen in the block.

Figure 3.40 View of the damage to the gel block.

Figure 3.41 Another look at the trajectory through the gel block.

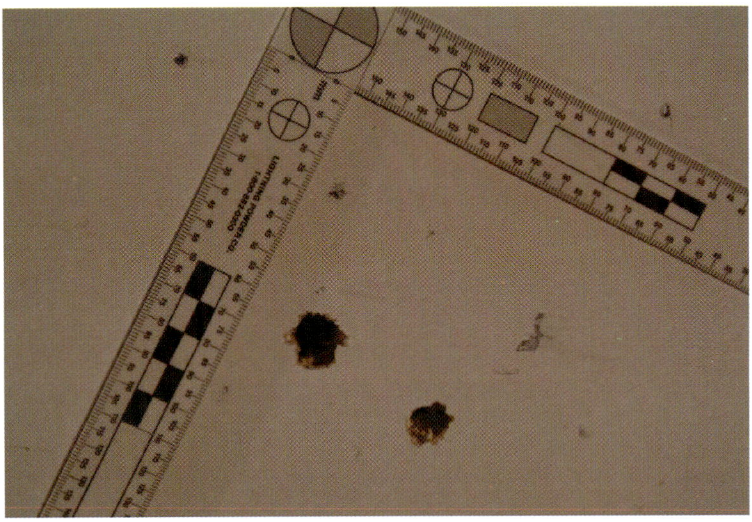

Figure 3.42 Damage to the sheetrock—note the "keyhole" entrance on the upper left—the bullet struck sideways.

After striking a primary target, the Nosler bullet, a JHP, exits and tumbles. If it enters a secondary target, that target shows evidence of tumbling as repeated tests confirmed. (See Figures 3.44 and 3.45.) The various powers to enter the sheetrock or to bounce off and land on the floor seemed to vary in a ratio of about 1-to-1. One variable seemed to be the most significant: the distance between the primary and the secondary targets. This data varied in the same ratio: the closer, the more likely the penetration into the wall. In no case did the bullet exit the building (Figure 3.46). The most penetration was documented on the exterior wall above when no gel block was used and the wall was the primary target (Figure 3.46).

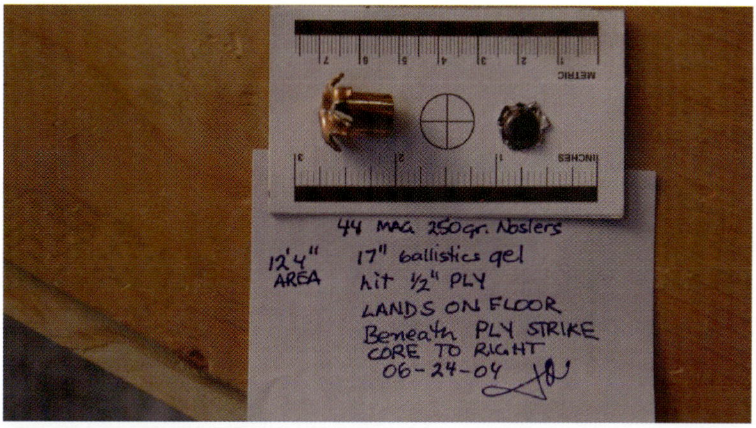

Figure 3.43 Damage to two parts of a Nosler.

Figure 3.44 The two parts of the Nosler that struck the sheetrock. They had no energy for a "tour of the neighborhood."

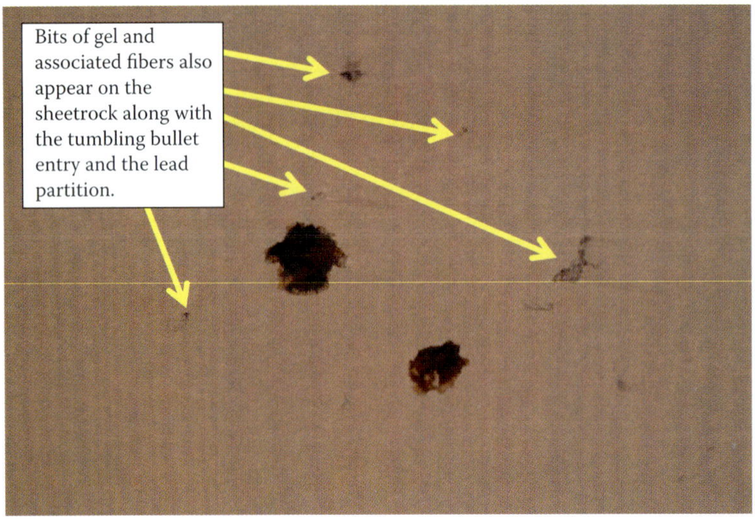

Figure 3.45 The characteristic shape and nature of these entry wounds show lead, gelatin "tissue," and then the keyhole entrance from both the bullet and its primary core as it enters the sheetrock.

Figure 3.46 Close-range shots still failed to perforate the walls of the building.

The Experiment's Conclusion

The four detectives' untested expectations predicted that ammunition powerful enough to kill big game, such as brown bears, at close range would simply perforate any human target in such a small 12 × 12 foot space and continue on its destructive path through the walls of a building. However, the results of this testing and controlled experiment overcame "pre-analytic intuitions": the 250-grain Nosler rounds do not behave as such intuitions predicted. Their initial destructive force remains great, but it largely dissipates when pursuing its course through a target (repeated gelatin block tests provided consistent results supporting this conclusion). The destructive range of such bullets becomes more limited when considering their power. Of particular interest is the behavior of these 250-grain Nosler bullets after perforating a human target when striking a secondary target.

Results

This testing with the gel block and the sheetrock helped develop an additional hypothesis to explain Maxalline's fatal injury. In our test room, I discovered that the autopsy documented more injuries than we had bullets, assuming that a single bullet caused each injury. But what if we formed a hypothesis holding that the bullet perforating Lesella's torso tumbled and then struck the top of Maxalline's head, eventually lodging in her left chest? This hypothesis was suggested by the shape of the tumbling bullet's entry into the sheetrock.

Often in what can be termed a form of scientific discovery, hypotheses are suggested by prior experiments or tests that have quite distinct objectives. Recall that the main purpose of the gel block and wall tests was to determine the penetrating and perforating power of the Nosler Dual Partition 44 cal ammunition used in the shootings. If we remain open to seeing and investigating both similarities and differences, then we can begin to develop an appreciation for exactly how hypotheses can be developed in the first place.*

Useful hypotheses are both risky and rewarding: In this case, for example, the reward includes developing an explanation for both the injury to Maxalline and an explanation for the lack of bullets corresponding with each injury. It also explains one injury to Lesella. In that sense, this is a rich explanatory hypothesis worthy of further analyses and tests. One risk, however, is that there are insufficient data either to support or to refute the hypothesis. We then have to settle for an informed ignorance: Science may not support the truth of any of the three related explanations. Without data, science may not be able to support the falsification of any of the explanations either. That, of course, also depends upon the data that we develop in the process of testing this hypothesis. The further risk rests with tying the three explanations tightly together. They may require separate hypotheses. And so it goes.

* Some texts explain the scientific method as if it were a great monolithic entity. The texts further define that method overly simply and inaccurately as a set of steps or rules: develop hypotheses, test hypotheses, and so on. Nowhere do we learn how to develop some hypothesis or how to test it once developed. The situation is similar to Steve Martin's famous quip about how to become a millionaire and pay no taxes. "First," he says, "get a million dollars. Then two simple words: I forgot!" That is no way to learn how to become a millionaire. Nor, in my view, is the above description the way to understand scientific methods.

Maxalline's Injury Hypothesis

The photo on the left in Figure 3.47 presents the entrance wound appearing on Maxalline's skull. Note that the head has been shaved and that the tissues have been moved aside in order to display the keyhole or sideways nature of the entrance defect on the skull itself that resulted from a tumbling bullet that had struck a prior target. The head X-rays showed that lead particles were present in the victim's scalp above the bone. The photo on the right presents, first, the keyhole entrance of a tumbling jacketed Nosler bullet (in the mushroom shape [uppermost hole] with the spread copper petals hitting the sheetrock sideways as we developed from our gel block testing) and the second hole produced by the Nosler's primary lead core (lower hole). We saw gel and fiber pieces (which resulted from the Nosler round striking the prior target) surround these two entrance holes in the sheetrock. The holes seen in the photograph to the right resulted when a 250-grain 44 cal Nosler bullet was fired through a ballistic gelatin block covered with a blue Russell Athletic sweatshirt similar to the sweatshirt worn by victim Lesella.

The JHP bullet fired through the gel also had blue fibers under the petals. Recall the blue fibers seen on the bullet removed from Maxalline at the autopsy.

The blue fiber on the recovered bullet strongly supports this hypothesis. Although blue fibers could possibly have originated from the perforated sweatshirt (Figure 3.48) and then a subsequent tumbling bullet could have struck Maxalline's head thereby picking up the fiber, this possibility does not account for the rest of the data. No other perforating shot could strike her head.

Figure 3.47 The entrance wound appearing on Maxalline's skull (left), and the keyhole entrance of a tumbling jacketed Nosler bullet and the second hole produced by the Nosler's primary lead core (right).

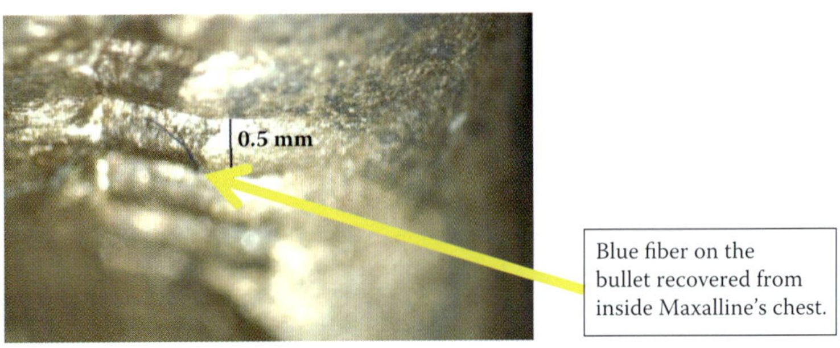

Figure 3.48 Blue fiber on the recovered bullet from Maxalline's chest.

Additional testing might help us further confirm the hypothesis. First, one might confirm that the blue fiber on the bullet recovered from Maxalline's chest is the same type of blue fiber that originated with Lesella's sweatshirt. That involves a search for additional sources of blue fibers. But this may not be a significant finding, for example, if Maxalline wore a cap made from the same material in the same color as Lesella's sweatshirt. This, however, was not the case.

Comparisons of fiber evidence are always a bit tricky. One must avoid the temptation to overstate the significance of any "match." The best we can get is a similarity in as many respects as possible, including the physical properties of both the known from the sweatshirt and the unknown from Maxalline's chest. One difference and the comparisons fail.

The fibers' properties can be developed and detailed by several methods, including polarized light microscopy (PLM) (Figure 3.49 to Figure 3.51). In this case, both the fibers from the bullet and the fibers from the sweatshirt shared a number of properties while exhibiting no differences among their properties.

These properties included morphology (shape), refractive index (Becke line movements), relief (visibility when the refractive index of the mounting medium differs from the sample), micrometry (size measured by a stage micrometer as well as by the microscope camera's software), and optical properties. The optical properties include being isotropic (one refractive index) or anisotropic (more than one refractive index), birefringence (the difference between an anisotropic sample's refractive indices), and pleochroism (absorbing

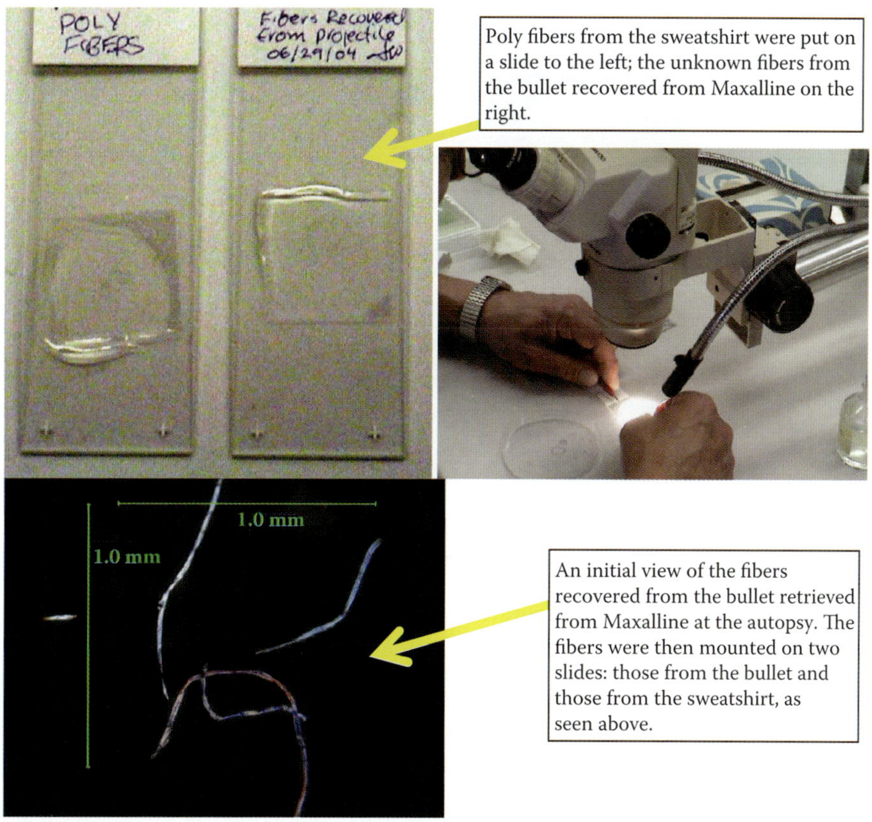

Figure 3.49 Viewing the recovered fibers.

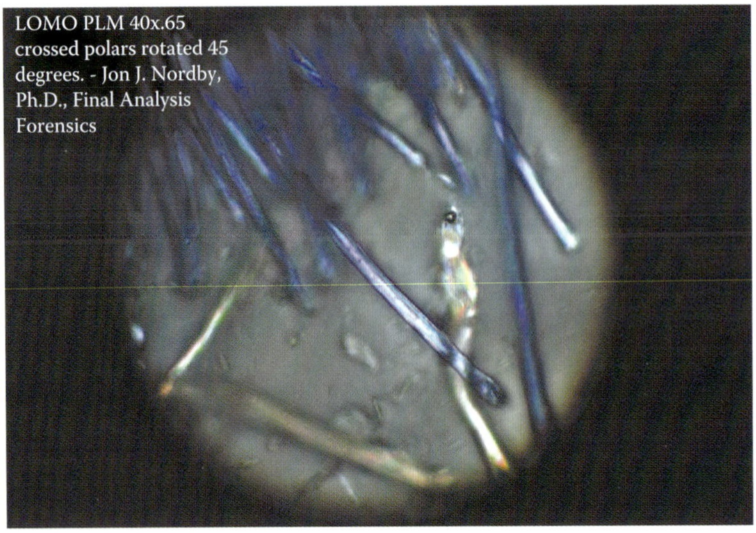

Figure 3.50 PLM photo of the blue fibers.

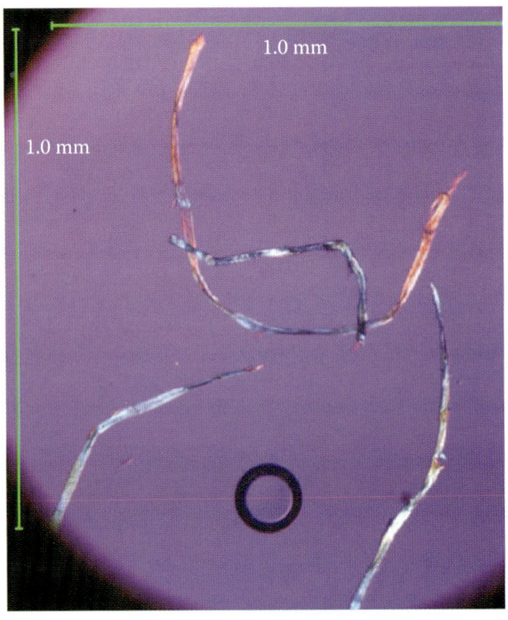

Figure 3.51 The fibers recovered from the bullet removed from Maxalline at the autopsy viewed under crossed polars with a full wave plate. Note the color changes.

different wavelengths of light when rotated under the PLM and therefore presenting different colors at different angles).* The fibers' chemical properties were also identical.†

* Any number of excellent references can provide the details of polarized light microscopy. One excellent reference is *Microscopy for Criminalists, Chemists, and Conservators* by N. Petraco and T. Kubic, CRC Press, 2004. Anything by Dr. Walter McCrone is also recommended.

† I am told by colleagues Petraco and Kubic that Chamot and Mason's *Handbook of Chemical Microscopy* is considered by practitioners to be the best reference work. I like to use the Fourier transform infrared spectroscopy (FTIR) for relevant samples as well.

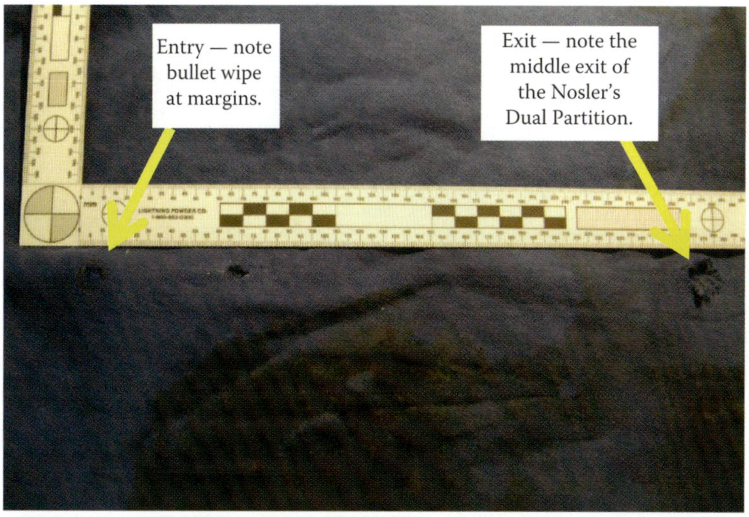

Figure 3.52 The blue sweatshirt in evidence, showing the entrance and exit.

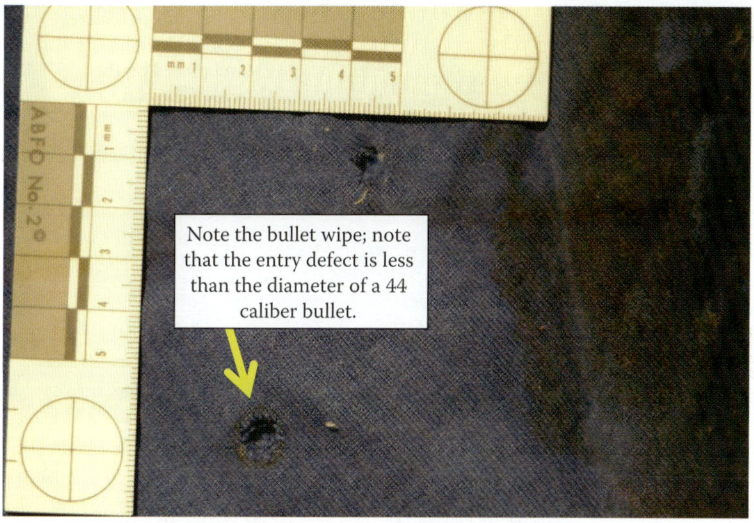

Figure 3.53 The blue sweatshirt seen in Figure 3.52 above.

Clothing Examination: The Blue Sweatshirt from Lesella

The blue sweatshirt has multiple bloodstains from the two gunshot wounds sustained by this victim (Figure 3.52). The perforating GSW can be seen in Figure 3.53.

Blanket Folded on Bed

This blanket was found folded at the end of the bed (Figures 3.54 and 3.55). Obviously, the bloodstains and the bullet entry and exit holes found on the inside folds did not occur with the item folded on the bed. One way to examine large items of evidence, such as blankets, for trace evidence and bloodstains is to use an alternate light source, or ALS. These are

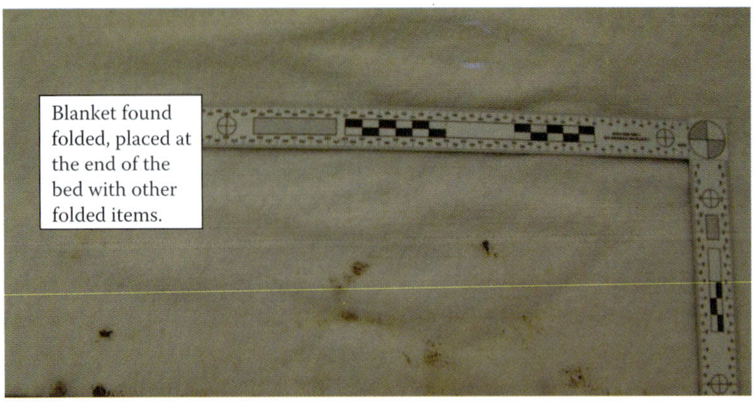

Figure 3.54 Numerous bloodstains appeared on this blanket and on all the items that were neatly folded and placed on the bed.

Figure 3.55 The blankets had to be in a position to get bullet holes—*not* folded.

variously configured but generally have full-spectrum lightbulbs and different filters that transmit different wavelengths of light from the ultraviolet through the infrared.

Similar filters on the camera allow the digital capture of items appearing on the material that may not be visible under ordinary incandescent or full-spectrum lighting. It is best to use several wavelengths to visualize as many different materials as possible. Some have evidential value; obviously, some do not. Often we do not know the difference at the time of the examination. Thus, it is best to photograph each, then document the complete search and examination.

Figure 3.56 to Figure 3.60 illustrate the use of an ALS in an effort to help visualize various items of trace evidence, such as metal particles. Infrared light at 676 nm is useful to visualize bloodstains when they are deposited on dark surfaces. Most sweeps of large items of evidence, such as blankets and sheets, involve beginning by using a light with about a 550 nm wavelength. That has proved to be the most useful wavelength to spot initial items or areas of interest for further analysis.

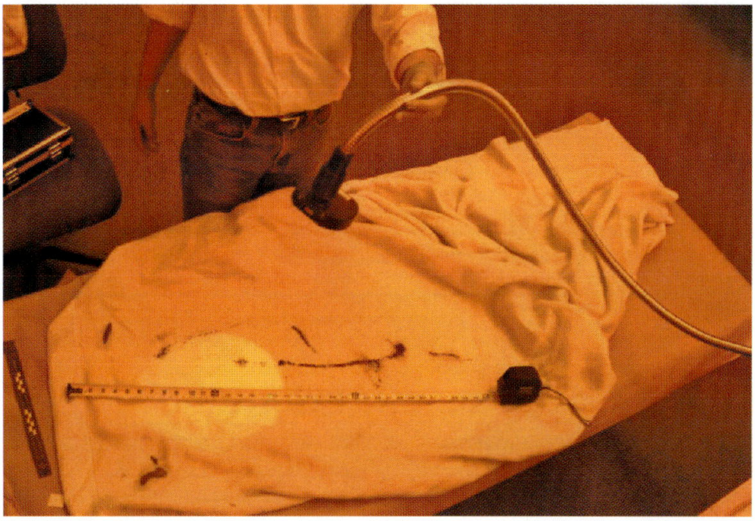

Figure 3.56 This photo is taken with an orange filter (#22) using a light source of about 515 nm. This filter is often used to increase contrast in blue preparations for later microscopy. It also helps visualize specific items, such as the heat produced by proximity to any heat source of sufficient temperature. (Michael J. Brooks of Brooks Photographic Imaging, LLC, has written extensively about the various camera filters available and the specific functions that they provide to aid in the effort to visualize items of potential evidence on various surfaces.)

Figure 3.57 This photo provides a good example of how otherwise invisible trace items can appear when using the appropriate wavelength of light. The iron used to smooth this sheet has left its unintended mark—not enough damage to see with full spectrum light and certainly not sufficient to photograph without the orange filter.

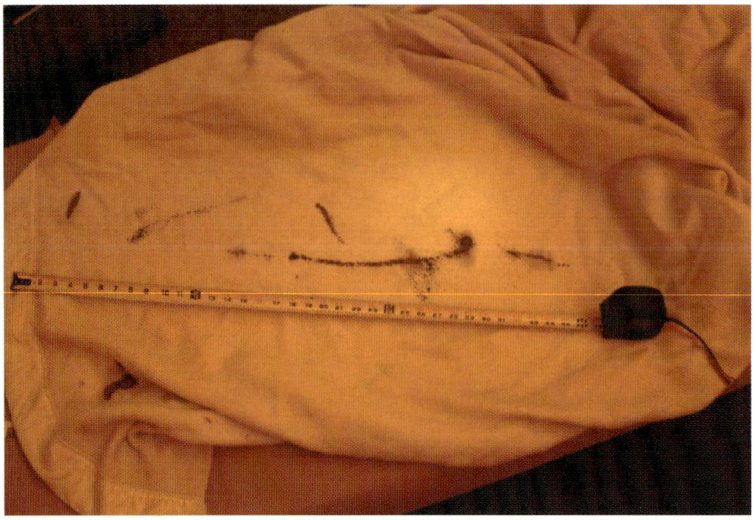

Figure 3.58 It appears that the bloodstains, which are visible, have muliple explanations for their presence. Folding the blanket proves to be the best hypothesis. Using a similar blanket with various mechanisms of blood contact and folding when wet helps to confirm that folding explains some of the complex patterns.

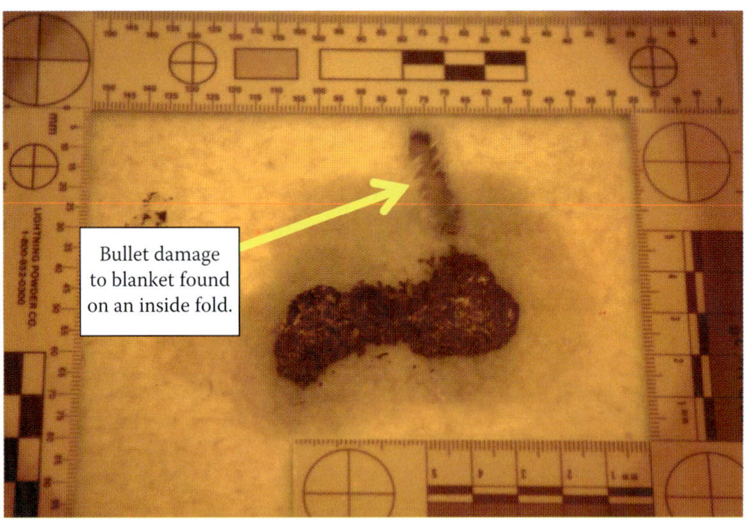

Figure 3.59 This item was also involved in the shooting events. Multiple bullet defects appear. The use of an XRF (X-ray fluorescence) spectrophotometer was used to confirm the presence of metals common to Nosler Partition bullets.

Figure 3.60 The complex bloodstains (multiple mechanisms of stain production) again provide evidence that the items in the room were moved after the shooting was completed. Transfer stains, along with bullet holes, and soaking and dripping stains, appear inside blankets that were found folded on the end of the bed. This blanket, along with the other bloody items found folded at the foot of the bed, provides strong support for the conclusion that the scene, as discovered, had been staged, or at least reorganized and sanitized.

Yellow Towel

A yellow towel found at the foot of the bed played some role in the shootings. Combinations of blood and gunshot primer residues as well as muzzle effluents (GSR refers to both) appear on the towel (Figures 3.61 and 3.62).

The towel was moved during the shooting or the clean up or both (Figures 3.63 and 3.64). Often, items of evidence do not clearly fit with one simple event or even one complex event, such as the three murders in this case. We do not know if the cleanup and the shootings were done by the same person or persons. Nor do we know the time interval between the two.

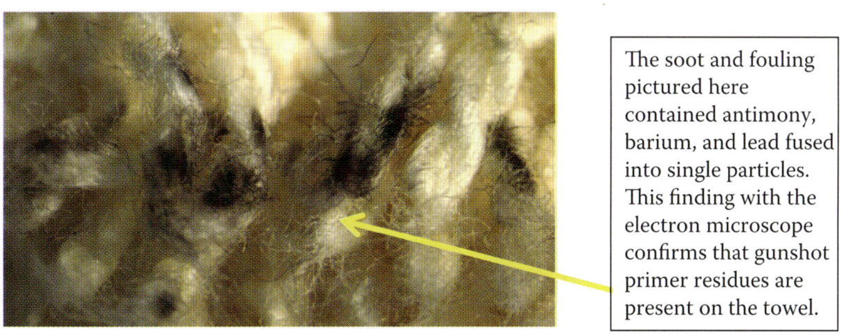

Figure 3.61 GSR on the yellow towel found at the foot of the bed.

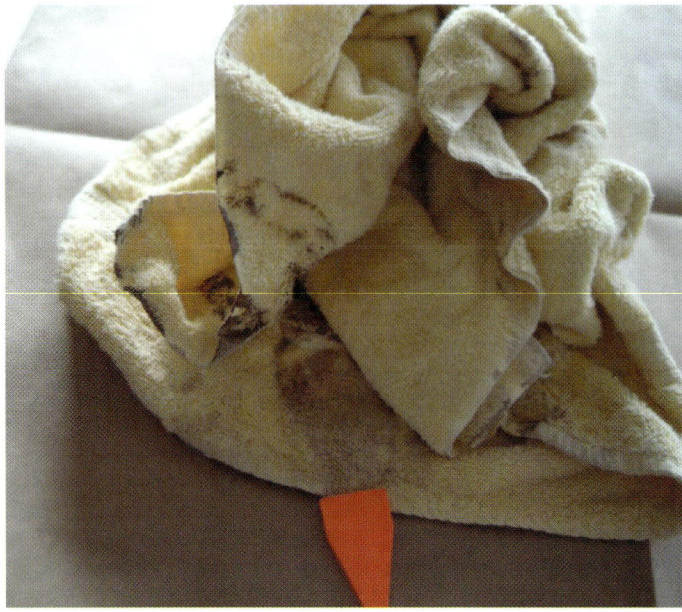

Figure 3.62 When folded in roughly this position, the 44 Magnum Smith & Wesson revolver fits in nicely with the crane and cylinder gap fitting in the circular area on the mid-left side of the upward fold.

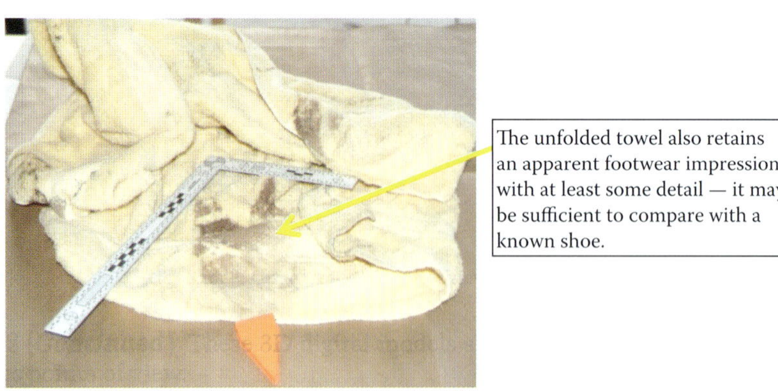

Figure 3.63 A footwear impression on the yellow towel.

Figure 3.64 The mirror-image property with blood, soot, and fouling. Note the GSR on the upper part of the towel indicated by the arrows. The initial screening is done with XRF and then if the elements appear (usually antimony, barium, and lead), samples are taken and put on scanning electron microscope (SEM) stubs to confirm that the elements are indeed fused into a single particle for the detection of primer residues.

Red Bedspread

Both impact spatter and some contact transfer stains appeared on the bedspread that was folded and put at the bottom of the bed (Figure 3.65). The bedspread had a number of different bloodstains (Figure 3.66). The DNA testing of the tissues mixed with the blood all came back to Lesella.

Figure 3.65 The red bedspread had impact spatter and some contact transfer stains.

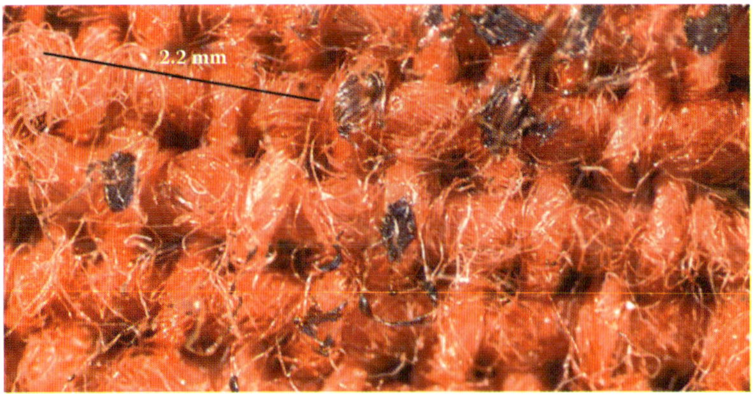

Figure 3.66 The bedspread had a number of different bloodstains.

White Pillow

Many items of evidence have more than one story to tell. Figure 3.67 captures at least two potentially interesting bits of data. One is the bullet hole; the other appears to be a footwear impression in blood.

The apparent bullet wipe and hole was examined under the microscope (Figure 3.68). Efforts to determine the shape of the pillow to accommodate a round or oval bullet wipe pattern developed more data. This proved significant when fibers similar to those making up the pillowcase were found on the yellow towel.

Note the apparent impact spatter visible in Figure 3.69. Obviously, this pillow and case had to be moved in order to explain the complex (more than one part) evidence that it displays.

Many pattern stains can be identified by their complex geometrical configurations. This contact transfer stain on the pillowcase (Figure 3.70) certainly suggests a footwear impression. Other areas also make it appear that the pillow had been stepped on at some

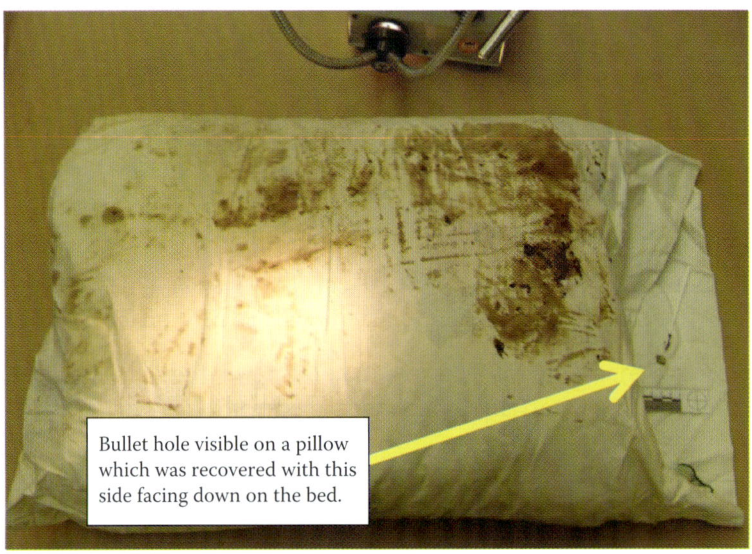

Figure 3.67 The white pillow had a bullet hole and an apparent footwear impression in blood.

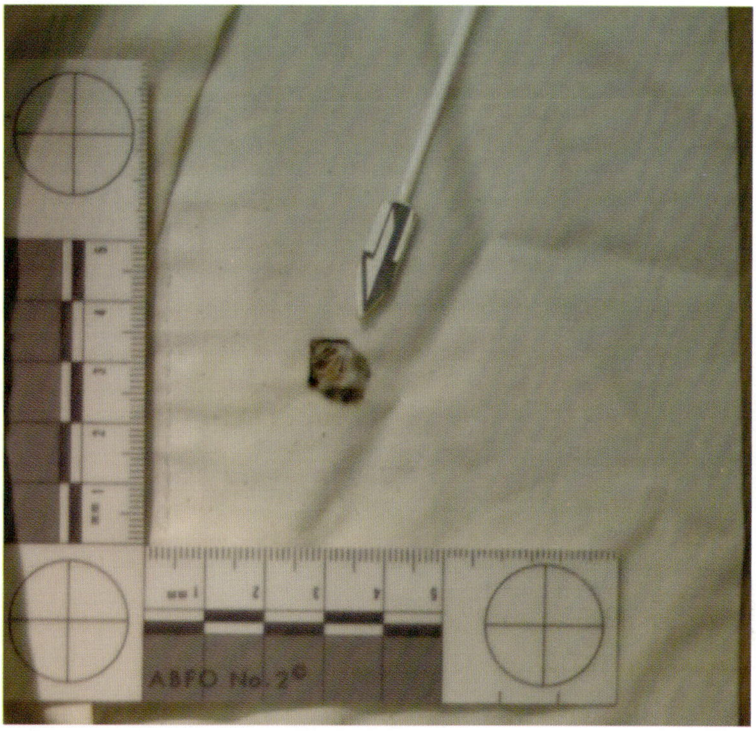

Figure 3.68 The apparent bullet wipe and hole was examined under the microscope.

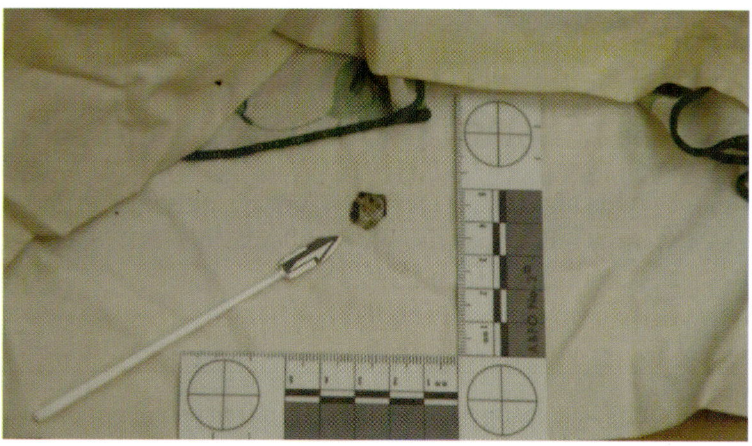

Figure 3.69 Note the apparent impact spatter visible on the pillow.

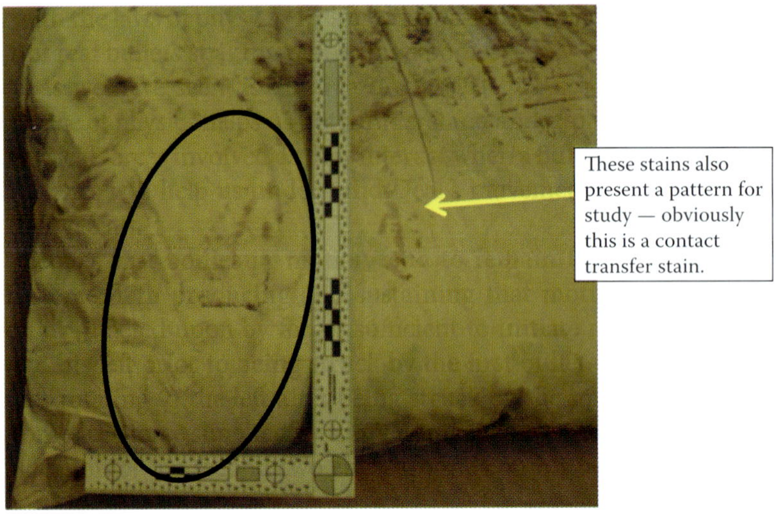

Figure 3.70 Contact transfer stain on the pillowcase.

point after the shootings. This could provide valuable information about the shooting even if the data is not sufficient to identify either the type of footwear or its individual properties.

Bedsheet

In a homicide scene as complex as this one, if apparent footwear impressions appear on one piece of evidence, and the scene has been cleaned or somehow altered, a good bet is that other such impressions will be found on other items. This was confirmed by one of the bedsheets (Figure 3.71), also found folded and placed neatly at the foot of the bed. In addition, learning which items have GSR and which items do not can provide data to help formulate hypotheses about the relative positions of these items during the shooting events.

Basic Progress

The shooting scene was staged or altered given the bloodstain and the GSR trace evidence. Blankets, sheets, and towels were folded, the pillows were moved, and the bed was put together. In addition, both Lesella and Maxalline were moved after sustaining their fatal injuries. It appears that the shooting and the cleanup could be separate events. That, of course, remains to be proven. The footwear impressions lead nowhere other than to suggest that a small shoe size left the impressions. It also remains unclear if the footwear impressions are from one shoe or from two different shoes.

In addition, one of the bullets was found on a table in the room. No trace evidence was uncovered beneath the petals (Figure 3.72). No specific data tying this bullet to a particular victim developed. Our experiments in the 12 × 12 foot model room showed that such a bullet could perforate a target and then merely bump into a wall, landing on whatever surface it first contacted. Yet, no specific evidence other than possibility supports that interpretation. The status of the bullet found on the table therefore remains unclear.

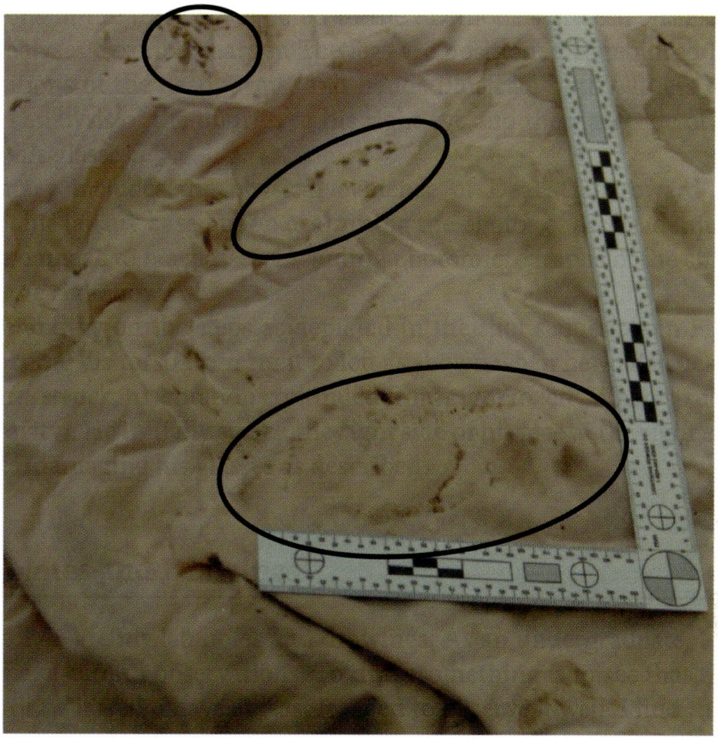

Figure 3.71 At least these three areas on the pillowcase proved useful in determining the size of the footwear impressions. The size was relatively small and could have been either a child or a woman's shoe size.

Figure 3.72 Bullet found on the table in the room.

Often, this is the best that we can do. We do not always get the answers that we need from an analysis of data. Science does not always provide the required answers. Only when this fact is forgotten and the data is modified just to fit some particular theory of the case do we abandon scientific methods. Doing so also abandons both ethical and legal methods. Bad things happen when this route is followed. If not right now, at least, they will happen later. That becomes a matter of faith in every sense.

So far, the physical evidence to help law enforcement find, arrest, and charge the murderer of these three women remains frustratingly beyond reach. This is not a television show or a movie. Real humans died and real families suffered. The search for a killer continues. Detectives will not give up—the good ones never do. And the good ones are not satisfied to just close a case. The good ones want to close the case by arresting the actual murderer(s). Nothing less and nothing more.

See No Evil

4

> If you tell the truth, you have infinite power supporting you;
> but if not, you have infinite power against you.
>
> **General Charles Gordon**

Introduction

On April 14, 2001, at approximately 4:30 P.M., Officer Tim Johnston of the Williamsville Police Department responded to the Curl Up & Dye hair salon in the downtown business district (Figure 4.1). There, the victim's mother, Mavis Martin, beauty shop owner and chief stylist, witnessed a struggle between her son, Percy Martin, and Johnston during which Johnston fatally shot her son. Percy Martin was pronounced dead at the scene.

Johnston had responded to the same scene earlier that day to quell another disturbance at the business also precipitated by Martin, who suffered from schizophrenia. Martin was, when off his meds, possessed by violent demons that neither mother nor son could control.

On his second visit to Curl Up & Dye, Johnston faced an even more belligerent Martin. Johnston used his baton. The struggle drew both combatants further into the salon past the styling stations, sinks, and blow dryers, toward the mirrored southwest wall. Johnston attempted to use his pepper spray, to no avail. Martin somehow seemed to possess superhuman strength. He had a firm grip on the officer's throat from behind. Johnston felt faint and dizzy. He tried to pull out his Glock.

The struggle now focused on the pistol. Martin let go of the officer's throat and began grabbing for the gun; Johnston continued struggling to keep it in his possession. Martin

Figure 4.1 The Curl Up & Dye Salon. My equipment appears in the foreground.

draped himself over Johnston's neck, pressed his chest into Johnston's back and reached over his shoulders with both arms, his hands pulling on the pistol. The struggle continued. The salon's magazines slid from the table, littering the floor, as the pair fell to the linoleum only to get up and struggle again.

In the course of this frantic battle for gun control, after about 15 minutes of intense combat, Johnston fatally shot Martin, double tap. Percy Martin slipped to the floor. His blood began soaking into the current issue of *People* magazine. Mavis Martin looked on in horror as her son died before her on the salon floor.

The next day, Howard Davis, M.D., conducted the postmortem examinations as the pathologist of record. The autopsy findings indicated that Martin died from two gunshot wounds, one to the right upper chest and one to the left chest. Both injuries showed evidence of stippling and tattooing, indicating close range, near contact, or close contact gunshot wounds. Internal affairs performed their standard investigation. In the course of that investigation they interviewed Mavis Martin. She told her story with great emotion, certain of what she saw.

She said that Johnston, in the course of the struggle, had gained the upper hand, and had pushed Percy away. Johnston had retreated, she said, about 6 to 10 feet away from her son. He held the Glock with both hands, she said, aimed at her son, and shot him twice. She had seen Johnston murder her defenseless son, and she would make the officer, the department, the city, the state, and everyone remotely connected to the case pay the price for his untimely death. She hired a lawyer. She would no longer speak with officers. She would see them in court.

In my independent practice of forensic science and forensic medicine, I always first assure my clients that I will reach the best, most reliable scientifically supported conclusions, regardless of their adversarial consequences. If they do not accept that fact, I do not accept the case. Assurances met by Mavis Martin's attorney, I agreed to look at the evidence and reconstruct the shooting based upon the physical evidence.

There were two conflicting scenarios offered to explain this officer-involved shooting death. Wisdom usually accompanies every effort to avoid the sort of "he said, she said" statements offered by witnesses of violent death. Focus on the evidence. But in this case, the conflicting description of events supplied two hypotheses to play against what the physical evidence might be able to show. The evidence might confirm one, confirm the other, or disconfirm both. So the focus remains the physical evidence.

The analyses of this evidence involved ballistics and trace evidence examination, as well as scientifically based and carefully limited shooting event reconstruction. Ballistics and gunshot residue (GSR) played a major scientific role.*

"Ballistics" Further Clarified

The first step, of course, is to determine what the issues are, and if and how the physical evidence might help settle the issues (or at least address them). All opinions must be based upon an analysis of data *as it relates to the physical evidence in a case*. Like other scientific endeavors, forensic science and forensic medicine are data driven. Evidence involving bloodstain patterns and ballistics, along with scene features and measurements,† must be developed. Since the issues seem to depend upon muzzle-to-target distances, the detection of muzzle effluents or gunshot residues also seems to be relatively important to help sort out the issues that physical evidence may help adjudicate. And since the target in this case involves clothing and a human body, forensic medicine and autopsy results also provide critical data.

Work with human tissues requires additional expertise in forensic medicine, which is not part of ballistics as a branch of classical physics. My own interests, education, training, and experience cover both ballistics as a branch of classical physics and terminal ballistics as an area of forensic medicine.‡

Ballistics as an area of forensic science, however, also involves a number of distinct tasks that more properly fall under the heading of "firearms examination." Risking abuse of the term from classical physics, as eventually lamented by the "father of modern ballistics," Calvin Goddard (Figure 4.2) when ruing his choice of the term "ballistics" to cover such activities, participants in both areas are sometimes referred to as "ballistics examiners" by the lay public. This can be confusing since ballistics is a branch of physics, and tool mark examination has yet to find any scientific home.

Besides conducting microscopic comparisons among bullets, cartridge cases, firing pins, gun barrels, and various gun parts, the firearms examiner also may study bullets and cartridge cases to identify the make and model of the weapon that fired them. This also may involve the disassembly and testing of various firearms and various ammunitions, restoring obliterated serial numbers, or conducting failure analyses when weapons malfunction, causing injury or death.

Scientists suitably trained in both ballistics (physics) and trace evidence (chemistry/microscopy) also significantly contribute to scientific crime scene reconstruction. This work may involve incorporating research conducted, for example, on the gunshot residue

* Of course the natural sciences of physics and analytical chemistry with microscopy cover these areas.

† Scene analysis includes, but is not limited to, various analyses of the relative locations of items at the scene, such as tables, chairs, and doors, that may become relevant to these events.

‡ Properties of bullets and cartridge cases and issues raised by attempts to distinguish individual versus class characteristics of marks from specific firearms on specific bullets and cartridge cases remains an active area of research for me with respect to 3D examination, documentation, and comparison issues.

Figure 4.2 Calvin Goddard, 1923.

patterns of specific guns or classes of either semiautomatic or fully automatic weapons. Results may help determine the possible positions of any shooters using this type of armament. The patterns of GSR or muzzle effluents must be investigated on clothing belonging to both a suspect and the victim.

Gunshot Residue as Trace Evidence Analysis and Analytical Chemistry

This GSR analysis is another area popularly associated with ballistics as well as with firearms examination on occasion. However, the techniques of gunshot residue analysis overlap the area of trace evidence examination. GSR analysis is most properly a branch of analytical chemistry. Trace evidence analysis involves the recognition, collection, examination, and identification of microscopic materials usually transferred from one item to another. One such item of trace evidence is gunshot residues or muzzle effluents. The techniques applied to analyze GSR derive from the practice of analytical chemistry.

Gunshot residues transfer to items that may be near a firearm during its discharge. The Association of Firearm and Toolmark Examiners (AFTE) defines GSR as any particulates resulting from the discharge of a firearm including burned and unburned gunpowder, primer residues, and metal particulates. They consist of burned and unburned gunpowder, lead or copper shavings from the bullet and its jacket (if present), and residues from the initiating primer. The latter are microscopic in size, usually between 0.5μ and 5.0μ with some from 10μ to 15μ. These materials exit from the muzzle and from any other opening in the firearm through which combustion gases may exit, depending upon the weapon's design and condition.

The Glock pistol, carried by Officer Johnston, is considered to be a high-quality gun and his was well cared for—cleaned, lubricated, and dried. In testing, I noted the locations of GSR exit. The GSR discharge for any gun can be visualized by a high-speed camera, such as the Keyence high-speed digital model, which shoots about 42,000 frames per second. The residues exit the Glock as seen in Figure 4.3.

GSR exiting from the muzzle usually exhibits a conical shape (Figure 4.4). When this conical pattern is not disrupted by environmental or situational factors, estimations of the distance between the muzzle and the target can be made by firing the gun at issue from specific distances into a clean white target, measuring the dimensions of the resulting residue pattern, and correlating the distances from the target with these dimensions of the resulting residue patterns. The size of this pattern from a measured distance can then be compared with the size of an unknown pattern found at the scene on the victim, and the actual distance from the shooter's muzzle to the victim then can be estimated. This may become impractical, for example, if the actual gun is not recovered, if it is unavailable or unable to be tested, or if potentially unknown environmental conditions, such as heavy rain or high wind, adversely affect these typical conical distributions.

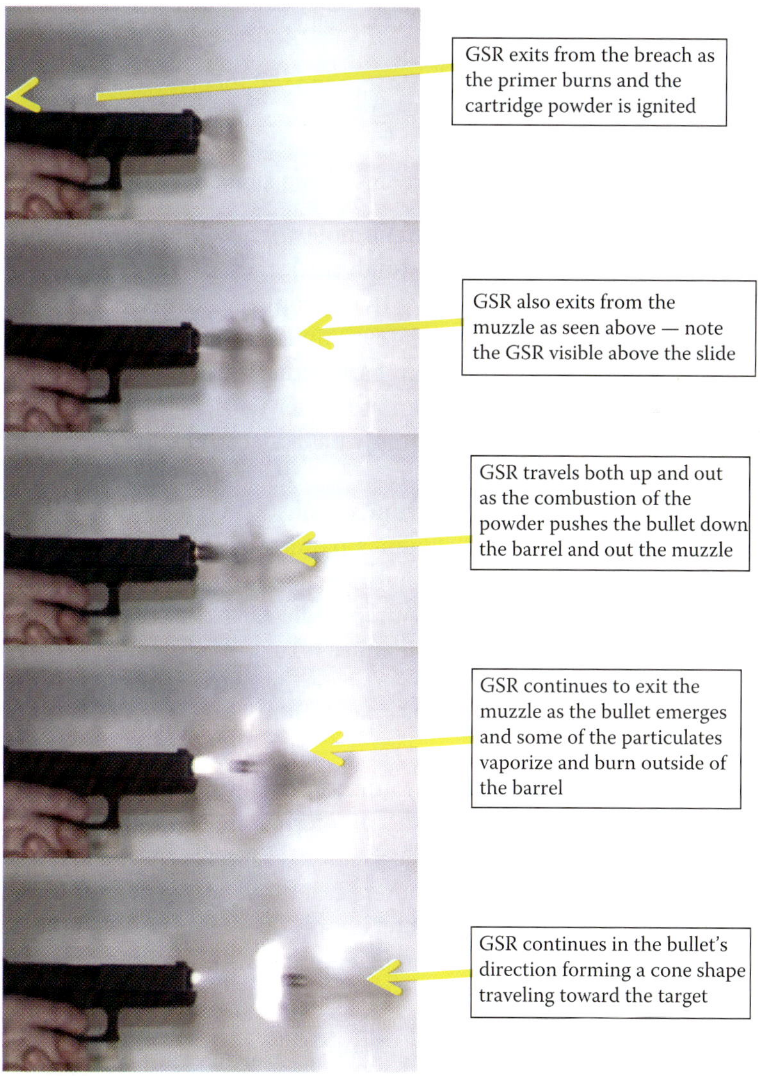

Figure 4.3 Shooting the Glock.

Figure 4.4 GSR behaves more like a mushroom-shaped cloud than like a cone-shaped cloud.

General considerations, however, become noteworthy. Most handguns (both pistols and revolvers, from the smallest 22 cal through the largest 50 cal Desert Eagle, distribute measurable conical GSR patterns no further than about 36 inches from the muzzle. Several factors influence this distance, including the type and load of the cartridge firing the bullet as well as the aforementioned environmental and situational factors. This does not mean, however, that no GSR can be found on a target farther than 36 inches from a muzzle. This means that such residues, when present, simply belie assessment as the result of a predictably measurable conical pattern.

Various factors may influence the distribution of such GSR on a target even though we may not be able to assess the muzzle effluent's pattern by estimating its dimensions. For example, transfer of GSR may occur through mechanisms other than from firing a gun, such as physical contact between a GSR source and a target. Excessive bleeding also may adversely affect the identification of such residues on any clothing target surface. Residues from muzzle effluent expected on flooring or carpeting may be affected by blood staining, redistributed by foot traffic, or may be disrupted by other factors, which, ultimately, are unknowable to the analyst.

GSR exiting the weapon can also be found on the face, hands, and clothing of anyone near a firearm's discharge, including the shooter himself. It may be transferred after a weapon's discharge through contact with the weapon. However, the identification of GSR on the hands of suspected or even known shooters can prove to be difficult in practice. Many laboratories no longer bother to test for such GSR on the hands of suspected shooters since the results, either positive or negative, often may present at-best ambivalent interpretations.*

* The FBI laboratory, for example, no longer tests hand swabs for GSR because the swabbing and collection techniques by various submitting agencies were haphazard at best and very sloppy at worst. In either event, the lab had no quality control over the collection of samples and therefore could not rule out false positives. This rendered such tests essentially meaningless.

Such residues easily can be lost from the hands through sweating and wiping, washing, or through almost any ordinary daily activity. Some shooters may wear gloves, protecting their hands from most such deposits. For this reason, hand swabs for GSR detection taken any time other than immediately after a shooting or taken from a deceased shooter who has not moved after potentially discharging a weapon are seldom productive. Additionally, problems of contamination through the collection process cannot be ignored.

The best evidence for individual GSR determinations on individuals remains on the clothing worn by either the victim(s) or the suspect(s) during any shooting. For other analyses, durable items present at the scene provide the best sources for testing. This gunshot residue may be detected through various scientific means, including chemical, microscopic, photographic, or other instrumental analyses. The choice of testing mechanism(s) ultimately depends upon the nature of the evidence and the need to preserve this evidence through nondestructive testing methods for additional or repeated analyses. It also depends upon the desire to explore any potential GSR patterns.

Posing the Questions

The account provided by Mavis Martin contrasted sharply with the account provided by Officer Johnston. He denied shooting Percy Martin from a distance while facing the decedent. He claimed that after getting control of the Glock, he pointed the gun along his left side toward the back and fired twice. He could not tell what position Percy occupied at the moment he fired, but he insisted that the decedent was behind him, choking him after failing to get control of the gun.

Thus, two competing scenarios could be stated. They helped to organize the physical evidence and to determine the priority of testing that evidence. The scenarios helped determine the potential evidential value of the data and guided the analytical approaches. There remain many things we can do in a shooting reconstruction: the question becomes what we should do and what we should omit in order to address the issues or develop leads.

Scenario 1

According to the first scenario, offered by Johnston, Percy Martin remained behind him during a struggle in which both men moved south and west into the hair salon operated by Mavis Martin. During this struggle, Percy Martin attempted to gain control of Johnston's service pistol with his right hand, which he placed on the gun as it rested in Johnston's holster. Johnston began to tire after the intense struggle, but managed to secure his service pistol after fighting off Martin for control of the holstered gun. With Martin still clinging to his back, Johnston reached under his left armpit holding the pistol in his right hand, and, fearing for his life, fired two shots toward Martin with Martin still in near contact with Johnston's own back.

Scenario 2

According to the second scenario offered by witness Mavis Martin, her son Percy and Johnston were engaged in a struggle in which both men moved south and west into her hair salon. In the course of the struggle, Johnston stepped backward and away to Mr. Martin's side. Then, she said, her son stepped forward, toward Officer Johnston. She said that when her son stood facing the officer from 3 to 5 feet away, the officer drew his service pistol and fired directly at him.

The processes for testing and analysis became clear. A starting point was to derive some measurements.*

Scene Visit

To help further clarify the scenario offered by Mavis Martin, I collected measurements at Curl Up & Dye. I visited the hair salon, and Mavis Martin indicated the relative positions of the two men as the shooting occurred. I photographed her depiction as follows, using V for victim to stand in for her son's location, and O for officer, to stand in for Officer Johnston's position (at the end of his Glock service pistol) (Figures 4.5 and 4.6).

Evidence Examinations

The first step, of course, is to obtain copies of the discovery and review any previous work including police reports, scene evidence collection, and resulting scientific reports. In some cases, this can be made needlessly difficult. While States Attorney Richard Cabeza stated clearly that no examinations or testing had been done on the evidence in this case, and that therefore no lab reports could be expected or would be forthcoming, I found evidence of prior analysis on virtually all the evidence I examined. (For example, photos of the officer's shirt show the numbers written in marking crayon by a previous examiner.)

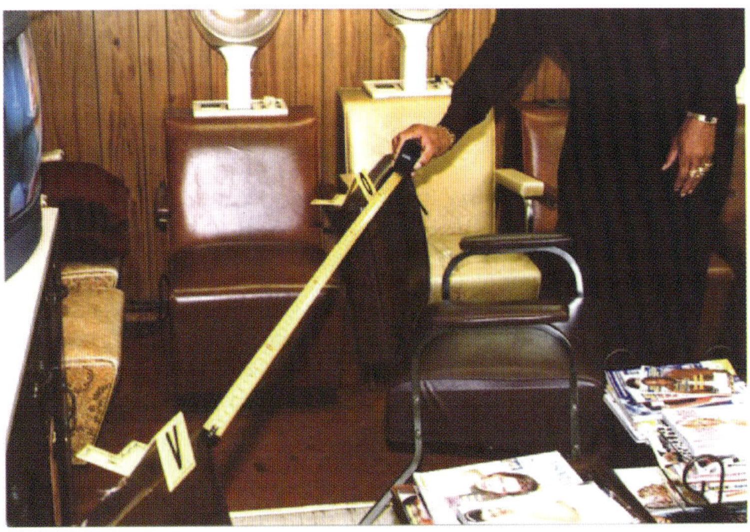

Figure 4.5 Illustrating the story.

* There are many who argue that unless some form of metrology is or at least in principle could be applied, then the scientific value of that work becomes questionable.

Figure 4.6 The tape indicates the distance to be approximately 3 feet 9 inches between the victim and the officer, according to Mavis Martin's account of the shooting.

In this case, the two opposing scenarios are clear. But even without the proposed scenarios, the exact same applications of science must be made to investigate this shooting. This includes, but need not be limited to, the following evidence:

1. The autopsy evidence of Martin's gunshot wounds
2. Gunshot residues present on both Martin's clothing and on Johnston's clothing
3. The physical position of the Glock pistol when fired toward Martin
4. The location of the ejected cartridge cases at the shooting scene
5. The particular service pistol's cartridge case ejection patterns
6. The particular service pistol's GSR patterns, including any leaks from the breach, as well as from the muzzle, through close contact to 4 feet intermediate range as seen in the photos documenting Johnston's Glock

Relevant Autopsy Results: The Victim's Injuries

Percy Martin was shot at near- or close-contact range, given the injuries sustained, their character, and their direction.

Examination of the Pistol

I tested the Glock both for GSR leakage as well as muzzle effluent volumes and distances. Using a Keyence Corporation high-speed digital camera, the following results were documented.

Leakage and Discharge

As seen in the earlier photographs, the Glock discharges GSR from the breach and from the muzzle (Figure 4.9). The volumes were not calculated but the deposition patterns were recorded.

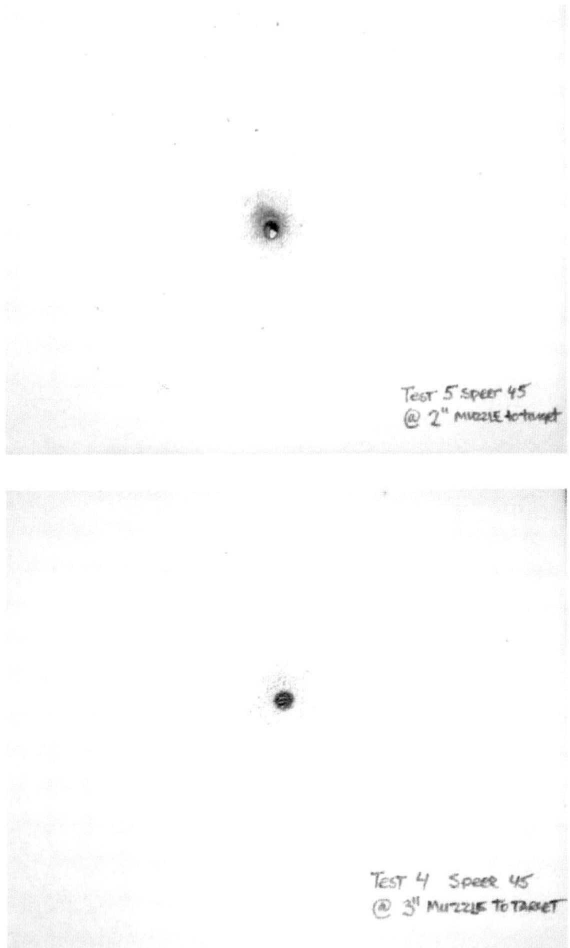

Figure 4.7 No GSR depositions were noted beyond 6 feet, but some should be present at between 3 and 4 feet, given the indoor scene.

Target GSR Depositions

To test Mavis Martin's account, some study of the distance that the Glock pushes the GSR was necessary (Figure 4.7).

Examination of Bloodstain Patterns and Trace Evidence

Officer Johnston's Glock

The pistol was in very close proximity to a blood source when fired. Photos were taken using 500-nm light with a 23 orange filter to display the presence of blood on the weapon, even after that weapon had been cleaned (Figure 4.8 and Figure 4.10). The weapon had been cleaned before the inside of the barrel could be examined.

Decedent's White T-Shirt

The decedent's T-shirt was examined visually and photographed. It was also examined using a forensic light source, and then analyzed using long-path infrared (IR) light, with

Figure 4.8 Officer Johnston's Glock.

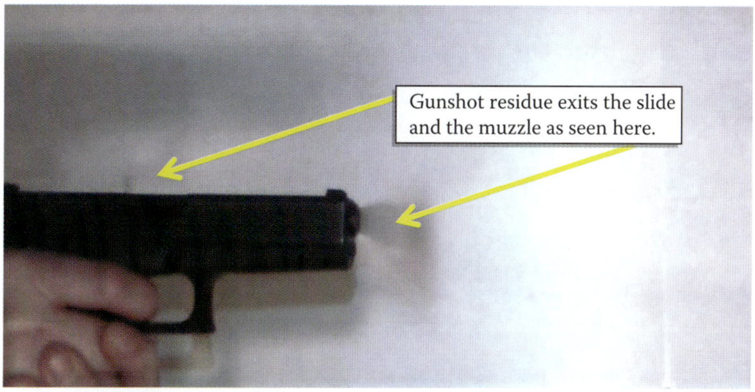

Figure 4.9 GSR exit points.

Figure 4.10 The blood shows signs of being wiped. Usually, such interpretable stains are simply called "smears."

Figure 4.11 Front of the victim's T-shirt.

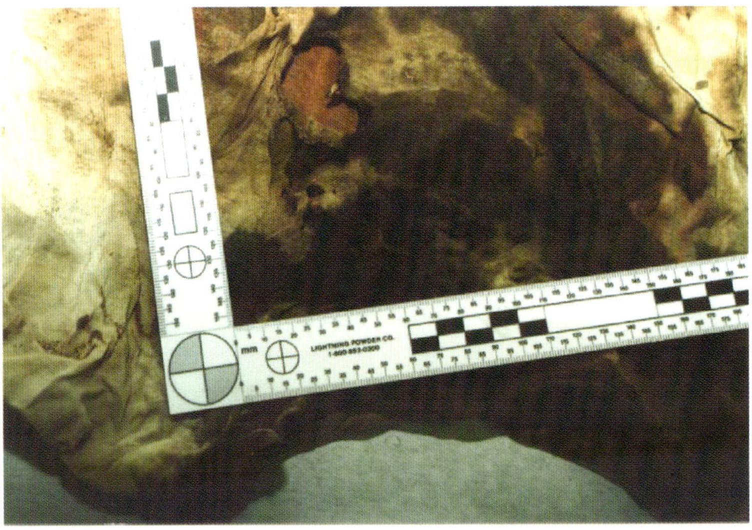

Figure 4.12 Defect in the front shows a great deal of burning and fouling.

a long-path IR filter on the camera. Finally, it was tested chemically for the presence of heavy metals characteristic of GSR primers found only close to the point of a weapon's discharge, with a positive result. The visual exam showed both gunshot entrance locations, with singeing around the margins. Despite the blood-soaked nature of the cotton material, GSR was present on the front of the victim's T-shirt, as indicated both by using the forensic light source and long-path IR light at 680 and 715 nm (Figure 4.11 to Figure 4.15).

Often using an infrared wavelength helps visualize GSR that might otherwise be invisible. This is especially helpful when examining a dark fabric, such as the officer's uniform, for GSR. IR light also helped visualize the evidence on the victim's white T-shirt as seen in Figures 4.15 and 4.16.

Officer Johnston's Dark Navy Blue/Black Uniform Shirt

The officer's uniform shirt was examined visually and photographed. It also was examined using a forensic light source, and then analyzed using long-path IR light, with a long-path

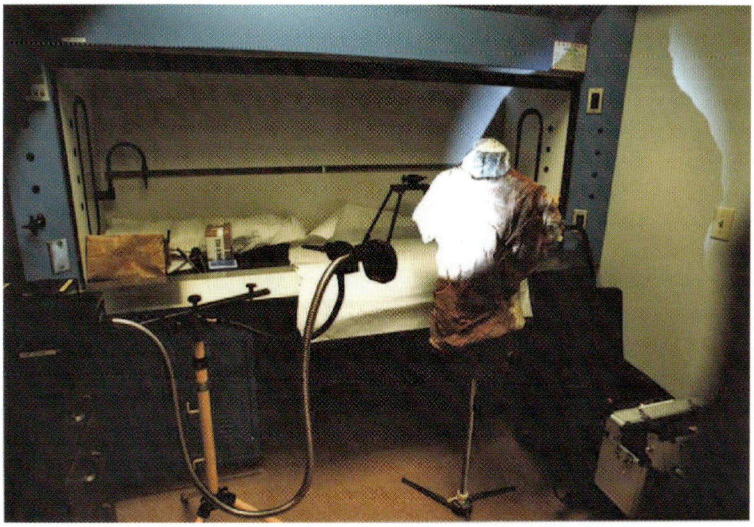

Figure 4.13 A fume hood, tripods, mannequins, and good light sources in a room that can be completely darkened are the necessary components for the various alternate light source (ALS) examinations of evidence.

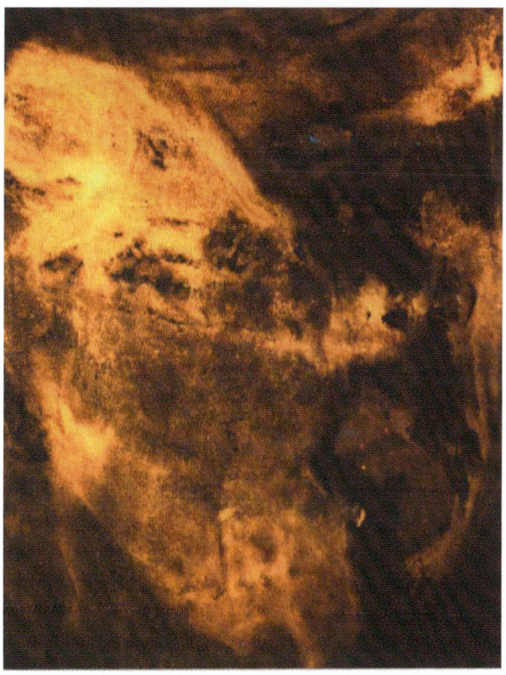

Figure 4.14 The blood and the powder appear as they do with normal full-spectrum lighting.

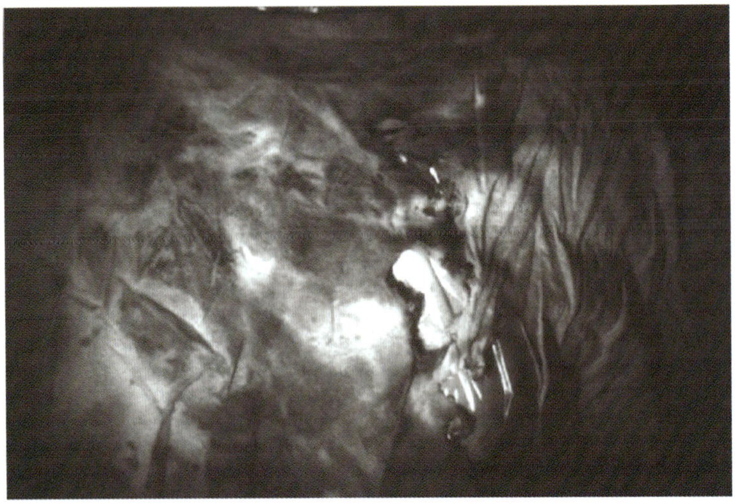

Figure 4.15 The victim's white T-shirt seen with a 715-nm light and IR film. Digital cameras require a custom chip to capture IR images.

Figure 4.16 A field microscope captured images showing blood mixed with dark flake gunpowder on the victim's T-shirt.

IR filter on the camera. Finally, it was tested chemically for the presence of heavy metals characteristic of GSR primers found within 3 to 4 feet from the point of his pistol's discharge. The visual exam showed apparent gunshot residues present on the garment's left side under the left arm as seen in Figures 4.17 and 4.18.

Despite the dark color of the officer's shirt, GSR was observed on the left side and back, as indicated both by using the ALS with long-path IR light at 680 and 715 nm as seen in Figure 4.19 to Figure 4.21.

I was also able to confirm the presence of blood on Johnston's uniform on the back by using the forensic light source, with confirmations made using my field microscope. The stains present as impact and wipe with swipe stains, meaning that a blood source came in contact with Johnston's back, depositing, transferring, and moving blood from that bloody source to several locations on his back.

Figure 4.17 Officer Johnston's uniform top.

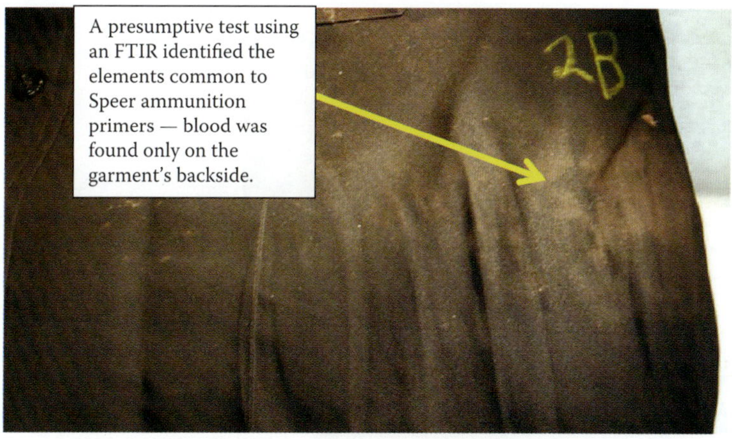

Figure 4.18 Uniform trace evidence.

Figure 4.19 Folds and voids.

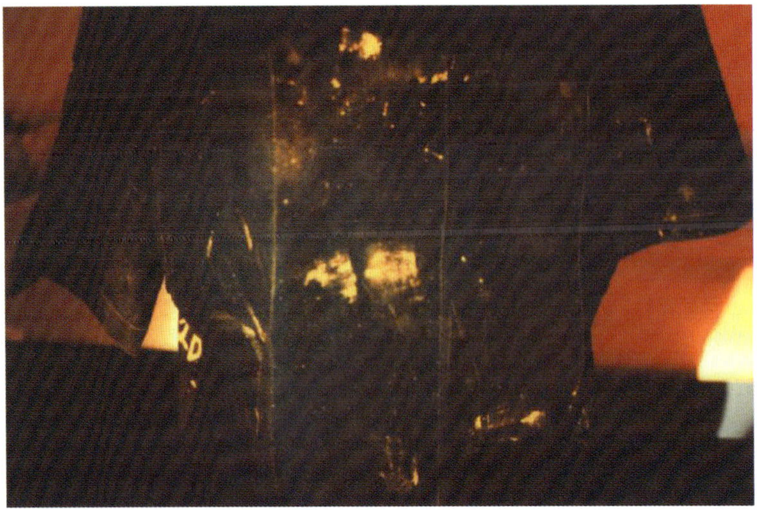

Figure 4.20 Blood appears more vividly with a series of filters for red light.

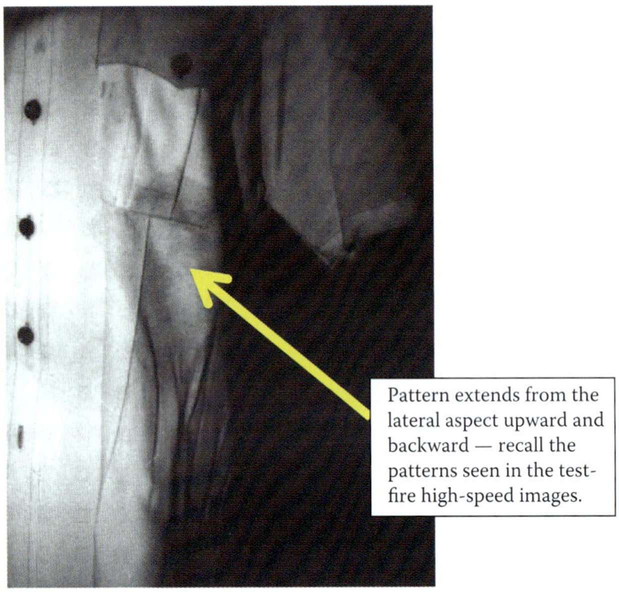

Figure 4.21 The left side of the uniform, indicating that a gun was discharged in close contact with the garment. The IR light with carefully chosen filters is a very good nondestructive method to visualize GSR in its full, accurate sense.

Crime Scene

Bloodstain Patterns at the Scene and Scene Drawing

Bloodstains visible in the scene photos show blood dripped onto the floor from a height at an angle, approximately indicating where the victim stood when initially shot. Bloodstains also appear on a blue-green, metallic-like shiny smooth salon chair and dryer hood seen in the pictures taken of the southwest corner of the business where the shooting occurred. I made a rough sketch of the scene to help me orient items of evidence (see Figure 4.22).

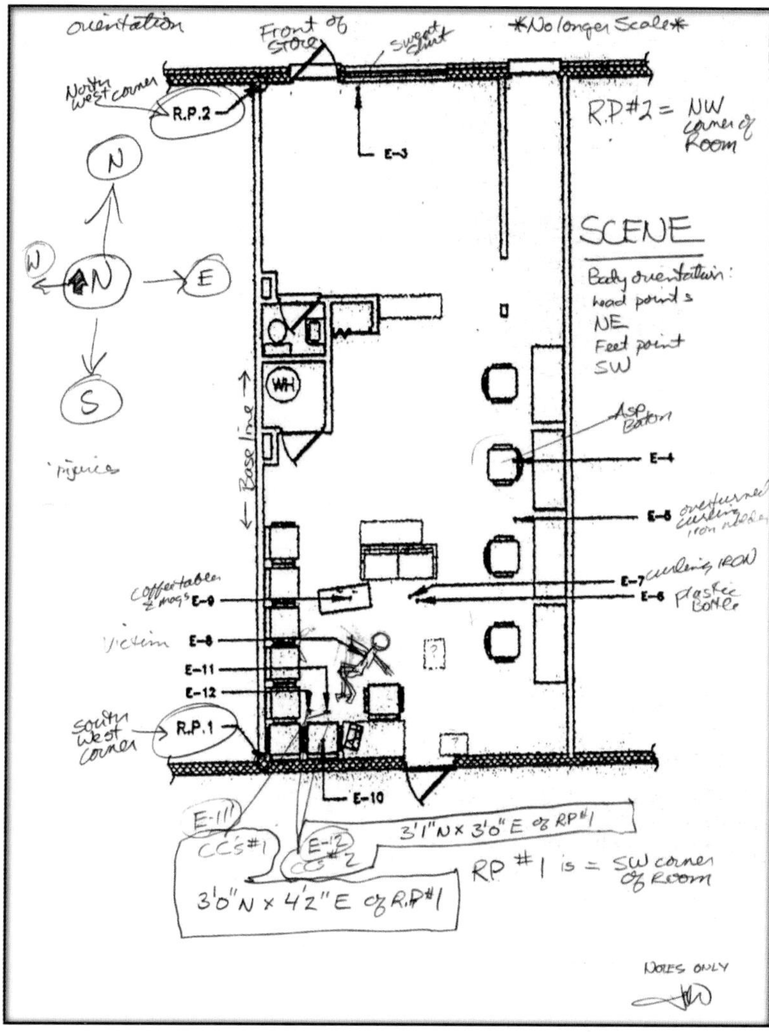

Figure 4.22 Very rough sketches can be developed into scale drawings if enough measurements are taken. Such drawings are also a useful tool to prompt questions that need answers.

I visited a storage building at Mavis Martin's mother's home to view the aforementioned blue-green metallic chair in order to further analyze the bloodstain patterns present. Although the photograph helps to understand the blood-shedding event, a further analysis is warranted. I located the chair, and did a phenolphthalein test for the presence of blood. I followed the positive result with a Luminol spray, lightly misted on the chair in the darkened storage building. Whereas luminescence appeared around the tacks holding the fabric onto the armrest and the seat, as well as along the cracks and folds in the upholstery, no bloodstains appeared on the seat's surface, or the surface of the salon chair's backrest. I could not locate the dryer hood pictured in the crime scene photograph (see Figure 4.24).

Despite my inability to locate this evidence, and despite the disappointing results with the aforementioned chair, all of the bloodstains visible in the photographs indicate that the decedent was standing approximately where he fell, facing the southwest wall. No bloodstain patterns indicate that the decedent moved independently after being struck by the two bullets fired by Johnston (see Figure 4.23).

Figure 4.23 Showing blood and cartridge cases at the scene, as well as the decedent's shoe.

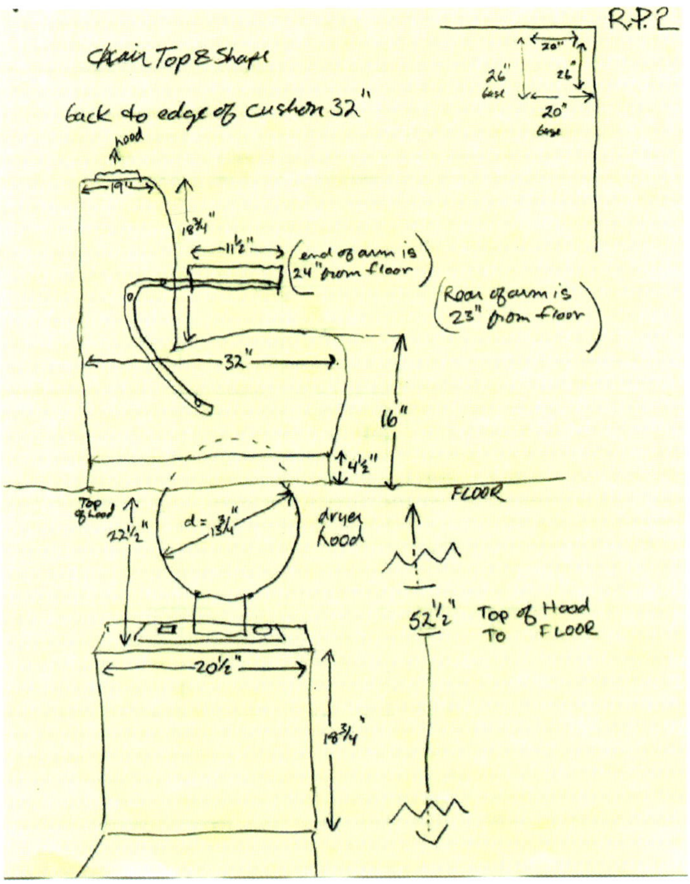

Figure 4.24 Sketch of the chair and hair dryer.

Scene Reconstruction Based upon Physical Evidence Examined

No reconstruction can account for every event involved in any shooting.* An attempt was made to reconstruct all movements involved in the struggle between Martin, the eventual decedent, and Johnston as such movements evolved through the Curl Up & Dye beauty shop. The evidence permits an account only of the mechanics of the shooting itself from the first to the end of the second shot.

1. Martin and Johnston struggle near the southwest corner of the building.
2. Johnston stands on the carpeted surface near the southwest corner, facing either west or slightly southwest (see crime scene drawing).
3. Martin also stands facing either west or slightly southwest in close contact with Johnston's back.
4. Johnston gains control of his service pistol, a Glock semiautomatic pistol.
5. Johnston, with Martin tight or very closely near his back, manages to twist slightly to his left as the decedent remains in close or near contact with his back (as indicated by wrinkles in his shirt by the areas of GSR); he points the pistol under his left arm just above waist high on his uniform (note that his legs may be bent, as may Martin's legs) and fires twice (double tap).
6. Martin is struck in the left chest by the first shot (see autopsy). As he slumps, he is struck in the upper right chest by the second shot (see autopsy).
7. Martin falls to the floor, as pictured in the crime scene photographs.
8. The two expended cartridge cases rest on the floor as ejected from the right side of the Glock as the pistol was pointed east or perhaps slightly northeast toward Martin (as pictured in Figure 4.23).

Conclusions

The physical evidence, including the victim's T-shirt, the officer's uniform shirt, the Glock pistol, the autopsy report and photographs, and the crime scene photographs, all support one conclusion: Officer Johnston fired at Percy Martin in the manner the officer described in his sworn statements. The physical evidence examined in this case fails to support any version of Mavis Martin's account. At most, 1 to 3 inches may have separated the victim from the muzzle of the weapon. Johnston's account, therefore, accurately describes events supported by the physical evidence in this case.

Another point deserves mention in this case. I was hired by Mavis Martin's attorney to examine the evidence in this case. Doing science does not involve any advocacy—save advocating for and supporting the practice of good, reliable science. Despite the beating I took from both attorneys, from one because I worked for the "enemy" and from the other because my results failed to support the client's story, my work product must stand on its own. Science does not play favorites. It is useful, in these stressful situations, to remember a

* For example, the reconstruction produced in the "Three Strikes and Out" case detailed in Chapter 1 accounts for only 1 second of an extremely complex event. However, that 1 second was judged to be the most important given the specific questions and the given evidence.

quote attributed to Thomas Paine: "When a man who is honestly mistaken hears the truth, he will either cease being mistaken, or cease being honest."

Brief Bloodstain Pattern Analysis Glossary

Arterial spurt or gushing: Pattern explained by pressurized blood pumped through a breached artery
Back spatter: Blood flying back toward the source of the force producing the spatter
Blood volumes: The average male has 6 quarts (5.6 liters) of blood
Cast-off stain: Pattern explained by blood released from a blood-bearing object in motion
Circulation: Arteries transport blood under high pressure away from the heart; veins transport blood under very low pressure toward the heart
Clot: The agglutination of blood producing a gelatinous mass through natural properties called "clotting factors" in red cells, fibrinogen, and platelets causing blood cells to clump together
Convergence point: The point representing the source of a bloodstain
Directionality: Features of a bloodstain's shape indicating its direction of flight
Drip pattern: Stains resulting from blood dripping into pooled blood
Drip stain: Pattern explained by blood dripped into blood
Expirated or exhaled blood: A pattern of small stains resulting from forceful exhalations through the nose or mouth (coughs or sneezes) that project blood from the airway to a target
Glancing angle: The angle between the line of bloodstain travel and a reference direction such as 90 degrees vertical; also called "directionality"
Impact angle: The angle between the direction of travel and the target surface defined, using trigonometry, as the arcsine of the stain's width divided by the stain's length
Impact site: Sometimes used to mean "point of origin"
Impact spatter stains or spatter: Patterns explained by blood dispersed from some force applied to a blood source, sometimes showing characteristics of the force that produced them
Misting or rouging: Blood reduced to a fine spray by applied force
Saturation stain: Pattern on an absorbent surface/area explained by a large quantity of blood soaking through the surface
Spatter: Small blood droplets resulting from the forceful projection of blood
Swipe: Stain created by a moving blood source contacting an unstained surface
Swipe stain: Blood transferred from a moving blood source onto an unstained surface
Target: Any surface that exhibits bloodstains
Transfer stains: Pattern explained by contact between a bloody surface and a second surface, producing an image of the first surface in the resulting pattern
Void: The absence of bloodstains in an otherwise continuous bloodstain pattern
Wipe stain: Blood pattern produced by an object moving through an existing stain altering its appearance

A Christmas Story

5

> Two Massachusetts eighth-graders who spent $200 for ballistics gelatin and seven months on a science project to show that BB guns can be dangerous were not allowed to present their project at the state's middle school science fair because BB guns can be dangerous.
>
> **"A Project on BB Guns Is Rejected in Amherst," Associated Press, *The Boston Herald*, June 12, 2005**

Introduction

Wrapping paper, torn and scattered, littered the carpet under the Christmas tree. Two small boys, one smaller than the other, played in the glow of the colorful lights. The older boy, Donny, awaited his "big present." The younger boy, Jimmy, had just gotten his: an official NFL jacket. He was a big fan. Well, a little one anyway. They lived just outside of Oakland in one of the few subdivisions with any sizable trees.

Then Donny had his turn: he finally got the big present. With repeated warnings from his parents not to "shoot his eye out," and, of course, not to shoot anyone else's eye out either, he got his wish granted: a new BB gun. It had dual capabilities. It shot either BBs or pellets and was a pump-style air rifle. Not a toy, his father warned.

Jimmy, dressed in his new Raiders jacket, and Donny, holding his new air rifle, played outside, running among the trees standing in the Pine Forest subdivision. Jimmy darted in and out behind the trees as Donny shot his rifle at a paper target tacked to one of the tree trunks. He worked his way backward, improving his aim as he went. By pumping the rifle up to 30 pumps, he found that both the range and his aim improved. About 90 feet from the target, Donny took aim and fired just as Jimmy popped out from behind the tree. Jimmy looked puzzled. Then he dropped like a sack of sand. Donny laughed and told him to get up, but when he didn't, Donny yelled for his mom.

At the hospital, Jimmy was pronounced dead on arrival. Someone, his folks decided, would pay. Their sights aimed at the air rifle manufacturer. They contacted a lawyer.

The lawyer asked me to review the available police, medical, laboratory, and additional follow-up reports concerning the physical evidence in the BB air gun shooting death of Jimmy DeLauro. After my review, I was asked to test two Daisy Air Rifles, both models 856: one manufactured for the U.S. market designed to fire pellets (data not stated for BBs) at a maximum of about 670 feet/second muzzle velocity at 10 pumps of the air chamber, and one manufactured for the Canadian market designed to fire pellets (again, data not

Figure 5.1 Unlike the Canadian model 856, the U.S. model 856 is apparently not designed to prevent overpumping and is said to be able to produce muzzle velocities greater than 670 ft/sec through overpumping the air chamber more than the 10 times recommended as the maximum in the instruction book.

Figure 5.2 Both rifles look very much the same. The boxes differ, as does the rifle's pumping design.

stated for BBs) at a maximum of about 490 ft/sec muzzle velocity at 10 pumps of the air chamber (Figures 5.1 and 5.2).*

Unlike the U.S. model 856, the Canadian model 856 is said to be designed to prevent overpumping of the air chamber. Regardless of the number of pumps over 10, the maximum muzzle velocity is said to be about 490 ft/sec. Of course, fewer than the recommended 10 maximum pumps in either model will also result in lower muzzle velocities. (Such data is provided in the instruction book published by Daisy accompanying each air rifle sold.)†

* I have had a Daisy Red Rider BB gun since I was about 7 years old. It remains one of my most cherished possessions. Behind the garage, I shot up my grandmother's Mason jars, jelly jars—anything I could find that broke when struck by a BB. I was trained by my parents and grandparents never to aim the rifle at anything that I did not want to break or kill, especially at things I could not see or places where I did not know what might pop out from behind the target. I accepted that responsibility. Personally, I find it difficult to blame an air rifle for anything, just as I would find it difficult to blame a pencil for a failing math grade. Others do not necessarily share my opinions.

† The instruction book states on page 8 that 2 pumps = 295 ft/sec; 4 pumps = 440 ft/sec; 6 pumps = 525 ft/sec; 8 pumps = 580 ft/sec. My testing holds very approximately to these values. Note the wide variations in testing; these values quoted by Daisy are rough approximations that cannot be expected with each and every shot meeting the pumping criteria.

Figure 5.3 Steel BBs.

Issues

In cases involving issues with a firearm, I find it useful to list what we know and what we do not know. I then try to develop a sound approach to any testing or experiment* designed to help us learn more about the unknowns that are relevant to the issues involving the firearm. Sometimes the testing and the experiment can occur simultaneously if the design is carefully crafted. We tested the rifles' expected behavior and investigated the relationships of pump numbers to distance and target penetration.

A similar U.S. model 856 fired a round steel BB (not a lead or alloy pellet)† striking 8-year-old Jimmy DeLauro in the upper left chest (see X-ray anteroposterior [AP] view and photo of entry wound). The BB perforated a poly-fiberfill NFL jacket near a seam, perforated a cotton T-shirt, and penetrated about 1½ to 2 inches of soft tissue, entering through an intercostal space, penetrating the mediastinum, perforating the pericardial sack, and entering the left ventricle of the decedent's heart. The injury sustained resulted in his death the same day.

Facts

The BB (Figure 5.3) perforated a poly-fiberfill NFL jacket through both external layers and the poly-fiber lining, perforated a cotton T-shirt and penetrated 1½ to 2 inches of soft tissue/muscle. The BB entered Jimmy DeLauro's upper left chest. The injury produced a cascade of events, including a cardiac tapenade, which resulted in Jimmy's death. The projectile apparently entered Jimmy's heart muscle and migrated to the right atrium after perforating the wall of the left ventricle.

* Writers often use the terms *testing* and *experiment* interchangeably. This use does not fit into the lexicon of the natural sciences as I learned them. Testing refers to an analysis of some object's performance—a machine, piece of equipment, or method with known or expected results. Often testing concerns issues of accuracy but, of course, not always. Experiment, on the other hand, refers to some set up of objects in order to develop consequences or results that inform us about an issue other than the accuracy of some apparatus. We usually, but not always, perform experiments to learn the outcome rather than to confirm that an outcome closely approaches a known or expected result.

† Pellets are usually soft lead or alloy, and have aerodynamic qualities, such as points. They also fit the muzzle tightly, sealing the projectile in the lands and grooves cut into the barrel. This results in both more power and greater accuracy due to projectile spin. Unlike pellets, BBs are round and have no such aerodynamic qualities. Shooting a pellet in a Daisy 856 is like throwing a football in a tight spiral, while shooting a BB in a Daisy 856 is more like throwing a curveball with a baseball.

Unknowns

The distance from muzzle to target is initially reported to be 20 feet, then revised to a distance of 90 feet. The actual muzzle-to-target distance, therefore, remains unknown for the purposes of these tests of the rifle and experiments about pumps, penetrations, and distances. The actual number of times the U.S. model 856 rifle was pumped before firing the BB that struck Jimmy also remains unknown for purposes of this experiment.

Protocols

In order to replicate the target, similar poly-fiberfilled material was sewn between two pieces of jacket material similar to the material in the jacket worn by the decedent when shot. An additional layer of T-shirt material was added between the jacket material and ballistics gel, mixed according to Federal Bureau of Investigation (FBI) protocols, and tested according to standards developed at the Armed Forces Institute of Pathology. Both rifles were pumped to the maximum recommended by Daisy: 10 times. Both weapons were also overpumped to 30 times in order to investigate the effect of overpumping the air chamber on both the U.S. and the Canadian market models 856.

Goals

The experiment is designed to help correlate the unknowns of the case (muzzle-to-target distance and number of pumps) with the known facts of the case as follows: Determine the possible muzzle-to-target distance(s) required to produce penetrations similar to those known to have resulted in the death of Jimmy DeLauro with the minimal number of charging pumps to the U.S. model 856 air rifle. With this accomplished, the identical shots are taken with the Canadian model 856 air rifle, and the effects compared with those produced by the U.S. model under identical conditions. I usually present the results of an experiment such as this one in table form.

Target Construction Based upon My Analysis of Jimmy DeLauro's Jacket

The Professional Sport Shop "Team NFL" Raiders jacket worn by the decedent exhibits a small tear approximately $5\frac{1}{4}$ inches down from the top of the left side of its snap lining, and $3\frac{5}{8}$ inches left of the left edge of this snap lining on the nylon material's external covering. The tear leads through the nylon shell, avoiding zipper fabrics, patches, and decorations. The resulting hole continues through the fiberfill lining and completes the jacket's perforation when exiting the nylon inner jacket lining. This perforation in the jacket appears to represent the entrance of the fatal BB into the victim. The BB then reportedly perforated the victim's cotton T-shirt, entering the decedent as described.

The jacket is very simply constructed, with a nylon fabric shell sewn over both sides of a thin polyester fiberfill lining. The only other material over the decedent is reported to have been a standard cotton T-shirt.

Target Material Selection and Design

To replicate the target characteristics for testing purposes, nylon fabric of the same type is sewn using a similar construction over both sides of a similar thin polyester fiberfill lining material. Cotton T-shirt material is sewn loosely over the back of the lining to represent the victim's T-shirt. Fabric straps are sewn onto the representational target to aid in securing the model of a jacket over the T-shirt, in turn, over the ballistics gel.

Ballistics Gel

After gel preparation according to FBI protocols, the blocks were tested using the standard penetration scale created by the National Firearms Unit. This helps ensure that the gel will adequately represent human soft tissues, including skin, fat, muscle, and other solid internal organs. (The gel represents only the decedent's soft tissues since no bones were struck.)

Target

The target consisted of a plywood platform approximately 3 feet from the ground with a perpendicular plywood support holding a ¼-inch steel plate 18 inches above the platform. An 8-inch thick block of ballistics gel 12 inches high (Figure 5.4) was then placed against the steel plate, and the jacket/T-shirt replica measuring about 15 inches was then affixed in front of the gel to the steel plate using the attaching straps (Figure 5.5). The targets, including the jacket with T-shirt material and gel, were changed at each distance—90, 80, 60, 40, and 20 feet—with the results documented for further analysis.

Shooting Range Design

Shooting design incorporated a measured course of 90 feet from target to muzzle, with increments established for muzzle-to-target distances at 80 feet, 60 feet, 40 feet, and 20 feet. A 10-foot mark was established for clarity, with no shots fired from that mark (Figure 5.6).

Both rifles were test fired from each set distance using two pumping values: one set of shots was taken with 30 pumps each of both the U.S. and Canadian models; another set of shots was taken with 10 pumps each of both the U.S. and Canadian models.

Figure 5.4 Jacket and T-shirt material to be installed over the gel.

Figure 5.5 Target.

Figure 5.6 A chronograph was used to document the muzzle velocity of each rifle as it was pumped and fired at the target.

U.S. Model 856 Air Rifle

Goals

The testing is designed to help correlate the unknowns of the case (i.e., muzzle-to-target distance and number of pumps) with the knowns and facts of the case as follows: Determine the possible muzzle-to-target distance(s) required to produce penetrations similar to those known to have resulted in the death of Jimmy DeLauro with the minimal number of charging pumps to the U.S. model 856 air rifle.

Method

The rifle will be tested with a fixed number of pumps at fixed distances from the target. At each target distance, the perforation of the clothing materials and the penetration of the ballistics gel will be measured to determine if the 1½- to 2-inch depth was reached.

A chronograph will be used to measure the muzzle velocity generated by each set of pumps (10 pumps and 30 pumps) at each distance from the target. (Note that muzzle velocity drops rapidly with such air-powered rifles firing nonaerodynamic projectiles such as round BBs, and muzzle velocity cannot accurately be correlated with target velocity.) The only reliable measure of target velocity and energy will be the actual perforation of the clothing material and the actual penetration distance of the ballistics gel. Therefore, chronograph values were used only to establish the integrity of the gel, not to measure the muzzle velocities of the Daisy model 856 weapons tested.

Results

The results provide a range of values including specific distances plus specific pumps (10 or 30) sufficient to perforate the clothing materials and penetrate the ballistics gel to a 1½- or 2-inch depth. I generally present the results in two forms: divided by rifle and put onto a single table.

Results Using U.S. Model 856

Distance	Perforate Jacket Material?	Perforate T-Shirt Material?	Depth of Ballistics Gel Penetration?
	Penetration Testing at 30 Pumps		
20 feet	Yes	Yes	3¾″
40 feet	Yes	Yes	3½″
	Yes	Yes	2½″ to 3″
60 feet	Yes	No	No
80 feet	Yes	Yes	1½″
90 feet	Yes	No	No
	No	No	No, bounced off
	Yes	No	No
	Yes	No	No
	Penetration Testing at 10 Pumps		
20 feet	Yes	No	No
	Yes	Yes	2½″
40 feet	Yes	Yes	3½″
60 feet	Yes	Yes	1¼″
80 feet	Tore outer	No	No
	Yes	No	No
90 feet	Yes	No	No
	Yes	Yes (stayed)	No

Canadian Model 856 Air Rifle

Goals

The goals of the testing are the same as stated previously for the U.S. model. The testing is designed to help correlate the unknowns of the case (i.e., muzzle-to-target distance and number of pumps) with the knowns and facts of the case as follows: Determine the possible muzzle-to-target distance(s) required to produce penetrations similar to those known to have resulted in the death of Jimmy DeLauro with the Canadian model 856 air rifle in which the propellant force generated by pumping is limited by the manufacturer regardless of the number of pumps over 10.

Method

The rifle will be tested with the same fixed number of pumps at fixed distances from the target as the U.S. model 856. Success will be measured by measuring the perforation of the clothing materials and the penetration of the ballistics gel to a 1½- to 2-inch depth.

A chronograph will be used to measure the muzzle velocity generated by each set of pumps (again, 10 pumps and 30 pumps) at each distance from the target. (Note that muzzle velocity drops rapidly with such air-powered rifles firing nonaerodynamic projectiles such as round BBs, and muzzle velocity cannot accurately be correlated with target velocity.) The only reliable measure of target velocity and energy will be the actual perforation of the clothing material and the actual penetration distance of the ballistics gel.

Results

The results provide a range of values including specific distances plus specific pumps (10 or 30, or a finding that the design successfully limits the power of the rifle regardless of the number of pumps) sufficient to perforate the clothing materials and penetrate the ballistics gel to a 1½- or 2-inch depth.

Results of Canadian Model 856

Distance	Perforate Jacket Material?	Perforate T-Shirt Material?	Depth of Ballistics Gel Penetration?	Notes
		Penetration Testing at 30 Pumps		
20 feet	Tore outside	No	No	Weak
	Bounced off	No	No	Bounces off fabric
40 feet	Dent fabric	No	No	Leaves mark, dent
	Bounced off	No	No	
60 feet	Bounced off	No	No	
	Bounced off	No	No	No mark left
80 feet	Bounced off	No	No	No mark left
	Bounced off	No	No	
90 feet	Yes	No	No	Tear outer fabric
	Tears fabric	No	No	
		Penetration Testing at 10 Pumps		
20 feet	Bounces off	No	No	Weak flight
	Bounces off	No	No	More power
40 feet	Bounces off	No	No	
	In seam	No	No	Hits, remains in seam
60 feet	Leaves hole; bounces off	No	No	Tore but bounces off lining
	Bounces off	No	No	
80 feet	Bounces off	No	No	Very weak
	Bounces off	No	No	

Results of Canadian Model 856

Distance	Perforate Jacket Material?	Perforate T-Shirt Material?	Depth of Ballistics Gel Penetration?	Notes
90 feet	Yes: tear	No	No	Still bounces off
	No: bounces off	No	No	

Conclusions to Be Drawn from the Data Sets

The overall interpretations allowed by the data sets include, but may not be limited to, the following:

1. Anecdotal characteristics of the Daisy model 856 over 100-plus shots
2. Correlations among *distances* and *number of pumps* producing clothing and gel penetrations sufficient for the reported fatal injury (with the U.S. model 856 and the Canadian model 856)

Summary of the Project

The experiments in this case resulted in a scientific investigation that remained limited to the physical evidence, photographs of the physical evidence, and the medical reports. One must resist the temptation to overinterpret the data or to advance conclusions totally unrelated to the data derived from the experiment's design. It is useful to list the bases for any conclusions thought to follow from the work. Then one may check to ensure that the conclusions presented were derived from the data produced.

The Bases for Conclusions

1. Review of police and medical reports
2. Analysis of the victim's jacket
3. Experiments with the two exemplar Daisy model 856 rifles using Daisy BBs as received, together with the results of these experiments
4. Photographs taken during laboratory examinations of the two Daisy models 856, one for U.S. and one for Canadian markets

With the bases for all the conclusions clearly in mind, one can check to ensure that they are in fact based upon reliable data. When offering observations unrelated to the experiments or the issues of the case—distance, pumps, and penetration—it is useful to label them as such: anecdotal* observations.

* Anecdotal observations in the sciences are still based upon evidence, but they are based on a very limited sample—in this case, the rifles examined (only two). The observation did not warrant any experiment and did not qualify as a test of the rifle's behavior in general. It is an observation made with those limitations in mind—no more.

Summary of Conclusions

I report these observations and give them the appropriate scientific status by noting their anecdotal nature. If lawyers do not understand the scientific uses of the word, it is always good to explain it in other words. One problem with the intersection of science and the law has always been, and always will be, using the same word to mean two different things, depending upon its context in science or law. Another example is "approximation"—a necessary tool in scientific contexts meant to confer upon some measurement (or physical feature, for example) that it is accurate enough to be reliable for the task at hand or even that it is more reliable than an alternative given the goal. I'm told that in law, an approximation means an unacceptable stand-in for the real, true point and, as such, is both weak and questionable. And so it goes. (With respect to communication, scientists are from Mars and lawyers are from Venus.) It pays to learn legal language.

So, I present the conclusions from the work according to its scientific type as follows:

1. Anecdotal characteristics of the Daisy model 856. Both rifles allow any BB once entered into the breach to roll out of the muzzle if the barrel is pointed less than 90 degrees to the ground. Both rifles also exhibit somewhat arbitrary performance characteristics; for example, some shots at 10 pumps exit the muzzle with authority (higher velocity as measured on the chronograph), whereas some shots at 10 pumps using the same rifle exit with temerity. The result is a very uneven performance with both models. The BBs fired from both models do follow a reasonably straight trajectory from 90 feet through 20 feet.

   In addition, overpumping the U.S. model did not necessarily improve the power of the BB measured by target penetration. For example, the U.S. model pumped 30 times at 60 feet failed to penetrate the jacket or the gel, while the same model pumped 10 times at 60 feet not only perforated the jacket and T-shirt material, but also penetrated the ballistics gel 1¼ inches.
2. Correlations among distances and number of pumps producing clothing perforations and gel penetrations sufficient for the reported fatal injury (with the U.S. model 856 and the Canadian model 856).

   The most notable finding appears to be that although the Canadian model consistently reached the target, it failed to penetrate the ballistics gel at any distance. On the other hand, the U.S. model penetrated the ballistics gel to potentially fatal distances at ranges from 80 feet to 20 feet with 30 pumps and from 60 feet to 20 feet with 10 pumps, noting the aforementioned unreliability throughout.

Medical Data

Note that the trajectory of the BB in the victim as described in this report is explained from the anatomical position.

At the hospital, the emergency room physician stated that it was "obvious what the child died from" so no medical examiner was contacted about the case and no autopsy was performed.

The ER report simply stated, "Shot in the left chest with a pellet gun; GSW in the left ventricle with another wound in mediastinum," that is, the area between the lungs. (The organs in the mediastinum include the heart and its large veins and arteries, the trachea, the esophagus, the bronchi, and lymph nodes.)

After a rib spreader was used, the pericardium was seen to be bulging, an incision released a large clot; an entry wound into the left ventricle was observed and sutured. An entry wound was also seen in the mediastinal tissues (medially).

A chest X-ray showed an endotracheal tube in the right main stem of the bronchus and "that a 5-mm round metallic density was seen over the right heart border."

Summary of Injuries: Victim J. DeLauro

GSW	In	Out	Found	Trace: Wound	Trace: Bullet	Trajectory
A	Left of midline in upper chest	No	Pericardium or right atrium; upper right heart	No trace noted	None recovered or examined/ none presented for analysis	L→R; F→B apparently lateral and slightly downward?

Ballistics and Firearms/Bullet Examinations

Sometimes charts are good summary formats but sometimes they are just confusing to the court. It is difficult to know how such presentations will be received. For advice, I talk with a good friend who works as a special education teacher in the public schools. They have a way of forcing one to simplify and make things understandable despite ourselves.

Chart of Relevant Air Rifle Data

Evidence Pellet Identified as a Daisy BB

Weight: .52±.02; Grams: 8.5 ±0.2; Grains (Σ measured)

Minimum Velocities

Skin perforation: 331 ft/sec

Bone penetration: 200±50 ft/sec

Bone perforation: 600 ft/sec

This is a subsonic round resulting in virtually no noise when fired. (Speed of sound = 1080 ft/sec at sea level: air rifles usually "pop" when fired due to the spring and the air pressure. (No sonic studies were done in this case.)

Lighter weight projectiles of any sort need higher velocities to perforate skin and bone than heavier projectiles. Upon penetration, the pellet does not lose all of this minimum velocity, but does lose up to 36%.

Summary of Test Data

Distance	Perforate Jacket Material?	Perforate T-Shirt Material?	Depth of Ballistics Gel Penetration?	Notes
		20 feet		
U.S.				
10 pumps	Yes	No	No	
	Yes	Yes	2½″	
30 pumps	Yes	Yes	3¾″	

Summary of Test Data

Distance	Perforate Jacket Material?	Perforate T-Shirt Material?	Depth of Ballistics Gel Penetration?	Notes
Canadian				
10 pumps	Bounces off	No	No	Weak flight
	Bounces off	No	No	More power this time
30 pumps	Tore outside	No	No	Weak
	Bounces off	No	No	Bounces off fabric
		40 feet		
U.S.				
10 pumps	Yes	Yes	3½″	
30 pumps	Yes	Yes	3½″	
	Yes	Yes	2½″ to 3″	
Canadian				
10 pumps	Bounces off	No	No	
	In seam	No	No	Hits, remains in seam
30 pumps	Dent fabric	No	No	Leaves mark, dent
	Bounces off	No	No	
		60 feet		
U.S.				
10 pumps	Yes	Yes	1¼″	
30 pumps	Yes	No	No	
Canadian				
10 pumps	Leaves hole, bounces off	No	No	Tore, bounces off lining
	Bounces off	No	No	
30 pumps	Bounces off	No	No	
	Bounces off	No	No	
		80 feet		
U.S.				
10 pumps	Tore outer	No	No	
	Yes	No	No	
30 pumps	Yes	Yes	1½″	
Canadian				
10 pumps	Bounces off	No	No	Very weak
	Bounces off	No	No	Very weak again
30 pumps	Bounces off	No	No	No mark left
	Bounces off	No	No	
		90 feet		
U.S.				
10 pumps	Yes	No	No	
	Yes	Yes; stayed	No	
30 pumps	Yes	No	No	
	No	No	No; bounced off	

Summary of Test Data

Distance	Perforate Jacket Material?	Perforate T-Shirt Material?	Depth of Ballistics Gel Penetration?	Notes
	Yes	No	No	
	Yes	No	No	
Canadian				
10 pumps	Yes; tear	No	No	Still bounces off
	No; bounces off	No	No	
30 pumps	Yes	No	No	Tear outer fabric
	Tears fabric	No	No	

Gel Block Testing

Figure 5.7 shows the perforation of the jacket material and T-shirt material after 10 pumps at 60 feet. One BB penetrated 3¾ inches, another 2½ to 3 inches. Perhaps the most surprising result was a penetration of 1½ inches at 80 feet with the U.S. model pumped 30 times (Figure 5.8). Figure 5.9 illustrates the perforation of the jacket and T-shirt material with the penetration of the gel block. Another shot appears on the left as noted and the shot corresponding with the clothing defect appears on the right. Another surprising experimental result occurred with 10 pumps of the U.S. model at 90 feet (Figure 5.10). The BB pushed the jacket insulation through the T-shirt.

Figure 5.7 Testing the rifle with clothing materials and the gel block.

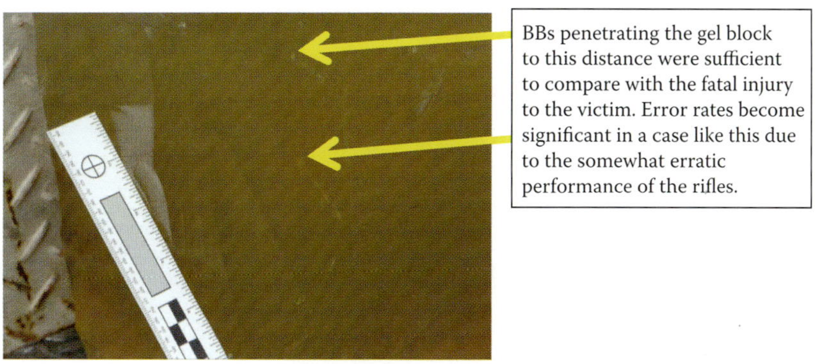

Figure 5.8 Penetration was sufficient for the fatal injury to the heart.

In Figure 5.11, note the lower BB's penetration of about 3½ inches, more than sufficient to kill a victim if striking soft tissue in front of the heart. The gel is mixed to approximate human tissues and organs for the purpose of measuring bullet penetrations, trajectories, and damage by observing and recording the temporary and the permanent cavity. Only the permanent cavity was considered in these experiments.

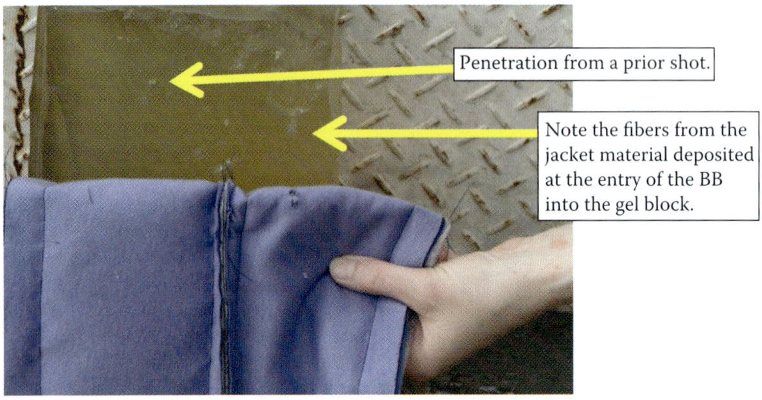

Figure 5.9 Gel block with clothing used to test the perforation of the clothing and the penetration distances into the gel block.

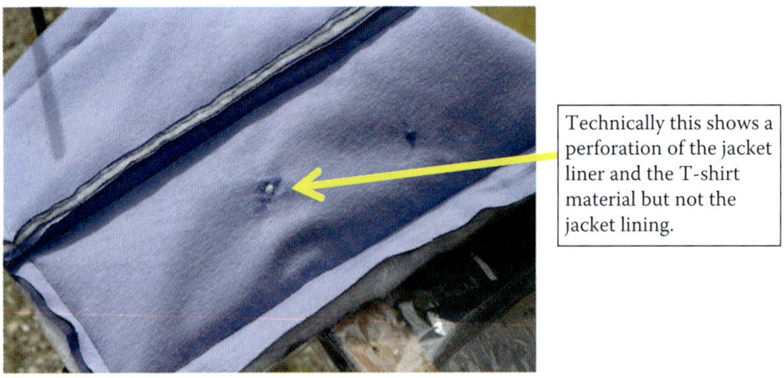

Figure 5.10 Clothing over the gel block—testing the distance penetrated by the BB into the gel block.

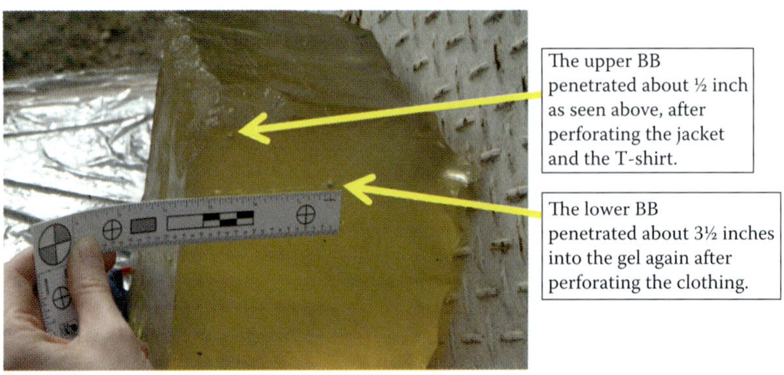

Figure 5.11 Clothing perforation and gel block penetration testing.

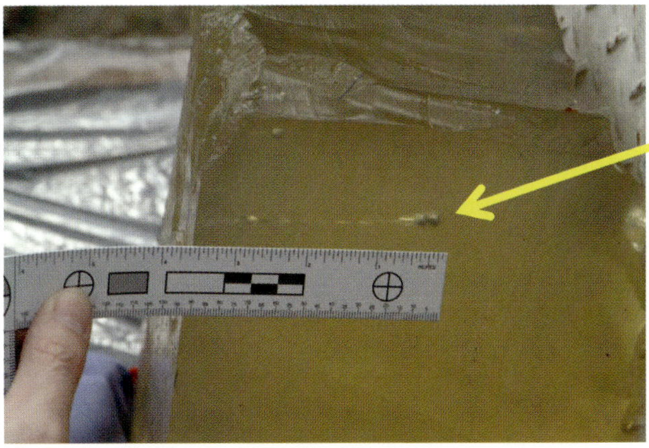

Figure 5.12 Evidence of an amazing (and coincidental) shot.

All my practice behind the garage shooting my Red Rider BB gun at my Grandmother's Mason jars as a child paid off. I was able to shoot two BBs through exactly the same permanent cavity, with the second BB striking the first, fired at a distance of 40 feet (Figure 5.12). I hesitated to tell my lab assistant that this event showed more luck than skill. I hate to think that as a kid I shot up all those valuable antique Mason jars for nothing.

Further Analysis of the Rifles

This testing was performed at 137.5 feet above sea level at 62°F with a barometric pressure of 1010 millibars and winds west-northwest WNW at less than 5 mph. All experimental conclusions based upon the data developed assume that these conditions are satisfied, unless explicitly stated otherwise.

As often happens in any shooting reconstruction, I was asked to disassemble both the U.S. Daisy model 856 and the Canadian Daisy model 856 rifles and document any design differences.

Both U.S. and Canadian versions of the Daisy model 856 initially appear and operate identically. They have the same exterior color, shape, and operator configurations, including safety, trigger, sight, and pump locations, as well as the same operational mechanisms. The first easily observable difference appears on the pump piston of each model when turning the rifles over. The Canadian model piston appears to be shiny bright metal, whereas the U.S. model piston appears to be black. Although both pistons exhibit similar rubberlike bushings and O-rings, this difference in appearance points toward the significant design difference distinguishing the two otherwise identical models.

The major components of both models are housed under left and right black plastic covers secured by plastic tabs and screws. Further disassembly reveals plastic, rubber, and metal components including those pictured later. Both models appear to have identical metal pumping chambers, with oiled rubber bushings and O-rings sealing the chamber and connecting it with the breach and the trigger mechanisms. The difference between the two models appears to rest with the design of the piston, which when pumped by the operator, fills the chamber with compressed air.

The Canadian model's piston has a 1.5-ml cavity cut into its end (see following photos). The U.S. model's piston appears flat on the end, having no such cavity (again, see following photos). The cavity in the Canadian model's piston functions to limit the capacity of the pump chamber. This ingeniously simple design will limit the amount of air compressible into the chamber, regardless of the number of times the piston is pumped by the operator. Whenever the chamber reaches its maximum fixed pressure level, the 1.5 ml chamber in the piston consistently withdraws 1.5 ml of air under greater pressure than the air that would be drawn into the chamber by the piston with each new pumping mechanism. This prevents additional air from entering the chamber and being compressed by the piston's action. The actual size of the cavity, the distance of piston travel, and the size of the pump chamber will determine the average pressures remaining in the chamber regardless of the number of times the piston is pumped by the operator.

The U.S. model's piston, however, has no such cavity. As a result, the piston will reach different and higher average equilibrium pressures (given indefinitely many possible pumps) since no action of the piston (no cavity) serves to reduce the available air supply into the chamber by presenting air from the chamber at a greater pressure than air from the atmosphere that continues to be pumped into the chamber by the piston's action.

As far as I can determine by making basic observations, the design and function of only one simple part distinguishes the two models.

Photos Comparing U.S. and Canadian Model Daisy 856 Air Rifles

See Figures 5.13 and 5.14.

Plastic housing removed from both models, the trigger pull data, and the comparisons for both models were made available if they were needed. One must be prepared to defend either the relevance or irrelevance of any data developed in the course of the analyses. Here with 20 tests each, the trigger pull data were virtually identical for both models. But that applies, as all the data developed, to these two rifles. One must examine a few more to reach any extended conclusions about the properties of the generic models. Further disassembly reveals components of trigger mechanisms (Figures 5.15 and 5.16).

The Canadian model (Figure 5.17) has a piston that appears to be different from the U.S. model (Figure 5.18), although both pump chambers appear to be identical with apparently identical components.

Figure 5.19 shows a piston present in the pumping chamber of the U.S. model.

Figure 5.20 shows the chamber from the Canadian model.

Figure 5.13 Canadian model disassembly.

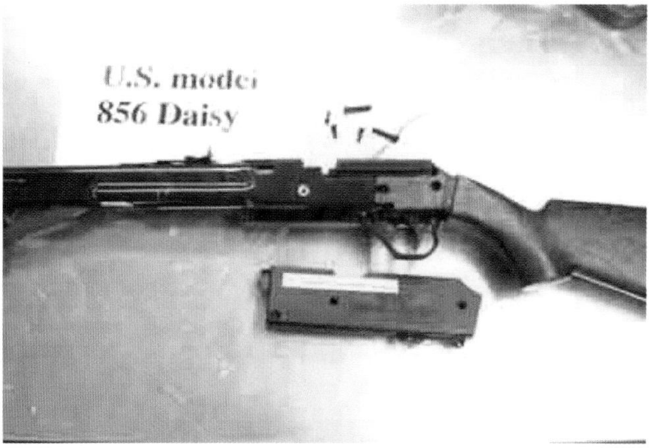

Figure 5.14 U.S. model disassembly shows similar components.

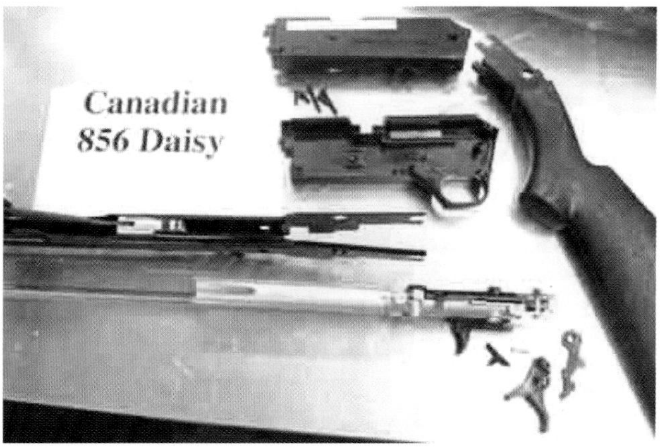

Figure 5.15 Trigger components and the Canadian rifle parts.

Figure 5.16 Trigger and rifle components of the U.S. model.

Figure 5.17 Canadian piston and pump chamber.

Figure 5.18 U.S. piston and pump chamber.

Figure 5.19 Piston in the pumping chamber of the U.S. model.

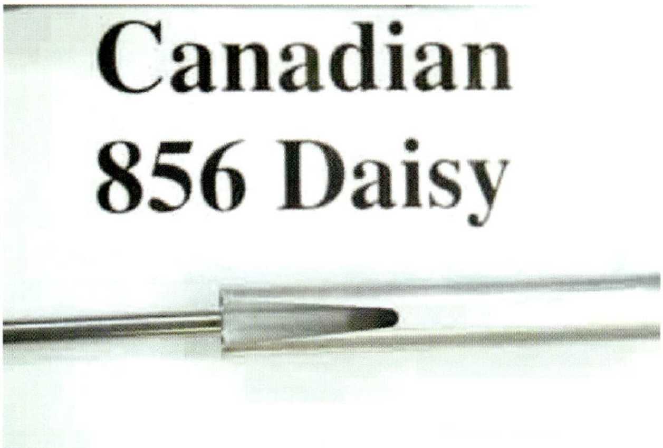

Figure 5.20 Pumping chamber of the Canadian model.

Figure 5.21 Pumping chamber.

The Canadian model's pumping chamber and the oiled rubber bushing and O-ring securing the chamber to the breach/trigger mechanism appear in Figure 5.22; the U.S. model appears in Figure 5.23.

The U.S. model's pumping chamber and the oiled rubber bushing and O-ring securing the chamber to the breach/trigger mechanism appear, upon observation alone, to be identical with the Canadian model.

The U.S. piston (Figure 5.25) and the Canadian piston (Figure 5.26), however, are different. The Canadian piston has a 1.5-ml cavity in the end. This end enters the pumping chambers.

Figure 5.27 best shows the design difference in the piston pump between the two models. Note the 1.5-ml "cup" cut into the piston on the Canadian model Daisy 856 Air Rifle.

Comments Regarding the Limits of Both the Data and the Conclusions

An error rate should, at least in principle, apply to most scientific projects. Informally stated, two sorts of error rate are at issue: error rates attributable to the method and error

Figure 5.22 Canadian model's pumping chamber and oiled rubber bushing.

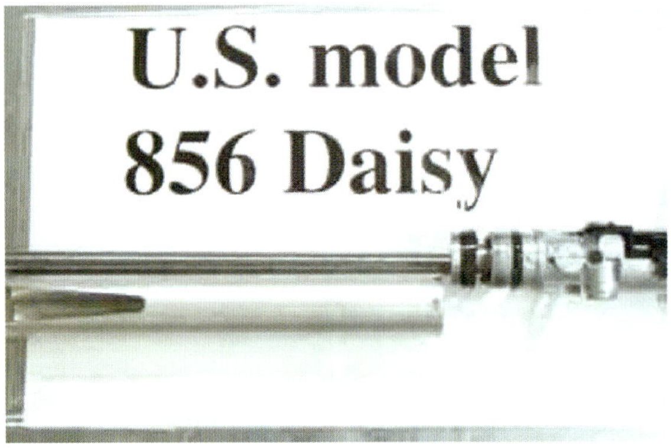

Figure 5.23 U.S. model's pumping chamber and oiled rubber bushing.

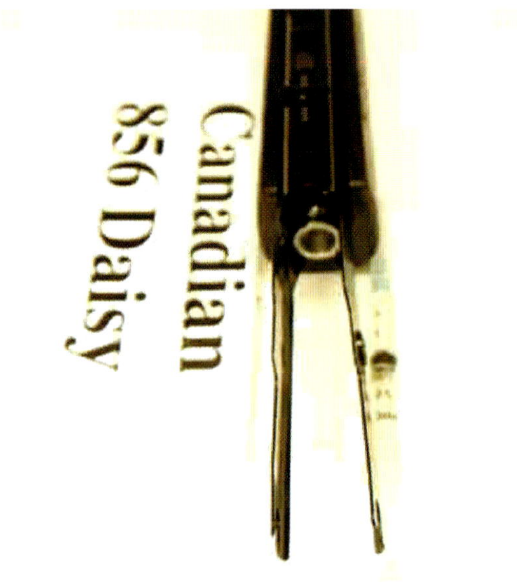

Figure 5.24 Canadian piston and pumping chamber.

Figure 5.25 The U.S. model piston.

Figure 5.26 The Canadian model piston with a 1.5-ml cavity in its end.

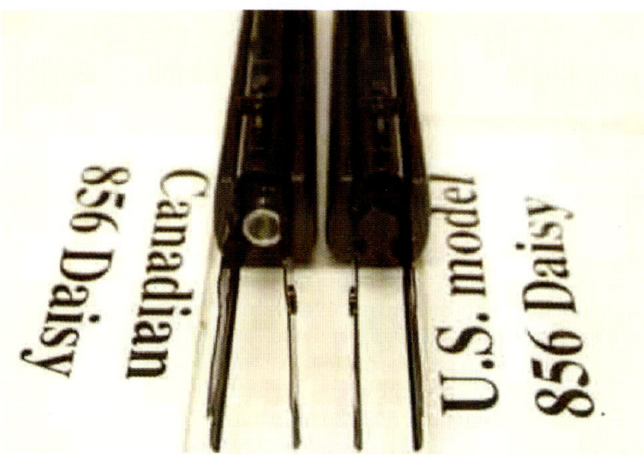

Figure 5.27 Design difference between the two models involves piston design.

rates attributable to the application of the method. A useful rule of thumb regarding error rates is to remember that anytime a measurement is involved in the work, error rates of both types apply. For example, using an American Board of Forensic Odontology (ABFO) ruler involves both types of error rate. The first error rate involves the error built into the manufacture of the ruler (or measurement apparatus) and its *accuracy*. The second error rate involves the error contributed by the user of the rule or measurement apparatus and the resulting measure of its *precision*.

Without a great deal of research and testing, determining the error rate built into the manufacture of a ruler, say, is very difficult to estimate. That error is not simply a statistical measure, but a measure involving the comparison of the ruler and its similarity with other rulers of the same design,* and their comparison with a known correct standard. Sometimes equipment manufacturers of various types provide this data. Sometimes they do not. It is important to stress that error rates are not mistakes; they are acceptable variance that becomes part of the data. They are, in short, one measure of our awareness of method and its application to the issues at hand. They become one more thing that is important to consider, regardless of error type.

In another form, scientific shooting reconstructions (even a very small part of one, such as the work in this case) require clarification so that the use of the report has some chance to be clear for the attorneys.† My work covered the function of both rifles, and a description of their designs, not an evaluation of their design qualities. Nor did it include an assessment of their lethality. (Of course in knowledgeable hands, even a pencil can become a deadly weapon.) The issue here involves an air rifle and not general "weapons." As far as the science applied here goes, the conclusions are limited to the following: The U.S. model can, under certain circumstances, perforate the described clothing and penetrate human tissues up to the measured distances, whereas the Canadian model cannot. Whether that makes one model *better* than the other is not subject to scientific analysis with the data provided. The issue appears to be a matter for ethical consideration, at least in the context that the question is posed.‡

Litigants usually attempt to push the scientist toward advancing opinions favoring their causes, whether or not the grounds for the opinion develop from the scientific work performed or not. It is important to stress that the work done on this case does not cover or establish the merits of any design or the defects of any product. It merely establishes the performance characteristics of two air rifle models of similar design while noting their differences. Whether someone chooses to find one design preferable to some other remains outside the scope of any shooting reconstruction. Despite the nudges and pushes to advance

* With a colleague, Eugene Liscio, I decided to evaluate the range of measurement differences and similarities among our laboratory's various ABFO scales as produced by various manufacturers. We found that among the 30 or so rulers used in the lab, the variation averaged about 1.5 mm either plus or minus among the 30 rulers, which is certainly an error when compared to a standard known to be true and correct. It is also classed as an "acceptable variance" when doing measurement work while using such an admittedly crude metrological instrument. Of course our study was informal and anecdotal. Nonetheless it makes a point. If accuracy is required, other tools are preferable.

† Of course, one must realize that often some attorneys see their role as making the clear into the unclear, the true into the false, and the reliable into the bogus. One recalls the activities of the Sophists in ancient Greece as described by Socrates throughout *Plato's Dialogues*. All the scientist can do is stubbornly stick to science with its honest limitations as well as its insightful advances based upon the data. More is impossible. Less is unconscionable.

‡ It is not a practical judgment—one is better for killing mice, rats, and rabbits than the other, for example.

a specific value judgment about each or about both products, the data remain to be the limit of the work. Goodness or badness does not lurk somewhere among the data alone.

Jimmy DeLauro is dead, either way. His family will suffer his loss forever, especially his brother. Time fails to heal some pains. They are just too severe. This is one of them. The debate will continue about who should bear the blame: the object, its manufacturer, its users, or those who provided the object to a child. I remember my Red Rider BB gun and the stern warnings that accompanied its delivery to me from under the Christmas tree. Perhaps accepting the suffering, and enforcing the warning are both parts of being a parent, that never get easier.

Hold Me Closer, Tiny Dancer

6

Truth forever on the scaffold; wrong forever on the throne.

James Russell Lowell, "The Present Crisis"

Prejudice is the reason of fools.

Voltaire

Introduction

Leroy Martin had a fulfilling Saturday. He reviewed the details, enjoying his weekend once again as he walked down the middle of Central Avenue on a cold, very early Sunday morning. It had been a good trip, a great experience. His much-prayed-for dream was finally coming true. He was on his way. The audition went very well, despite his handicap. Yes, he was overcoming! And spending time with Clarence in his small apartment had always been special. He continued walking west, his hands in his pockets, carrying his black bag over his left shoulder, leaning forward with his cap pulled over his ears and his eyes down. A cold wind blew steadily in his face. Snow still covered the ground. The pavement was dry. He continued talking to himself, walking down the road's double yellow line.

Unlike many of the young black males that he had grown up with from his neighborhood, Leroy Martin had no criminal history, no wants, and no warrants. His much-loved family was important to him. School remained important to him. His dream stayed even more important. His faith in the Lord was the most important of all. He wore his heavy orange jacket and black nylon warm-up pants. Both garments covered his lithe, slight body.

The location on Central Avenue is a mixed residential and commercial area, not heavily populated at the time of the shooting: a railroad track appears to the north and a garage is to the south. It remains unclear who called the police, reporting a "young black man walking suspiciously down the middle of the street talking to himself." Four squad cars responded within minutes.

No less than eight officers established tactical positions, both in front of and behind Martin. One of them, Officer Bud Stevens, positioned himself in the road just to Martin's left. Stevens repeatedly requested that Martin comply with police instructions to stop and to take his hands out of his pockets. But Martin continued to walk down the middle of Central Avenue, and he kept talking to himself. As the distance between the two closed, Stevens pulled his Glock 40 from his service holster. The double tap took everyone by surprise. Martin dropped. His dreams died with him.

Official Stories

The department and the prosecutor's office both concluded that the shooting was justified. Officer Bud Stevens said that he feared for his life as Leroy Martin continued to approach him. Stevens further stated that Martin had two glass bottles, one in each hand, which he continued to wave "in a threatening manner." The bottle in his left hand, Stevens said, had its bottom broken out and became a jagged sharp weapon. He said that Martin refused to comply with official police instructions. The seven other Pond Valley officers confirmed his story. Leroy Martin was a threat. He was taken down. End of story.

Well, not quite.

My Assignment

An attorney acting on behalf of the Martin family contacted me. The family members simply wanted to know what had happened to their son, brother, nephew, cousin, and friend, Leroy Martin. They could not understand the official pronouncements. Martin's interest had been focused on ballet, not on the violence accompanying gangs, drugs, and guns. I was asked to develop independent answers and a scientifically based account of the shooting. The request included a nonscientific project: an assessment of the police department's use-of-force policy and its implementation in Martin's shooting death.

To begin, I reviewed police, autopsy, laboratory, and follow-up reports, including scene and autopsy photographs, which might specifically help develop evidence in this gunshot homicide. Specifically, first, I was asked to determine the decedent's position relative to the shooter, specifically his arms and hands, when he was shot. Second, the nonscientific component of the request, I was asked to evaluate the department's use-of-force policy as applied to any reconstruction that I developed. Of course, the ability to answer these or any questions directly relates to the ability to recover information and to develop data for analysis.

I received items of documentary and physical evidence through the attorney's efforts. These items were read, examined, or analyzed, as appropriate. When an event has already occurred and the scene and body are long gone, all one has to work with are any documentations of the event, the collected physical evidence, and, of course, photographs usually of varying quality and significance. The material eventually finding its way to the laboratory included the typical fair for such cases. I usually list each item when presenting any written report.

Files, Transcripts, and Reports

Approximately 6000 pages of documents related to this case were provided and considered. I specifically note the most significant documents in most instances since much of the testimonial evidence provided in transcript form provides little value for a scientific reconstruction. Of course, there are always exceptions and one must certainly consider each document with those exceptions in mind.

- Autopsy report
- Police reports

Autopsy and Scene Photographs

- Autopsy photographs
- Police scene photographs
- Scene measurement notes and scene drawings
- Scene videotape

Physical Evidence

Items of physical evidence organized by a simple numbering system make references to the item in written reports and in courtroom testimony somewhat easier. But often, collected items receive a number from the police department, another number from the crime lab, and yet another number from the prosecutor's office. Nonetheless, one must be familiar with both the item and its associated numbers.

Item Number and Item Name

1. Received with physical evidence (white envelope) with photos and documents
2./3. One bag marked "Items 2 & 3"
4. Box with E&J Brandy bottle
5. Box with Old English label bottle, broken
6. One bag with "Black bag containing misc. items"
7. One Mommi & Me candies box with unknown brown/green substance
8. One plastic bag of "unknown green substance" from the black bag in Item #6
9./14. One bag labeled "Items 9 & 14"
10. One bag labeled "One (1) orange jacket"
11. One bag labeled "One (1) pair black nylon pants"
12. One envelope labeled "One (1) set of inked fingerprint impressions"
13. One envelope labeled "Two (2) test tubes"
14. One Glock 40 pistol from PD Officer Bud Stevens

Decedent's Clothing as Follows

1. Burgundy-colored sweatshirt (appears under open orange jacket)
2. Undershirt
3. Underwear
4. Shoes, left and right
5. Socks, one pair

Scene of Shooting

The area was not heavily populated, provided no areas of cover or concealment, and the weather was good, as can be seen in the photographs of the scene (Figure 6.1 to Figure 6.3). The items of evidence appear with markings in Figures 6.2 and 6.3. Martin has been turned from lying on his chest to lying on his back. I review the autopsy first, and then make an effort to understand the injuries and apply that understanding to an examination of the decedent's clothing. I began with the orange jacket, the outermost upper garment.

Figure 6.1 The decedent on Central Avenue with the lens capturing the view looking north toward the railroad tracks.

Figure 6.2 The decedent on Central Avenue with the lens pointed toward the southwest and the buildings.

Figure 6.3 The decedent on Central Avenue with the lens pointing east down the road in the direction from which Leroy Martin came.

Evidence Examination: Item #10 Orange Jacket

Taking the time to make an initial review of the evidence prepares one for the examinations, in their myriad forms, yet to come. Without attempting to make artistic statements, drawing the items focuses observation and demands a careful and critical eye from the observer. I resist the temptation to use generic template-style representations of items: coat, pistol, T-shirt, jacket with front snaps, and so on. I have noted many features of the item by drawing my own representation, one item at a time, for each item relevant to a case.

But no single way to develop such representations is the right way. Such a claim would reduce a complex scientific step—making observations—to a ridiculously simple level. Drawing is merely one way to help accomplish the task.

Most recipe, flowchart, or prescriptive process approaches to developing crime scene reconstructions are best viewed as postdevelopment descriptions of complex processes that provide little instruction for their creation or purpose. These approaches seem to me, at least, to confuse producing a list of steps with applying a method.*

Multiple bullet holes appear in the orange jacket (Figures 6.4 and 6.5). A bullet entrance appears in the left side of the chest at the zipper about 18⅞ inches from the waistband at the left side of the zipper. The bullet damages the zipper, showing that the zipper was zipped when the bullet entered. This bullet was recovered in the decedent's lumbar spine at the autopsy.

A bullet entrance appears in the lateral left forearm about 5 inches from the cuff, exiting about 6 inches from the cuff medially; a secondary bullet entrance appears on the left side of the jacket about 3 inches above the waistband. A single projectile produced these exits and entries. Articulating these exists and entries with the nylon pants entry and exit (see Item #11 later) allows us to estimate the trajectory of this shot and the approximate position of the left arm when struck by Officer Stevens' bullet.

Elemental Analysis Methodology

The orange jacket was photographed, examined microscopically, photomicrographed, analyzed with infrared (IR) light at 715 nm, photographed with IR film, and then subjected to X-ray fluorescence (XRF) spectroscopy analysis, as described next.

Gunshot residue analyses were conducted on Item #10, the orange jacket. Three nondestructive methods were used: optical microscopy for visual particle identification and comparison with other possible trace items, such as cigarette ash and automobile emissions, together with microscopic measurement of the particles; XRF for both qualitative and quantitative elemental analyses; and infrared photography for qualitative detection and potential gunshot residue (GSR) pattern analysis. All three testing mechanisms are well established in the field of gunshot residue analysis and are considered nondestructive, allowing for the repetition of testing as well as for additional testing in the future.

* I do not mean that such recipe, flowchart, or prescriptive processes lack value. I mean that they are not to be confused with what we mean by a method, particularly when such lists are meant to instruct students in the nature and application of scientific methods. They seem to lack any instructive power or purpose for the application or instantiation of the items composing the list. So drawing should not be viewed as just an item on a list but one way of developing the ability to make careful observations.

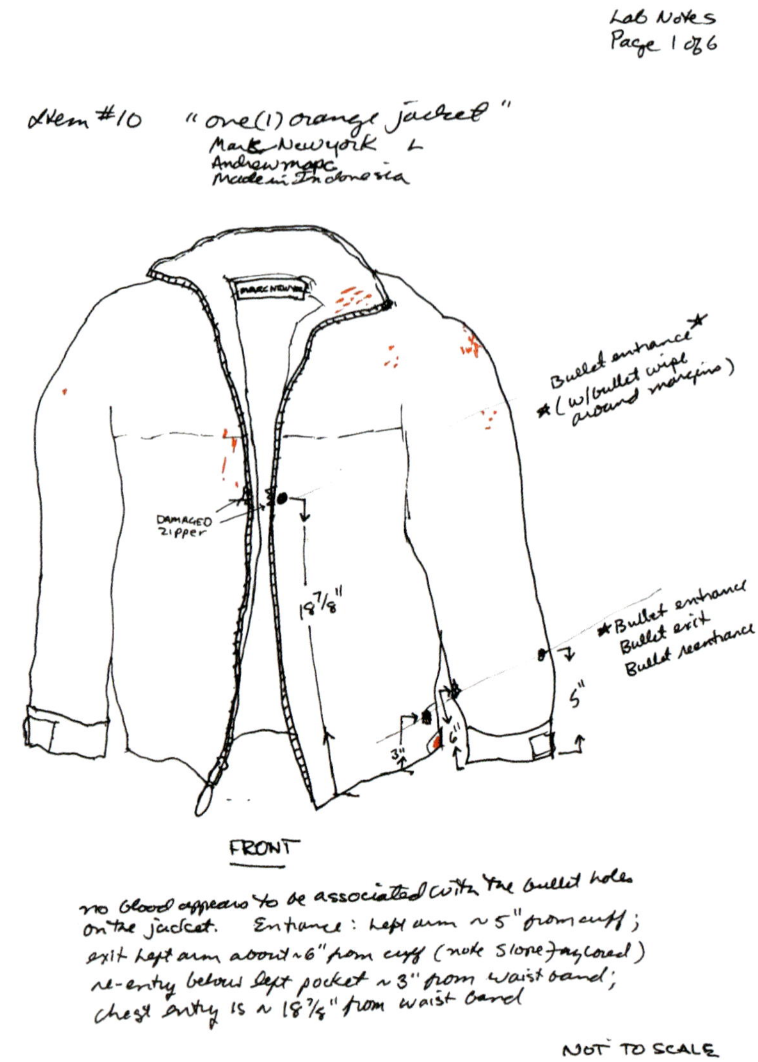

Figure 6.4 I think that to draw something is to see that thing—my way to help make observations—not great art, not to scale, just a way to look and to capture what you see. My drawing of the front of the orange jacket.

XRF analysis is similar to energy-dispersive X-ray (EDX) spectroscopy analysis typically used in scanning electron microscopy (SEM). One relevant difference includes the ability to visualize particles with an SEM, while XRF provides only qualitative spectra via graphs and quantitative analyses based upon determinations of the parts per million (PPM) concentrations of selected elements.

XRF analyses provided both quantitative detection (how much of each statistically detectable element) and qualitative detection (presence or absence) of primer residues, which form part of any GSR distributed during a firearm's discharge (Figure 6.6). But the tests are considered presumptive only. They cannot be used to establish the presence of GSR. Thus, to establish known elements for GSR identification on the aforementioned clothing items, known samples were taken from the primers of two fired

Figure 6.5 My drawing of the back of the orange jacket.

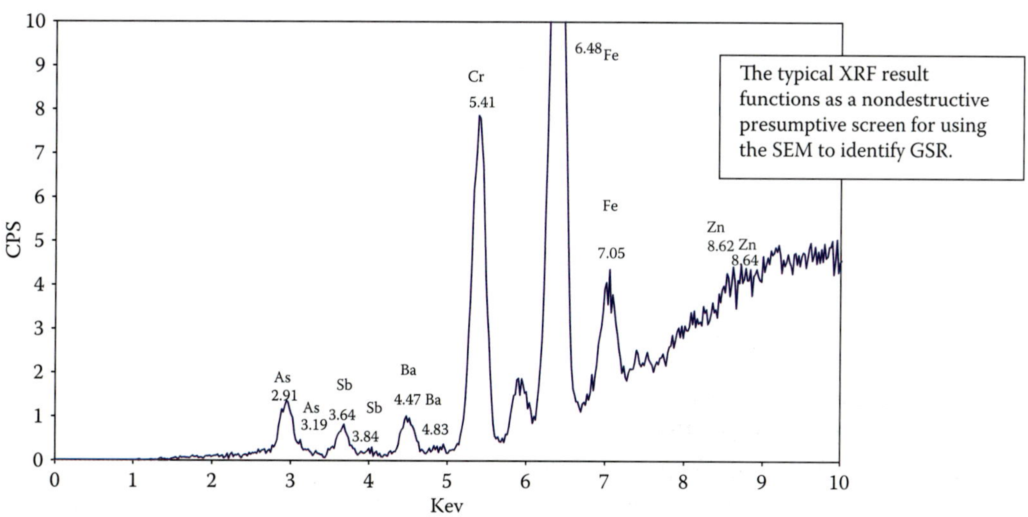

Figure 6.6 XRF spectra as a presumptive test for metals.

Glock 40 caliber cartridge cases (cartridges identified as those having been fired at the decedent). These knowns were then tested both qualitatively and quantitatively using XRF to identify the chemical elements in question. The only reliable way to determine the presence of GSR to date remains the electron microscope. The samples were then run on the SEM.

The orange jacket was tested both qualitatively and quantitatively by XRF for the presence of any of the relevant chemical elements. With the significant elemental components from both primers identified via XRF, the results of the XRF analyses on the clothing can then be assessed and potentially interpreted using the SEM. These assessments, together with results from both microscopic and IR analyses, allow one to determine if the clothing items were present when the relevant firearm or firearms were discharged. Given the identification of certain GSR patterns via IR photography (if the GSR is confirmed as such by the SEM), the mechanism of some GSR distributions may be revealed in the patterns on the clothing. In this case, the IR photos did not reveal an apparent pattern sufficient for scientific interpretation. The SEM confirmed its presence, but the particles were randomly spaced and few in number.

XRF Item #10 Orange Jacket

Analysis using XRF and the SEM shows that the jacket was exposed to GSR, but neither the concentrations nor the IR images are sufficient to help determine muzzle-to-target distances. Therefore, firing the Glock 40 (which fired the fatal shots) at known distances from fixed targets to determine range-of-fire distances proves unnecessary in this case. The shots were fired from Officer Stevens' service pistol from a distance greater than 3 feet muzzle to target. The GSR could have been deposited on the orange jacket by any number of mechanisms. More cannot be said with any scientific reliability.

Bloodstain Pattern Analysis

Bloodstains on the jacket consist of relatively small stains on the front, including stains on the chest, collar, and both shoulders, which I tried to capture in my sketches and notes. Martin was wearing multiple layers of clothing underneath the jacket and he had on a pair of gloves. This clothing would soak up any blood exiting through the chest wound as well as any bleeding from the left arm. The small stains on the coat were probably produced when he was turned from lying face down to face up, or otherwise moved. Back spatter and forward spatter from a perforating wound could play a role, but layers of clothing often mitigate the effects. His left glove would have been important evidence, as blood that might have soaked into that glove might have produced some stains when his arm(s) were moved. No stains appear on the inside of the jacket's left arm. The gloves were "lost" by the police department. His arm position was one important key to the case. Where were they when he was shot? Holding a glass weapon? Or holding nothing?

The left side and rear left side of the jacket have soaking stains along the waistband. These soaking and wicking stains resulted from the jacket's contact with the blood on Central Avenue. The bloodstains appeared only near the double yellow lines on the road.

Evidence Examination: Item #11 Black Nylon Pants

Physical Description

The black nylon pants are described as 100% nylon shell, with 65% polyester and 35% cotton lining. They have a Chicago Bulls logo on the front.

A bullet entrance appears on the front left thigh area about 6$^{11}/_{16}$ inches from the waistband and about 2⅜ inches from the midline seam (pictured in both drawings [Figures 6.7 and 6.8] and photos). The bullet entrance measures slightly less than 1.0 cm in diameter. A yellow mark appears on the right leg near the knee. This yellow mark probably originated when Martin's right knee struck the double yellow line on Central Avenue.

A bullet exit appears on the left rear thigh area about 12⅝ inches from the waistband and about 3$^{9}/_{16}$ inches from the midline seam. The bullet exit measures about 1.5-cm vertically and about 0.8-cm horizontally, further indicating a tumbling or otherwise distorted round that struck a prior target (arm). This bullet was not recovered at the scene after it exited from Martin's left thigh.

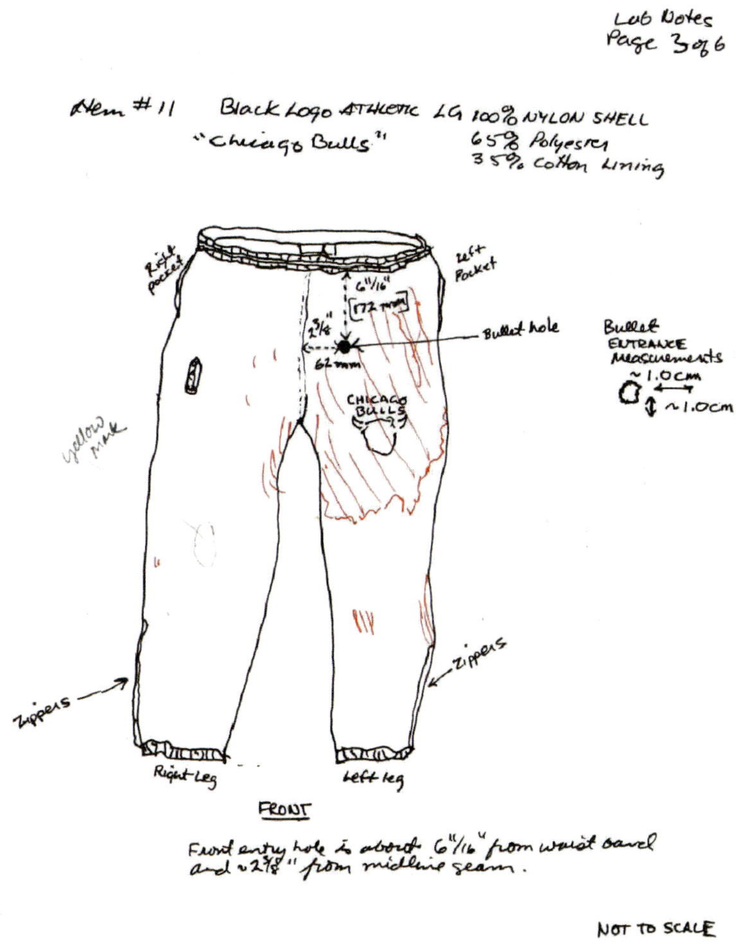

Figure 6.7 Drawing of the front of the black nylon pants.

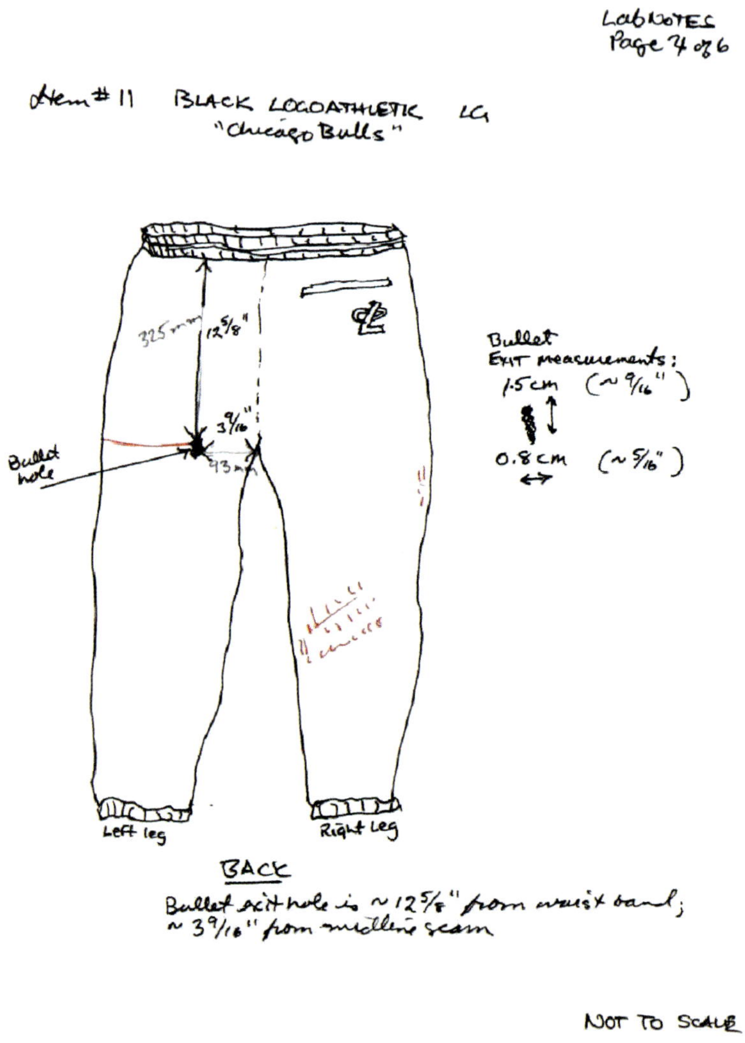

Figure 6.8 Drawing of the back of the black nylon pants.

Bloodstain Pattern Analysis

Blood soaks through the front upper left thigh area onto the right thigh midline area. This large stain resulted when the decedent lay on his face on Central Avenue while bleeding. This indicates that Martin lived for some undetermined period after suffering both gunshot wounds since this bleeding requires some continued heart activity. The other bloodstains observed on the pants resulted from moving the decedent for some purpose, such as EMT activity or transport to the medical examiner's facility or some other purposes that remain undiscoverable. We can determine some movement, but we cannot determine the intent behind the movement. It pays to remember the limits of scientific reconstructions. Intent might easily be inferred, but that is not what we need to state as a scientific inference.

Trajectory Analysis: Item #10 Orange Jacket, Item #11 Black Pants, and Autopsy Data

I use mannequins of various statures to display such items in the lab. I used a 6-foot articulated mannequin as a stand-in for Martin since he was 6 feet tall. It is not perfect, but it produces reliable scientific results with a range of acceptable variance for the purposes at hand. To set up the mannequin, I often refer to scene photos as well as the autopsy photos. If there is a great deal at stake in providing accurate and precise measurements, I will take the time for the activities described in Chapter 1. Scientific methods require data, but just because a tool in available in our scientific tool belt does not justify its application in every case to develop the data. One must continually monitor the purpose at hand and the time available to complete the work.*

Scene Photographs

Evidence of Shots Fired

Two shots were fired at and struck Martin as described in the autopsy report (Figure 6.9). Using the officer's Glock for an ejection pattern test could be slightly helpful, but the patterns are not regular and an exact conditions test in this case proved impractical. That happens often. The lives of others must go on—traffic must flow, blood must be cleaned up, and the scene must disappear. A crime scene can be viewed as a tiny slice in time, captured by photographs, descriptions, and measurements. After a short time, that is all there is left.

Bloodstain patterns on Central Avenue show that Martin fell to the street striking his right knee on the right yellow line, with his left thigh resting on the left yellow stripe.

* On the stand in an infant bathtub drowning case once, the infant was left alone in a bathtub splashing about in 1 inch of water. The faucet continually leaked a small stream of water. The caregiver said that he left her alone "for a few minutes." When I arrived, the infant was face down and the tub was overflowing onto the floor. Time left alone was an issue. I described to the court how I took an empty orange juice pitcher with volume marks, held it under the stream of water exiting the faucet, and looked at my watch. Using the great tool of calculus, I calculated the volume of the tub. The rest of the math told me that the infant was left alone in the tub for at least 3½ hours. The attorney asked, "But Dr. Nordby isn't science supposed to be exact?" I answered, "No. Its methods must be reliable to accomplish the task at hand." That threw him a bit.

Figure 6.9 Items labeled 2 and 3 are the two cartridge cases from Officer Stevens' Glock 40.

The black bag was removed from its position over his left shoulder by cutting the shoulder strap. The bloodstains show that Martin was rolled over from lying face down to lying face up. Note that his arms remain at his sides (Figure 6.10). Once moved, all bets are off, but the clothing always has some story to tell.

Limited Trajectory Analysis

Chest Shot

Figures 6.11 and 6.12 show the location of the entry wound described and measured at autopsy, and the bullet's recovery point in the lumbar spine. The distance from entrance to recovery point is about 6 inches and Martin's torso was approximately 5 inches thick from front surface to back. Using simple geometry, then, the angle of the shot can approximately be determined.*

Figures 6.13 and 6.14 show the approximate trajectory of the bullet striking Martin in the chest. The measured angle is about a 40-degree declination, the same as the 38 to 40 degrees calculated geometrically. One must remember that this angle appears in the anatomical position: Martin may have been bending about 50 degrees forward when struck in the chest if Officer Stevens fired his weapon approximately parallel with the road surface.

* Error rates apply, of course, to all measurements including angle determinations. For the purposes here, they are reasonable scientific approximations sufficient to address the questions about the decedent's position when shot.

Figure 6.10 Victim on the street.

Figure 6.11 Autopsy measurements (once verified) are added to the mannequin with clothing in place.

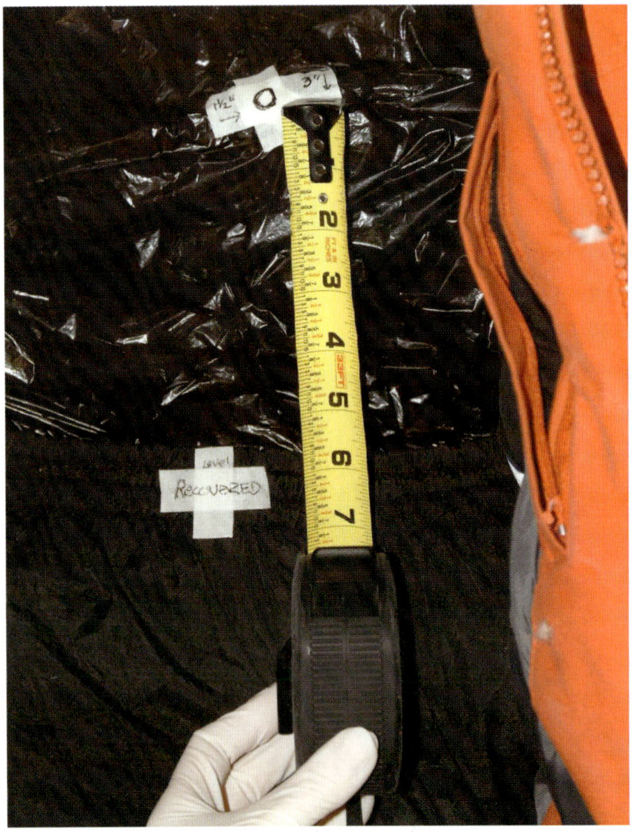

Figure 6.12 Measurements here must be regarded as approximations.

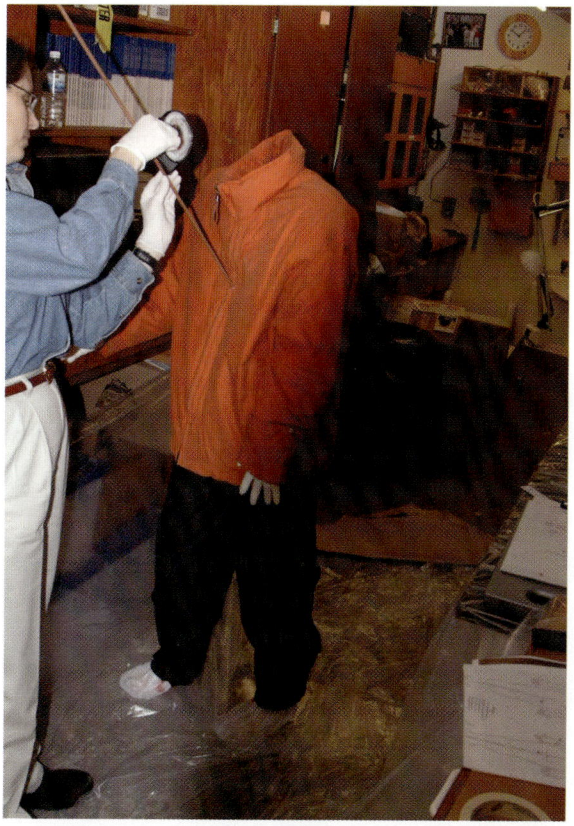

Figure 6.13 Here, a simple angle gauge was used on a trajectory rod to measure the approximate angle. Another digital gauge was also used. In addition, simple geometry was applied to derive a third value. Each value fell within ±3 degrees of the other values.

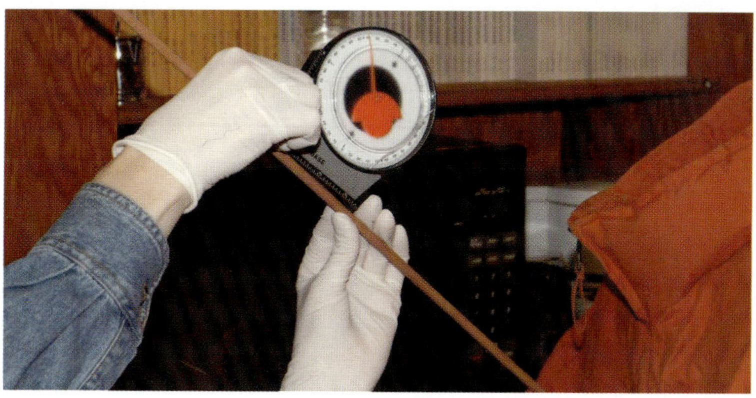

Figure 6.14 Showing the angle gauge.

Left Arm, Left Leg Shot

Figure 6.15 to Figure 6.20 show the trajectory of the shot striking Martin in the left arm and left thigh. Again the trajectory shows approximately a 40-degree angle of declination. This is the same trajectory as the chest shot. More important, regardless of the error rate and acceptable variance of these measurements, the results show that Martin had his left arm and hand down, perhaps in his pocket when he was struck by this bullet. This is directly contradictory to the officer's account of the shooting. He claimed that he was threatened by the decedent, who was waving bottles in both hands above his head. This work is sufficient to call the officer's story into serious question.

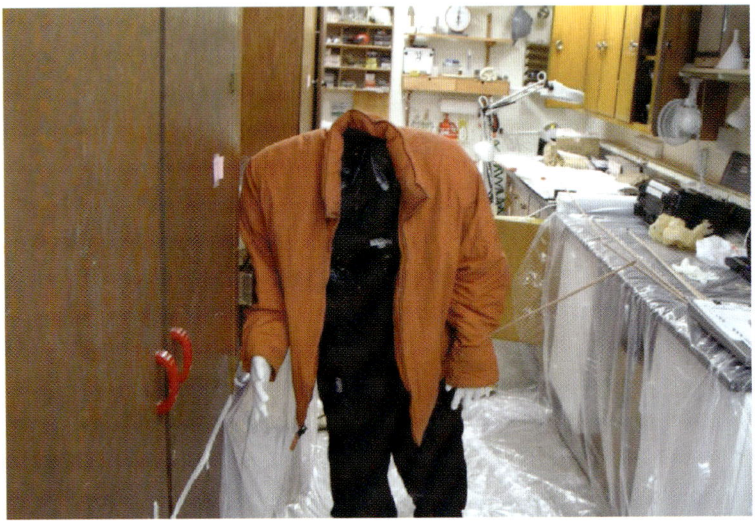

Figure 6.15 Both the autopsy measurements, as well as the bullet holes in the jacket, were used to help approximate the trajectories of the other shot perforating the body.

Figure 6.16 Trajectory rods and angle gauges help document the two relevant angles—the incident and the azimuth angles.

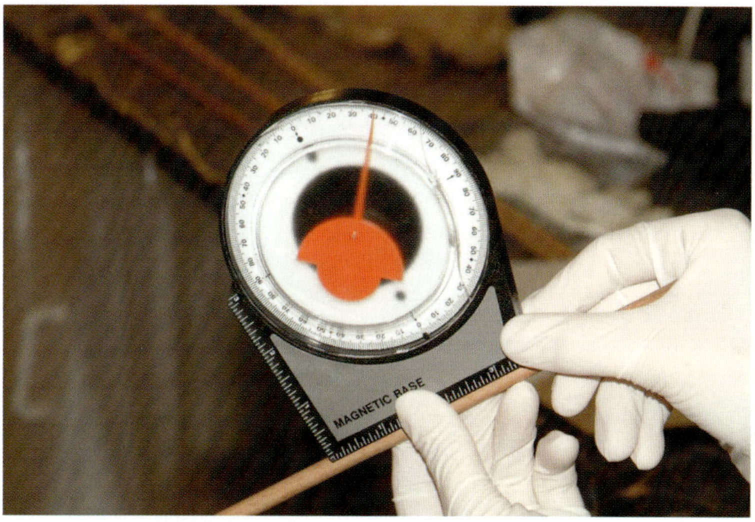

Figure 6.17 The same three methods were applied to this shot to measure its trajectory. Again, all the values fell within the ±3-degree range, more than acceptable for the approximations developed to account for the bullet's trajectory.

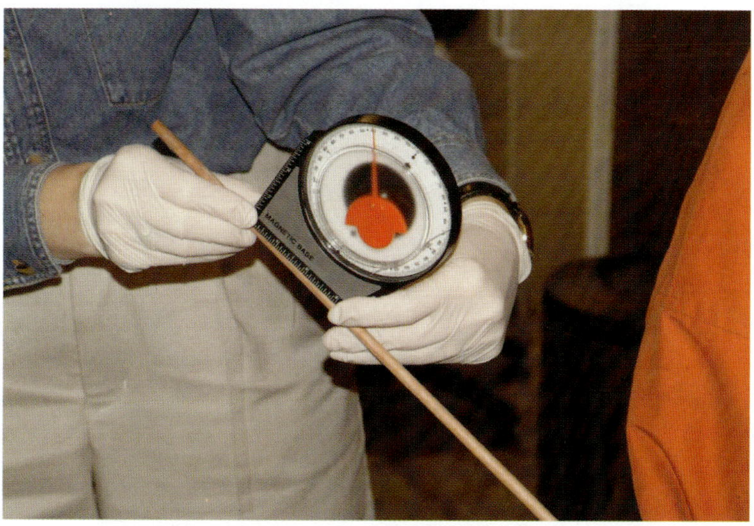

Figure 6.18 The angle of incidence and the azimuth angles can be measured by several useful methods.

Figure 6.19 I use the angle gauge as the last measurement to confirm the other two results.

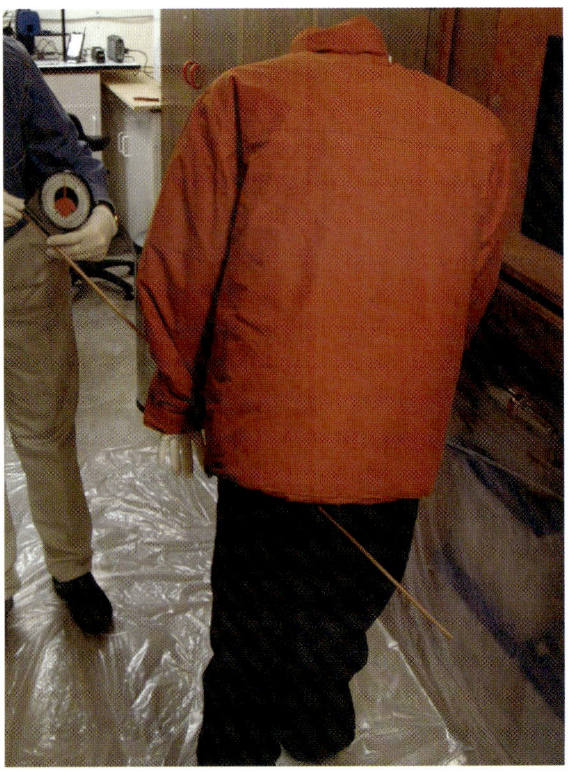

Figure 6.20 Trajectory rods must be rigid and screw together.

Conclusions

The aforementioned trajectories must be interpreted by considering that human bodies move in multiple ways. We know that Martin was about 6 feet 2 inches tall and that Officer Stevens is about 6 feet 4 inches tall. If Officer Stevens fired his pistol with the barrel roughly parallel with the road surface (as one would normally fire a pistol), then Martin was bent forward about 50 degrees from vertical (or 140 degrees from horizontal) when struck, thus accounting for the 40-degree angle of declination for each shot trajectory with the bullet fired roughly parallel with the road surface. Although other relative positions are possible, none fit the available evidence as well as this estimation.

Evidence Examination: Item #5 Old English Label Bottle—Broken Pieces of Glass

Scene Photographs

Figure 6.21 shows the broken bottle alleged to have been in Martin's left hand. One can clearly see that a piece of the bottle is to the left of the number 5 and that it is a relatively large piece.

Figure 6.21 Broken Old English Label bottle at the scene.

Description: Broken Pieces

Figures 6.22 and 6.23 show all the bottle pieces received in evidence. There is a piece missing that would have had to break off when the rest of the neck was broken. If Martin carried the bottle with only the bottom broken off and the rest intact, then the missing piece should appear both in the scene photos and with the collected evidence. The piece appears in neither.

Figure 6.22 Pieces of the bottle.

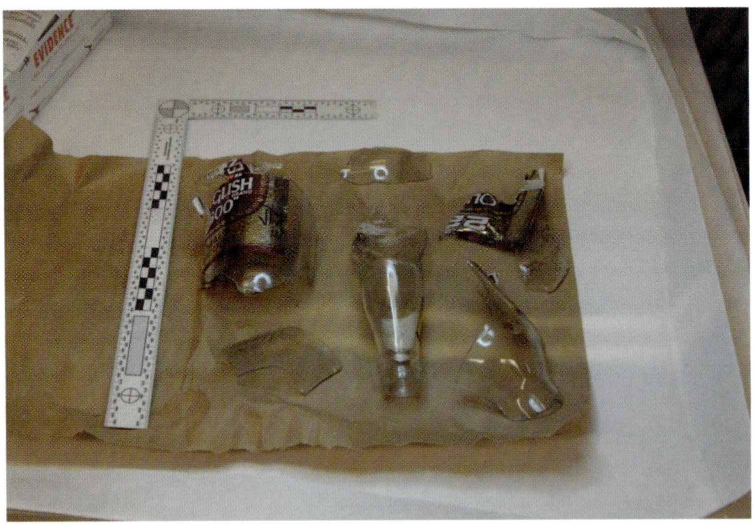

Figure 6.23 The bottle.

Broken Pieces Reassembled

The pieces were numbered and reassembled after examination of the glass fracture patterns (Figure 6.24 to Figure 6.31). Although it sticks to latex gloves as easily as it sticks to skin, Super Glue works well. The pieces were assembled in order to determine the shape of the bottle and to learn if the fracture patterns explained each of the existing pieces. Given the fracture patterns, if the bottle broke near the body on Central Avenue, then all pieces should have been recovered at the shooting scene where it was photographed. Alternative explanations rest with loss of the item, or failure to collect a "small" item.

Figure 6.24 Labeling the parts.

Figure 6.25 Assembling the parts.

Figure 6.26 Use "super glue," since that is part of most laboratory fuming chamber chemistry.

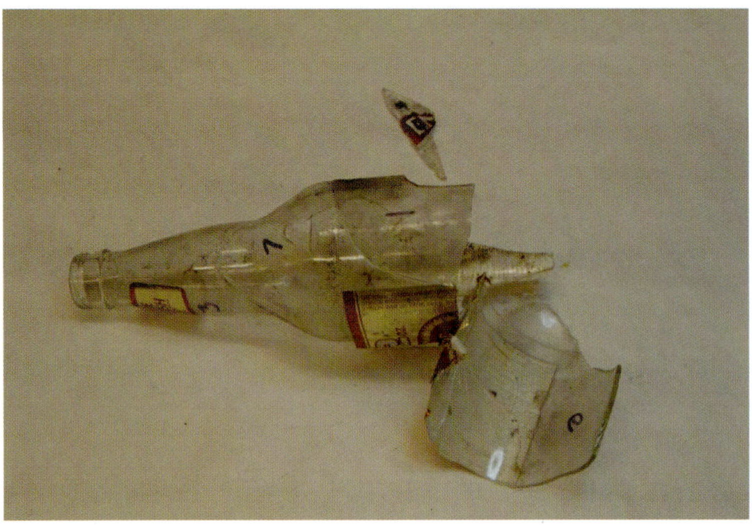

Figure 6.27 Assembly of the pieces.

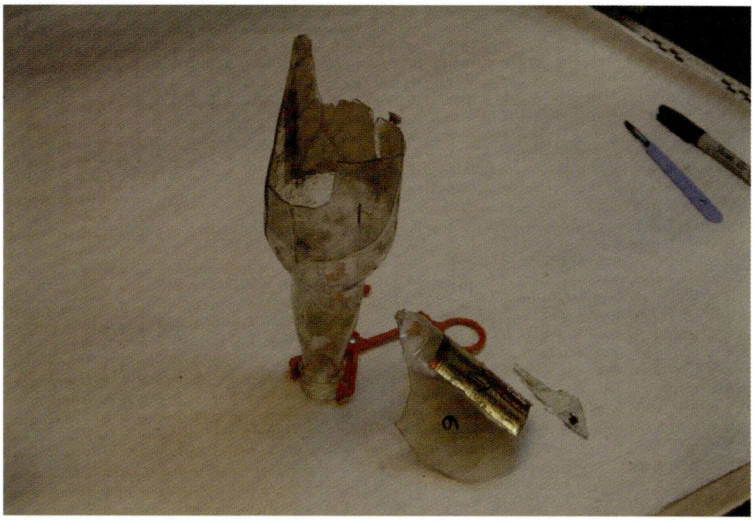

Figure 6.28 The numbers fit together.

Figure 6.29 The result.

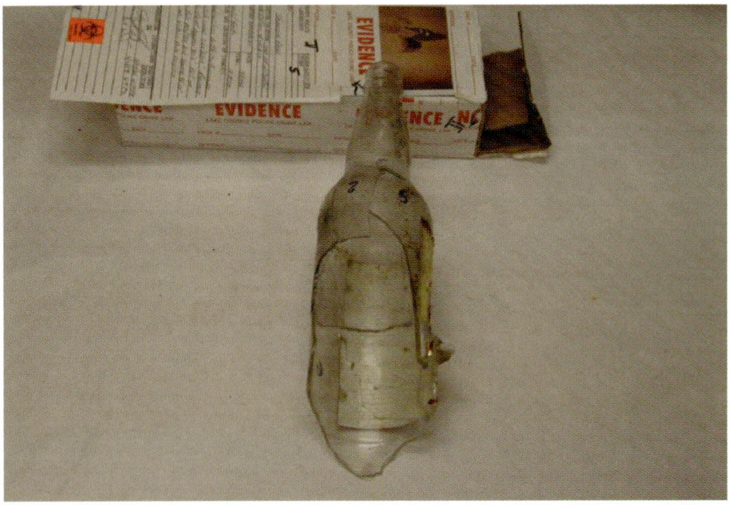

Figure 6.30 Determining the fracture lines.

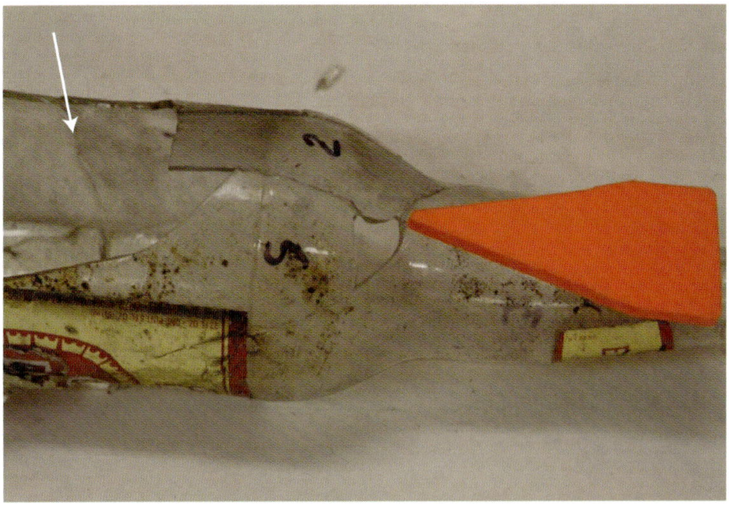

Figure 6.31 The white arrow indicates the missing piece, which should have been recovered with the rest of the broken pieces.

Conclusions

The absence of the glass piece that formed part of the bottle's neck also might indicate that the pieces appearing on Central Avenue in the evidence photograph and as collected by crime scene investigators were not fractured and broken apart at that location. All the fracture lines indicate that the missing piece would have broken when the other radiating fractures to the bottle neck occurred. Again, the other possible explanations for the missing piece should be considered: careless or improper evidence collection by the crime scene officers responding to the case or loss of the piece in the evidence room. But either explanation appears suspect: I got a similar bottle and tried to break off the bottom while leaving the rest of the bottle intact. I found that this is not easy to do. In fact, the bottle either failed to break at all or broke into pieces with a similar pattern as the evidence bottle.

Shooting Evidence Examination: Item #6 "Black Bag Containing Miscellaneous Items"

Physical Description

The black bag containing books, a tape player, tapes, floppy discs, school materials, as well as personal notes, writings, journals, and family papers weighed 27 pounds.

Brief Observations

Photocopies of four pages from one of the notebooks are typical of the more personal, religious writing found among these effects.

One book in his collection concerned certain types of "mind-expanding drugs," such as marijuana, and their use in "spiritual quests." It appears from toxicology reports that Martin had THC metabolites in his urine but none in his bloodstream. It is therefore possible that he had experimented with such drugs in an attempt to further this perceived spiritual goal (most medically trained health practitioners find this dubious advice at best). Some green vegetable material was collected in evidence as noted but was not tested. One could reasonably assume that it is marijuana or a related drug, although tests must be done to confirm such a conclusion if much depends upon its accurate identification. It seemed, now at least, to be irrelevant.

Use of Deadly Force Issues

It is useful to label the nonscientific extras such as police procedures. Authorities in police procedures should be prepared to defend their experience with the issues, even though the facts in a specific case should be allowed to speak for themselves. Arguments from authority, when offered without supporting reasons, are weak.

This case demonstrates the importance of proper training for law enforcement officers. Police officers are sworn to "protect and serve." This includes service to citizens who are guilty of crimes as well as to those who are innocent of any wrongdoing. Since officers are forced by their profession to deal with people alleged to have committed crimes, the notion of protect and serve must be established through practical training to cover all such cases. This oath must certainly cover those deemed less than desirable by civil society as well as those for whom the individual officer may have little personal regard or sympathy, whether or not such lack of regard or sympathy is justified.

Leroy Martin had not committed any violent act nor was he acting violently when he was confronted. According to witnesses, Martin may have exhibited behavior typical of individuals with mental defects sufficient to render them unable to follow a simple instruction to stop walking down the middle of Central Avenue, but he posed no threat. International Association of Chiefs of Police (IACP) policy clearly states:

> Police officers have been given the grave responsibility to protect life and to use their firearms, if necessary, in order to fulfill their responsibility. Therefore, officers must use utmost prudence and restraint in exercising their authority to use firearms. They should discharge their firearms only when shooting is necessary to protect a life, or when it is the only effective

means to apprehend an individual who poses a significant threat of death or serious bodily injury to officers or to citizens. (IACP Order #92-04, section I)

Officers have legal as well as moral obligations to seek the truth, to work effectively to this end as a team, and to protect and serve themselves. The choice by Officer Stevens to place himself in front of an apparently harmless person walking down the middle of a quiet street and to both draw and fire his weapon at that person does not demonstrate a proper grasp of this concept.

This case, while egregious in its departure from accepted IACP guidelines and generally accepted police practices, even appears to violate the extremely general and brief standard operating procedures of the Pond Valley Police Department regarding the use of deadly force.

"Section IV: Reporting, Arrest, Prisoners" of the Police Department Manual, Subsection 4.10 reads as follows:

4.10 Use of Force

Officers shall not use more force in any situation than is reasonably necessary under the circumstances. The extent of force used must be in compliance with law and with departmental policy.

Refer to Policy to Rule 4.10

Policy Rule 4.10

An Officer may use all the force reasonably necessary to arrest a person whom the Officer has probable cause to believe has committed a crime, even a misdemeanor. He can overcome resistance with all reasonable force, short of taking a life, if the officer reasonably believes that force is necessary to effect a lawful arrest.

A law enforcement officer is justified in using deadly force *only if* the officer [emphasis added]:

Has probable cause to believe that deadly force is necessary;
To prevent the commission of a forcible felony; or
To effect an arrest of a person who the officer has probable cause to believe poses a threat of serious bodily injury to the officer or a third person; and
Has given a warning, if feasible, to the person against whom the deadly force is to be used.

Regarding the use of weapons, "Section V: Conduct—General, Subsection 5.06" of the Pond Valley Police Department Manual reads as follows:

5.06 Use of Weapons

Officers shall not use or handle weapons in a careless or imprudent manner.

Although the Pond Valley Department Manual may require serious revisions to detail explicit policy on the use of nondeadly force versus the use of deadly force, and to correct grammar and spelling, it remains clear that Officer Stevens' decision to use deadly force against Martin fails to meet the required necessary condition stated in Policy Rule 4.10 for its use.

It is not clear that walking down the middle of Central Avenue is even a misdemeanor; however, failing to respond to a lawful request from Stevens does violate the law. Therefore, Martin had committed a misdemeanor offense. However, Martin was in no way resisting

Stevens; at most, any claim of resistance must construe it as merely passive. The use-of-force policy states that Stevens "can overcome *resistance* with all reasonable force, *short of taking a life*, if the officer reasonably believes that force is necessary to effect a lawful arrest" [emphasis added]. This entails the use of reasonable force short of taking a life to make the arrest. Such force might include pepper spray, a net, baton, physical constraint, and so forth.

It cannot be assumed that Martin committed a prior violent crime, nor could he reasonably be thought to be committing or be planning to commit a forcible felony. However, according to Stevens, Martin was brandishing two bottles, which the officer construed as weapons. One, said to be in his left hand, was a broken bottle with a jagged edge. Stevens reported that Martin was swinging the bottles over his head in a threatening manner.

The scientific and the witness evidence available in this case both cast considerable doubt on the position and role of the bottles in this shooting. However, even if it is true that Martin had the bottles in his hands as Stevens states, namely, the jagged bottle in his left hand and the intact bottle in his right, it does not follow that Stevens had either a reasonable belief or probable cause to believe that Martin posed a threat of serious bodily injury to any of the officers or to any third person. Martin had a 27-pound bag over his left shoulder, rendering the role of any broken bottle in his left hand ineffective at best. The only risk was introduced when Stevens took an immobile position in front (or to the left) of Martin and stopped walking backward. In no reasonable sense can an officer take such a position, only to claim that the person advanced toward him and therefore posed a threat necessitating deadly force. No such circumstance justifies the use of deadly force.

That is not to say, however, that no use of force was justified. It is to say that the use of *deadly* force in this case was not justified. It is, in other words, to say that the Pond Valley Police Department Manual's Section IV, Subsection 4.10 does apply: "Officers *shall not use more force* in any situation than is reasonably necessary under the circumstances" [emphasis added]. Appropriate force in this case was not spelled out in the department's manual and, at least according to Stevens, no effective training was provided by the department in the use of nondeadly force.

The Pond Valley Police Department Manual in Subsection 4.10 also states that "the extent of force used must be in compliance with law and with departmental policy." As discussed earlier, it is not clear that this condition is satisfied by the circumstances of Martin's shooting.

Nor does Stevens' tactical position appear in keeping with the goal of protecting innocent bystanders from injury or death. Considering the location of the cartridge cases documented in the scene photos, along with the location of Martin's body and his injuries, it appears that Stevens shot toward the southeast. He shot toward the side of Central Avenue with the closest buildings, potentially risking civilian injury. Indeed, the shot eventually perforating Martin's left thigh was never recovered. If Stevens had taken a tactical position facing the other way, any shots he fired would have gone toward the open area by the railroad tracks with its greater distance to any potentially occupied buildings.

This decision raises serious questions about the training provided regarding "Section V: Conduct—General, Subsection 5.06" of the Pond Valley Police Department Manual. It appears that Stevens, at the least, failed to consider the requirement that "Officers shall not use or handle weapons in a careless or imprudent manner."

The IACP has prepared policy guidelines for police departments to consider, to adopt or adapt, or to follow. It is important to note that these are guidelines and are not meant by the IACP to be used as criterion for measuring any department's existing policies and procedures.

The IACP guidelines on the use of force dictate that deadly force be used only as a last resort. They require, among other things, that "the officer must reasonably believe that all other means of effecting the arrest are futile" (IACP, G. O. 92-04 I, C.2). None of the officers at the scene had by any means evaluated the situation accurately enough to determine that firearms were their only tactical option. In fact, in common police practice, the first option in the face of such behavior would be to adopt a nonlethal tactical response—something for which, by his own admission, Officer Stevens, and perhaps other members of the Pond Valley Police Department, had no effective training.

Both reasonable police procedures and common sense dictate that standing immobile in front of an approaching, unresponsive subject and firing, even if the subject is armed with bottles, is contraindicated, unnecessary, and clearly avoidable. Such an approach merely provokes tactically avoidable violence. Procedures and common sense require that officers who face situations for which they are untrained, unprepared, or unequipped, call for additional appropriately trained, prepared, and equipped backup. This would help to defuse any situation if the responding officers feel threatened or unable to control the situation. There is nothing to be lost in this case by waiting.

According to IACP guidelines, and as written in many police departments' operational orders, officers will not needlessly place themselves in situations of great danger and use this as justification for the use of deadly force. Stevens' first obligation was to avoid increasing tensions. At the very worst, Martin was a danger to himself. Approaching such a person with weapon in hand is a classic mistake, the result of poor training, limited experience, or both. Stevens' actions needlessly escalated a potentially resolvable situation, and violated stated use-of-force guidelines written by his own department.

General Concluding Observations

In a dissenting opinion, *United States v. Wade*, 388 U.S. 218 (1967), Justice White states what he takes to be the legal obligations of law enforcement officers:

> Law enforcement officers have the obligation to convict the guilty and to make sure they do not convict the innocent. They must be dedicated to making the criminal trial a procedure *for the ascertainment of the true facts surrounding the commission of the crime* [emphasis added]. To this extent, our so-called adversary system is not adversary at all; nor should it be.

My work on this case highlights specific shortcomings that prevented "the ascertainment of the true facts surrounding the commission of the crime." These shortcomings include the lack of officer training in use-of-force options as well as the absence of a fruitful shooting reconstruction. The latter, in my view, results from a shooting scene that was not presented accurately or clearly to the crime scene investigators who eventually documented the evidence. Nor was all the relevant physical evidence collected or examined. These shortcomings, regardless of their true explanation, contribute to the enduring suspicion that must surround any purported justification for Martin's shooting.

Crime Scene Shooting Reconstruction Summary

Important caveat: This sequence of events is not meant to represent a fixed time line but only a rough structure of the events that could occur within a few minutes of each other or that could occur with a significant time lapse between items. Some of these events may have occurred simultaneously or in an alternate order.

1. Leroy Martin walked west down the middle of Central Avenue leaning forward, carrying his 27-pound black bag over his left shoulder, his hands in his pockets, when he was intercepted by Pond Valley Police officers, including Officer Bud Stevens.
2. Martin failed to respond to Stevens' repeated commands to take his hands out of his pockets and to stop walking down the middle of the street.
3. Stevens took a position in the street to the front and slightly to the left of Martin.
4. As Martin walked forward, Stevens walked backward with his Glock 40 pistol pointed at Martin.
5. As Martin walked, officers took positions behind him, armed with pepper spray.
6. Stevens said that he fired two shots at Martin from his Glock 40 cal pistol, after Martin ignored his repeated commands to stop.
7. These two shots were fired at Leroy Martin from *at least* 3 feet or *more* muzzle-to-target distance (conservative).
8. The two shots striking Martin both were fired from his left side. So either Stevens stood to Martin's left as Martin walked straight while both facing and bending forward, or Stevens stood in front of Martin and Martin turned to his right facing away from the shooter, exposing his left side while bending forward.
9. The two shots striking Leroy Martin both entered his body at about a 40-degree angle of declination anatomically. So if Stevens fired his weapon with the muzzle parallel to the road surface, then Martin was bent forward about 50 degrees from standing perpendicular to the road surface when struck by each of the two shots.
10. Martin's left arm was down by his waist when struck by a bullet.
11. This bullet perforated his left forearm, exiting through his jacket, reentering through his jacket below the left jacket pocket, entering his black nylon pants in front of his left thigh, perforated his left thigh from front to back, and left to right, exiting to his right rear, and was not recovered.
12. Martin's orange jacket was zipped when he was struck in the chest by a bullet.
13. The bullet striking his chest pursued a 40-degree angle of declination as described in the autopsy, and was recovered in the lumbar spine.
14. The "broken bottle" was either moved to the location where it was photographed, remains unrelated to the shooting scene, or both.
15. The unbroken bottle was either moved to the location where it was photographed, remains unrelated to the shooting scene, or both.
16. The shoulder strap on Martin's black bag was cut, and the bag moved.
17. Martin was turned over onto his back, his jacket was unzipped, and his arms were moved away from his body.

Laboratory Notes and Evidence Photographs

The function of lab notes and drawings are to stimulate thought and aid memory as much as to document evidential observations. Cameras can do that, but lab notes aid in both looking at and describing the items and their properties. I always include my lab notes, sketches, and data in any report that I produce (Figure 6.33 to Figure 6.52). Of course, all my drawings are approximations and none are drawn to exact scale. Despite that, I make every effort to draw items of evidence both clearly and accurately. In addition, the notes reflect the fingerprint examination of the bottles recovered at the shooting scene. No identifiable friction ridges appeared on the glass or label surfaces of either bottle.

My laboratory notes are often followed by my evidence photographs, any XRF spectra, SEM photographs, or charts, tables, and scene photographs. All items of evidence are opened under a fume hood (Figure 6.32) and examined for biohazards and any relevant transitory trace evidence prior to further handling and analysis—both the orange jacket (Figure 6.33) and the black pants were so examined. The latter section includes photos digitally cropped according to FBI Scientific Working Group (SWG) protocols for the use of digital evidence. When photos are cropped for a report, I always include a copy of the original uncropped photograph for reference and examination.

When running samples in the XRF, printouts from the unit as well as notes help to establish the results, which may then be graphed. It must be recalled that this use of XRF is a presumptive test for GSR: the SEM must be used if the sample passes the threshold by revealing the requisite elements. Only the SEM can identify a particle, for example, as having the morphology, size, and composition of three elements—Sb, Pb, and Ba—fused to form the single particle. Only then do we entertain the possibility that this is GSR. We still may be uncertain of its source.

Figure 6.34 to Figure 6.53 illustrate one use of lab notes for recording basic measurements, and organizing presumptive test results. Sometimes among the evidence, a very personal item will appear. While not scientific, the journal excerpt appearing in Figure 6.54 to Figure 6.57, provides information about the victim in this case that in some fundamental way reveals an insight into his character and personality. Finally, Figure 6.58 to Figure 6.110 are typical evidence photos used to illustrate the examination and to document observations and analyses.

Figure 6.32 The fume hood is designed to protect users from any chemicals or biohazards that may accompany the evidence.

Figure 6.33 I document packages with evidence items before, during, and after opening. Too much documentation is seldom a problem. Too little is often a problem.

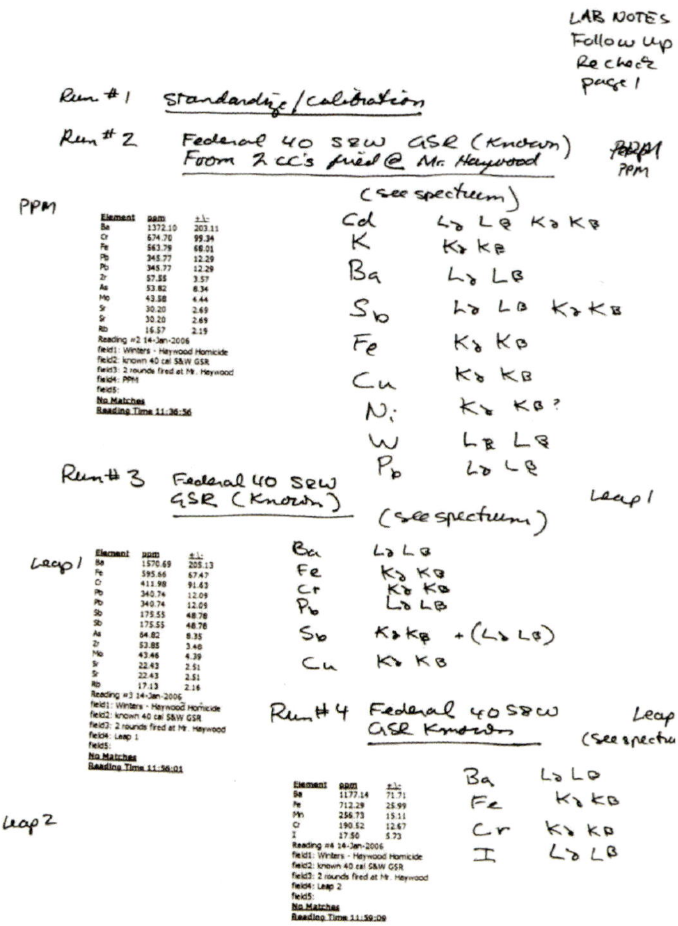

Figure 6.34 Lab notes (see also Figure 6.35 and Figure 6.37 to Figure 6.4.)

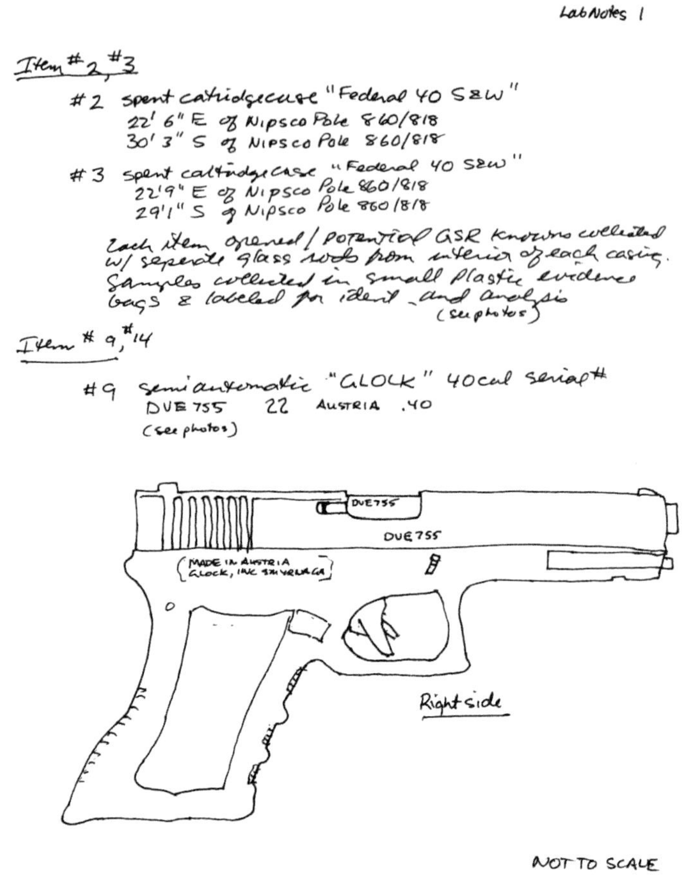
Lab Notes 1

Item #2, #3

#2 spent cartridge case "Federal 40 S&W"
22' 6" E of Nipsco Pole 860/818
30' 3" S of Nipsco Pole 860/818

#3 spent cartridge case "Federal 40 S&W"
22'9" E of Nipsco Pole 860/818
29'1" S of Nipsco Pole 860/818

Each item opened / potential GSR knowns collected w/ separate glass rods from interior of each casing. Samples collected in small plastic evidence bags & labeled for ident. and analysis (see photos)

Item # 9, #14

#9 semiautomatic "GLOCK" 40cal serial#
DUE 755 22 AUSTRIA .40
(see photos)

Figure 6.35

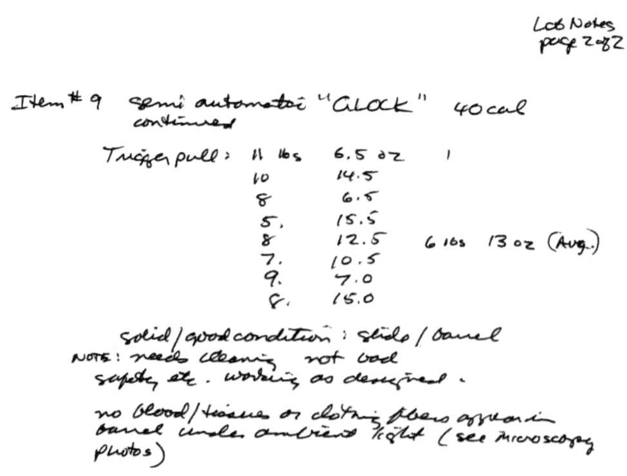
Lab Notes
page 2 of 2

Item # 9 semi automatic "GLOCK" 40 cal
continued

Trigger pull: 11 lbs 6.5 oz 1
10 14.5
8 6.5
5. 15.5
8 12.5 6 lbs 13 oz (Avg.)
7. 10.5
9. 7.0
8. 15.0

solid/good condition: slide / barrel
NOTE: needs cleaning not load
safety etc. working as designed.

no blood/tissue or clothing fibers appear in barrel under ambient light (see microscopy photos)

Figure 6.36 Trigger pull measurements supply good material to practice calculating error rates for the actual measurements that one makes using any number of devices.

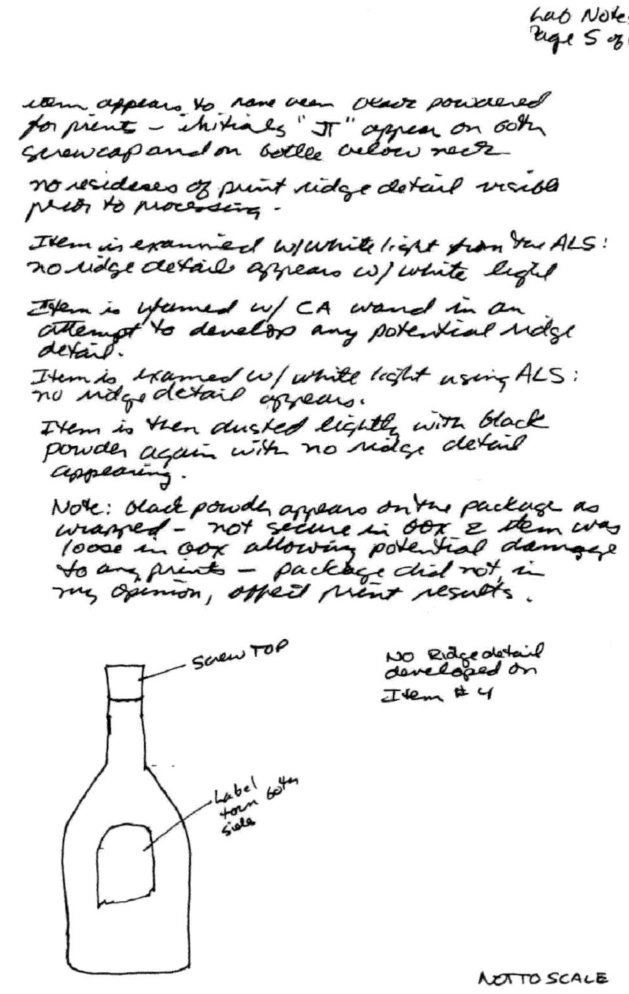

Lab Notes
Page 5 of 6

item appears to have been black powdered for print – initials "JT" appear on both screwcap and on bottle below neck

no residues of print ridge detail visible prior to processing.

Item is examined w/white light from the ALS: no ridge detail appears w/ white light

Item is fumed w/ CA wand in an attempt to develop any potential ridge detail.

Item is examed w/ white light using ALS: no ridge detail appears.

Item is then dusted lightly with black powder again with no ridge detail appearing.

Note: black powder appears on the package as wrapped – not secure in box & item was loose in box allowing potential damage to any prints – package did not, in my opinion, affect print results.

Screw TOP

Label torn both sides

NO Ridge detail developed on Item # 4

NOT TO SCALE

Figure 6.37

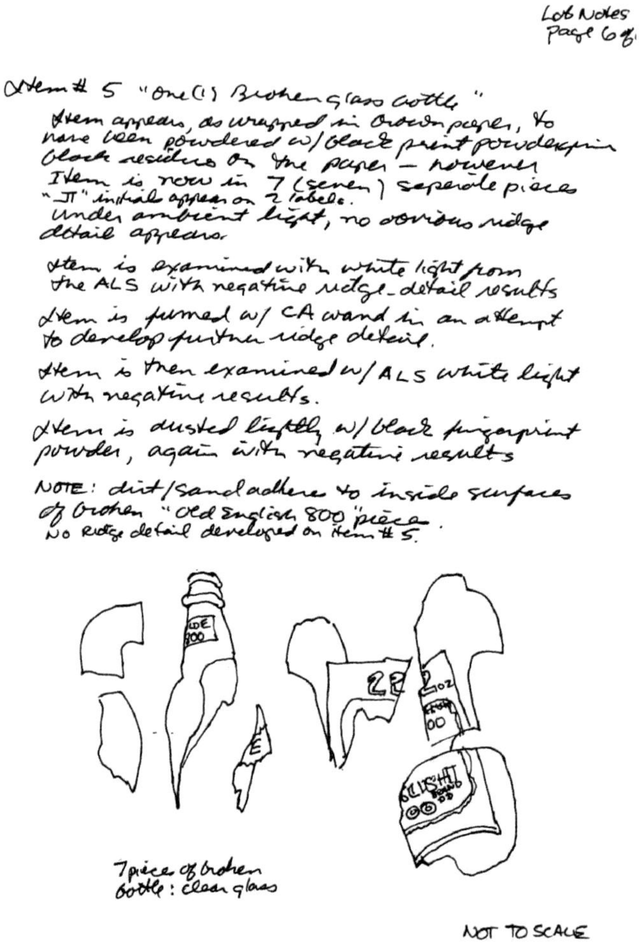

Lab Notes
Page 6 of

Item # 5 "One (1) Broken glass bottle"
Item appears, as wrapped in brown paper, to have been powdered w/ black print powder given black residue on the paper — however
Item is now in 7 (seven) seperate pieces
"JT" initial appears on 2 labels.
Under ambient light, no obvious ridge detail appears.

Item is examined with white light from the ALS with negative ridge-detail results

Item is fumed w/ CA wand in an attempt to develop further ridge detail.

Item is then examined w/ ALS white light with negative results.

Item is dusted lightly w/ black fingerprint powder, again with negative results

NOTE: dirt/sand adhere to inside surfaces of broken "Old English 800" piece.
No ridge detail developed on Item # 5.

Figure 6.38

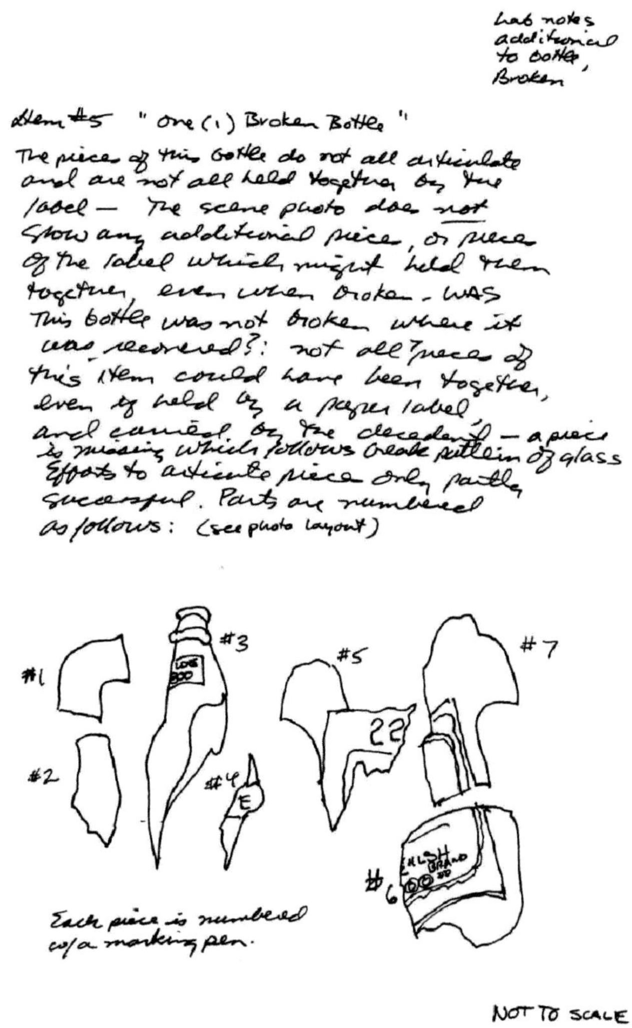
Lab notes additional to bottle, Broken

Item #5 "One (1) Broken Bottle"

The pieces of this bottle do not all articulate and are not all held together by the label — The scene photo does not show any additional piece, or piece of the label which might held them together, even when broken. WAS This bottle was not broken where it was recovered?: not all pieces of this item could have been together, even if held by a paper label and carried by the decedent — a piece is missing which follows break pattern of glass Efforts to articulate piece only partly successful. Parts are numbered as follows: (see photo layout)

Figure 6.39

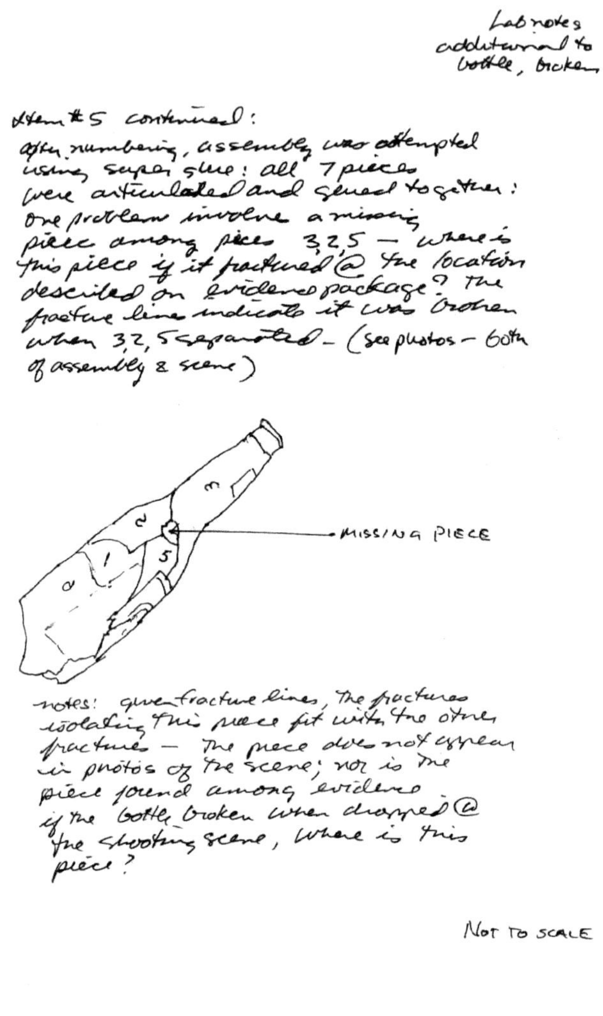
Lab notes
additional to
bottle, broken

Item #5 continued:
after numbering, assembly was attempted
using super glue: all 7 pieces
were articulated and glued together:
One problem involve a missing
piece among pieces 3,2,5 — where is
this piece if it fractured @ the location
described on evidence package? The
fracture line indicates it was broken
when 3,2,5 separated — (see photos — both
of assembly & scene)

notes: given fracture lines, the fractures
isolating this piece fit with the other
fractures — the piece does not appear
in photos of the scene; nor is the
piece found among evidence.
if the bottle broken when dropped @
the shooting scene, where is this
piece?

NOT TO SCALE

Figure 6.40

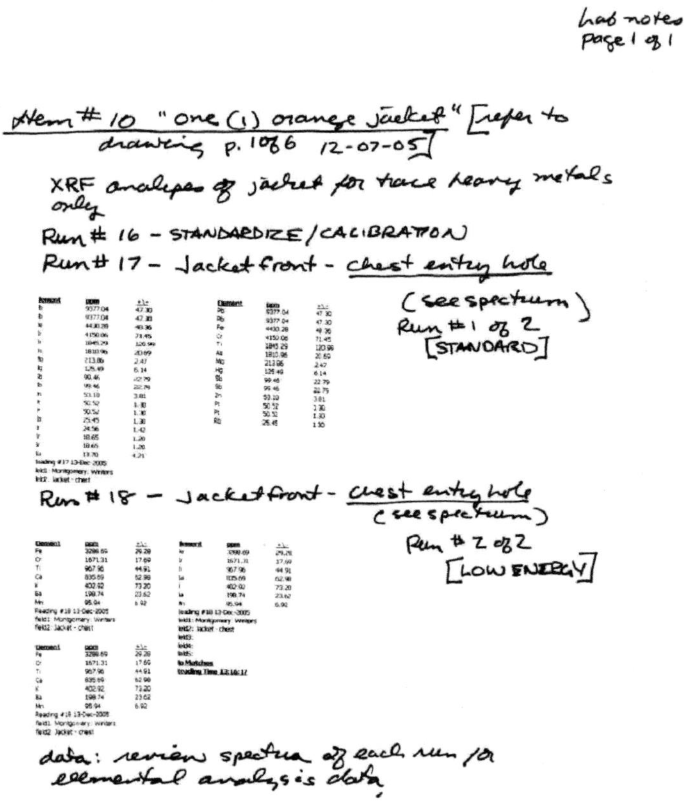

Figure 6.41 XRF results help determine the presence of metals—the analyses provide a presumptive test for the presence of GSR—if the components are found, then the SEM must be used to determine if the elements are fused into one particle and to determine its morphology.

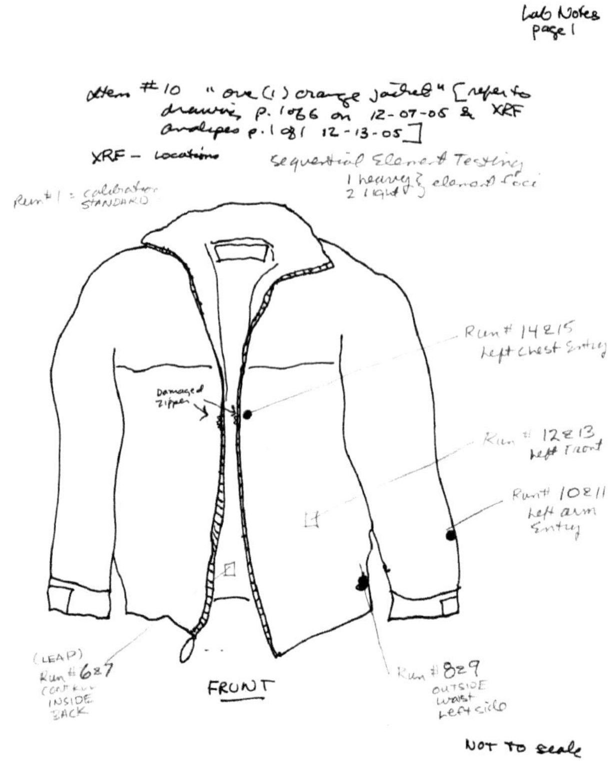

Figure 6.42 Lab notes continued (see also Figure 6.43 to Figure 6.45).

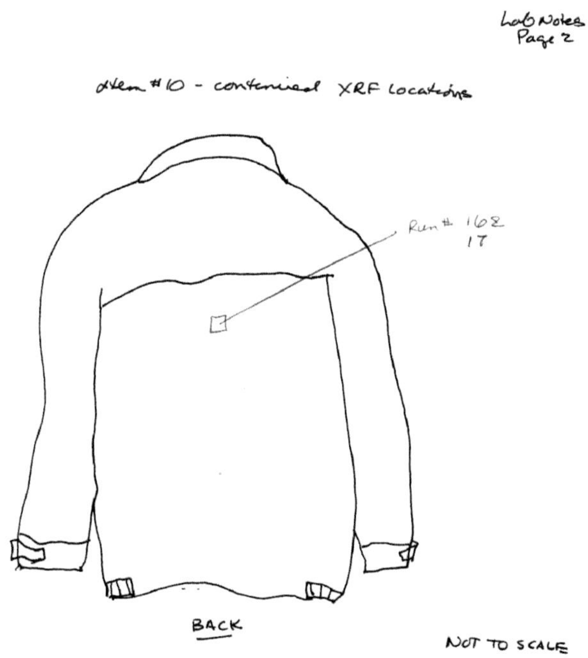

Figure 6.43

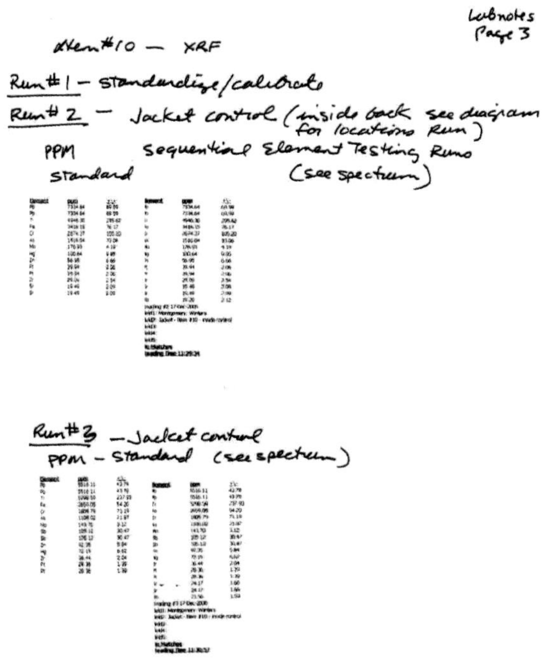

Figure 6.44

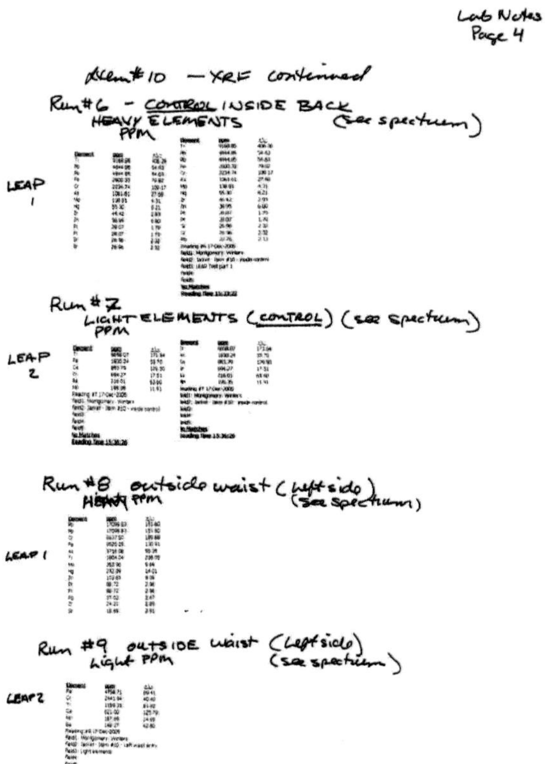

Figure 6.45

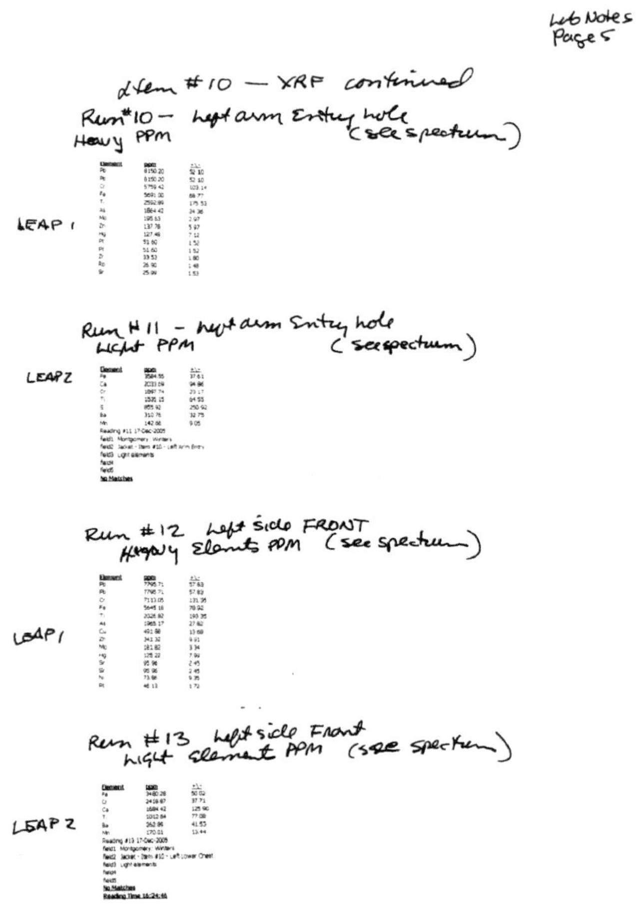

Lab Notes
Page 5

Item #10 — XRF continued

Run #10 — Left arm Entry hole (see spectrum)
Heavy PPM

LEAP 1

Element	ppm	+/-
Pb	8150.20	52.10
Pb	8150.20	52.10
Cr	5759.42	103.14
Fe	5691.00	68.77
Ti	2592.89	175.53
As	1864.42	24.36
Mo	195.63	2.97
Zn	137.78	5.97
Hg	127.48	7.12
Pt	51.60	1.52
Pt	51.60	1.52
Zr	33.53	1.80
Rb	26.90	1.48
Sr	25.99	1.53

Run #11 – Left arm Entry hole (see spectrum)
Light PPM

LEAP 2

Element	ppm	+/-
Fe	3584.55	37.61
Ca	2033.69	94.86
Cr	1897.74	23.17
Ti	1535.15	64.55
S	855.92	250.92
Ba	310.76	32.75
Mn	142.68	9.05

Reading #11 17-Dec-2005
field1 Montgomery : Winters
field2 Jacket - Item #10 - Left Arm Entry
field3 Light elements
field4
field5
No Matches

Run #12 Left side FRONT
Heavy Elements PPM (see spectrum)

LEAP 1

Element	ppm	+/-
Pb	7795.71	57.83
Pb	7795.71	57.83
Cr	7113.05	131.35
Fe	5645.16	78.92
Ti	2026.82	193.35
As	1965.17	27.82
Cu	491.88	13.69
Zn	341.32	9.91
Mo	181.82	3.34
Hg	125.22	7.99
Sr	95.96	2.45
Sr	95.96	2.45
Ni	73.68	9.35
Pt	46.13	1.72

Run #13 Left side Front
Light Element PPM (see spectrum)

LEAP 2

Element	ppm	+/-
Fe	3480.28	50.02
Cr	2418.87	37.71
Ca	1684.42	125.90
Ti	1012.84	77.08
Ba	362.86	41.53
Mn	170.01	13.44

Reading #13 17-Dec-2005
field1 Montgomery : Winters
field2 Jacket - Item #10 - Left Lower Chest
field3 Light elements
field4
field5
No Matches
Reading Time 16:24:46

Figure 6.46 The absence of elements that are typical of primer residues, such as antimony barium and lead is significant: no SEM follow-up is necessary since GSR primers have been excluded.

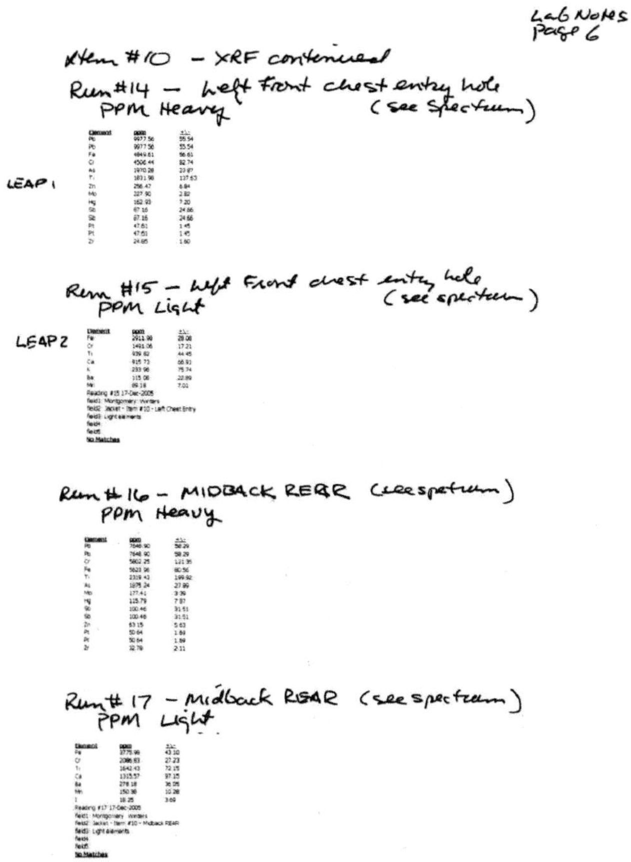

Lab Notes
Page 6

Item #10 – XRF continued

Run #14 – Left Front chest entry hole
PPM Heavy (see Spectrum)

LEAP 1

Element	ppm	+/-
Pb	9977.56	55.54
Pb	9977.56	55.54
Fe	4849.61	56.61
Cr	4506.44	82.74
As	1970.28	23.87
Ti	1831.98	137.63
Zn	256.47	6.84
Mo	227.90	2.82
Hg	162.93	7.20
Sb	87.16	24.66
Sb	87.16	24.66
Pt	47.61	1.45
Pt	47.61	1.45
Zr	24.65	1.60

Run #15 – Left Front chest entry hole
PPM Light (see spectrum)

LEAP 2

Element	ppm	+/-
Fe	2911.98	28.08
Cr	1491.06	17.21
Ti	939.62	44.45
Ca	915.73	66.93
K	233.96	75.74
Ba	115.06	22.89
Mn	89.18	7.01

Reading #15 17-Dec-2005
field1: Montgomery: Winters
field2: Jacket - Item #10 - Left Chest Entry
field3: Light elements
field4
field5
No Matches

Run #16 – MIDBACK REAR (see spectrum)
PPM Heavy

Element	ppm	+/-
Pb	7648.90	58.29
Pb	7648.90	58.29
Cr	5802.25	121.35
Fe	5623.96	80.56
Ti	2319.43	199.92
As	1875.24	27.89
Mo	177.41	3.39
Hg	115.79	7.87
Sb	100.46	31.51
Sb	100.46	31.51
Zn	63.15	5.63
Pt	50.64	1.89
Pt	50.64	1.89
Zr	32.78	2.11

Run # 17 – Midback REAR (see spectrum)
PPM Light

Element	ppm	+/-
Fe	3775.99	43.10
Cr	2086.83	27.23
Ti	1642.43	72.15
Ca	1315.57	97.15
Ba	278.18	36.05
Mn	150.38	10.28
I	18.25	3.69

Reading #17 17-Dec-2005
field1: Montgomery: Winters
field2: Jacket - Item #10 - Midback REAR
field3: Light elements
field4
field5
No Matches

Figure 6.47 It remains important to identify GSR as being more than simply primer residues—any muzzle effluent counts as GSR, so one must consider the elements involved in the production of such residues.

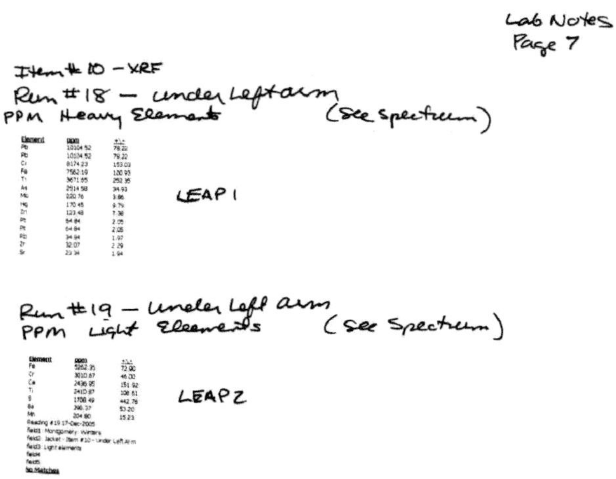

Lab Notes
Page 7

Item #10 – XRF

Run #18 – under Left arm
PPM Heavy Elements (See spectrum)

LEAP 1

Element	ppm	+/-
Pb	10104.52	79.22
Pb	10104.52	79.22
Cr	8174.23	153.03
Fe	7562.19	100.93
Ti	3671.85	252.35
As	2914.58	34.93
Mo	220.76	3.86
Hg	170.45	9.74
Zn	123.48	7.38
Pt	64.84	2.05
Pt	64.84	2.05
Rb	34.94	1.97
Zr	32.07	2.29
Sr	23.34	1.94

Run #19 – under Left arm
PPM Light Elements (See Spectrum)

LEAP 2

Element	ppm	+/-
Fe	5262.35	72.90
Cr	3010.67	46.00
Ca	2436.95	151.92
Ti	2410.87	108.61
S	1708.49	442.78
Ba	390.37	53.20
Mn	204.80	15.23

Reading #19 17-Dec-2005
field1: Montgomery: Winters
field2: Jacket - Item #10 - Under Left Arm
field3: Light elements
field4
field5
No Matches

Figure 6.48 The spectra require knowledgeable assessments to eliminate the phantom peaks and accept the correct indicators of those elements present.

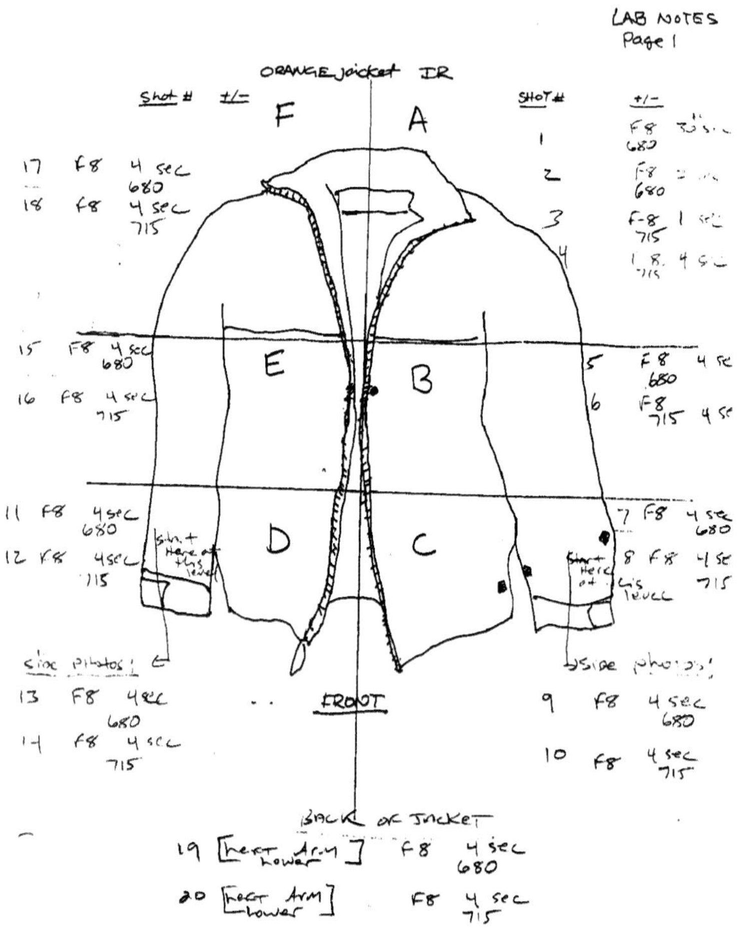

Figure 6.49 Mapping the item allows a better insight into where IR photos should be taken to cover the item and recover the locations captured in the photographs.

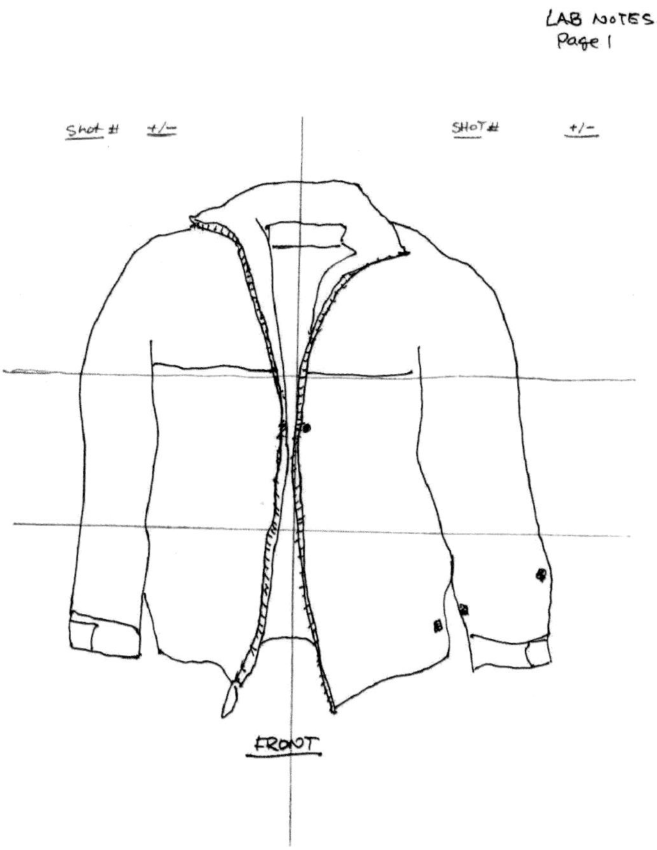

Figure 6.50 The jacket divided into sections for IR or XRF or FTIR analyses—the sections can then be identified, and the results determined for each section, front and back.

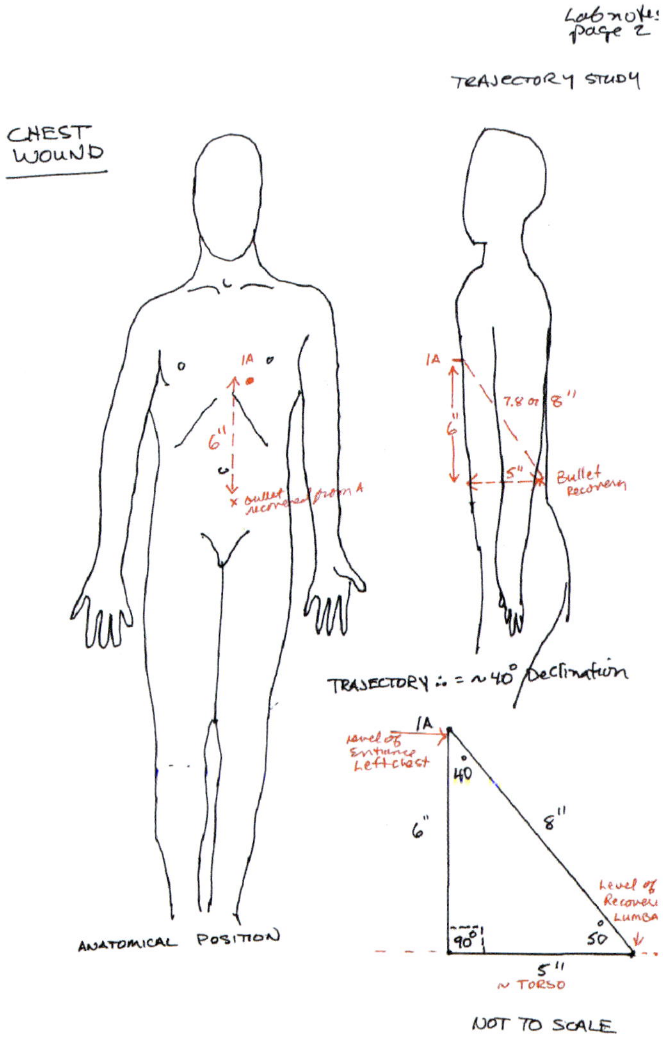

Figure 6.51 Additional lab notes (see also Figures 6.52 and 6.53).

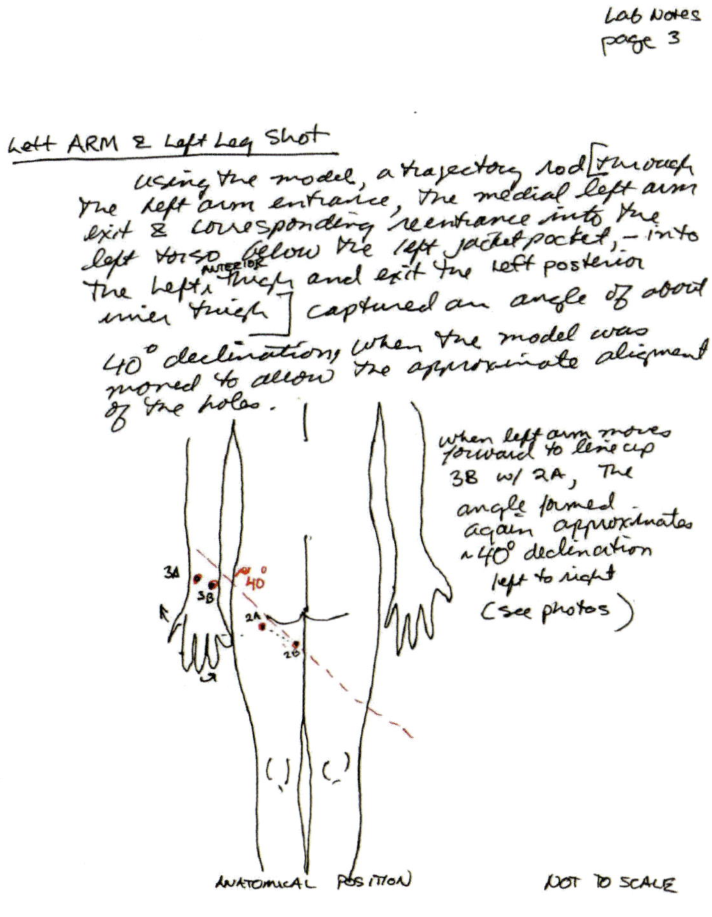
Lab Notes
page 3
Left ARM & Left Leg Shot
Using the model, a trajectory rod [through the left arm entrance, the medial left arm exit & corresponding reentrance into the left torso below the left jacket pocket, - into the left anterior thigh and exit the left posterior inner thigh] captured an angle of about 40° declination, when the model was moved to allow the approximate alignment of the holes.
when left arm moves forward to line up 3B w/ 2A, The angle formed again approximates ~40° declination left to right (see photos)
3A
3B
40°
2A
2B
ANATOMICAL POSITION
NOT TO SCALE

Figure 6.52

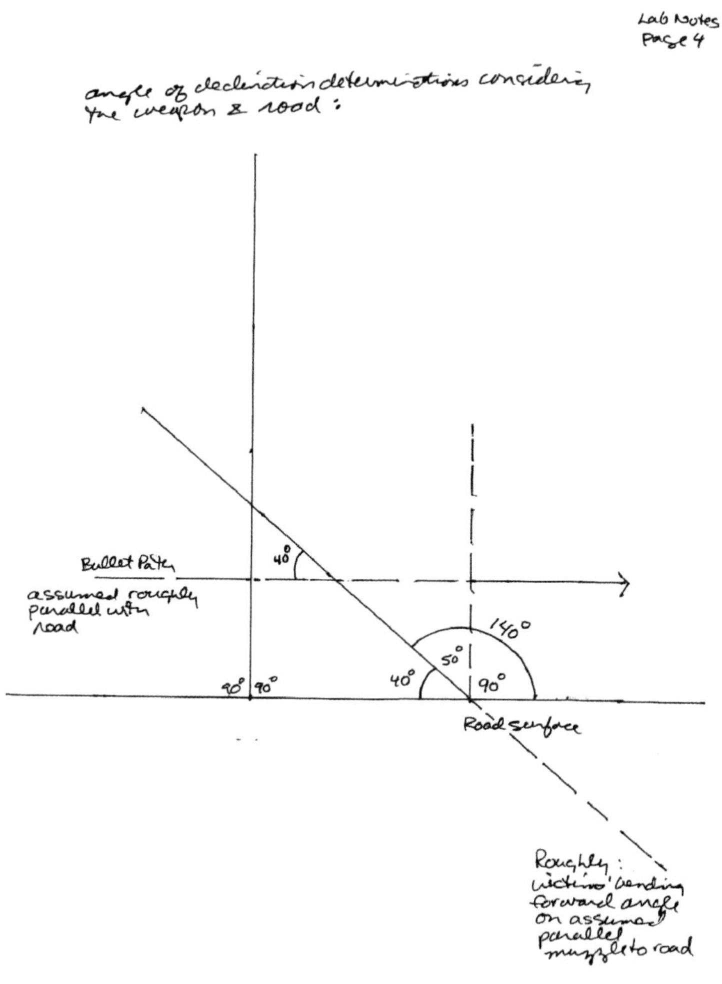
Lab Notes
Page 4
angle of declination determinations considering the weapon & road:
Bullet Path
assumed roughly parallel with road
40°
140°
50°
40°
90°
90°
90°
Road surface
Roughly: victims' bending forward angle on assumed parallel muzzle to road

Figure 6.53

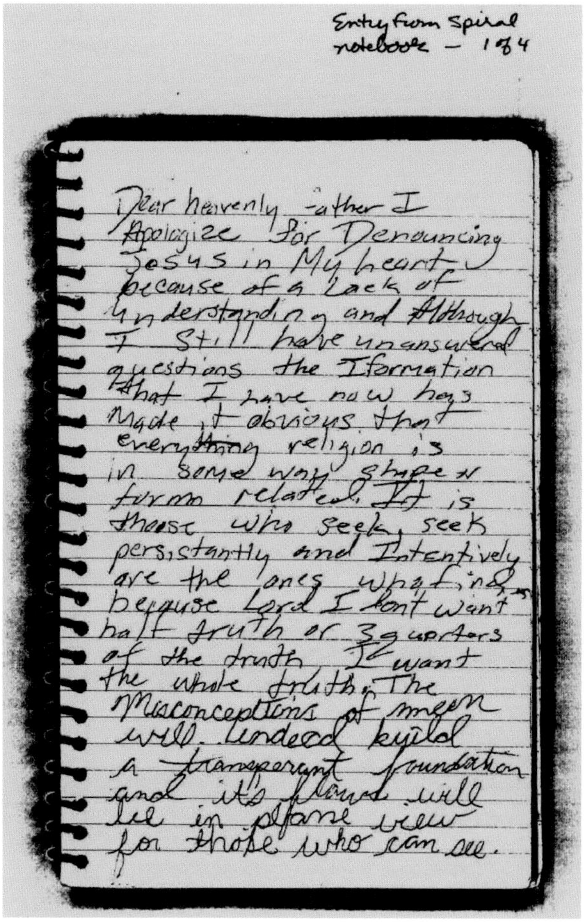

Entry from Spiral notebook – 1 of 4

Dear heavenly father I
Apologize for Denouncing
Jesus in My heart
because of a Lack of
understanding and Although
I Still have unanswered
questions the Information
that I have now has
Made it obvious that
everything religion is
in some way shape or
form related. It is
those who seek, seek
persistently and Intentively
are the ones who find,
because Lord I dont want
half truth or 3 quarters
of the truth. I want
the whole truth. The
Misconceptions of men
will indeed build
a transparent foundation
and it's flaws will
lie in plane view
for those who can see.

Figure 6.54 Spiral notebook notes (see also Figure 6.55 to Figure 6.57).

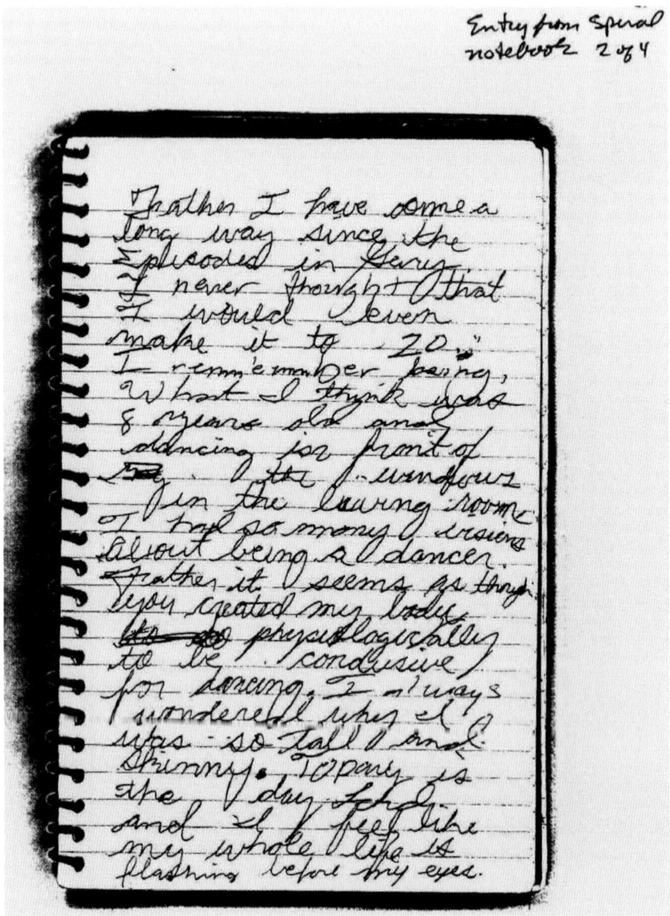

Entry from spiral
notebook 2 of 4

Father I have come a
long way since the
Episodes in Gary.
I never thought that
I would ever
make it to 20.
I remember being,
What I think was
8 years old and
dancing in front of
the windows
in the living room
I had so many dreams
about being a dancer.
Father it seems as though
you created my body
physiologically
to be conducive
for dancing. I always
wondered why I
was so tall and
skinny. Today is
the day
and I feel like
my whole life is
flashing before my eyes.

Figure 6.55

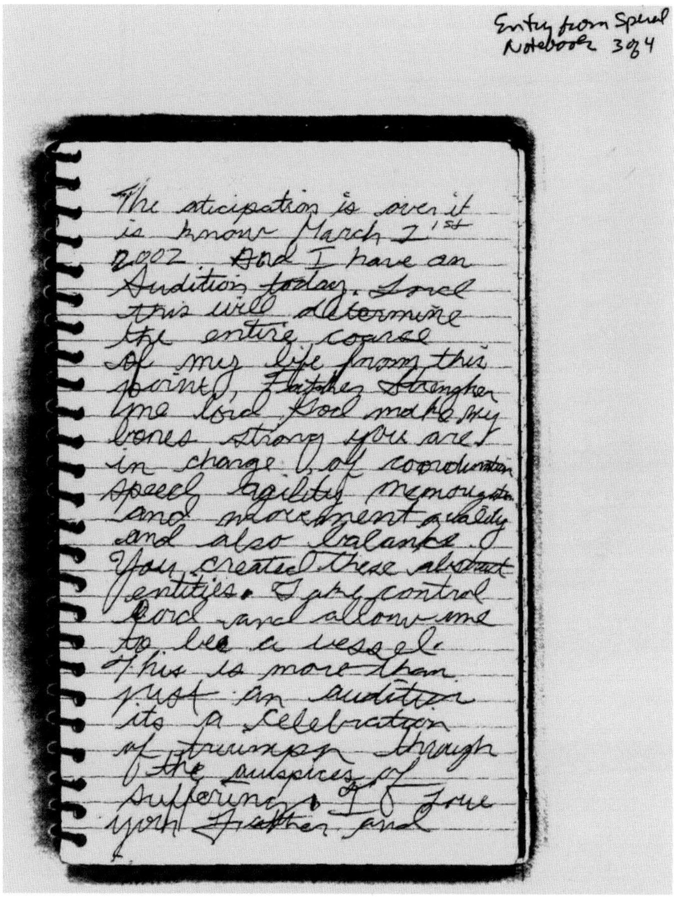

Entry from Spiral Notebook 3 of 4

The aticipation is over it is known March 1st 2002 And I have an Audition today. Lord this will determine the entire course of my life from this point, Father Strengthen me Lord God make my bones strong you are in charge of coordination speed agility memory and movement quality and also balance. You created these abstract entities. Take control Lord and allow me to be a vessel. This is more than just an audition its a celebration of triumph through the auspices of suffering. I love you Father and

Figure 6.56

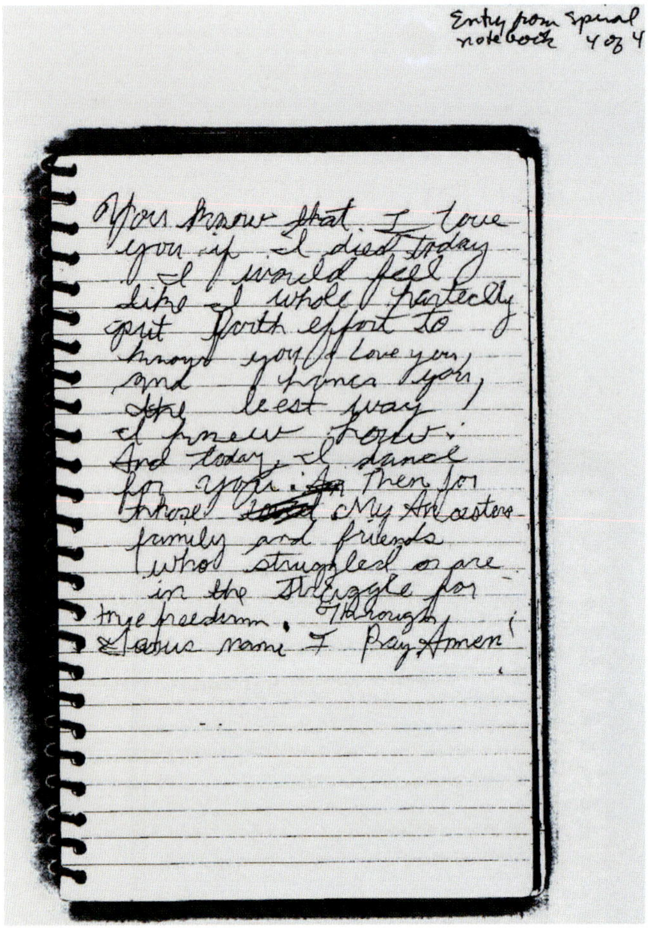

Entry from spiral notebook 4 of 4

You know that I love
you if I died today
I would feel
like I whole heartedly
put forth effort to
know you, Love you,
and [illegible] you,
the best way
I knew how.
And today I dance
for you. Then for
those My Ancestors
family and friends
who struggled or are
in the struggle for
true freedom. Through
Jesus name I Pray Amen!

Figure 6.57

Figure 6.58 The orange jacket (see also Figure 6.59 to Figure 6.70).

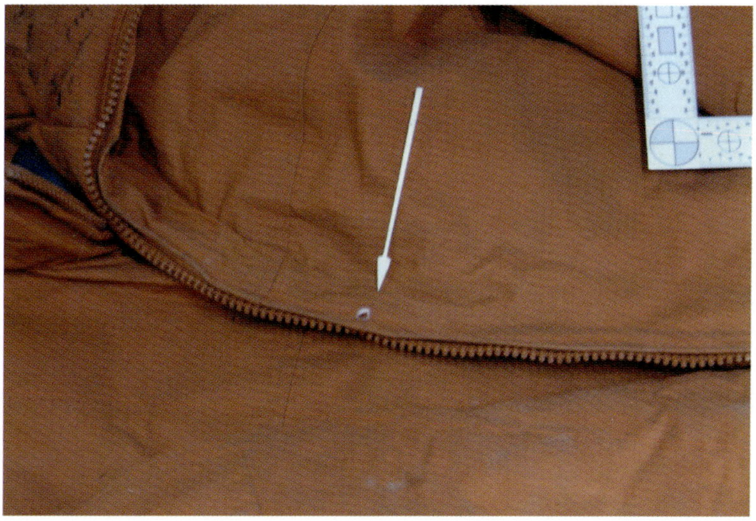

Figure 6.59

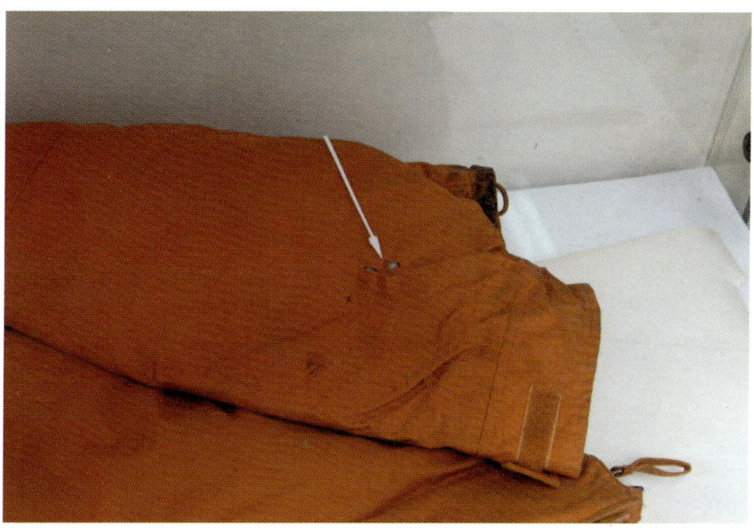

Figure 6.60

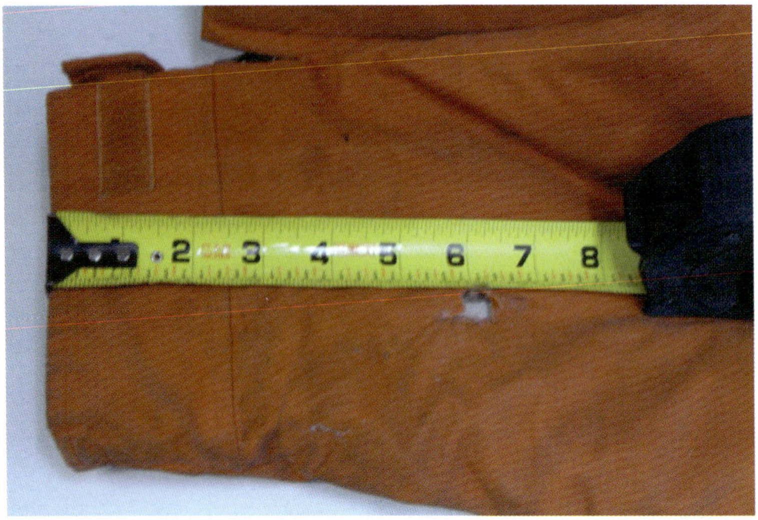

Figure 6.61

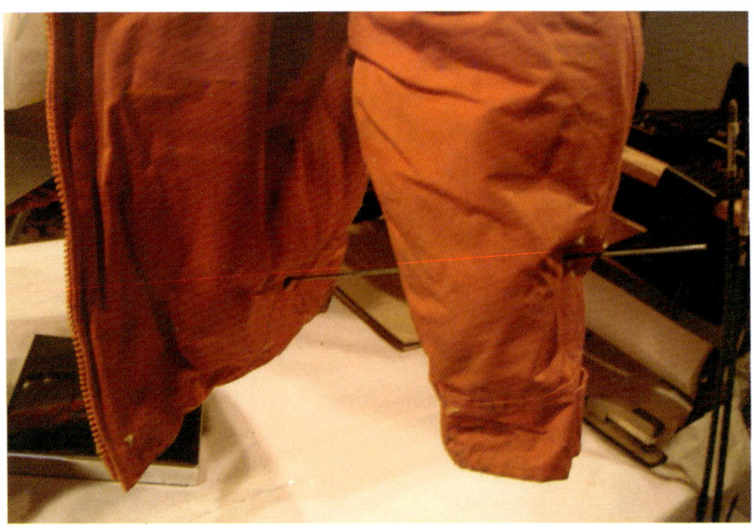

Figure 6.62

Figure 6.63

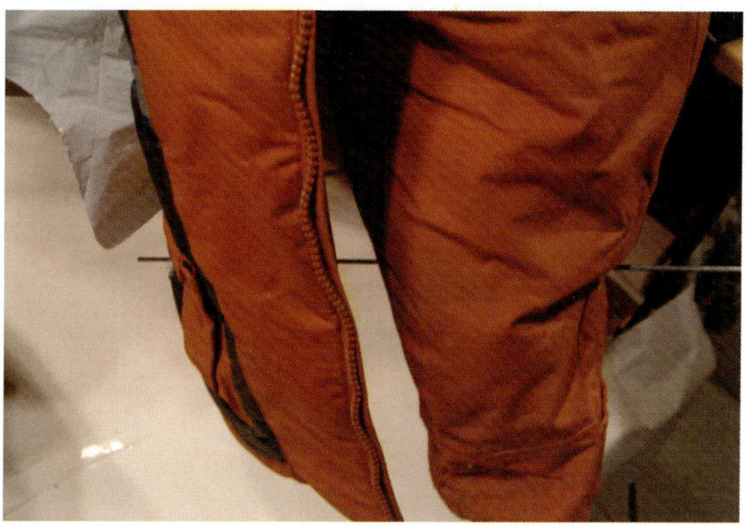

Figure 6.64

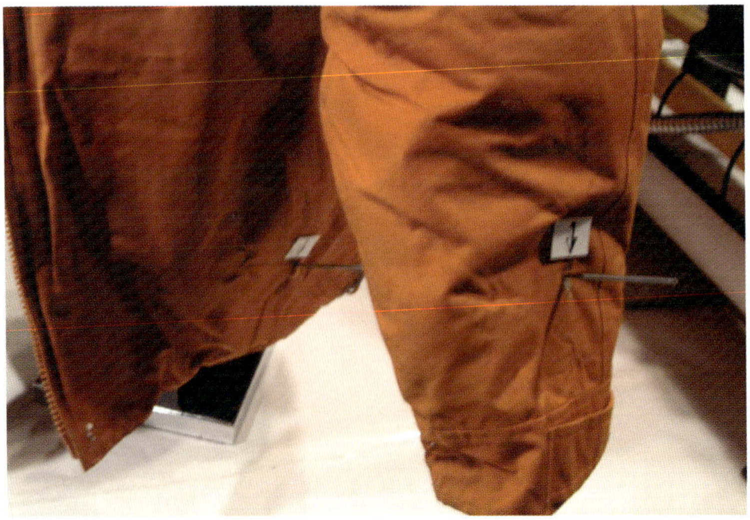

Figure 6.65

Figure 6.66

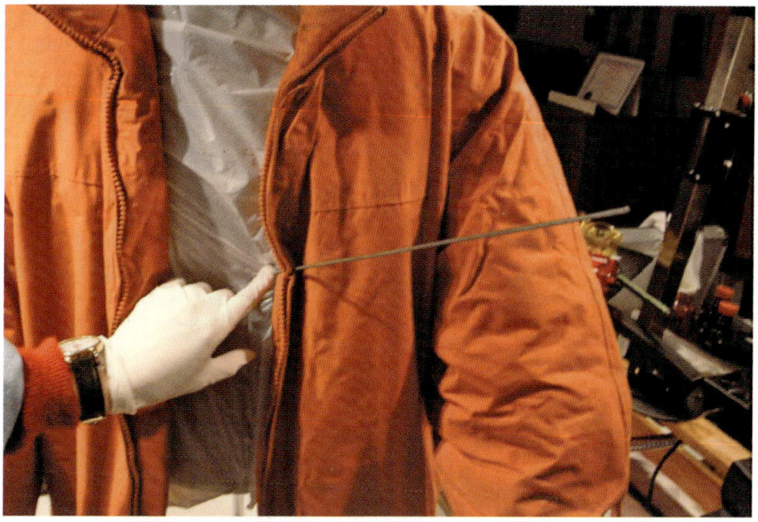

Figure 6.67

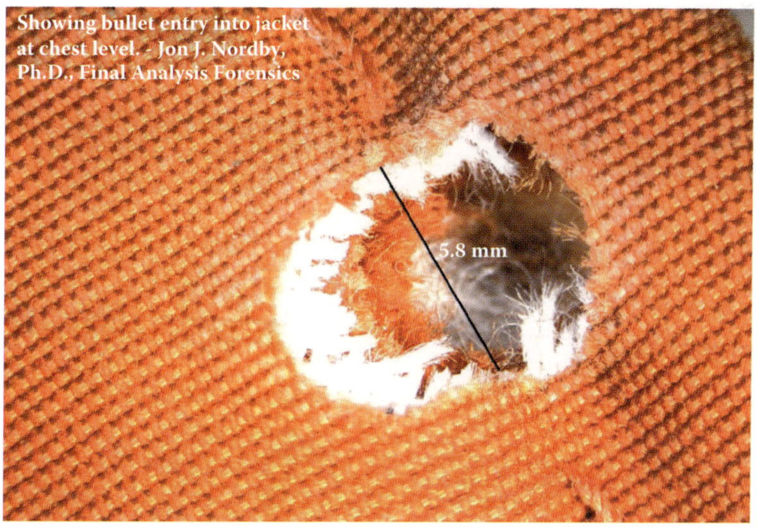

Figure 6.68

Figure 6.69

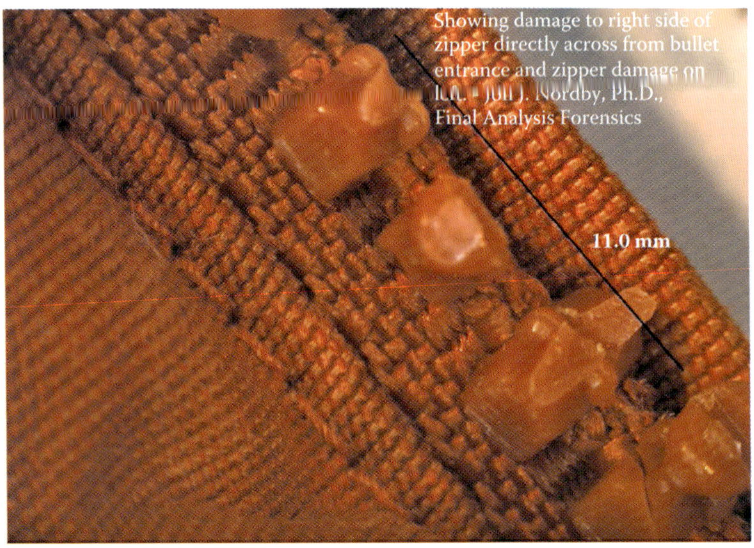

Figure 6.70

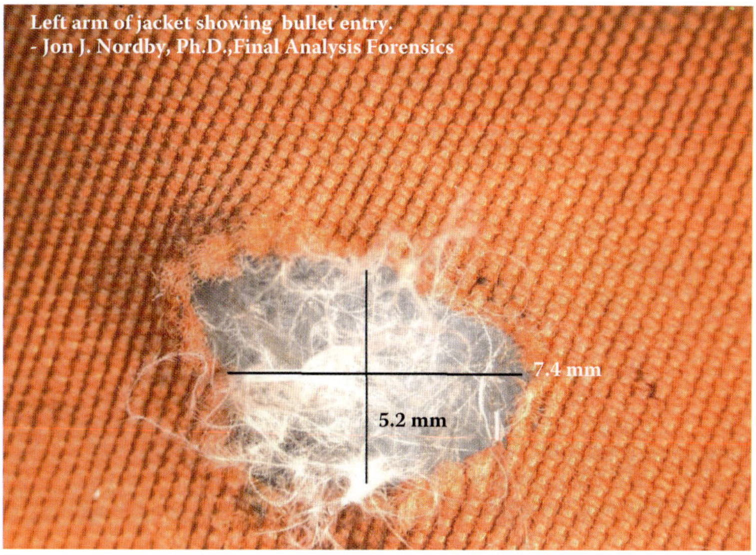

Figure 6.71 Left arm bullet entrance.

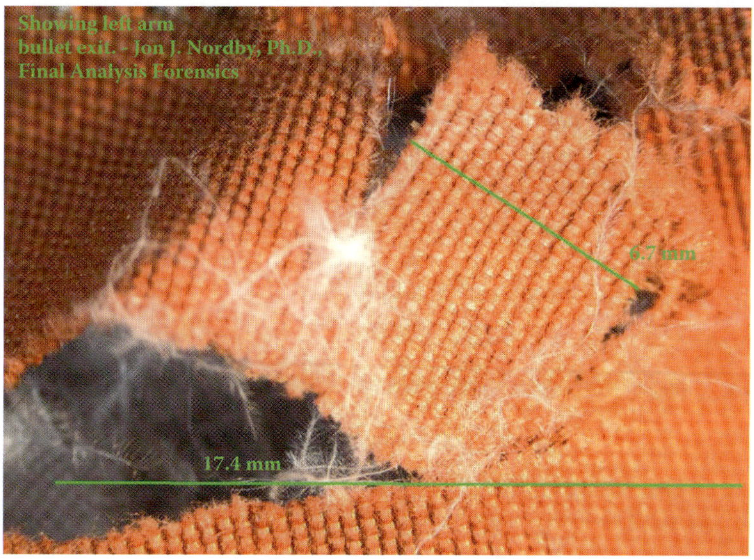

Figure 6.72 Left arm bullet exit.

Figure 6.73 Secondary entrance.

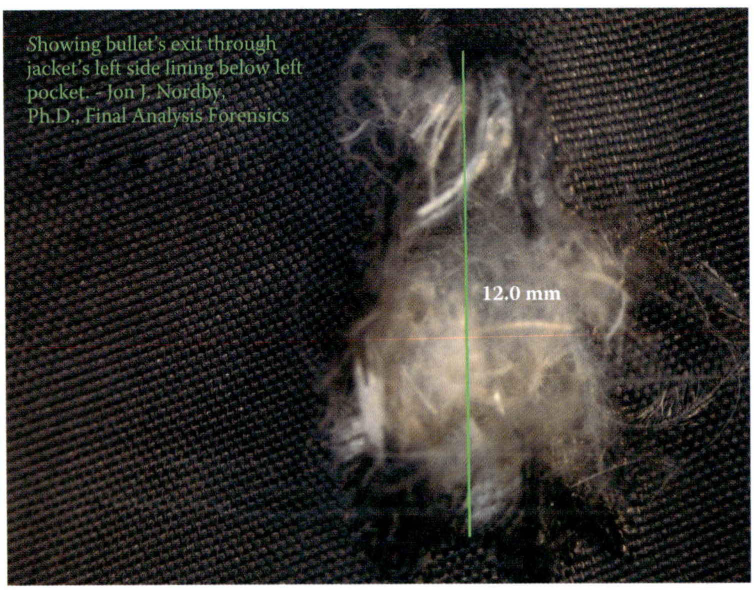

Figure 6.74 Exit through the left side lining, below the left pocket.

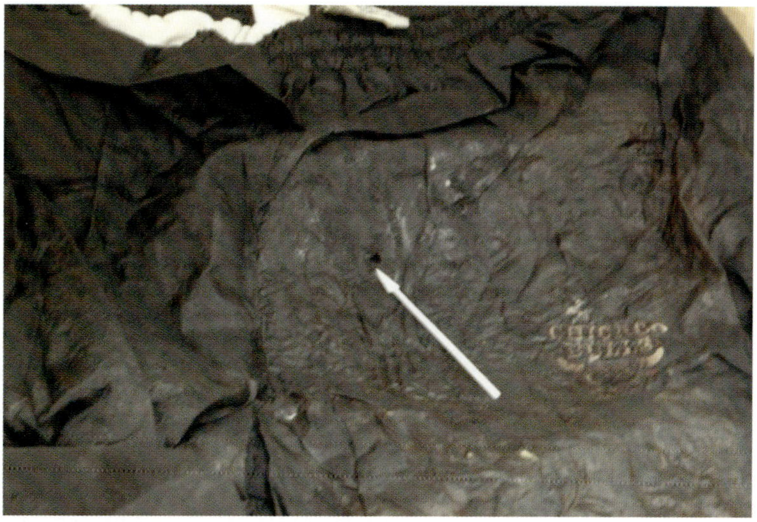

Figure 6.75 The black pants showing the entrance.

Figure 6.76 The entrance.

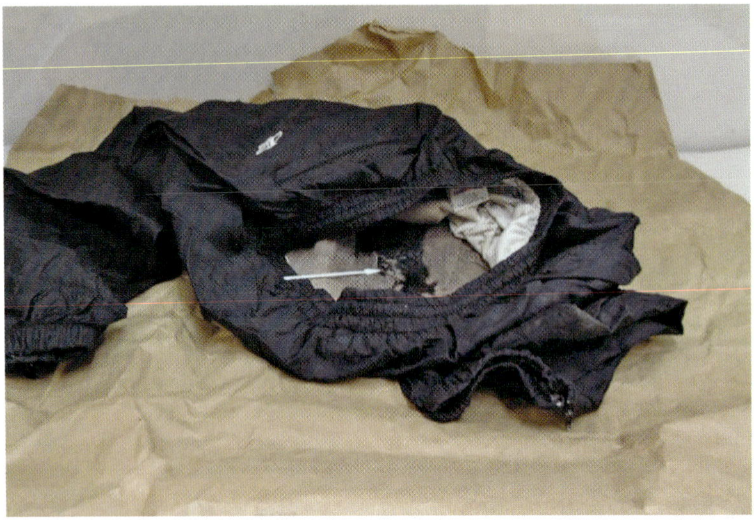

Figure 6.77 Exit from the body, then entrance on the pants.

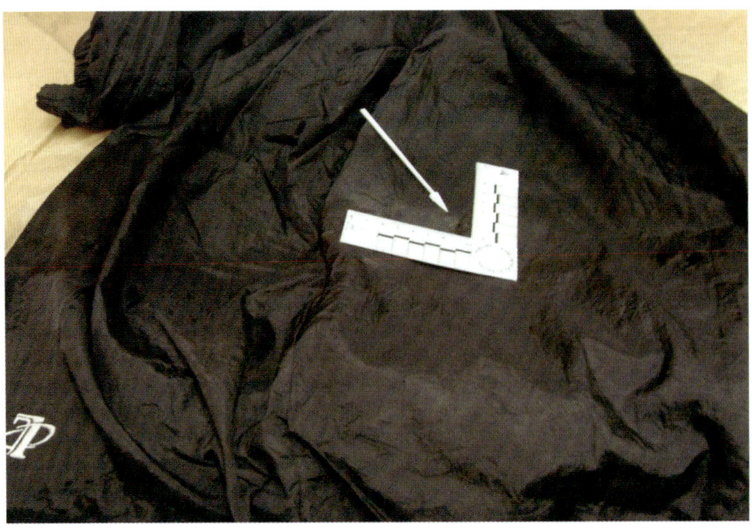

Figure 6.78 Exit from the outside surface of the pants.

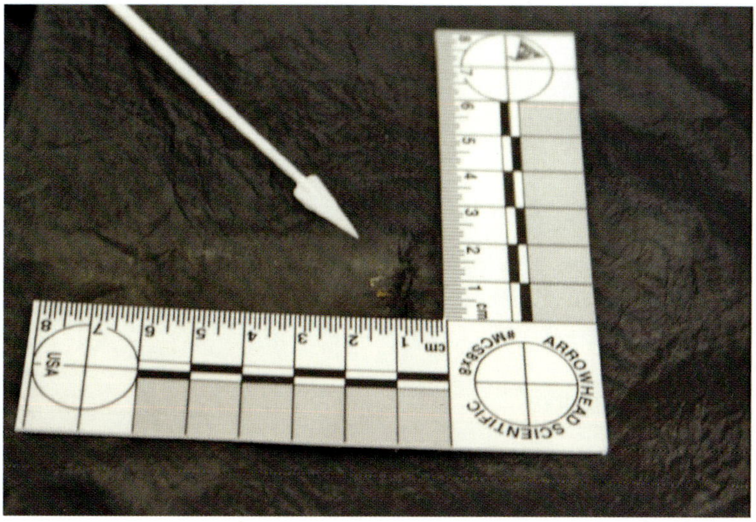

Figure 6.79 The exit.

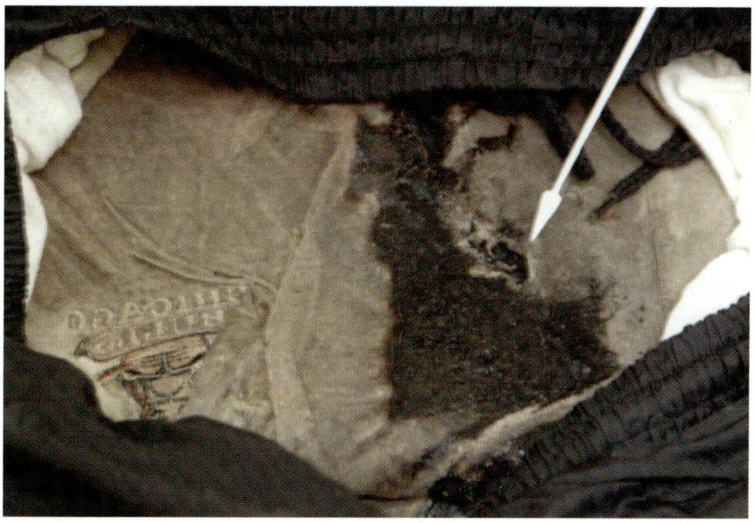

Figure 6.80 Exit from the body into the inside of the pants.

Figure 6.81 Intact bottle.

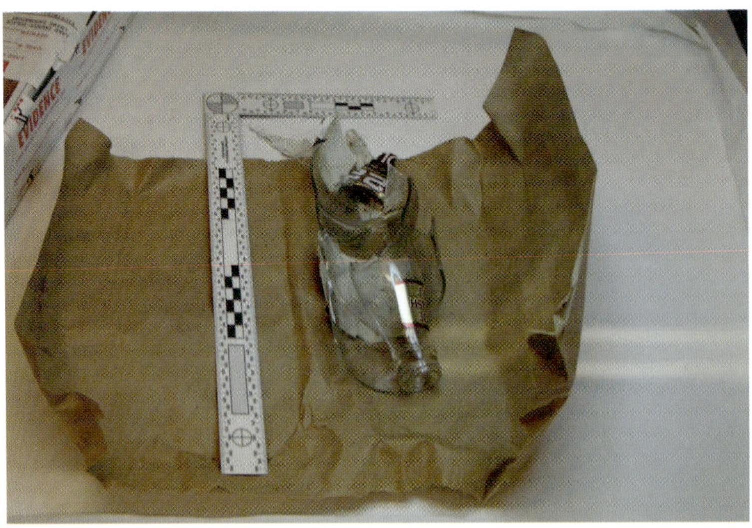

Figure 6.82 Fractured bottle.

Figure 6.83 40 cal Glock.

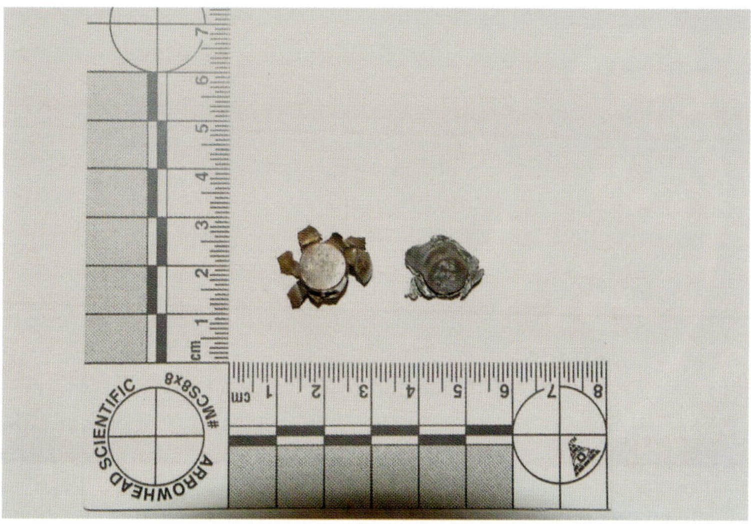

Figure 6.84 Recovered bullets.

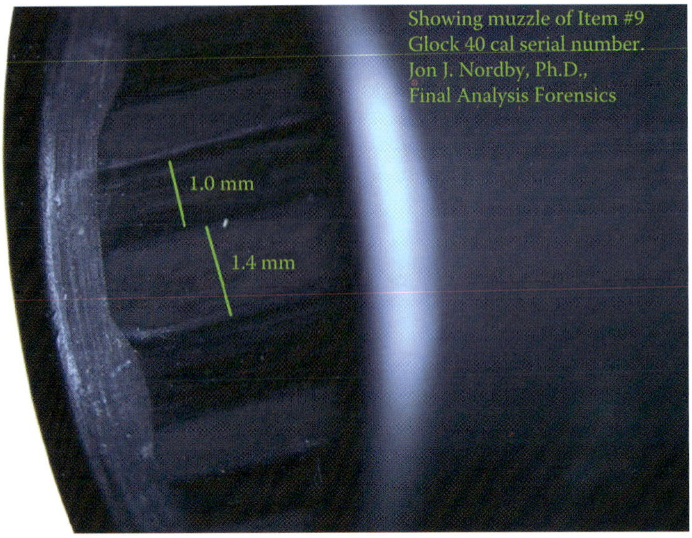

Figure 6.85 The muzzle of Item #9—the Glock—lacks typical lands and grooves with its polygonal barrel.

Figure 6.86 Recovered cartridge cases from the scene were compared in the hopes of establishing the number of shooters. Despite stories, the story must be confirmed or rejected, given the available evidence, if possible.

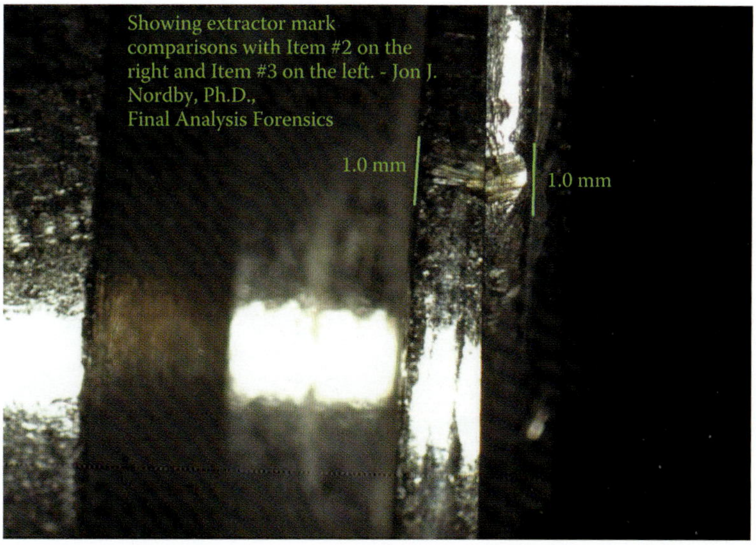

Figure 6.87 Extractor or ejector marks are also examined and compared.

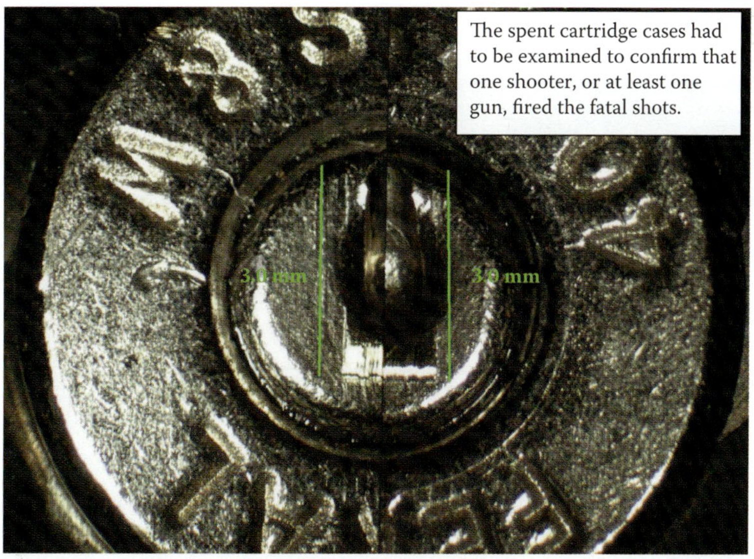

Figure 6.88 Comparisons involve measurements at several different levels, given the depth-of-field limitations of most optical microscopes.

Figure 6.89 Showing the initial measurements.

Figure 6.90 The cartridge case—basic measurement of the firing pin and breach block impressions.

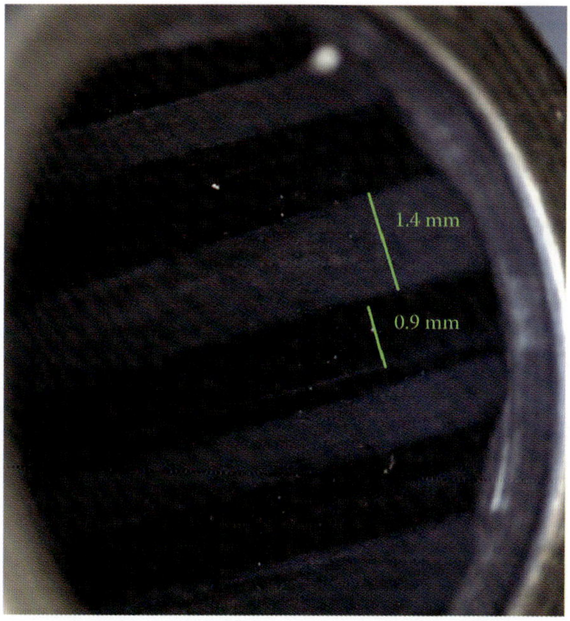

Figure 6.91 Measurements include those from a known pistol's barrel.

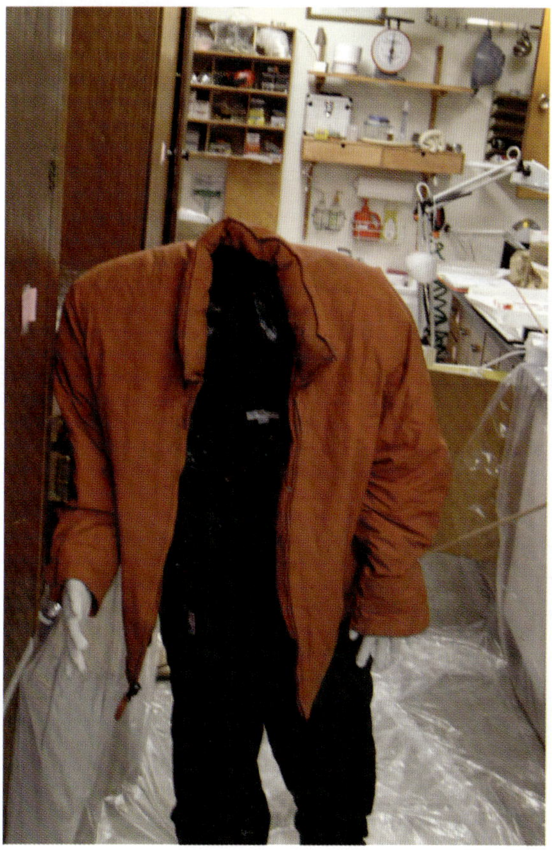

Figure 6.92 The jacket.

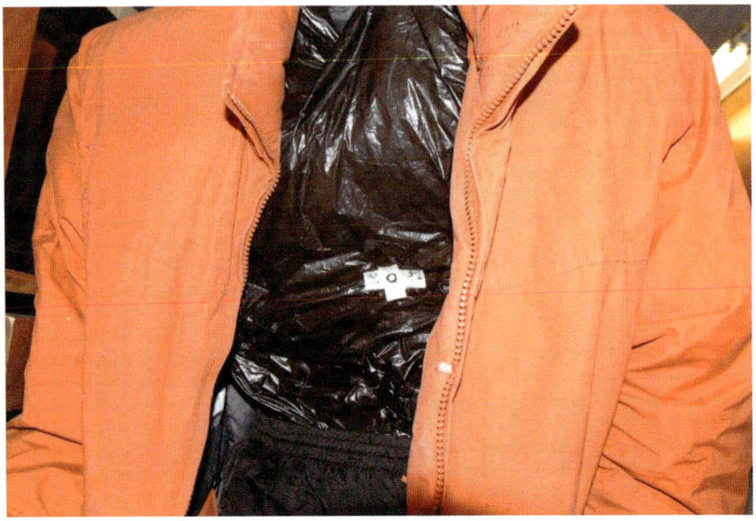

Figure 6.93 Close-up of top of jacket.

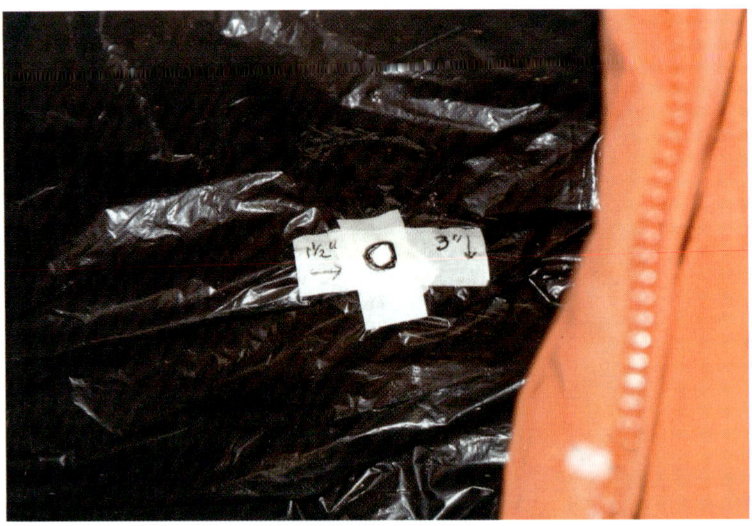

Figure 6.94 Various measurements inside the jacket (see also Figure 6.95 to Figure 6.105).

Figure 6.95

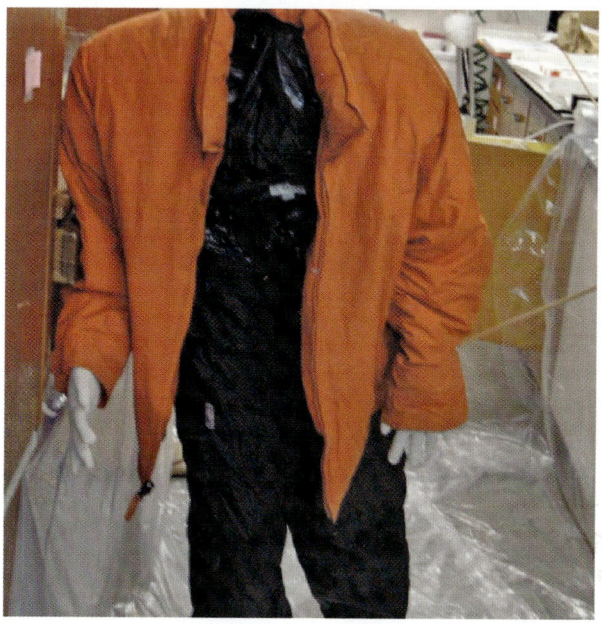

Figure 6.96

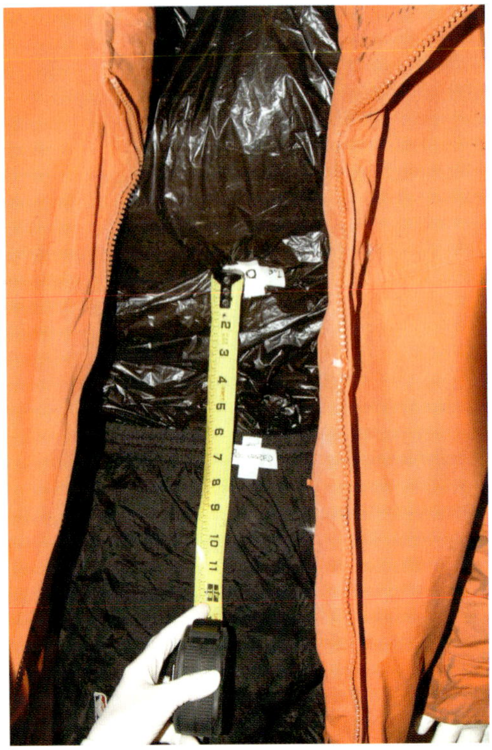

Figure 6.97

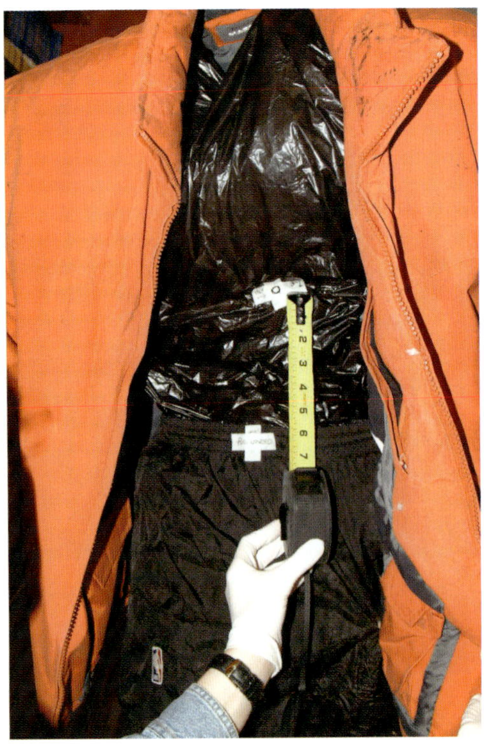

Figure 6.98

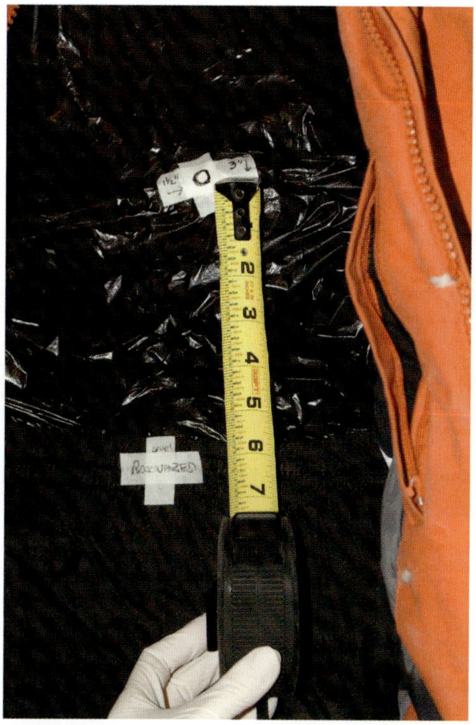

Figure 6.99

Figure 6.100

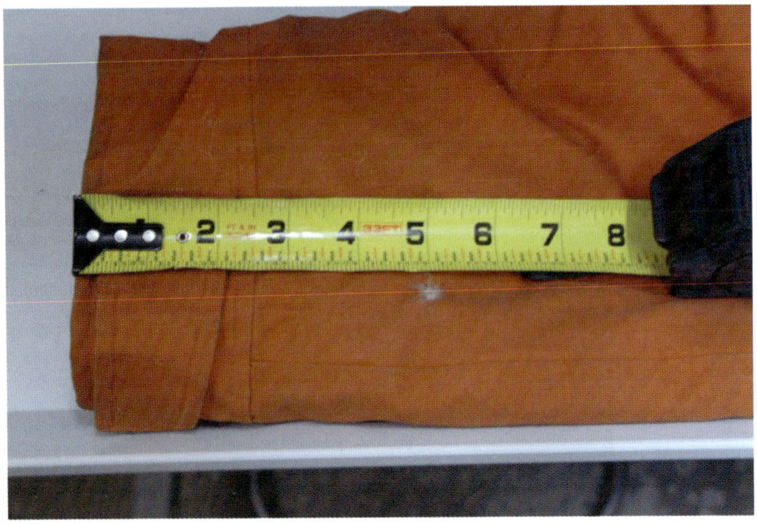

Figure 6.101

Figure 6.102

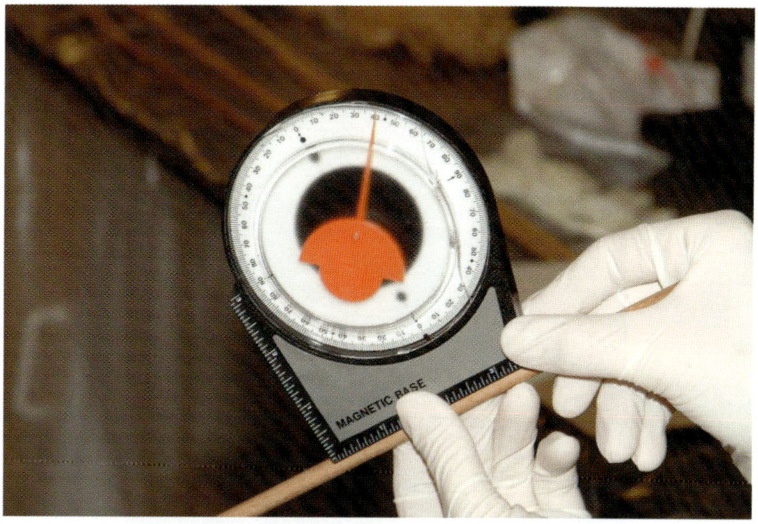

Figure 6.103

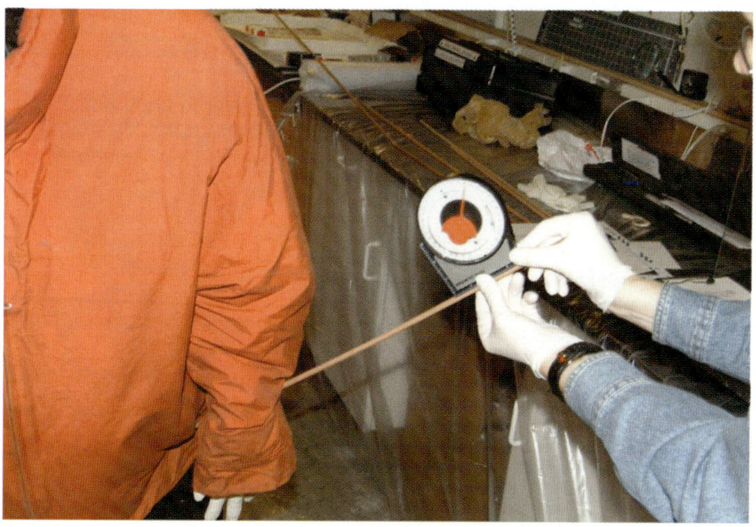

Figure 6.104

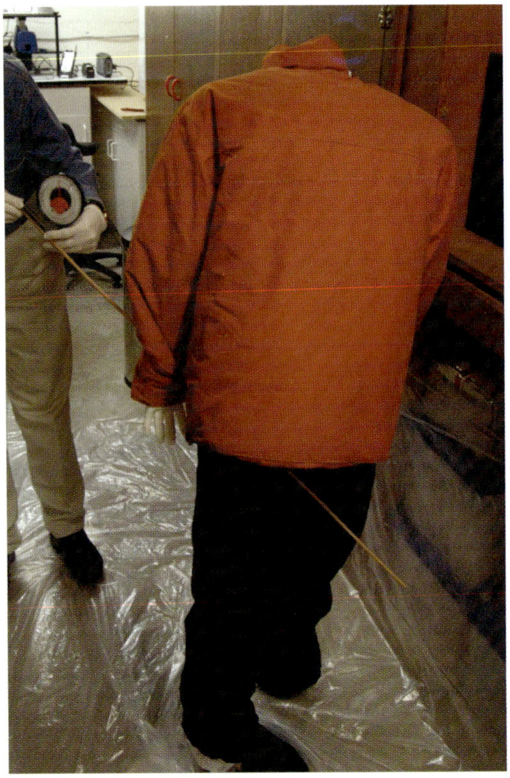

Figure 6.105

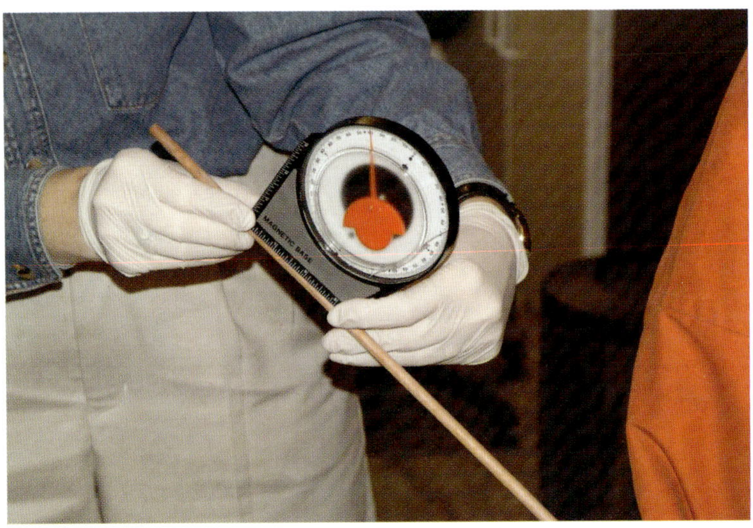

Figure 6.106 An angle measurement.

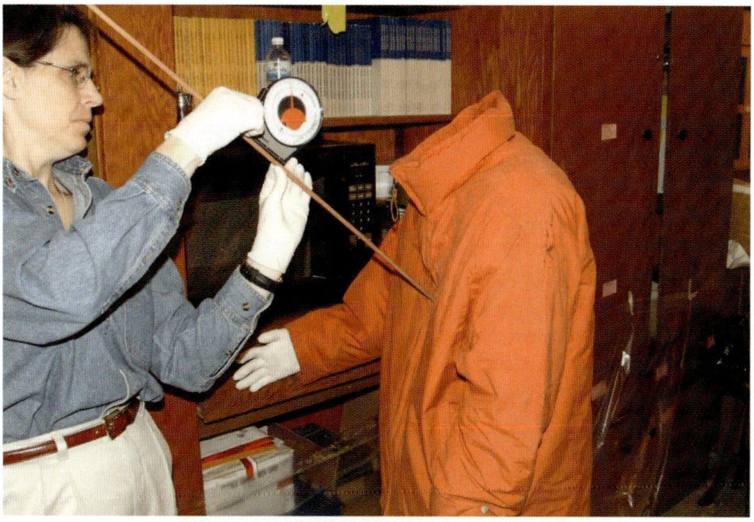

Figure 6.107 Lab assistant Peg Foley uses the angle measurement device on a trajectory rod.

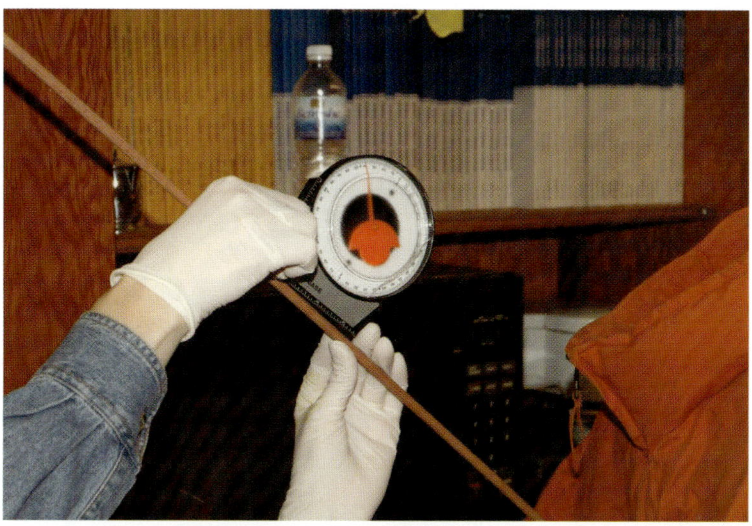

Figure 6.108 Photo documentation of measurements can also be useful when presenting results.

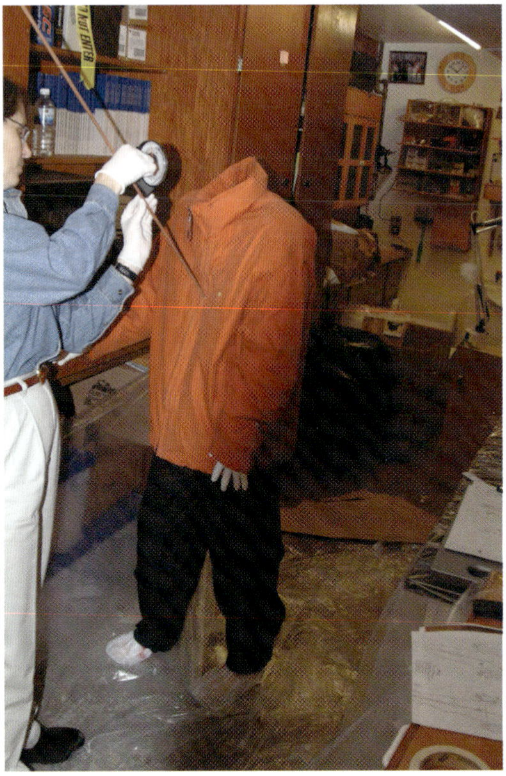

Figure 6.109 All clothing is examined on the appropriate-sized model.

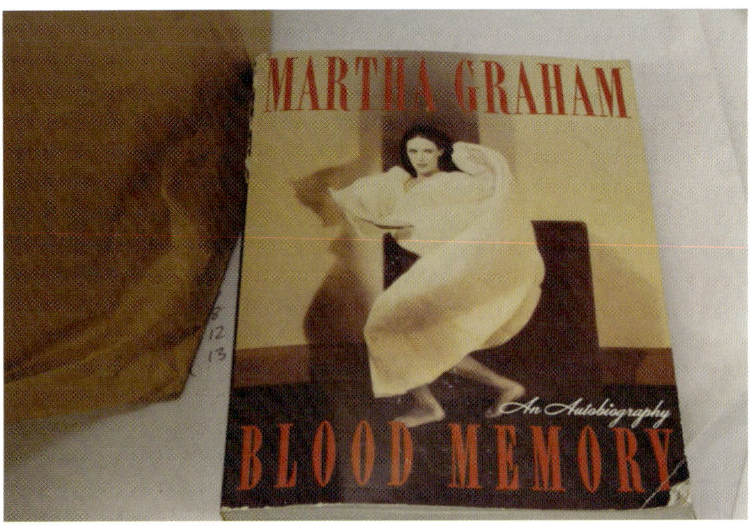

Figure 6.110 A book found in the black shoulder bag recovered on the pavement.

XRF Charts Showing Elemental Spectra

Figure 6.111 to Figure 6.130 show the XRF charts of elemental spectra.

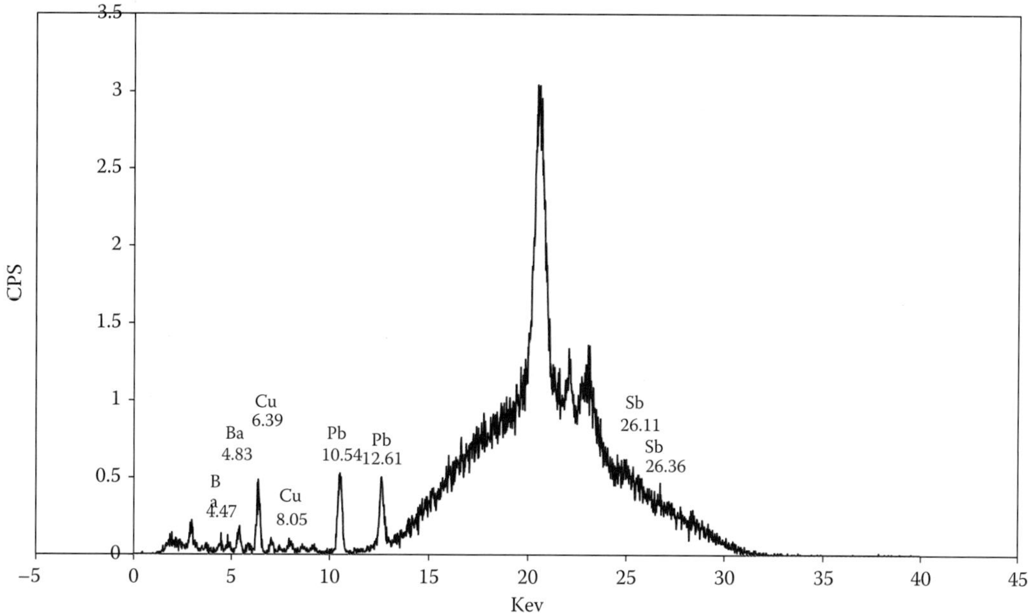

Figure 6.111

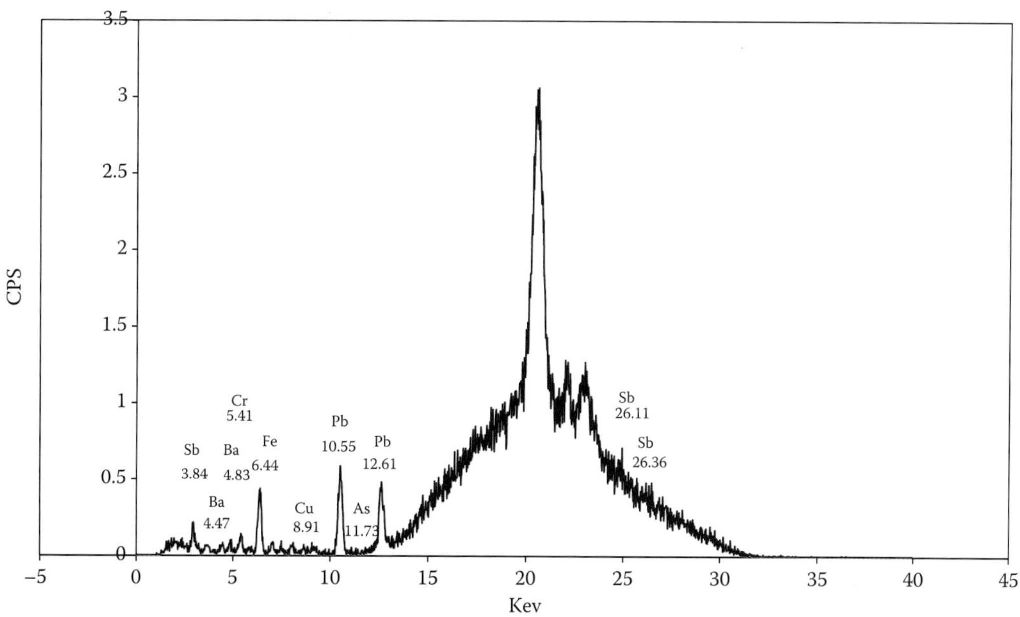

Figure 6.112

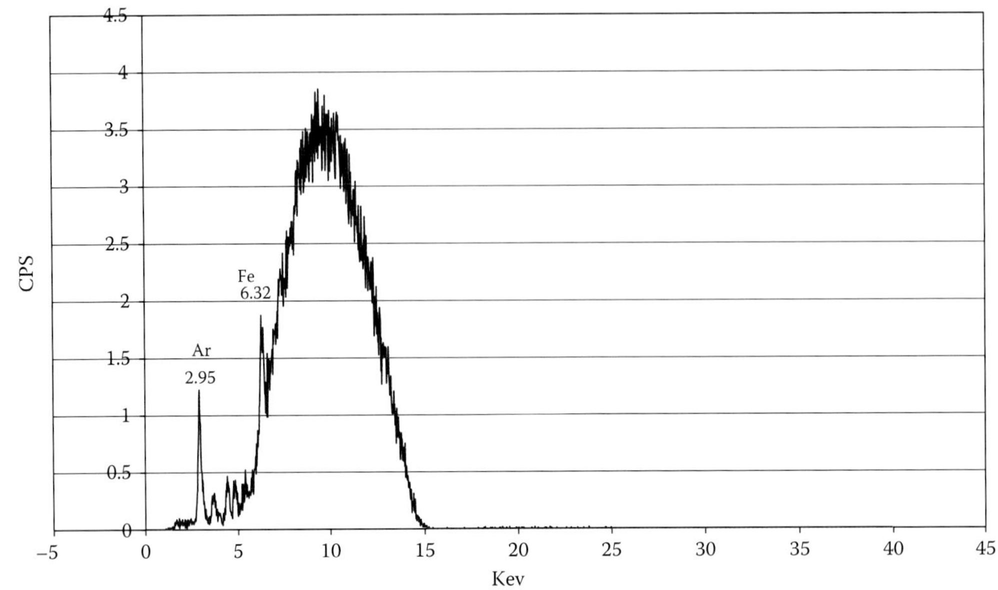

Figure 6.113

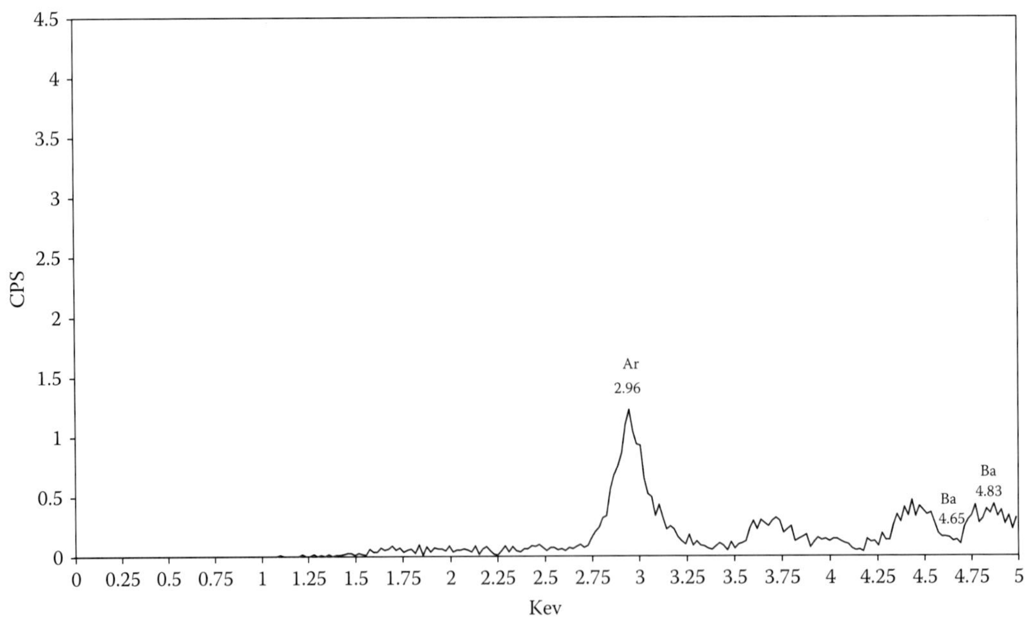

Figure 6.114

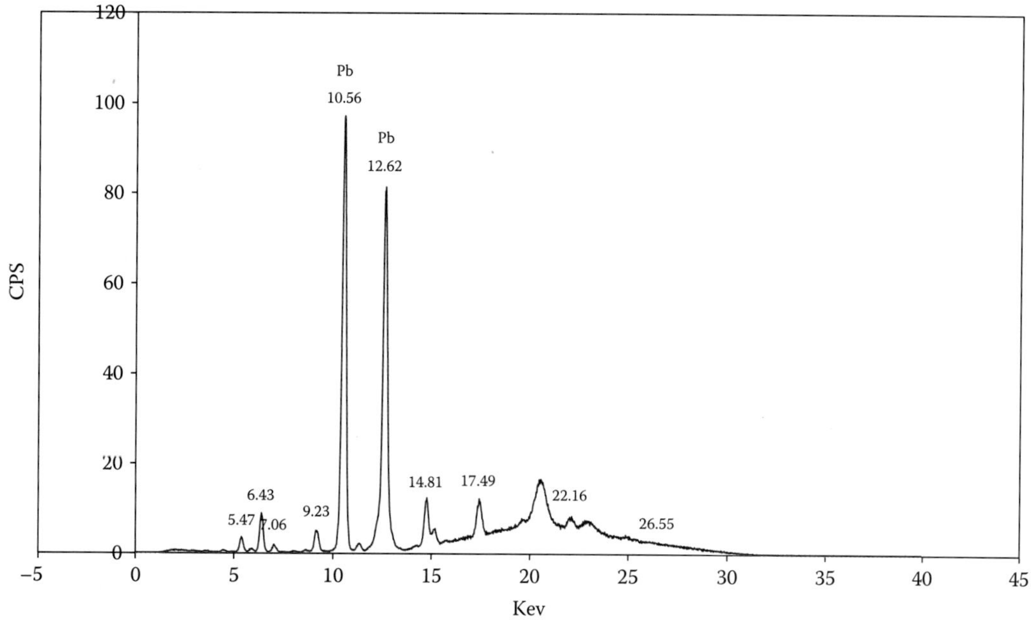

Figure 6.115

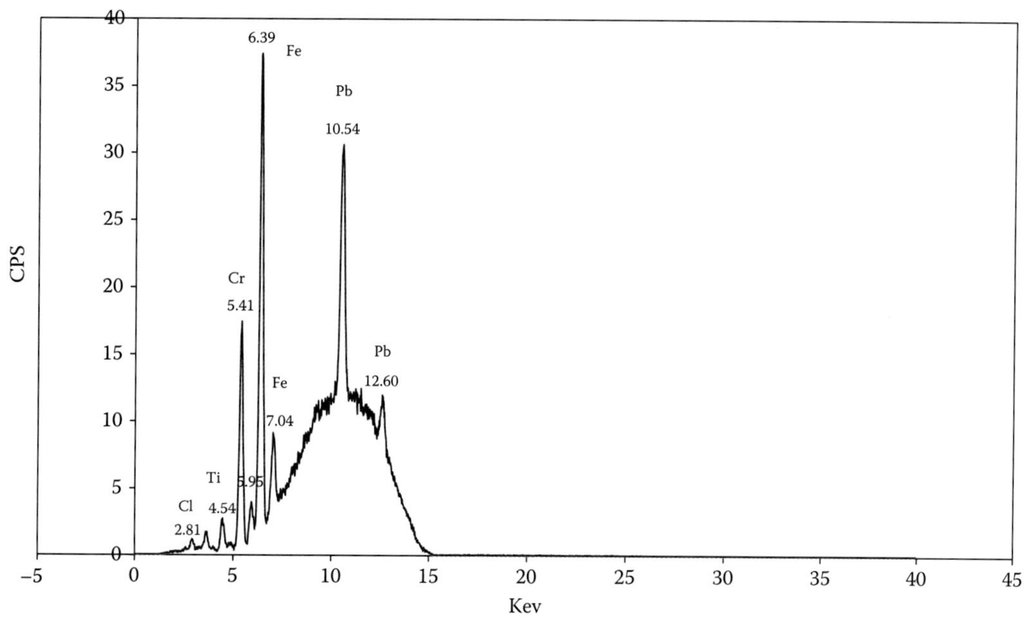

Figure 6.116

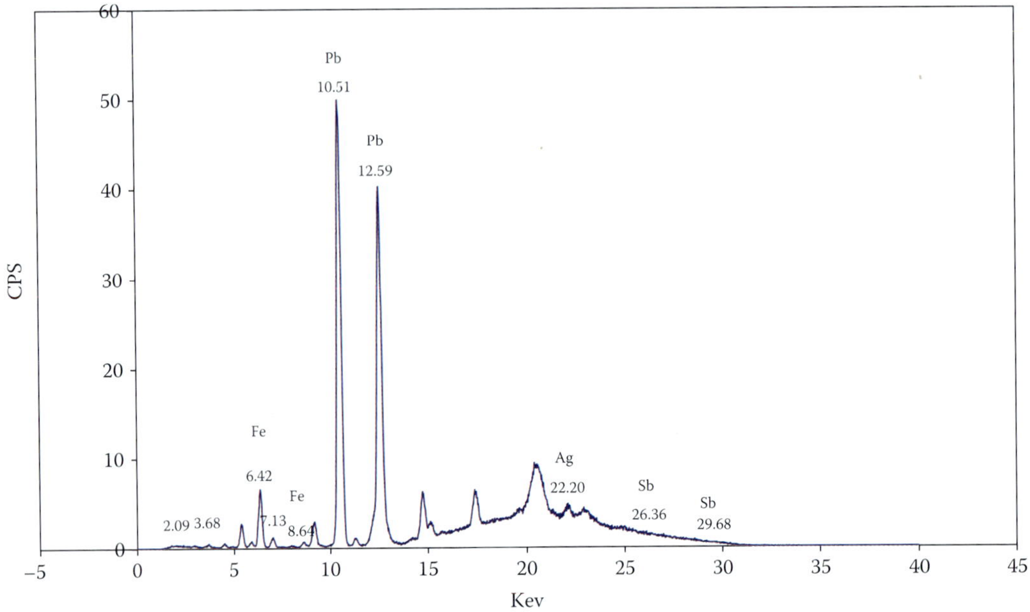

Figure 6.117

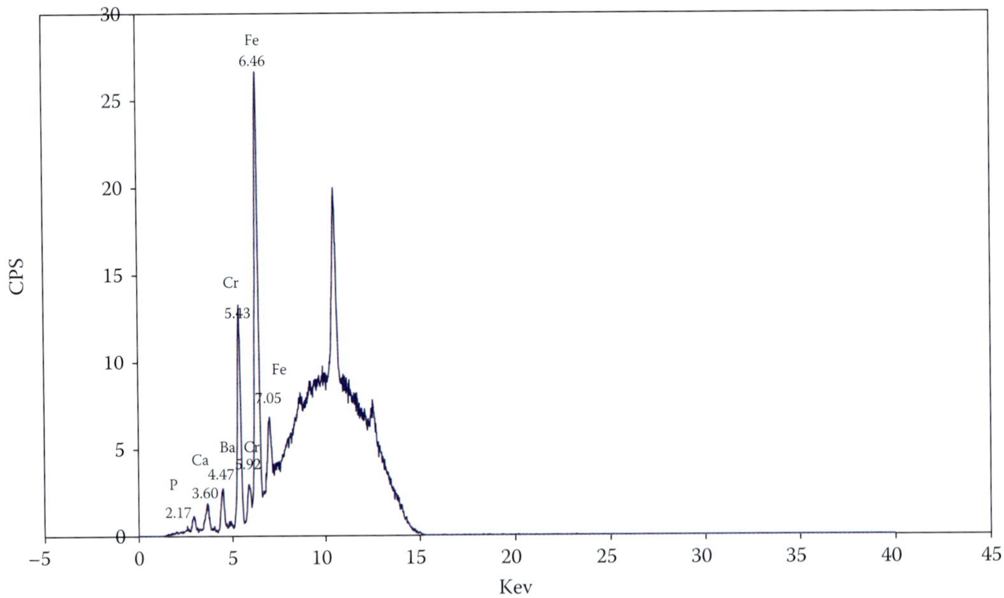

Figure 6.118

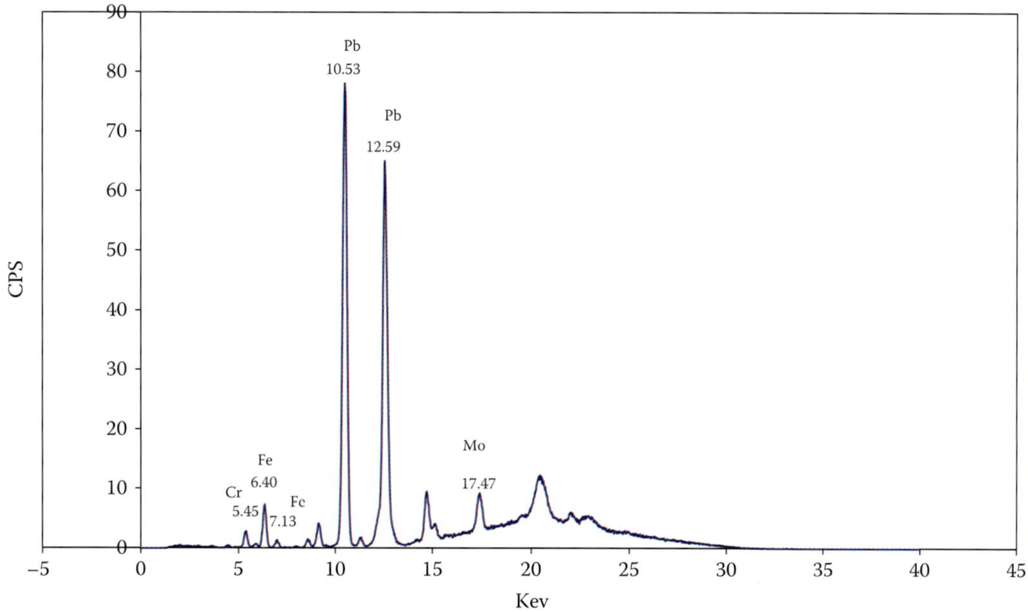

Figure 6.119

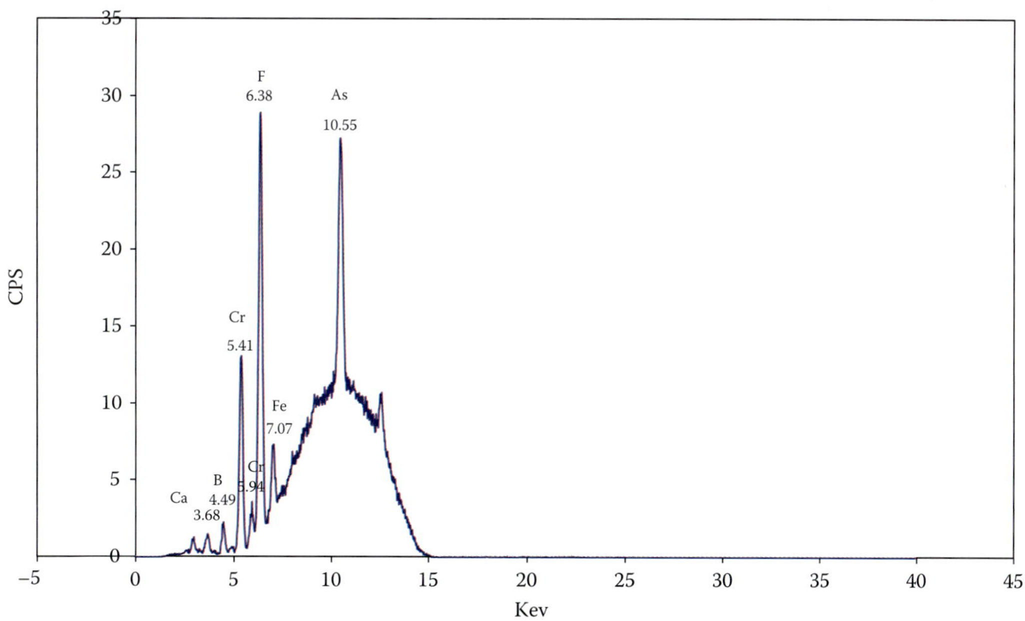

Figure 6.120

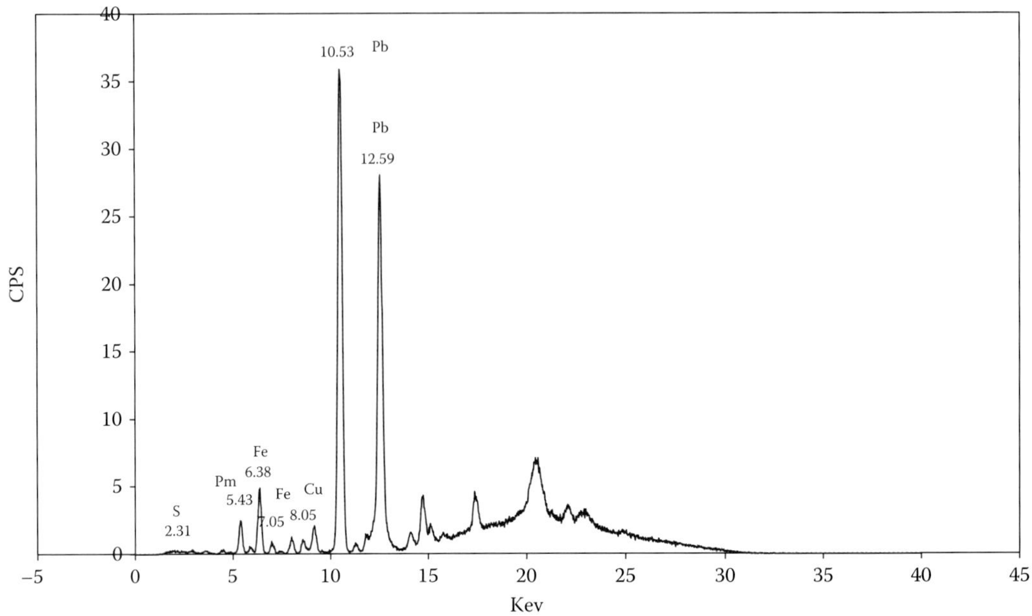

Figure 6.121

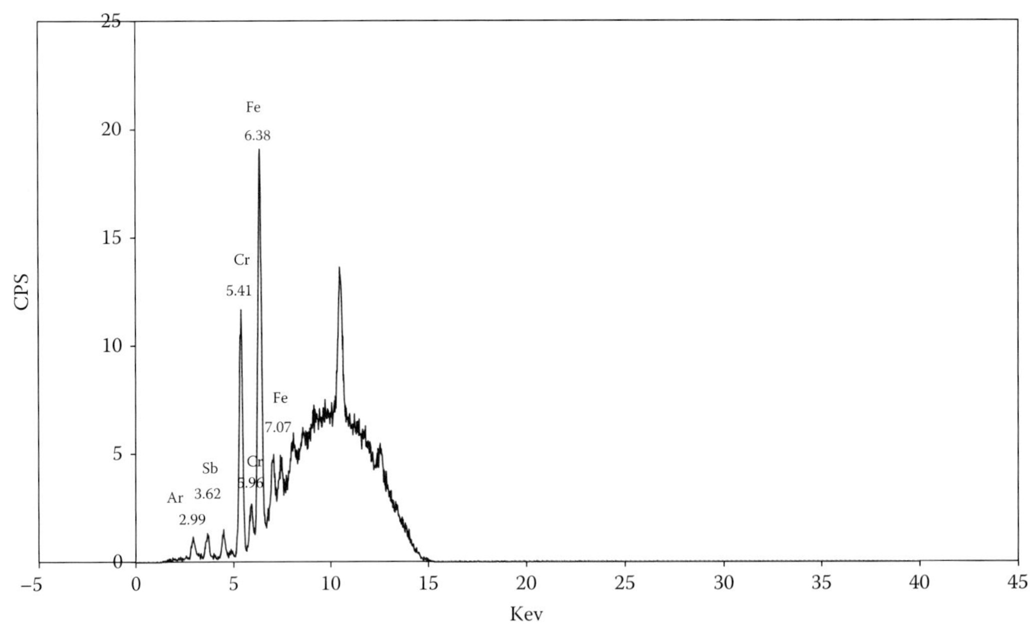

Figure 6.122

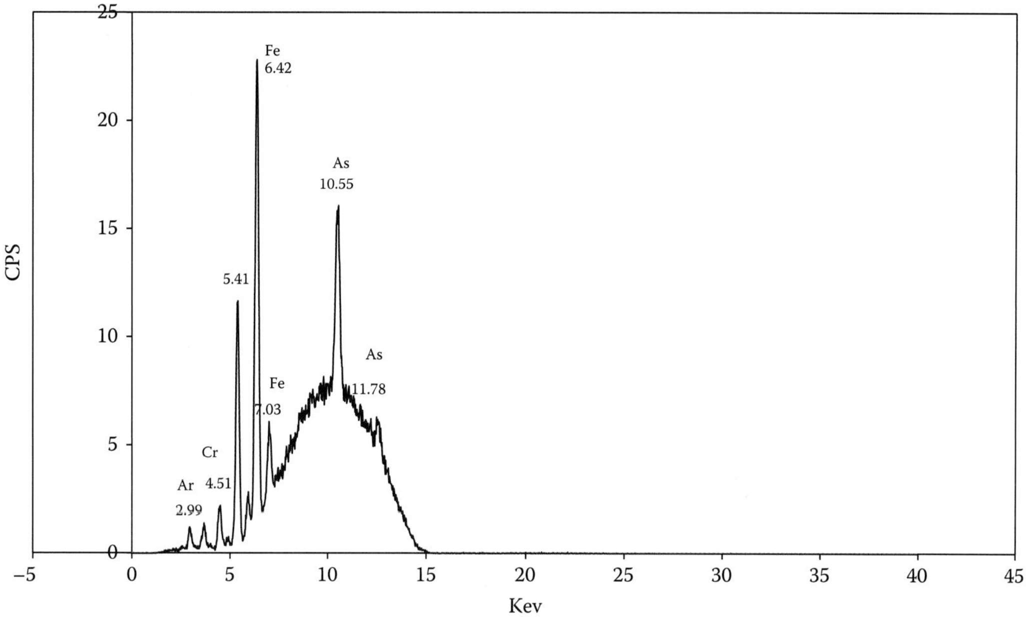

Figure 6.123

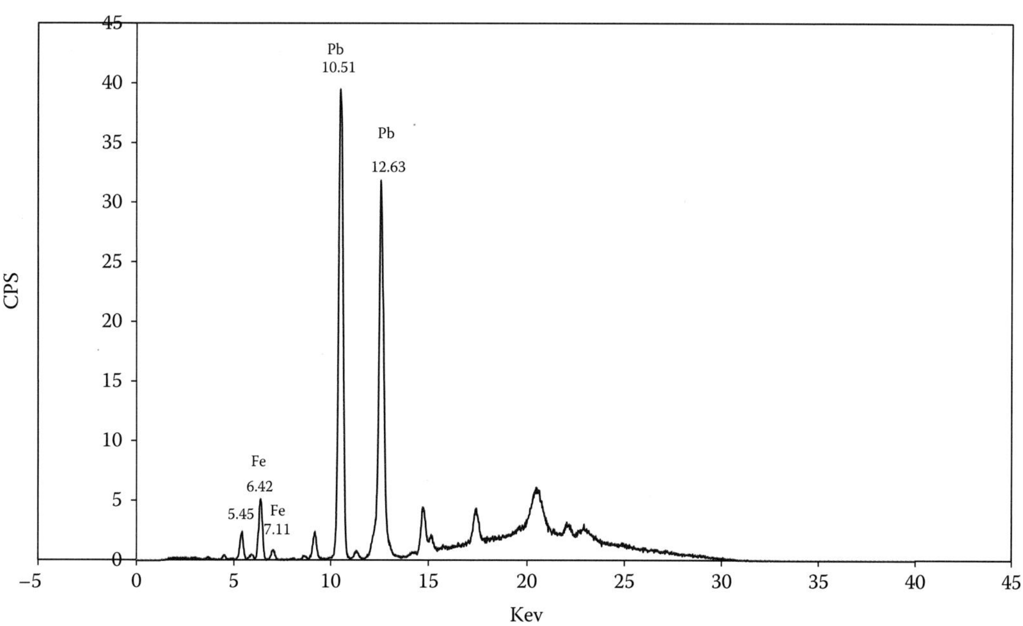

Figure 6.124

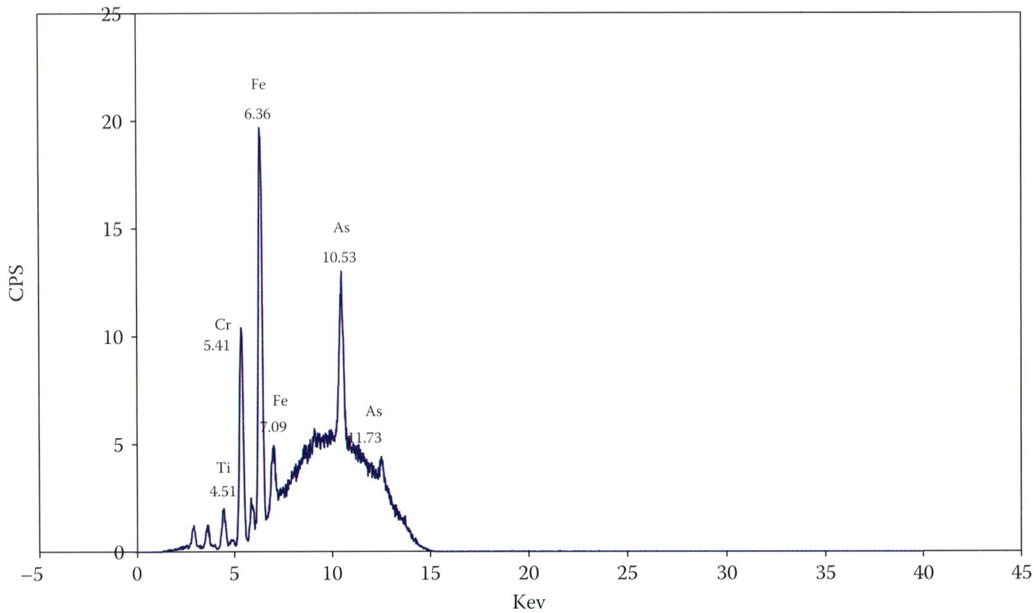

Figure 6.125

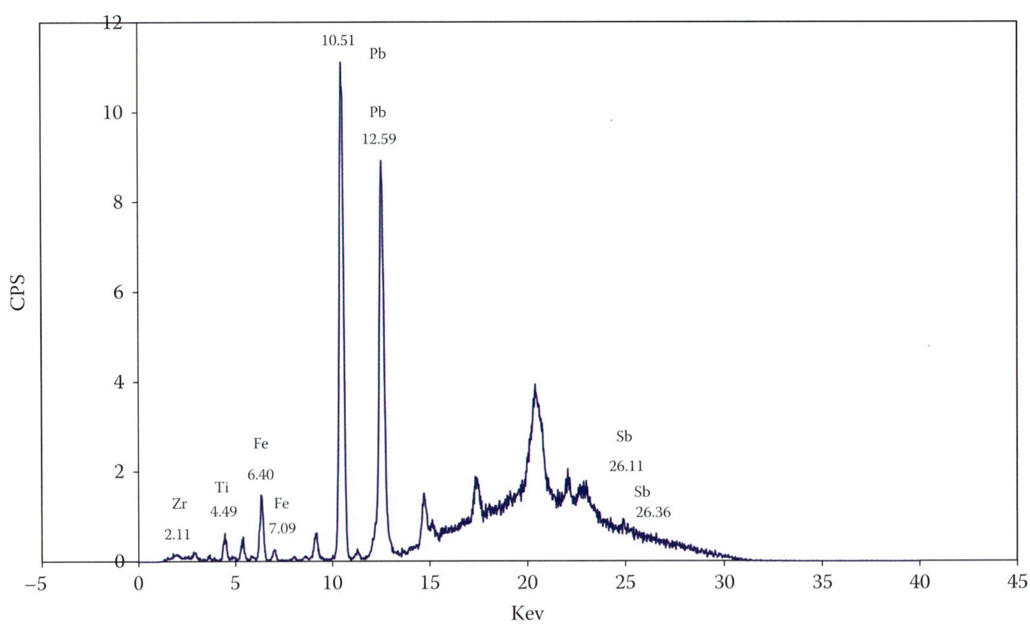

Figure 6.126

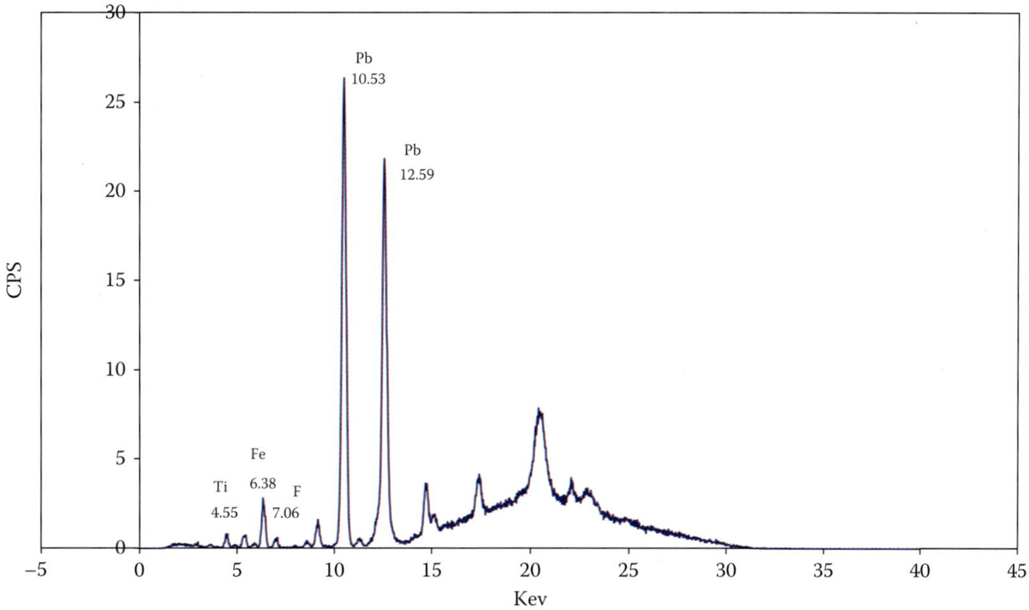

Figure 6.127

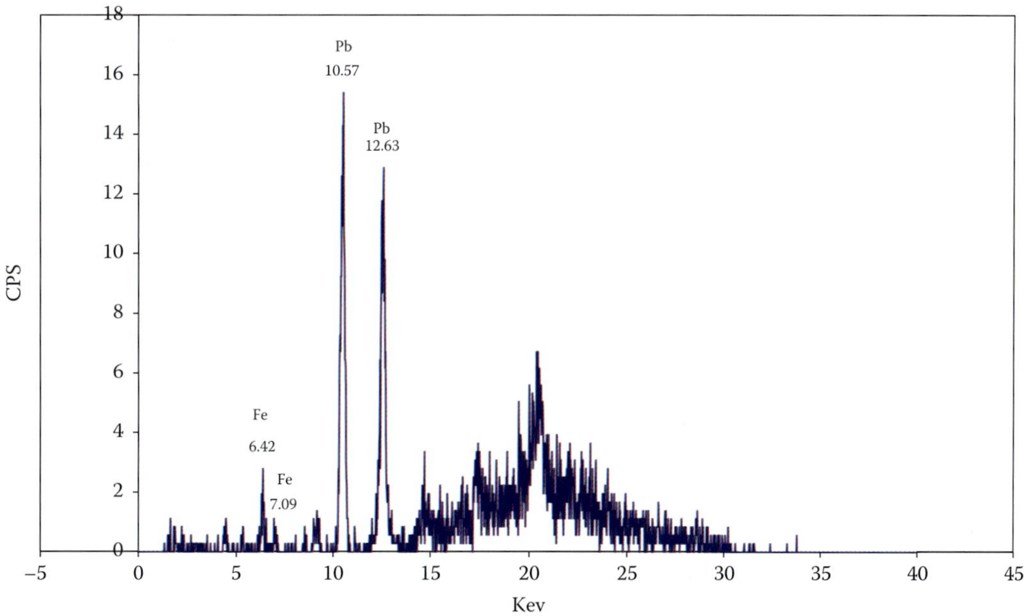

Figure 6.128

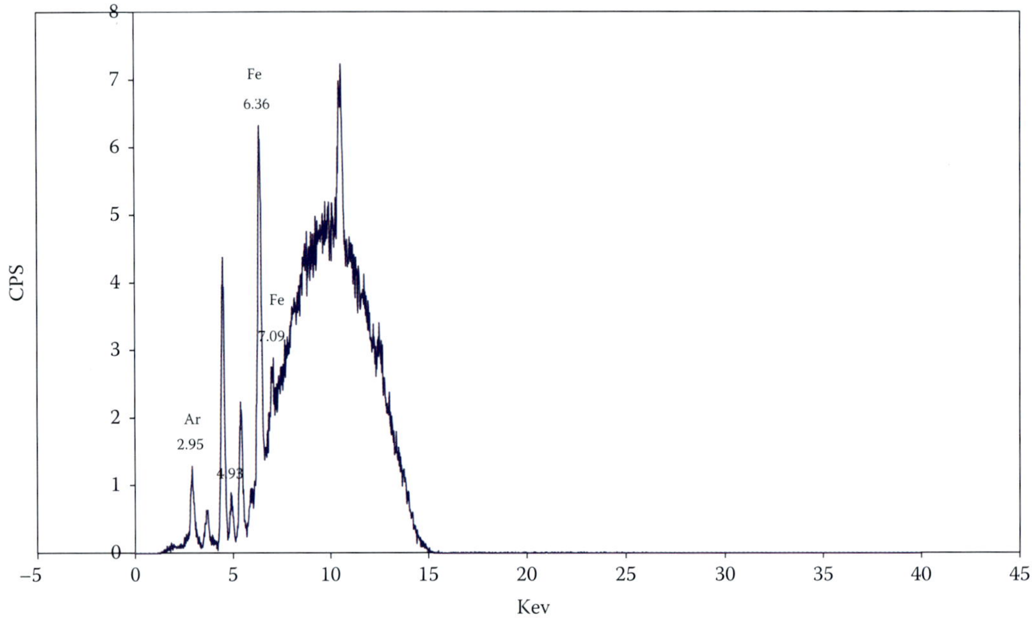

Figure 6.129

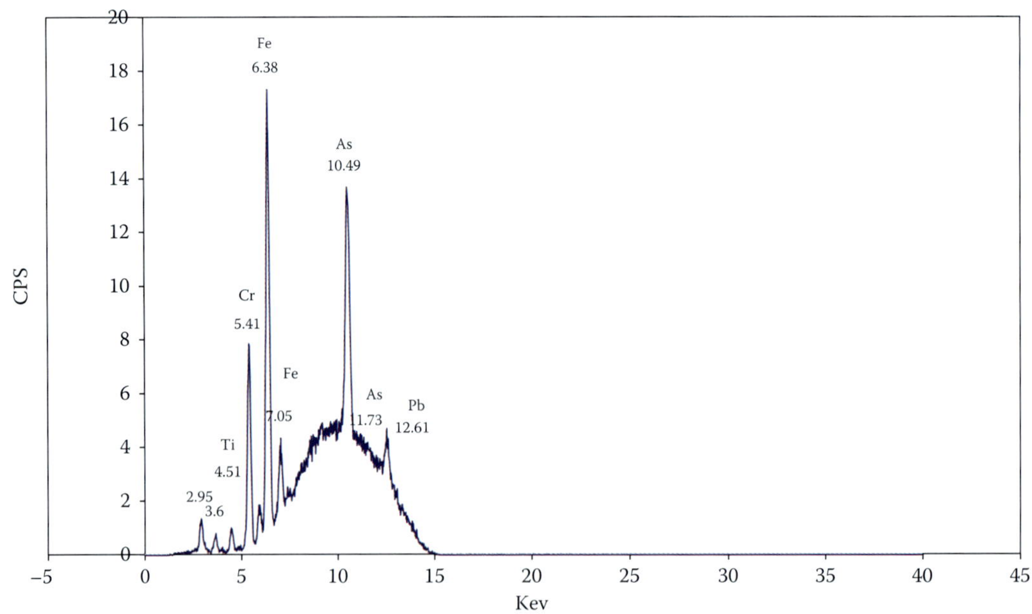

Figure 6.130

Relevant Scene Photographs

Figure 6.131 to Figure 6.143 are relevant scene photographs showing the location of the evidence items. It is always necessary to think through the photos, to see if the location of the items makes physical sense.

Figure 6.131 This photograph shows the bottles, bag, and Martin. The unbroken bottle was said to be in the decedent's right hand, and the broken bottle was said to be in the decedent's left hand. Martin was rolled over toward the north (toward the background, or his right) from face down to face up.

Figure 6.132 This photo shows the relative positions of the bottles and the black bag.

Figure 6.134 shows the broken bottle. The bottle was broken into pieces when recovered and taken into evidence. If Martin held the "jagged bottle" in his left hand and dropped it here, then the bottom was broken out earlier (as stated by witnesses). But if Martin dropped it here when shot, then the bottle had to break into these pieces (as shown) when it fell from Martin's left hand.

Figure 6.133 The bloodstains from Martin's left leg appear on the pavement near the center line. This shows his position when he lay on the pavement bleeding.

Figure 6.134 The broken bottle.

Figure 6.135 Photo of the unbroken bottle on the pavement.

Figure 6.136 Police lab photo of the bottle pieces with fingerprint powder.

Figure 6.137

Figure 6.138 Another view showing the improbable relationships among the decedent and bottles, even if they were somehow "moved."

Figure 6.139 Various scene photographs (see also Figure 6.140 to Figure 6.143).

Figure 6.140

Scene photos have a story to tell. Once told, the science must be applied to investigate those stories. Sometimes the investigation can go forward, other times it languishes in a stagnant state.

Clearly in the photos, the neck of the bottle is broken. If the bottle broke in this position, then a piece from the neck remains to be missing from the scene. The neck and bottle fractured at the same time, otherwise the "jagged bottle" would be nothing more than a pile of debris in his left hand. But since there is a piece missing from the neck, it follows that the bottle did not break into the pieces seen above at this scene. It follows that the bottle broke, as seen above, when the bottom was broken out since all the fracture lines appear to indicate that the missing piece could not have been broken out without the entire bottle being broken.

The scene pictured in Figure 6.141, if the decedent were simply turned over where he fell, makes no sense. The bottles would not appear in their present locations if they were in his hands. The broken bottle was said to be in his left hand with the unbroken bottle in his right hand. No bloodstains appear anywhere between the two bottles. Officers did not say that they moved him other than to "turn him over where he fell."

Note that Mr. Martin was turned over. Yet the officers claimed that he held the broken bottle in his left hand. If he landed face down and was turned onto his back, then the broken bottle, as it lies on the pavement, would have to have been in his right hand. This problem apparently never occurred to the officers, who seemed to create their accounts well after the fact. Further, the unbroken bottle is farther to the left in Figure 6.138 than a bottle in his right hand, landing where it fell. Yet, the bloodstains only appear on the pavement under Martin. No evidence supports any claim that he was moved from the left of the photo between the bottles to his present resting place. The only claim is that he was turned over where he fell. Bottles move much more easily than bloody bodies. Bottles do not bleed onto the pavement.

Sometimes it is best just to list the point you want to cover in a report. Be prepared to provide all the reasoning to support each point.

1. No physical evidence supports any statement that Martin carried bottles in his hands at the time he was shot.

Figure 6.141

Figure 6.142

Figure 6.143

2. If Martin had a bottle in his left hand, then the 27-pound bag over his left shoulder would help render any such implement ineffective as a life-threatening weapon.
3. The scene remains inconsistent with any claim that Martin had a broken bottle in his left hand.
4. Steps to force Martin to comply with instructions include nonlethal options.
5. One option: The officer behind Martin reportedly had the opportunity to put both arms around him, pinning Martin's arms to his sides; the officer in front of Martin could then have assisted by helping to take him to the ground—no weapons required.
6. Another option: Use of a baton, net, Taser, pepper spray, or other nonlethal mode, if available.
7. Another option: Simply wait since no obvious threat to the public or to responding officers appeared imminent.
8. It is never acceptable police practice deliberately to put one's own life at risk, when such a risk is not the only available option to protect life, simply to help justify the use of deadly force.
9. The foremost goal and the fundamental value of modern policing is to preserve human life; this includes the lives of the public, the suspect, the officer himself, as well as his fellow officers.
10. That foremost goal and this fundamental value were violated by shooting Martin and firing from the tactical position taken by the shooter.
11. Shooting Martin was unnecessary, excessive, and inconsistent with customary police practices designed to protect officers and citizens from harm.

Conclusions

In my report, I first described a scientific investigation that remained focused on the physical evidence as depicted in supplied photographs of this specific physical evidence, the autopsy report and police reports, and the evidence itself. But second, I was asked to research the police department's policies and procedures manual and comment separately on the use of deadly force in this case. Here, the reasoning from the premises was clear and unambiguous. So I offered my unscientific yet reasoned views. Often the two projects become entangled. This can be a problem that one must be prepared to untangle. Unfortunately, I never had that opportunity.

First, I provide the basis for any conclusions and opinions:

1. Review of police, medical examiner (autopsy) and laboratory reports, and principal interviews
2. Analysis of crime scene photographs, autopsy photographs and video, as well as those photographs clarified for my laboratory examinations and analyses following SWG protocols for digital evidence
3. Elemental XRF analysis (GSR) of Item #10 orange jacket
4. Microscopic analysis of Item #10 orange jacket
5. Bloodstain analysis of Item #10 orange jacket and Item #11 black nylon pants
6. Trajectory analysis of Item #10 orange jacket and Item #11 black nylon pants placed upon a 6-foot articulated mannequin
7. Glass fracture analysis and reassembly of Item #5 "Old English Label" bottle-broken pieces
8. Examination of Item #6, "Black bag containing misc. items"
9. Review of threat assessments based upon what responding officers knew about Martin before the shooting used to justify the use of deadly force
10. Review of threats to the safety of the public and to the responding officers posed by Martin's location and actions used to justify the use of deadly force
11. Review of alternative actions open to officers in lieu of the use of deadly force in these circumstances
12. Review of the Pond Valley Police Department Manual regarding use-of-deadly force; review of other generally accepted guidelines on use-of-force policy

Then, it is easiest and sometimes best in a lengthy work product to list the conclusions one by one in summary fashion as follows.

Shooting Evidence

1. The scene as photographed does not accurately reflect the shooting event in terms of the body's position, his jacket's zipper position, the black bag's location, or the two bottles' positions.
2. Two shots were fired at Leroy Martin from a Glock 40 cal pistol.
3. These two shots were fired at Martin from at least 3 feet or more muzzle-to-target distance.
4. The two shots striking Martin were both fired from his left side. So either the shooter stood to Martin's left as Martin walked straight while both facing and bending forward, or the shooter stood in front of Martin and Martin turned to his right facing away from the shooter, exposing his left side while bending forward.
5. The two shots striking Martin both entered his body at about a 40-degree angle of declination (anatomically). So, if the shooter fired his weapon with the muzzle parallel to the road surface, then Martin was bent forward about 50 degrees from standing perpendicular to the road surface when struck by each of the two shots.
6. Martin's left arm was down by his waist or in his left pocket when struck by a bullet.
7. This bullet perforated his left forearm, exiting through his jacket, reentering through his jacket below the left jacket pocket, entering his black nylon pants in

front of his left thigh, perforated his left thigh from front to back, and left to right, exiting to his right rear, and was not recovered.

8. Martin's orange jacket was zipped up when he was struck in the chest by a bullet.
9. The bullet striking his chest pursued a 40-degree angle of declination as described in the autopsy report, and was recovered in the lumbar spine.
10. Martin carried his 27-pound black bag over his shoulder using the shoulder strap.
11. The black bag's shoulder strap has been cut with a sharp blade further showing that the bag was hanging over his shoulder.
12. The photos showing the bag, body, and bloodstains support the opinion that the black bag was hanging over his left shoulder.
13. The broken bottle taken into evidence is missing a triangular-shaped curved piece from its neck, which would have broken off together with the pieces present in evidence when the bottle was broken.
14. The broken bottle did not break in the position photographed. It was either moved to the location where photographed, remains unrelated to the shooting scene, or both.
15. The unbroken bottle was either moved to the location where it was photographed, remains unrelated to the shooting scene, or both.
16. Martin's personal writings found in his 27-pound black bag (along with his possessions) profess and express his Christian religious faith and spiritual quests.

Use-of-Deadly-Force Opinions

1. Martin had no criminal history, no wants, no warrants, and none were known to the responding officers.
2. Martin was reportedly walking west down the middle of Central Avenue with his hands in his pockets, carrying a black bag, leaning forward with a hood up and eyes down, talking to himself.
3. The location on Central Avenue is mixed residential and commercial area, not heavily populated at the time of the shooting. A railroad track appears to the north, and a garage and another building appear to the south.
4. The temperature was reported to be 34°F, the weather was sunny and clear, and the pavement was dry.
5. Martin walked down the middle of the road leaning forward, probably with his left hand in his left nylon pants pocket and probably his right hand in his right nylon pants pocket with his jacket zipped up.
6. He walked with his 27-pound black bag over his left shoulder (bloodstains on pavement, location of bag, reported movement of body).
7. No physical evidence supports any statement that he carried bottles in his hands at the time he was shot.
8. If he had a bottle in his left hand, then the 27-pound bag over his left shoulder would help render any such implement ineffective as a life-threatening weapon.
9. Officers established tactical positions both in front of and behind Martin repeatedly requesting that he comply with instructions to stop and to take his hands out of his pockets (scene photos, trajectory analysis, witness reports).
10. Steps to force Martin to comply with instructions include nonlethal options.

11. One option: The officer behind Martin reportedly had the opportunity to put both arms around him, pinning Martin's arms to his sides; the officer in front of Martin could then have assisted by helping to take him to the ground—no weapons required.
12. Another option: use of a baton, net, Taser, pepper spray, or other nonlethal mode, if available.
13. Another option: simply wait since no obvious threat to the public or to responding officers appeared imminent.
14. It is never acceptable police practice deliberately to put one's own life at risk, when such a risk is not the only available option to protect life, simply to help justify the use of deadly force.
15. The foremost goal and the fundamental value of modern policing is to preserve human life—the lives of the public, the suspect, the officer himself, as well as his fellow officers.
16. That foremost goal and this fundamental value were violated by shooting Martin and firing from the tactical position taken by the shooter.
17. Shooting Martin was unnecessary, excessive, and inconsistent with customary police practices designed to protect officers and citizens from harm.

Postscript

No charges were ever filed in this case. The department's Internal Review Board and the citizen-member Shooting Review Board both cleared Officer Bud Stevens, concluding that shooting Martin was justified. No one held a news conference. No media appeared. No independent organizations took an interest—not the NAACP, not the ACLU, not even Martin's church. No one asked for any explanations. Nancy Grace stayed home. Apparently everyone ignored my conclusions.

Later, after all reports were completed, I was made aware of one interesting fact. Martin had not been ignoring officers' commands as he walked down Central Avenue. He had been totally deaf since birth.

Stevens remains on patrol. He was quoted in the county's weekly newspaper as saying, "I'd rather shoot a suspect than arrest 'em." He now carries a Glock 45.

Elton John still sings "Tiny Dancer"* to those of us who do hear:

> But oh how it feels so real
> Lying here with no one near
> Only you, and you can hear me
> When I say softly, slowly
> Hold me closer tiny dancer
> Count the headlights on the highway
> Lay me down in sheets of linen
> you had a busy day today

* Music: Elton John; Lyrics: Bernie Taupin; Piano and vocals: Elton John; Drums: Roger Pope; Bass: David Glover; Electric guitar: Caleb Quaye; Steel guitar: B.J. Cole; Acoustic guitar: Davey Johnstone; Backing vocals: Lesley Duncan, Sue Glover, Sunny Leslie, Barry St. John, Liza Strike, Roger Cook, Tony Burrows, Terry Steele, Dee Murray, Nigel Olsson. © 1971 Dick James Music Limited. Available on the album *Madman Across the Water.*

Man Down

7

It takes few words to tell the truth.

Chief Joseph of the Nez Perce (1840–1904)

Introduction

Carl Whitehorse was no angel. Nor was he a devil. He was a laborer; a son to loving parents, a boyfriend to one, and a friend to many. Yes, he did get into a bar fight one night at a local watering hole, The Bare Bar,* in his adopted hometown of Del Mar. And in the course of that altercation, a fellow sheetrocker did lose an ear. Not merely an ear lent to a countryman, mind you, but an ear rent from its very roots. Blood flowed everywhere, dripping in a trail from the bar toward the parking lot. Whatever else he was, Whitehorse was neither weak, nor faint of heart—his now earless barroom opponent was 6 feet 6 inches and weighed at least 275 pounds. The one-eared sheetrocker ran away, crying out in terror. Whitehorse finished his beer, got in his van and drove home.

The Del Mar police were called and responded to the bar, arriving about 2 hours after Whitehorse departed. Their response time was typical for the department; the officers often had trouble finding the locations of their calls. Thomas Brother's maps got a real workout (Figure 7.1).

At the bar, Officers Zwank and Dom got a name for the suspect from patrons—Carl Whitehorse—and the address of his digs on Vidalia Road. They were anxious to respond and to solve this bloody crime. After all, 2 weeks prior they were both unloading palettes at a local big box store. Now they got to drive fast and carry guns. Life was good. But a few minutes after their arrival on Vidalia Road, Whitehorse's life was over, and everything changed. An investigation into the death of Whitehorse ensued; it highlighted numerous shortcomings in the Del Mar Police Department.

As mentioned in Chapter 6, in a dissenting opinion in *United States v. Wade*, 388 U.S. 218 (1967), Justice White states what he takes to be the legal obligations of law enforcement officers: "Law enforcement officers must be dedicated to making the criminal trial a procedure for the ascertainment of the true facts surrounding the commission of the crime."

This case highlights specific failures of the Del Mar Police Department that prevented "the ascertainment of the true facts surrounding the commission of the crime." These shortcomings affected the two officers' actions leading up to and including the shooting. This remained the focus of the administrative review.

* A not-so-clever reference to a weightlifter's hangout where on occasion athletic females exercised on floor-to-ceiling poles.

Figure 7.1 The Bare Bar appears to the left of the light pole between the two furthest cruisers.

The police investigation therefore covers the officers' actions during the shooting, and the department's actions in the shooting's aftermath. This administrative investigation was conducted by independent police agencies and high-echelon policy makers. It certainly went well beyond the bounds of a traditional shooting event reconstruction. The review in a fundamental way relied upon the scientific results.

The administrative investigators found that the department's work continued "embodying a pattern of conduct inconsistent with IACP* guidelines and generally accepted police practices." Failures included deliberately ignoring the recommendations of a grand jury previously impaneled to investigate overzealous use of force.

My work in this case certainly was not to evaluate the department's conduct, nor was it to assess the conduct of the two officers who ended Carl Whitehorse's life. The request differed from the Leroy Martin shooting (Chapter 6). The two projects remain distinct and have entirely separate principles to apply. A shooting reconstruction ought to remain a scientific project. The administrative and legal components remain separate projects. A shooting reconstruction is based upon the physical data eventually assembled and the analyses of that data developed by reliable applications of the relevant natural sciences. The moral dimensions of the scientific work product remain obvious only to morally developed humans.† The administrative and disciplinary aspects lay with these high-echelon policy makers as they do in all such cases, even when reasoning is offered to stimulate decisions. As in the Martin shooting, the powers are free to ignore any inferentially derived advice from any outside source. They often do.

* The International Association of Chiefs of Police, an organization of police chiefs from around the world, who, among many other things, makes policy recommendations. Most such recommendations are accepted if they fit within a department's organizational structure.

† As Aristotle put it, one cannot teach ethics to immoral people—at least not in the sense that ethical judgments are properly made and correctly applied.

The administrative investigation concluded that the lack of proper training, the failure to follow both mandated and recommended procedures, and the lack of corrective supervision, explains this shooting and its investigation from the policy maker's point of view. The policy makers appealed to or referenced some aspect of the scientific shooting reconstruction of Whitehorse's death. In this case, the administrative review could not proceed without the foundation provided by a scientifically reliable reconstruction of Whitehorse's shooting. Thus, the two distinct projects can overlap in an instructive manner: the administrative component often depends upon the scientific component. Or at least in many cases it ought to. We just cannot escape the moral dimensions involved with the application of science to practical problems. The interface of science and policy often becomes an important consideration. But that never ought to affect the scientific project.

Administrators' Use of the Shooting Reconstruction

First, they concluded that the shooting at 1948 Vidalia Road, which resulted in the death of Carl Whitehorse, was unnecessary, excessive, and inconsistent with customary police practices designed to protect officers and citizens from harm.

Officer Zwank stated that he was following up on a shooting incident at The Bare Bar allegedly involving Carl Whitehorse. Officer Dom joined Officer Zwank as his backup when Zwank searched for Whitehorse. Zwank wrote no report on the incident's investigation with his work starting at the restaurant's bar and ending up in Whitehorse's work van. He says that he did "no report on any incident leading up to and including the ear-tearing event."

Zwank was hired in 1993 after working as a laborer. He had 124 hours of training in basic law enforcement. He was a new patrolman. Dom had been a police officer for only a week. He was trained as a corrections officer for 536 hours and had 124 hours training in basic police skills. Dom states that "Officer Zwank asked me to assist him on a call."

After learning Whitehorse's address through interviews, the two officers approached the residence at 1948 Vidalia Road. After interviewing an upstairs tenant without taking the tenant's name or documenting the visit, the two officers approached Whitehorse's apartment (Figure 7.2).

According to the information given to these officers, Whitehorse was armed, had been drinking, and was involved in a shooting earlier that evening (actually it had been a fight and not a shooting). In a case of stolen identity confusing Whitehorse with another man named Marvin Roberson, they also believed that Whitehorse had a felony warrant for his arrest. When Whitehorse's female companion, Kathy Bates, exited the Vidalia Road address in tears, she managed to tell Dom that Whitehorse had a gun and was wrecking the place in despair, in the belief that he had killed his friend by smashing him in the head while the two argued at the bar. The victim had bled all over, after all.

It was Zwank's call, yet Dom approached the door while Zwank supplied "backup" by positioning himself near the northeast window. He said that he had the following "facts" in mind:

1. That Whitehorse was alone in the apartment
2. That he had been drinking
3. That he was emotionally unstable given the earlier fight and Ms. Bates's statements

Figure 7.2

4. That he was armed and should be considered dangerous
5. That he had an outstanding felony warrant (false as it turns out)

Officer Dom approached the front door with his weapon, a 9-mm Glock, drawn. IACP guidelines* on the use of force dictate that deadly force be used as a last resort. The guidelines require, among other things, that "the officer must reasonably believe that all other means of affecting the arrest are futile" (IACP, G. O. 92-04 I, C.2). Neither officer had by any means evaluated the situation sufficiently in order to determine that firearms were their only tactical option. In fact, in common police practice, the first option would be to call for backup, as indicated in the Del Mar Police Department's* own Operational Order 6.1.0.

Both reasonable police procedures and common sense dictate that immediately approaching a suspect with a drawn pistol pointed in his direction is contraindicated. Such an approach merely provokes what may have been avoidable violence. Procedures and common sense require that officers call for additional backup, preferably trained negotiators, to defuse any potentially volatile situation. There is nothing to be lost by waiting. Whitehorse was not going anywhere.

According to Operational Order 6.1.0, "officers will not needlessly place themselves in situations of great danger and use this as justification for the use of deadly force" (6.1.0, III, D, 4). The two rookie officers' first obligation was to de-escalate tensions. At the very worst, Whitehorse was a danger only to himself. Approaching such a suspect with weapon in hand is a classic rookie mistake, the result of poor training, limited experience, bad judgment, or all of the above. Both officers' actions needlessly escalated a potentially resolvable situation and violated stated use of force guidelines (Operational Order 6.1.0).

From this point on, the situation deteriorated rapidly. Zwank took a position by the northeast window while Dom first approached the front door (Figures 7.3 and 7.4). Dom

* These are guidelines only and the IACP is clear in pointing out that these guidelines are not to be used as the basis for any disciplinary action: only the specific department's SOP, or standard operating policies, as stated in the policies and procedures manual actually apply to the issues. Of course, most departments produce policies very close to the IACP guidelines.

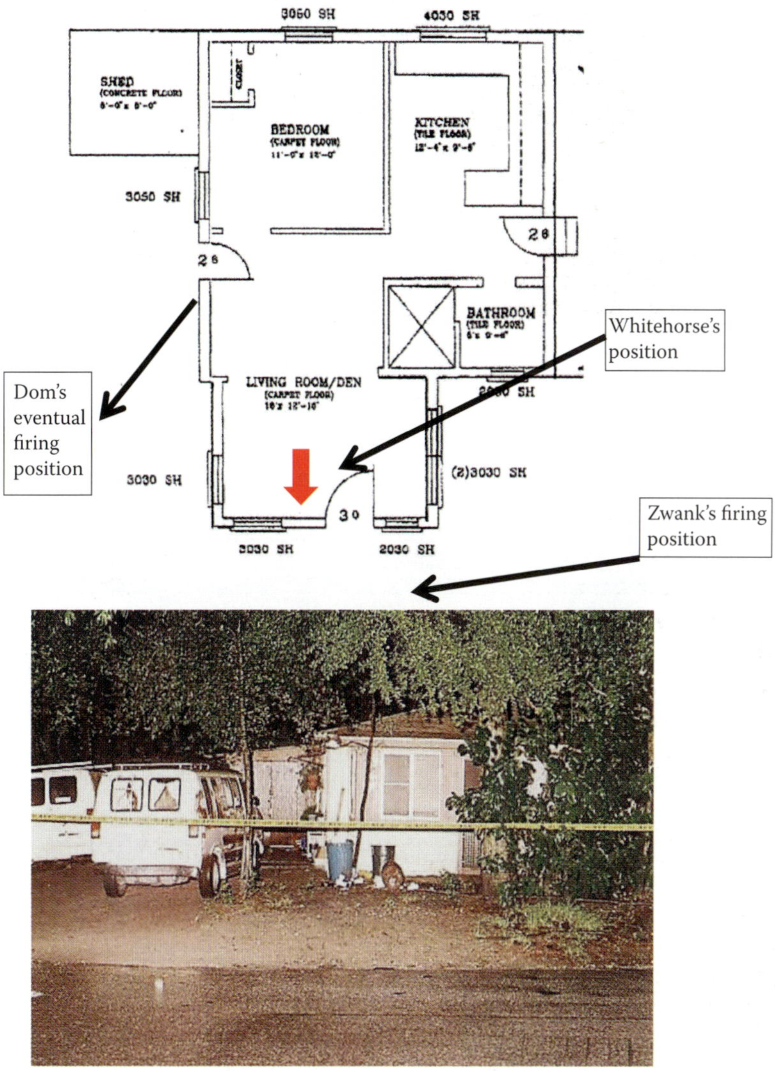

Figure 7.3 A view from the northeast showing the positon of the vans and the location of the window through which Zwank fired at Whitehorse (on the right behind the greenery).

Figure 7.4 The window through which Officer Zwank fired. Note the evidence markers locating the position of each cartridge case ejected from his pistol then landing below the window.

states that Whitehorse was standing in the middle of the room facing the doorway "about 5 to 10 feet away with his right hand behind his back." Dom then states that Whitehorse "pulled his gun out and fired a shot." "After he fired his shot, I fired back. I started exiting the doorway—took cover and—continued hearing shots" (Figure 7.5). Dom states that he did not know Zwank's position.

Dom reports, "I see the muzzle flash and hear the shot." He thinks that he fired two shots at Whitehorse from the doorway. He did not continue to return fire until he repositioned himself behind a van parked in the driveway. "I went to another position between the van, and I could see him in one of the windows here on the side, and I fired through that window." During this interval, he says that he does not communicate with Zwank (Figure 7.6).

He says he remembers seeing Whitehorse through that window (Figure 7.7). "I know he was looking around but I remember seeing him facing more toward south." He claims to see him moving and he "could see a gun in his hand."* (The photos obviously raise serious doubts about the truth of these statements.)

* The reconstruction and the scene documentation convincingly refute this statement, especially photos of the window visible to Dom as seen in the photograph.

Figure 7.5 Dom's position after leaving the front door; cartridge cases are marked with tents and flags. Apparently, the scene processing van was not carrying an adequate supply of numbered tents.

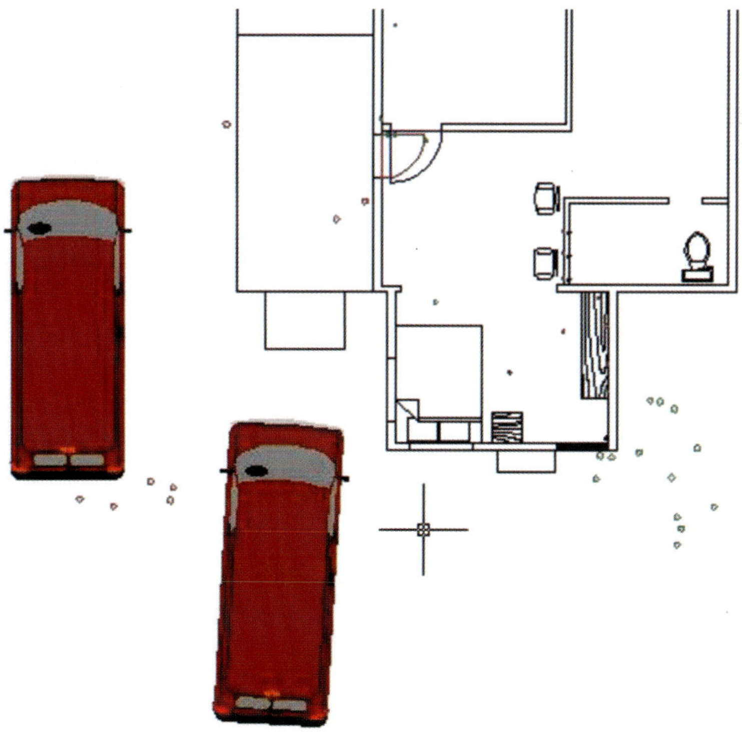

Figure 7.6 In some shooting scenes, a top-down, two-dimensional computer-generated image can show the related locations of the bullets and cartridge cases.

Figure 7.7 The window through which Dom claims that he could see Whitehorse. The blinds are not transparent and do not show shadows when the outside is dark and the room is lighted.

Meanwhile, Zwank was stationed outside the northeast corner window. He states that he remembers the curtain being pulled on the northeast window. He claims that he saw, "the male take an aggressive stance towards the door with his right hand behind him." Zwank assumes that Dom is by the door (he cannot see him) and claims that he screams several times to Dom "he's got a gun." Zwank says he saw Whitehorse "come around with his right hand, comes up in a shooter stance position, fires one shot, comes up to fire the second shot." He says that he *saw* Whitehorse fire two shots. "He was very fast. I was going as fast as I could, but simultaneously as his second shot was going off, I opened fire."

It would seem that Officer Zwank might be responding to Dom's shots, misidentified as Whitehorse's shots. Yet, Dom claims to have seen a muzzle flash and heard a shot *before* he fired. Obviously, this confusion escalated into an undisciplined shooting frenzy. IACP guidelines clearly state that "deadly force should not be used if its use would threaten the safety of innocent persons" (IACP 92-04ID). Since Dom had no idea of Zwank's position, and vice versa, both officers recklessly endangered each other's lives by firing blindly into the windows of this building (Figures 7.8 and 7.9). The windows, as it turned out (and as we can see in the photo of the west window), had the Venetian blinds both drawn and closed. The window Zwank fired through also had a curtain drawn—it was opened to the side for photography during the investigation (Figure 7.10).

Zwank states that his "first two (shots) seemed to have absolutely no effect at all. As I fired, he just turned towards me, gun in hand. He seemed to get madder and madder from his original aggressive stance. I fired as fast as I could. It seemed to have no effect, and he went for cover right at my feet." Zwank claims to have seen Whitehorse's facial expressions, as well as experiencing auditory exclusion.*

* Meaning, of course, the physiological blocking of sound as an incoming stimulus. One might find it difficult to explain his claim of hearing some things but failing to hear others due to auditory exclusion.

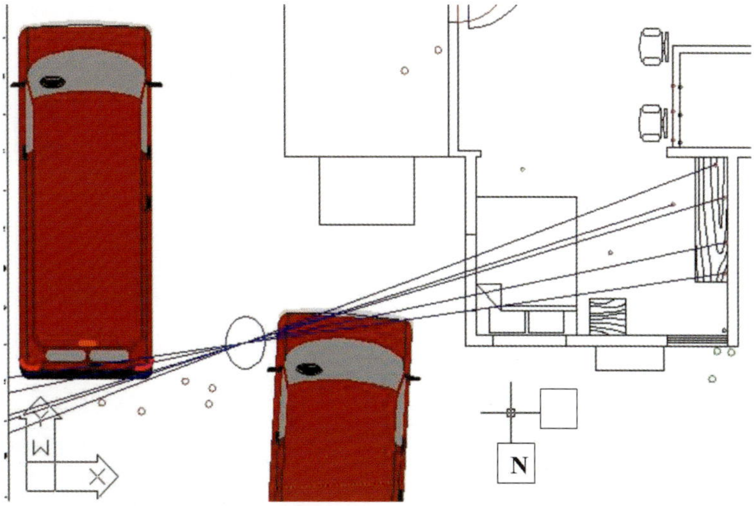

Figure 7.8 This scale drawing documents the angles of Dom's first five shots into the window with the blinds closed. Zwank's position remained by the northeast window. Dom eventually fired three more shots for a total of eight shots that we can document.

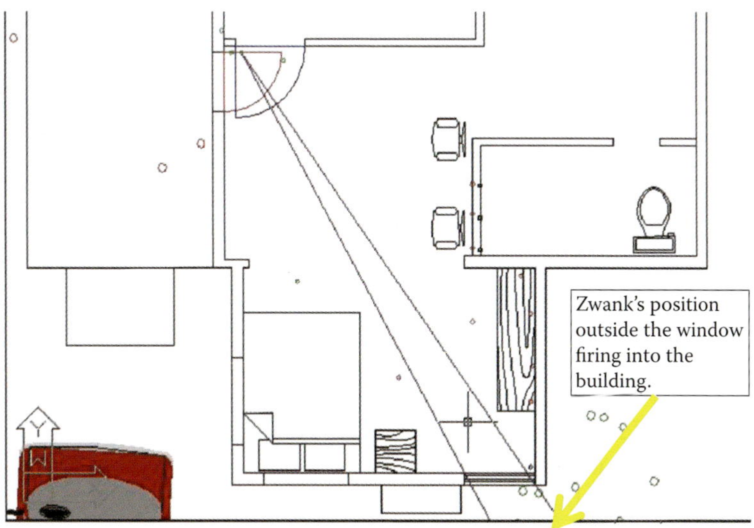

Figure 7.9 This diagram illustrates the angle of Zwank's shots. Zwank fired 15 shots through the northeast window with 11 shots striking Whitehorse in the back as he lay prone on the floor.

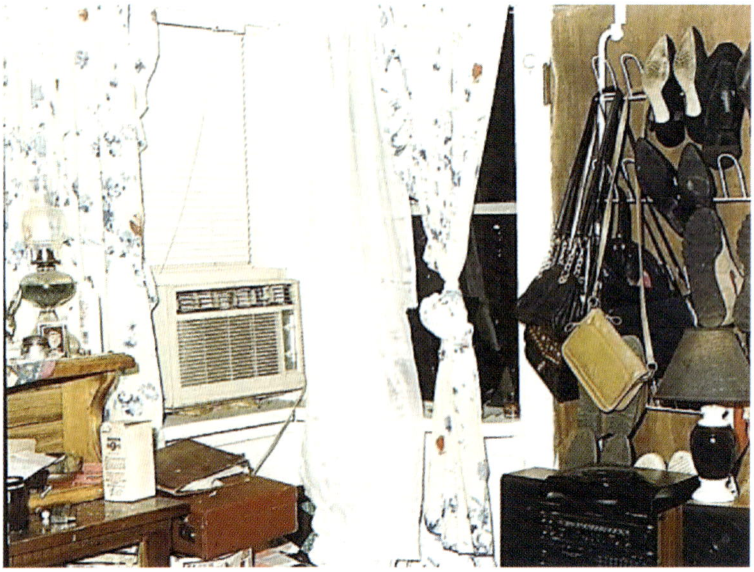

Figure 7.10 Photo taken of the northeast window after the blinds and curtain were opened. Evidence that they were closed during the shooting sequence consists of damage to the blinds left by the bullets fired by Zwank.

It appears that Zwank may have also suffered from hallucinations. For example, he says he saw Whitehorse fire and even says that he saw Whitehorse's gun slide moving (to expel spent cartridge cases). "I could see—I could actually see the action of the gun working, and I could see the smoke coming out, but I never heard the shot." Later, he adds that, "His slide—I could see his slide move, and I could see the smoke come out almost in slow motion." Zwank testified that he had target acquisition all the time he was shooting, and that he never lost sight of Whitehorse through the entire episode. He says that when Whitehorse dove for cover, "gun in hand at my feet, his back was facing me, his lower back." He states that he was maybe 2 to 3 feet away from Whitehorse at this point, and that he continued shooting. "I never stopped." That part, at least, is probably correct.

At some point during these events, Dom, firing his pistol, retreated from the front door to a position behind one of the vans parked in the southeast side of the driveway. Learning that his partner, Zwank, had taken a position at the northeast corner of the building by a window, Dom commenced firing into the southeast window. The Venetian blinds were closed (again, we note the bullet damage to the closed blinds). Yet, with his partner out of his sight, directly opposite from his line of fire, Dom fired blindly in the direction of his partner through a window obscured by closed blinds. He can see nothing through the window.

Zwank apparently fires through the northeast window, with the curtain down (see Figure 7.11). He fires first, apparently as Dom is retreating from the door. Both officers claim to have seen Whitehorse fire his 9-mm Star. The investigation revealed, however, that Whitehorse never fired his pistol at the scene or anywhere else recently. It was clean, the magazine was full, and no cartridge cases were found inside that had been fired from a 9-mm Star. With Dom firing blindly into the southeast window, Zwank fires at the decedent lying face down on the concrete floor of the garage apartment.

Obviously, inexperience and adrenalin fueled the tactical errors made by the rookie officers and accelerated the violence. While Whitehorse was on his stomach on the ground

Figure 7.11 The window through which Officer Zwank fired. Note that the blinds are closed and the curtain is drawn.

and no longer a threat, Zwank repeatedly fired at the decedent. In fact, all rounds hitting Whitehorse were fired into his back while he lay face down on the floor (see Figure 7.12). This evidence raises profound questions about the degree of threat that Whitehorse actually posed at the time he was shot by Zwank.

Also, given the injury to the right wrist (Figure 7.13), it is difficult to imagine Whitehorse holding on to anything, let alone a gun. His position raises a question: Where were his hands when the shots were fired from Zwank's position? When was he struck by this bullet?

Clearly, shooting from these positions becomes extremely dangerous to the officers themselves and anyone else nearby (Figure 7.14). Officer Dom proved to be a greater threat to Officer Zwank than did Whitehorse. It is a wonder that Dom did not shoot and kill his own backup.

Further Administrative Review

The reviewers concluded from their interpretation of the shooting reconstruction report that the original police investigation of this shooting at 1948 Vidalia Road was inadequate,* incomplete, and improperly directed—at the best exhibiting gross incompetence, at the worst exemplifying an attempted cover-up of events under investigation.

Procedural irregularities and inadequacies continued when the department's crime scene unit documented the Vidalia Road crime scene. Referring to the measurements from a baseline to an item of evidence, Crime Scene Officer Henry Broderick states that "these measurements are approximate." Certainly scene sketches and even some finished drawings may not be to scale and show the approximate or merely relative positions of objects. However, the measurements themselves must be accurate, and never merely approximate in this sense of the term. In the scene documentation done in the Whitehorse shooting, outside walls are used as baselines for inside items of evidence. Such measurements inaccurately approximate the length of an inside wall. For example, wall D is measured to be 13 feet 6 inches on the

* An inadequate investigation is defined as an investigation both raising and ignoring additional questions, while either ignoring or otherwise failing to answer the initial questions giving rise to the investigation in the first place.

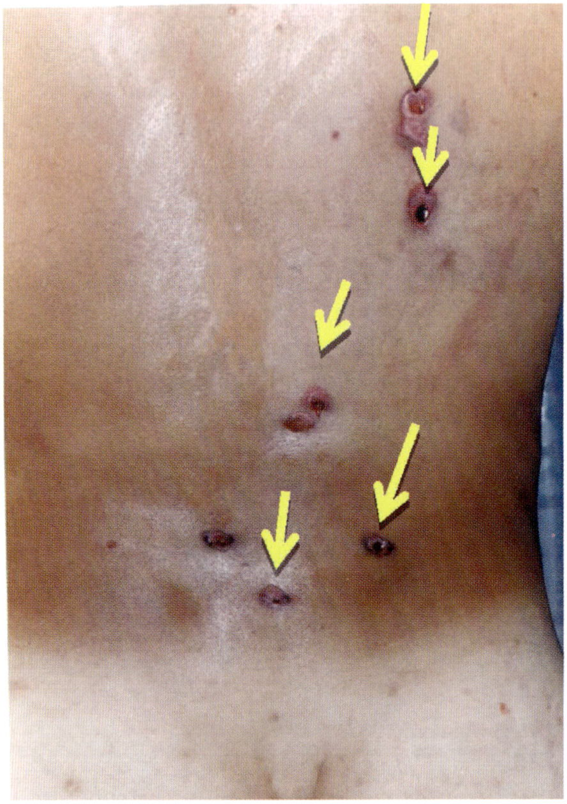

Figure 7.12 All of the 11 rounds that struck Whitehorse struck him in the back while he was lying face down on the concrete floor. All the bullets enter at approximately the same angle as seen in the arrows—at about a 40-degree declination and top to bottom. If we position Whitehorse prone on the floor with at least one hand behind his back or on top of his head, the bullet through his wrist lines up with the trajectory of a bullet fired from Zwank's position, striking him in the back. On the other hand, the injury may have resulted when Dom fired from the doorway. But no bloodstains were found to support this interpretation. Additionally, a 9-mm bullet has a muzzle velocity of about 1160 feet/second. When striking a secondary target after perforating a primary target, the entrance wound can be difficult to distinguish from the wound in the primary target.

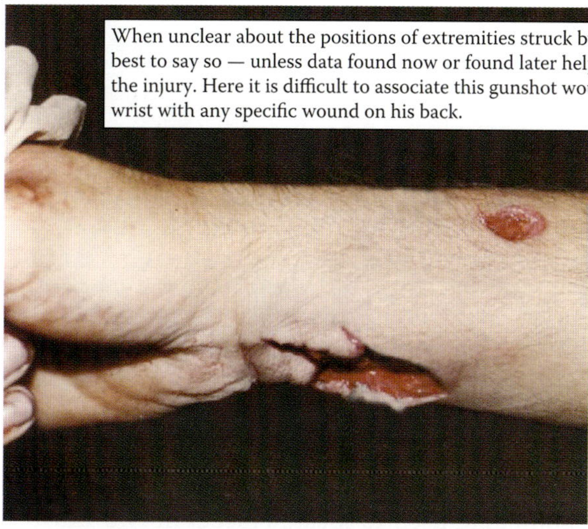

Figure 7.13 Injury to Whitehorse's right wrist.

Figure 7.14 The van provided cover for Dom as he shot into the window with the blinds closed; Zwank shot into the window on the right, standing up to the sill—it is a wonder that Dom missed hitting Zwank.

outside but the inside measurement of the same wall is 12 feet 10 inches. In some exceptional cases, this may be adequate, but most investigations will not tolerate errors of this magnitude simply built into the protocol. Furthermore, such errors invite suspicions about the accuracy of other areas of evidence identification, collection, and documentation.

Crime Scene Officer Stanley Farnsworth seemed to misunderstand the ultimate purpose of crime scene sketches and measurements. When asked about a sketch and measurements done by Broderick, he is unable to decipher its significance: "No, ma'am, that's his—that's his diagram." "OK, so you have no idea what this refers to at all?" He replied, "I'm not sure." The purpose of such diagrams is to allow anyone to locate items of evidence, visualize a specific scene, or to undertake specific calculations of their own based on the recorded facts in the diagram (see calculated positions of Zwank and Whitehorse that follow). Such sketches cannot be individually idiosyncratic and totally subjective exercises.

Figure 7.15 Note the cartridge case inside the window showing that Zwank fired directly into the room at Whitehorse as the decedent lay on the floor.

When two reference points are used to locate objects in two-dimensional space, the lengths and exact nature of reference points must be clearly stated. Although such reference points are available in some form in the crime scene sketches prepared by the Del Mar department, they are not obvious or immediately useful as presented. When walls are used for such reference points, outside wall measurements must be distinguished from inside wall measurements. They are never identical. For example, 13 feet 6 inches is the stated length of wall D. This data appears on the crime scene unit's drawing labeled "Outside scene of 1948 Vidalia Road." Yet this measurement is referred to as a baseline used to locate objects inside the residence. Measurements such as (D to 1 = 4 feet) appear to pose no problems. However, the thickness of D is ignored in this measurement. When D must be used to calculate, say, the location of the northeast window, we run into problems. Here the thickness of the wall (i.e., whether we use inside or outside measurements) is crucial for accuracy. We also risk errors of at least the thickness of wall E when locating the objects, such as the gun in the room, using E as a baseline. It would get only worse for bullet entries and exits; the error range could be greater than the target size. That is one more value of being careful to calculate error rates involved in any measurements. That exercise alone should compel practitioners to adjust their protocols and, more important, to clarify and critically assess their own reasoning processes.

There are no measurements documented by the crime scene unit of the northeast bedroom window through which Zwank fired. This is a key omission since Zwank's rounds were the only shots fired that met their target: Whitehorse's back (Figures 7.15 and 7.16). (The northeast bedroom window has been measured and found to be approximately 2 feet 6 inches wide and 3 feet 2 inches tall.)

Administrators found that specific Del Mar documentations of the bullet trajectories in the Whitehorse shooting scene revealed a serious lack of relevant knowledge and experience.* When bullet entrances and exits are investigated, the use of numbers to reference

* It is interesting to note that the least experienced Del Mar crime scene unit officer had been working at his craft for over 35 years. That seems to support the view that years of practice do not necessarily teach one anything. If one performs a task incorrectly, it does not matter how long they continue the performance. Doing any activity for a long time does not guarantee that the activity is correctly performed. So much for experience and authority as criteria for truth. Look at the work first and last; the work should speak for itself. The rest, despite what lawyers would have juries believe, is irrelevant window dressing.

Figure 7.16 Another photo of the northeast window shows that while firing downward at Whitehorse, who lay at his feet, one of Zwank's bullets hit the windowsill.

distinct marks, protractors to find azimuth and incident angles, and strings or lasers to identify bullet tracks, intermediate targets, and so forth are common. Newer techniques including total stations (see Chapter 1) are valuable tools, but only if the basics are first understood (Figure 7.17).

Basic concepts—and by basic I mean fundamental, not simple—such as a set point must be understood before the total station can even begin to generate useful data. I find that making crime scene drawings and locating the items of evidence can be taught more easily by doing the work manually (Figure 7.18). Once that has been mastered, the benefit of technological developments can help speed up and improve both accuracy and precision, as long as we do not forget to factor in the error rate and to validate the specific contrivance put to the task. We used a total station as described in Chapter 1.

Such basic investigations involve detailed sketches, photographs, and measurements, which help to determine angles from certain fixed points (i.e., from a wall or some other fixed target to the muzzle of a firearm). Two points are OK; three points are much preferred, depending on materials. When multiple shots are fired, the effort is tedious, complex, and involves attempts to identify the origins of all shots fired, unless specific, unusual circumstances require the documentation of only a select few shots. There are no such trajectory reports or analyses among the Whitehorse shooting scene documentation provided by the Del Mar department. Investigators, apparently, did not conduct the most thorough of searches.

Scene photographs taken by the department show a somewhat clumsy attempt to discover and document trajectory assessments (Figure 7.19). The Del Mar photos demonstrate a lack of principled trajectory investigation. A single point is insufficient to determine a trajectory, except under unusual circumstances.* Which wall is studied and why it is studied and photographed to the apparent exclusion of others must be explained. One naturally assumes that investigations are meant to illuminate and answer some question or solve

* Those circumstances include the material composing the target. Techniques familiar to bloodstain pattern analysis can be applied in exceptional circumstances; I use them only as a check or comparison measurement.

Figure 7.17 My colleague Stuart James looks on as I discuss the use of and the data from a total station with an officer working with me on an unrelated case. The problem was that the Del Mar CSI unit ran the machine without knowing the most basic principles for documenting items of evidence at a crime scene. Regardless of its user, no technical device will take the place of the careful, critical intelligence so central to the practice of natural science.

Figure 7.18 Doing things the old fashioned way is sometimes best. It helps to develop an understanding of concepts as basic as documenting the location of evidence items. They exist in at least a three-dimensional (3D) space. The documentation requires an X, a Y, and a Z axis. The fourth dimension, if one considers the evidence, is time. That, in my view, is an important and basic concept to consider with evidence as well as the three-dimensional axes. That notion probably developed from my medical work.

Figure 7.19 Following trajectories from at least two or three fixed points often helps me to find a bullet's final resting spot. Lining up lasers can be time consuming, but they are extremely useful tools to document trajectories and to locate bullets. Once the bullet is found, then an investigator has a better chance to determine the bullet's trajectory or to identify some intervening target and any deflection angles.

some puzzle. What is puzzling here is why these numbers were written on the wall, circled, doweled, and why these photos (see Figure 7.20 to Figure 7.22) were taken in the first place. Shots from Zwank's location to Whitehorse's location, lying on the ground at his feet, are much more relevant here. The numbers have no reference to any item of evidence collected and never appear in any document, except for these photos.

Some indication of how many shots were fired from Dom's and Zwank's stated positions would be important to know, as would any evidence that Whitehorse fired his 9-mm Star, since officers Dom and Zwank repeatedly made that claim. The following table reflects the results of analyzing my own data, as well as the Del Mar department's original crime scene documentation of the shots fired by each officer.

Name	# Shots Fired per Del Mar Officer	# Shots Accounted For	# Casings Recovered at Scene
Dom	8	8?	8
Zwank	13 or 15?	14	14
Whitehorse 9 mm	5 or 6 fired per officers' statements	0; No bullets	0; No cases
Whitehorse .25	0	0	0
Whitehorse .22	0	0	0
Totals	21 or 23	21 or 23	22

Although the Del Mar department's investigation provided no proof that Whitehorse fired a gun at the scene of his death, Dom's and Zwank's written statements apparently were enough to convince investigators that the decedent had fired a gun.

There are no new documents developed as part of that internal affairs investigation. If no other documents were developed and no new data was developed by a later scene investigation, then the results of this subsequent investigation (either by the state or by Internal Affairs) must be based on the same contradictions that they were supposed to resolve in the first place.

It becomes obvious that investigators ignored evidence documented and collected by Crime Scene Officer Broderick. For example, they apparently ignored the absence

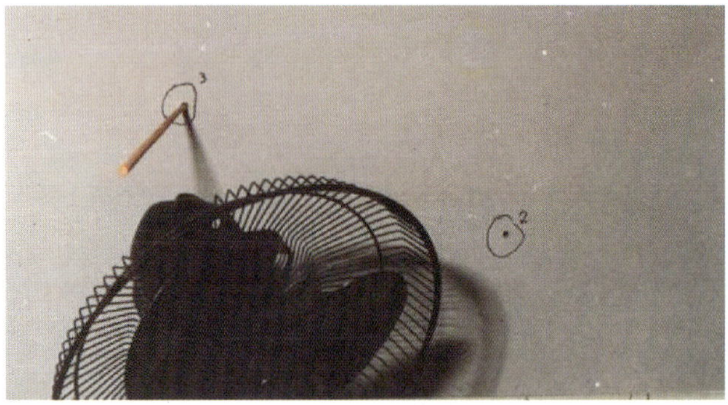

Figure 7.20 Two holes circled, numbered 3 and 2 with a dowel protruding from 3. What does the fan demonstrate?

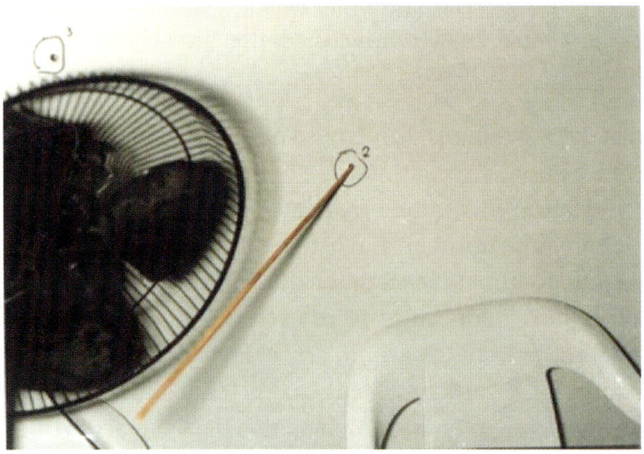

Figure 7.21 These same two holes, 3 and 2, with the rod now protruding from hole 2. A chair appears, along with the fan, but these photos require some documentation to explain their significance. No photo log accompanied any of the photographs.

of cartridge cases found at the scene consistent with the ejection characteristics of Whitehorse's Star semiautomatic pistol. Detective Billy Ray Blind underscores this point. He states that Broderick "came to me at one point and advised me that there were … did not find some shell casings inside and that was a concern, and at that time I went in to assist looking for them." The fact that no casings from Whitehorse's gun were recovered at the scene he says, "Needs to be clarified." When asked if it was clarified, he says that he does not know. Indeed, as far as he knows, no one ever followed up on the lack of shell casings. That fact alone certainly vitiates the entire investigation of Dom and Zwank's actions during the shooting—at least that is what the administrative reviewers concluded.

Detective C. Charles Hershey stated that the purpose of such an investigation is "to document the facts and gather together the facts in regards to the circumstances around why officers Bret Dom and Joel Zwank fired their weapons that night and the circumstances surrounding that." He says that he informed his supervisor, Sgt. C. D. Goldfinch, about the lack of any 9-mm Star cartridge cases at the scene, but he does not recall telling State's Attorney Sheldon Baxter.

Figure 7.22 An as yet unidentified curtain now associates with the trajectory rod and the fan, but the purpose needs to be documented so that in the future the effort here can be interpreted in some meaningful manner. If the trajectory rod perforates the window and the wall, then an account of those holes must be provided somewhere. Also, one needs to consider any possible deflection angle when using a trajectory rod between only two points. Sometimes two points can be adequate for the purpose of determining trajectory, but often more data are needed. It pays to know which case we have at the scene.

From these statements, it appears that no one has taken responsibility for the investigation of Whitehorse's death. Information is collected and documented, and yet nothing appears to be done with it. At best such indifference to the facts is irresponsible and unprofessional; at worst it smacks of a criminal whitewash.

It is odd that Dom and Zwank's statements after the shooting were written by Detectives Hershey and Blind. Each detective signed as a witness for the other's handiwork. It is unusual for two allegedly busy detectives to write down each of the rookie officer's statements, crossing over to sign as witnesses for each other's statements. IACP guidelines require that officers involved in shootings must initiate an offense report "and detail the facts regarding the use of deadly force or the discharge of a firearm. The report must be completed prior to the end of the officer's shift" (IACP 92-04; VII. A.1a, 3). It also requires "all officers present to submit reports detailing their actions and observations." Many patrol rookies would be very happy to have detectives write their reports for them, but most detectives would refuse, and not too politely.

Both statements assert that the officers witnessed Whitehorse firing his pistol, one saw a flash from the barrel, the other saw smoke from the same barrel. The statements assert that Whitehorse fired first. It appears that the detectives may have coached these officers when collecting their statements. Perhaps this coaching might suggest some sort of attempt to cover up the rookie errors noted earlier. In any event, the statements contradicted the facts.

Both the scene and the nature of Whitehorse's injuries documented in the autopsy report and autopsy photographs blatantly contradict the officers' recorded statements (Figures 7.23 and 7.24). There is no evidence to support Dom's claim that he could see Whitehorse, or anything for that matter, through the window with drawn blinds and pulled curtains. There is no evidence to support the claim that Whitehorse was struck by any shots at all until he lay face down on the floor literally at Zwank's feet. The injury to

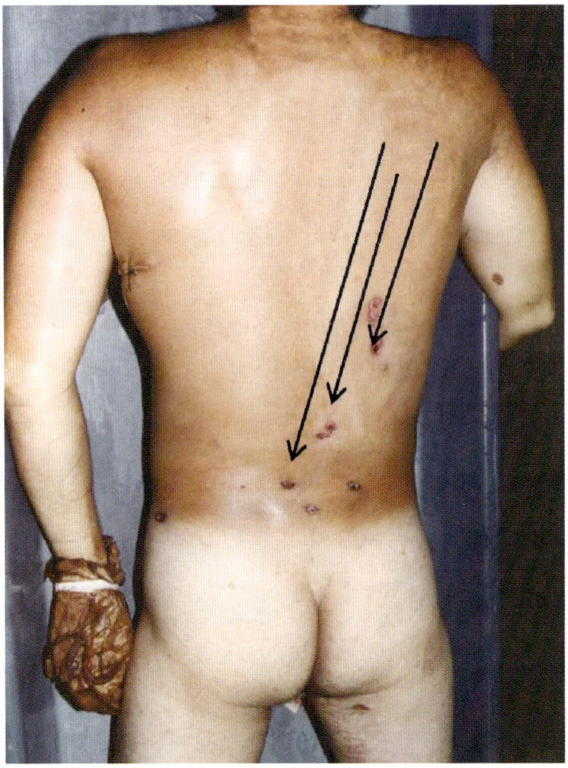

Figure 7.23 Whitehorse's entrance wounds show the angle of entrance. Dissecting the wound track at autopsy also establishes the connection of the entrance wounds with the "exits." This, in turn, establishes their trajectories through the body.

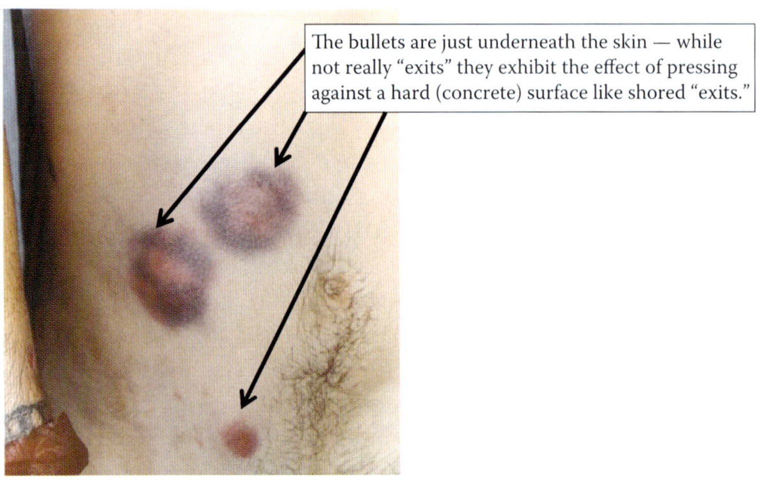

Figure 7.24 The exits are not, of course, really exits at all. This photo indicates that the body was pressed against a hard surface, such as the concrete floor of Whitehorse's apartment, preventing any actual exit. The bullets are just beneath the skin. In that sense, they are shored trajectory end points.

Whitehorse's arm raises questions about his ability to hold a gun as well as the position of his arms. Obviously, these inconsistencies raise questions about police department actions leading to Whitehorse's death as well as the conduct of his death's subsequent investigation. It may be that Whitehorse lay on the floor with his arms, or at least one arm, behind him when he was repeatedly shot.

Not only were these contradictions ignored throughout the alleged investigations, but evidence also suggests that no further investigation was attempted. At the very least, such an investigation must:

- Determine the approximate positions and actions of Whitehorse, Dom, and Zwank using all case data including photographs, drawings, evidence collected, together with recognized methods of crime scene reconstruction.
- Account for all the bullets from each of two Glock 9-mm handguns that struck Whitehorse, those that missed, and any rounds fired from Whitehorse's 9-mm Star to account for each shot fired.
- Note that two police Glock 9-mm handguns were fired at Whitehorse. Due to unique characteristics of the Glock, individual shots cannot be linked to one gun versus the other through standard comparison methods unless one has unusual damage leaving unique marks on a bullet. (However, bullets fired from either of the two Glock pistols can be distinguished from bullets fired from Whitehorse's Star pistol—had any been fired using class characteristics alone; the Star lacks a polygonal barrel.)
- Distinctions between Zwank and Dom's rounds must be based upon their stated positions during the shooting and crime scene drawings, cartridge case locations, damage to the surroundings, and injuries to the victim, as well as all recovered cartridge cases and bullets.
- Account for Whitehorse's actions by analyzing the crime scene data as well as the officer's statements. Whitehorse reportedly had his Star in his right hand; the gun was reportedly found under his body and inside a plastic bag. This places Whitehorse's head toward the head of the bed and his feet toward the door. Mr. Whitehorse was face down on the floor when shot by Zwank.

None of these things were done either before, during, or after the scene was released.

A sloppy job of documenting this crime scene resulted when technicians failed to understand the ultimate purpose of such documentation: to answer any future questions that arise about the evidence, and to present data in a clear and orderly fashion, both to investigators and to the court. Some obvious concerns include the following:

- Failure to sketch the scene accurately, including the bathroom and the north bedroom, indicating their relative positions with other rooms
- Failure to photograph evidence with standard measuring devices or informative arrows identifying the evidence in place
- Failure to use trajectory rods correctly and meaningfully
- Failure to document each photo, and to keep a photo log describing the content of the photo, the number of rolls, and the number of photos taken
- Failure to account for each of the officer's bullets, their terminus, as well as each of their ejected cartridge cases

- Sloppy documentation of measurements
- Sloppy photo archiving, including multiple cases on a single roll of film
- Failing to identify film rolls with a photo identifying photographer, location, time, and case number
- Failing to identify and photograph Whitehorse's 9-mm Star, its magazine, and the number of rounds chambered with the number in the magazine
- Failing to address the absence of a 9-mm Star cartridge case at the scene
- Failure to identify intermediate targets at the scene
- Failure to document the plastic bag around Whitehorse's gun, recovered under the decedent's body (Figure 7.25 to Figure 7.27)

The body was neither photographed nor sketched at the scene. Its position may have been very helpful in interpreting events leading up to the shooting. Detective Petigone Hisome states that the crime scene techs were at work not only before detectives arrived, but also before the emergency medical team had responded. This raises serious questions not only about the investigation's quality but also about its ultimate purpose: Did the body show something at the scene that investigators would rather conceal? Their actions, unfortunately, lend support to sinister conclusions. The problem here is that questions were never asked or answered. That simply invites speculation.

Detective Hershey also seems perplexed about who controlled the scene when they arrived. Hisome refers to the crime scene techs when he calls it "their scene" and describes how he is just trying to keep out of their way. This is certainly contrary to the IACP guidelines

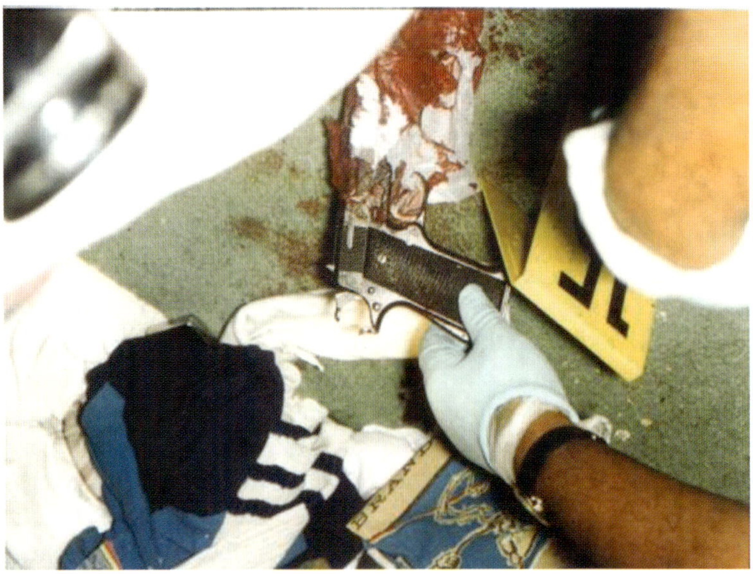

Figure 7.25 Whitehorse's gun, presumably as it was originally found on the floor under his body (see the blood and void patterns in the subsequent photos). The original position of the plastic bag remains undocumented. The decedent's female companion said that the pistol was stored in and stayed in a plastic bag, at least when she left the apartment. The gun's original position should have been more carefully documented. Objects around it should have been identified and examined. The carpet should have been taken into evidence as well. These tasks become even more important when the data could help to answer questions about the pistol's role in the shooting events.

Figure 7.26 The gun appears to have been moved when this photo was taken. Note the carpet lacks bloodstains to the left of the gun. The original position appears perhaps, in the earlier photo within the void: the arrow represents the pistol's handle. The muzzle was pointed away from Zwank's position in the window.

Figure 7.27 In this view, the fact that the gun has been moved can be clearly seen. Again, note the bloodstain pattern and void on the rug, and the bloodstains indicating that the decedent was dragged from his position on the floor while still face down.

(92-04, VIII, A. 8), which state: "Any Commissioned Officer arriving on the scene prior to the arrival of the Patrol Watch Commander will be responsible for the direction of subordinate personnel until the arrival of the Patrol Watch Commander. The control of the scene will then be handed over and will remain the responsibility of the Patrol Watch Commander."

The failure to use any crime scene data for reconstructing and analyzing the basic events of that night is a stunning investigative failure. This error is compounded when no one seems to be in charge of this case, let alone the scene. The evidence, once collected, was never used in the homicide investigation itself, the state's investigation, or the Internal Affairs' investigation. There quite simply were no such investigations. Scene processing became a simple cleaning operation rather than an investigation charged to exonerate or to condemn the two officers' actions. Crime scene drawings were not analyzed, nor were the various reports examined and compared with the collected evidence to draw out and resolve obvious contradictions.

Such contradictions between the evidence and both the statements of the officers and the reasoning applied by officials to determine that this was a justifiable homicide raise more questions. A proper investigation is supposed to answer questions or at least explain why answers are not forthcoming. Even a minimal effort to document Zwank and Whitehorse's relative positions during the shooting raises questions. They certainly cast a shadow of doubt on the reliability of Zwank's statements.

Inconsistencies among statements and physical evidence were ignored. Every effort must be made to resolve such contradictions, documenting the attempts to resolve them, and noting the efforts' success or failure. The important inconsistency involves the officers' statements to the effect that they saw Whitehorse fire his weapon at Dom (muzzle flash, puff of white smoke), together with the conclusion established by scene investigation that Whitehorse did not fire his weapon in his apartment. Despite this inconsistency, potentially resolved by the available physical evidence, the shooting is ruled "a justifiable homicide."

Reconstruction Data

The first step to understand and sort out the evidence in the shooting events variously described, of course, was to search the scene and to locate all of the bullets. We know the location of those 11 recovered at the autopsy. We can determine Whitehorse's approximate position when shot from the bloodstains on the floor. After locating all the bullets, any evidence of their paths must be discovered and evaluated. Only then can one hope to develop an assessment of their trajectories. We may not be able to account for them all.

The cartridge case count indicates that 23 shots were fired into the Whitehorse residence. With 11 striking Whitehorse, that leaves 12 bullets to find. Dom's location provided 8 cartridge cases; Zwank's location provided 14 cartridge cases. So, there are two firing locations or, more properly, two shooters providing trajectories to develop.

Location of the Bullets

There is no easy or best way to search for bullets; each scene provides its own challenges. This scene was mercifully small. That, however, provides its own challenge: associating bullet damage with specific bullets and distinguishing one damaged area from another. Direction can be misleading. Bullets may behave unpredictably when striking primary

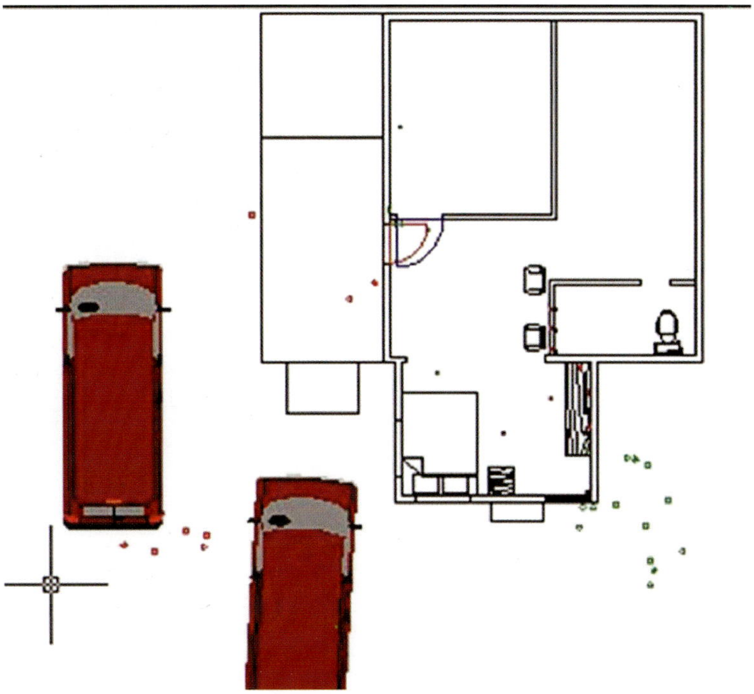

Figure 7.28 The red dots represent cartridge cases from Dom's pistol; the green dots represent cartridge cases from Zwank's pistol.

and secondary targets of different materials. Eventually the bullets were found and associated with the positions of the two shooters. Some of the bullet and cartridge case locations appear in Figures 7.28 and 7.29.

If the location or the complex trajectory of a bullet becomes a puzzle or raises issues, identify the materials composing the primary, secondary, tertiary, and so forth targets and design some experiments. Deflection angles for many materials can be measured. Deflection angles for combinations of multiple materials, similar or distinct, are also subject to experiment. That is the best way to answer questions about the trajectory followed through multiple targets. It is wise to rely on some type of experiment rather than merely to speculate about a complex trajectory, especially if there are multiple bullets in or near the same final target. Of course, multiple perforations may correspond to one bullet's trajectory.

If human tissues occur somewhere along the trajectory, I suggest using a ballistics gel block of the relevant size (Figures 7.30 and 7.31). If the bullet strikes bone, add bone to the block. Several types of gel are now available; they still require validation. But the clear gel* allows for somewhat better photographs and can be reused.

Multiple target setups must consider methods to measure the property at issue. With bullets fired through various materials, I find that a Keyence imaging system taking about 42,000 frames per second is more than adequate (Figure 7.32). It allows an accurate measurement of the bullet's path and captures the deflection of the bullet as it travels through

* Perma-Gel, Inc. (Albany, Oregon), developed and provides a clear, reusable material with an indefinite shelf life that requires no refrigeration. In my experiments comparing Perma-Gel with regular ballistic gelatin, the air rifle penetrations of the BB at 600 ft/sec were within .75 inches of each other over 10 shots on each type block.

Figure 7.29 All the bullets were located and all were photographed for the report on this case.

Figure 7.30 Testing the effects of bullet strikes to various materials at specific distances are part of most shooting experiments.

Figure 7.31 Gel blocks approximate the properties of soft tissues. Bone can be added if needed for testing.

Figure 7.32 Michael Graham of the Keyence Corp. of America adjusts the computer element of the high-speed camera in our shooting range at Final Analysis Forensics.

various materials. Setting up the camera's location can be time consuming but is critical to the development of good data. Another virtue of this approach remains the fact that the images can be shown in court if the issues demand the consideration of a jury.

Experiments with vehicles can be conducted in a similar manner. In Figure 7.33, a van was used to evaluate known trajectories and their passage through the vehicle's multiple material types. High-priced tools are convenient but not necessary. I took this photograph with my pocket digital camera that cost less than $100. A time exposure function remains the only requirement.

Experiments also inform us about the impact angles of either similar or dissimilar materials at fixed distances (Figure 7.34). By abstracting* relevant features, experiments are possible.

* To abstract anything is literally to cut off or separate. Abstraction, then, is eliminating irrelevant aspects and including only those aspects thought to apply to the issues on the table.

Figure 7.33 Laser light showing the trajectory established for each shot into a van.

Figure 7.34 Targets set at 45-degree angles.

Figure 7.35 to Figure 7.39 illustrate the search and recovery efforts for several of the 12 bullets that missed Whitehorse. (Recall that 11 struck Whitehorse. If 22 or 23 shots were fired, then we should look for 11 or 12 bullets.) Often, the bullet itself tells us about its targets along the way once it is recovered.

Any of the techniques or all of them may be applied in the search for bullets at the scene. A field microscope, either optical or digital, is useful for examining a recovered bullet for trace materials to help learn what its targets may have been.

The bullet path through materials in a room can provide information about the position of the item when struck. In this case, two doors: one a closet door, the other the front door of the apartment (Figure 7.37 to Figure 7.39).

Sometimes the bullet one seeks at a scene may have gained new positional properties along its course through the various targets. Figure 7.34, for example, shows the entrance of bullets at a 45-degree angle and Figure 7.40 shows the primary target.

Figure 7.35 Bullet recovery at the Whitehorse shooting scene.

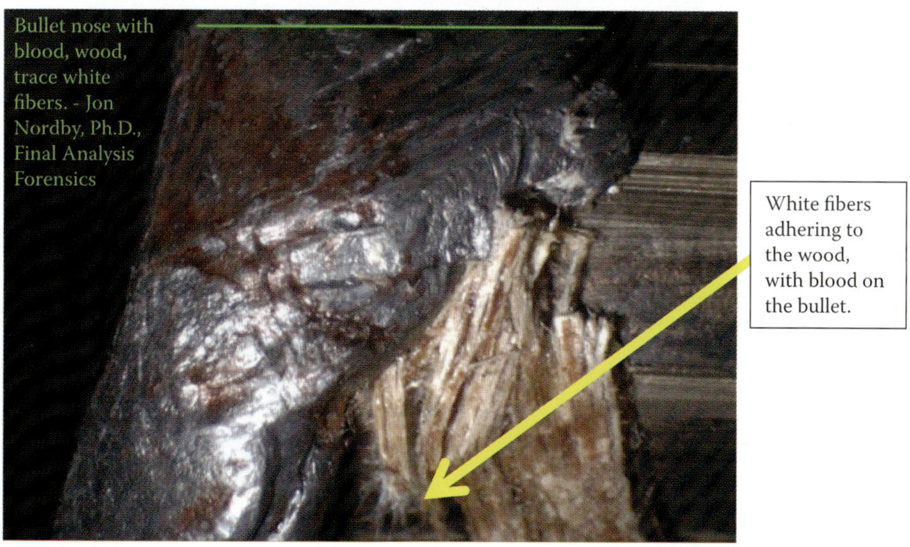

Figure 7.36 Trace evidence on a bullet is often both useful and important.

Figure 7.37 Bullet recovery points allow one to investigate and discover any other targets involving that bullet.

Figure 7.38 Specific targets such as doors can provide information about their position—open or closed.

Figure 7.39 The front door.

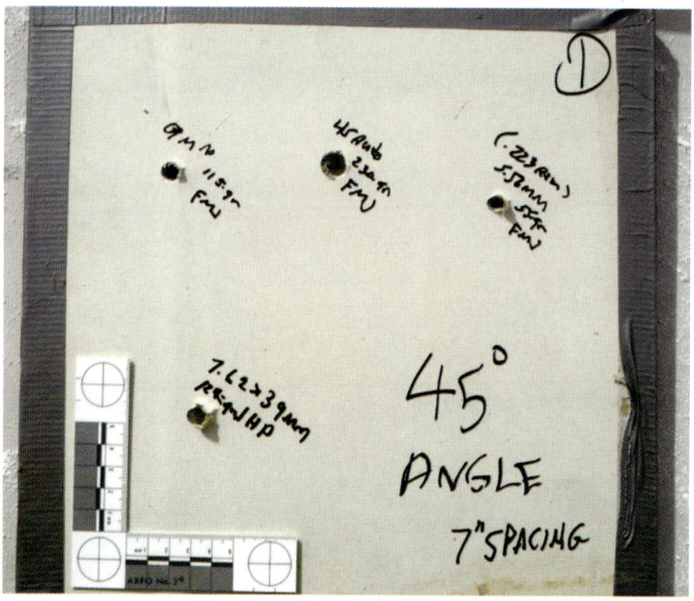

Figure 7.40 Primary sheetrock target.

On a trajectory through a target series of six pieces of sheetrock, for example, some bullets exhibit tumbling while others continue without tumbling (Figure 7.41). Multiple tests with the relevant caliber and the same ammunition at issue in the case are required for any such experiments through multiple or even through any target(s). It remains preferable to use ammunition from the shooting incident, but often that is neither practical nor possible. The same brand, the same properties, and the same firearm are sometimes the best that we can do.

One useful skill to develop involves understanding what physical features distinguish bullet entrances from bullet exits in different materials (Figure 7.43). The physics of terminal ballistics, therefore, involves more than simply understanding the effects of a bullet fired into

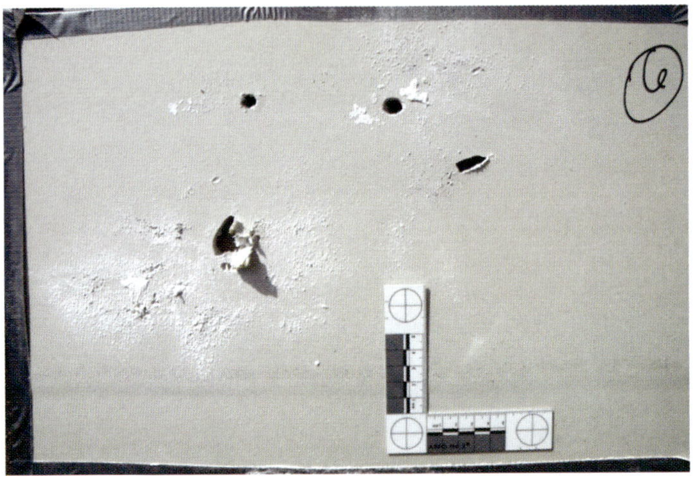

Figure 7.41 Differing entrances on the time the bullet strikes the sixth target at 45 degrees.

Figure 7.42 Lighting and camera positions, regardless of tools, remain to be one of the most important steps in gathering data. Michael Graham adjusts the Keyence high-speed camera to generate measurements from the images showing the bullet deflections through materials.

human tissues and bone. Sometimes, as I mentioned, a laser is helpful in suggesting alternative trajectories to follow (Figure 7.42). Sometimes one finds a bullet at the end of the line.

The discovery and documentation of bullet damage can be very time consuming. Although interesting, it is hardly what one would call exciting work. It must, however, be done the best one can within the limits imposed by both scene and circumstance (Figure 7.44). Finding bullet fragments is difficult as well, since often the fragments pursue

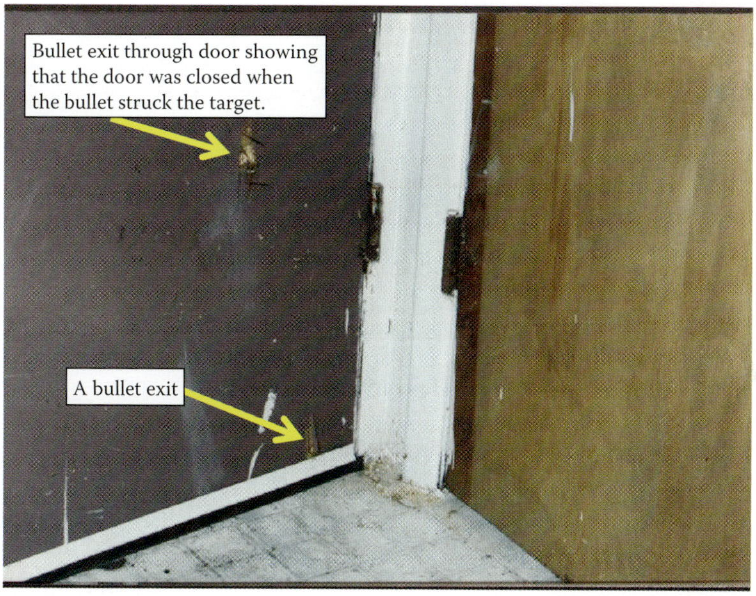

Figure 7.43 Adjacent doors—the door on the right is a hollow-core door with a thin, plywood face.

Figure 7.44 Testing can sometimes help distinguish bullet entrances from exits—it is not a simple matter if an entrance appears in the fourth or fifth target (see Figure 7.41). This photograph shows an entrance.

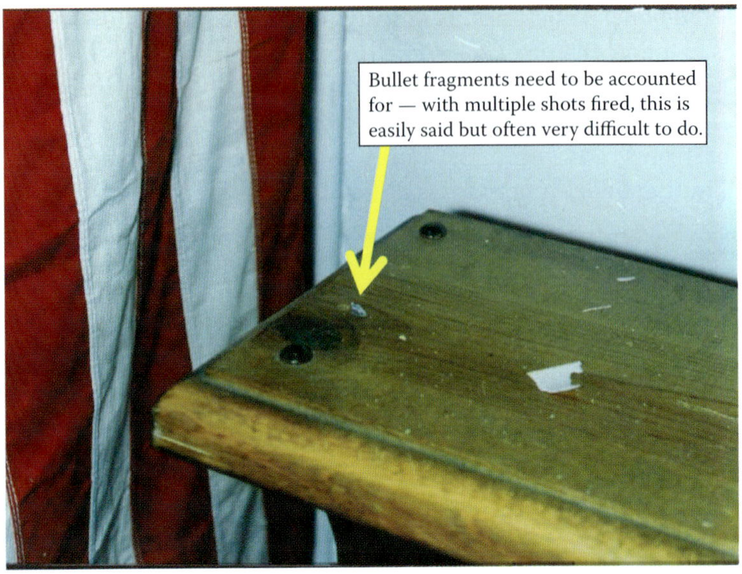

Figure 7.45 The search for bullets includes bullet fragments.

independent trajectories after a bullet perforates a target (Figure 7.45 to Figure 7.47). I have seldom been able to associate one fragment with one bullet unless the pieces articulate like those in a puzzle. With multiple independent targets, this is unlikely.

Various methods can be used to note bullet locations. The favored method is to either use numbers or letters printed on yellow or brightly colored plastic tents. Some systematic principle to organize the type of tent proves useful when recording the photo log.

A close look can sometimes help explain damage. The most difficult items to evaluate with either bullet damage or bloodstain patterns are those that can be moved easily. Doors, walls, floors, ceilings, and any other fixed items provide the best, most reliable evidence. Often, movable objects cannot be reliably evaluated.

When possible, a series of photographs or a video of the undisturbed scene should be taken (Figure 7.51 to Figure 7.54). This will help determine the initial position of moveable items, but this does not mean that this position replicates their position when the shots were fired. It is seldom that easy.

Cartridge cases are another item to be located and documented. Their ultimate positions can sometimes be explained by multiple mechanisms—being kicked, stepped on, or striking another object when ejected from the pistol. At best, their locations provide some general information about the relative position or positions of a shooter. Caution must attach varying and modest significance to these observations (Figure 7.48 to Figure 7.50).

The broken glass (Figure 7.55), the nick in the sill (Figure 7.57), and Whitehorse's position when struck by the 11 bullets help fix Zwank's position: he was shooting through the window. Cartridge cases were found outside to the right of the window when facing in (Figure 7.56).

Figures 7.58 and 7.59 document the window through which Dom fired. The damage to the Venetian blinds (previously seen) shows that they were closed with louvers shut when Dom fired into the building from his position in front of the van. If they were closed, then how could Dom see Whitehorse, with his gun's slide moving and the cartridge cases ejecting? Add to that the lack of any cartridge cases from Whitehorse's Star pistol, and the fact

Figure 7.46 When multiple shots are fired from multiple locations by multiple shooters, there is no shortage of entrances and exits, as well as fragments.

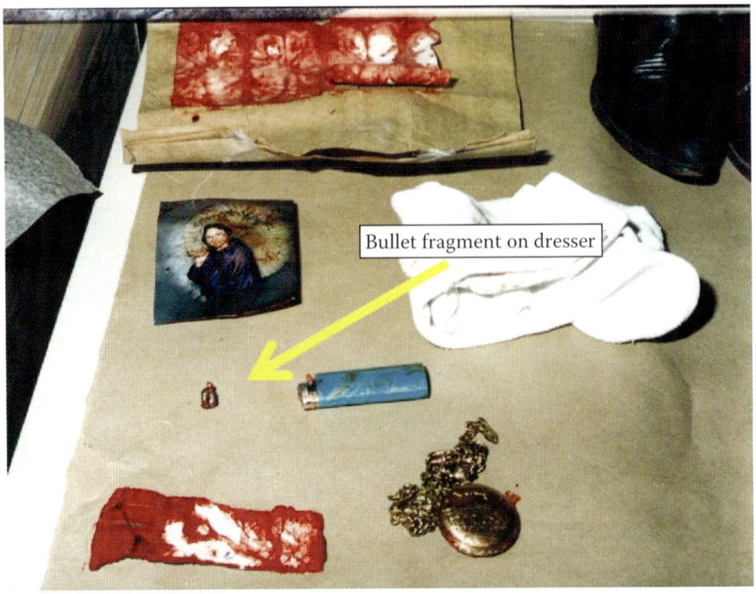

Figure 7.47 Sometimes bullets can be overlooked when they seem to blend in with ordinary objects.

Figure 7.48 The cartridge case in this hanging plant was found by trying to associate bullets with cartridge cases.

Figure 7.49 Cartridge case in hanging plant.

Figure 7.50 This cartridge case provided another bit of data raising questions about the shots fired by Dom, as well as accurate documentation by the Del Mar investigators. If Dom did not enter the residence, then he shot through a window or a solid wall leaving no damage to either. How did one of his cartridge cases get into the flowerpot hanging near the front door? And without his entry, or without shooting through a window or wall, at least one of his shots which was reported to be from his position outside is not possible.

Figure 7.51 Bullet in the dresser drawer.

that the pistol was clean, oiled, and had not been fired. Five of Dom's eight bullets perforated this window. One even ended up toward the west side of the apartment (Figure 7.60).

Officer Dom's Shooting Positions and Trajectories

The bullets originating from Dom's direction, according to his sworn account, have him firing from just behind the front of the van parked in the driveway. If his statements are true, then the trajectories look like the two-dimensional representation in Figure 7.61.

It is obvious that these trajectories are impossible—bullets do not make sharp angles through sheetrock. Any doubts should be satisified by testing—shoot into a sheetrock wall, constructed to replicate this wall, from the same angles claimed by Dom and see what

Figure 7.52 Photographs such as this one remain profoundly mysterious and useless for most elements of a scientific reconstruction.

Figure 7.53 Identifying bullet damage to items may involve identifying and examining relevant trace evidence.

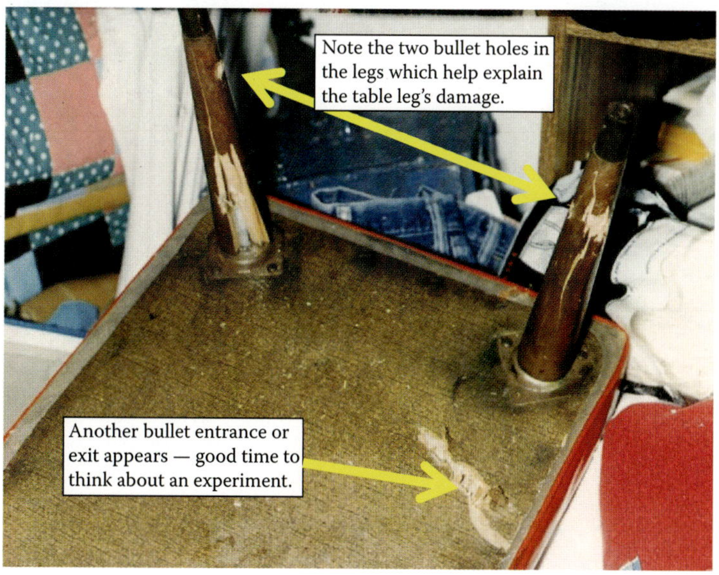

Figure 7.54 Bullet damage to movable items must be examined and carefully evaluated—we may not know the position they occupied when they were struck by the bullets.

Figure 7.55 Additional data are provided by damage to the windowsill and the location of the cartridge cases.

Figure 7.56 One of Zwank's cartridge cases, in a closer view, landed on the window ledge.

Figure 7.57 Zwank fired downward into the room while his target lay face down on the floor—one bullet struck the window ledge on the way in.

Figure 7.58 Dom fired into the window when the blinds were down—he could not see his target—the damage to the blinds is clear.

Figure 7.59 Dom's bullet entered this window—the blinds are raised.

Figure 7.60 This bullet was recovered in the west side of the building behind some furniture.

happens. Bullets do not make such overlapping angular turns. The spent casings also supply additonal evidence that Dom is not telling the truth.

Five cartridge cases lay by the van indicated by the small circles in the drawing, which is roughly to scale. The other three bullets attributed to Dom must have been fired by the apartment's front door since three cartridge cases rest on the porch. Their positions are noted by the small circles. If he fired eight shots, then the five plus the three on the porch total eight (Figure 7.62).

These trajectories developed from the four perforations of the walls (two for each wall: entrance and exit, reentrance and reexit) by each 9-mm bullet together with the three cartridge cases show that Dom fired at least two shots into the apartment from either just inside, or just outside, the front door. The third shot, closest to the van, raises questions. If that was his first shot, then he fired it from inside the apartment. He can not shoot through

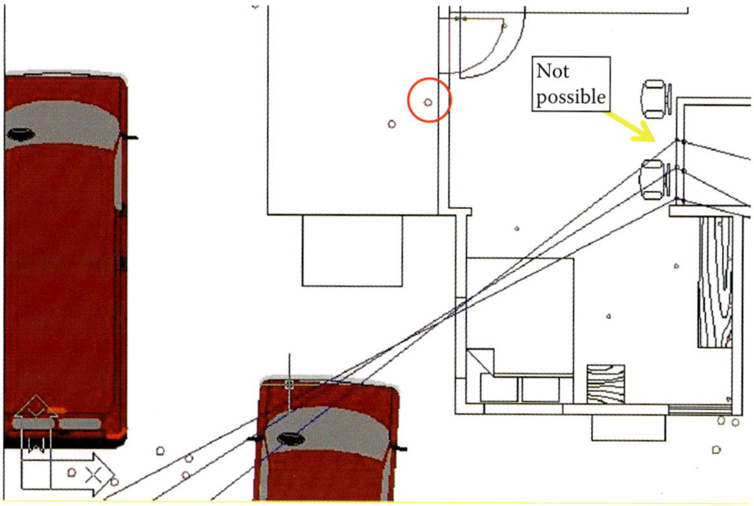

Figure 7.61 Crime scene drawings of the shot trajectories (see also Figure 7.62 to Figure 7.69).

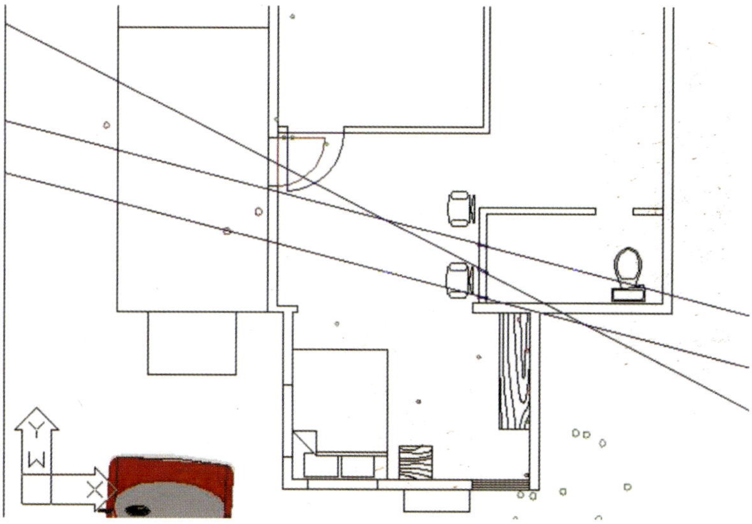

Figure 7.62

a solid wall or a closed window, leaving no trace of a bullet perforation. If he fired from inside the apartment, then one cartridge case should be found in the apartment. Since it was not, then the one cartridge case from inside must have been moved outside by some unknown agent. The cartridge case found in the planter seems to be a likely candidate (the red circle closest to the building just to the right of the front door when facing the door). The alternative is that Dom has discovered a mechanism for shooting through solid walls or closed windows and leaving no trace. This trajectory and the missing cartridge case is one of the concerns voiced in the administrative review charging official misconduct at the crime scene.

An analysis of Dom's shots provided the explanation for three of the bullet perforations of the wall into the bathroom, and into or through the bathroom's exterior wall (Figure 7.63).

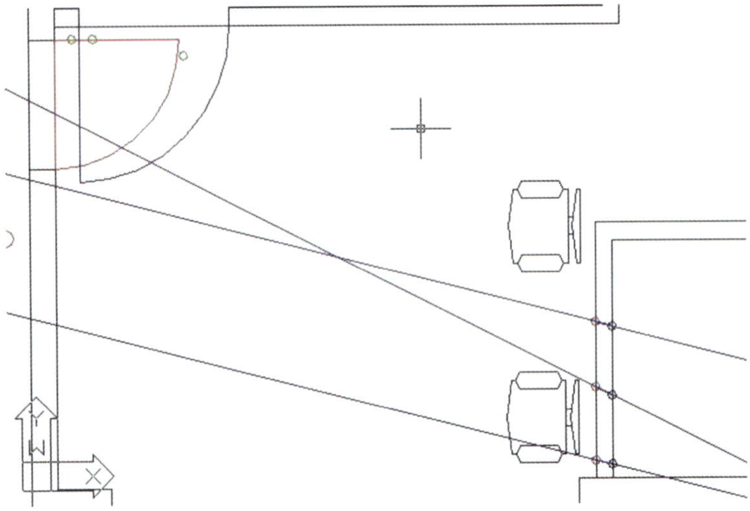

Figure 7.63

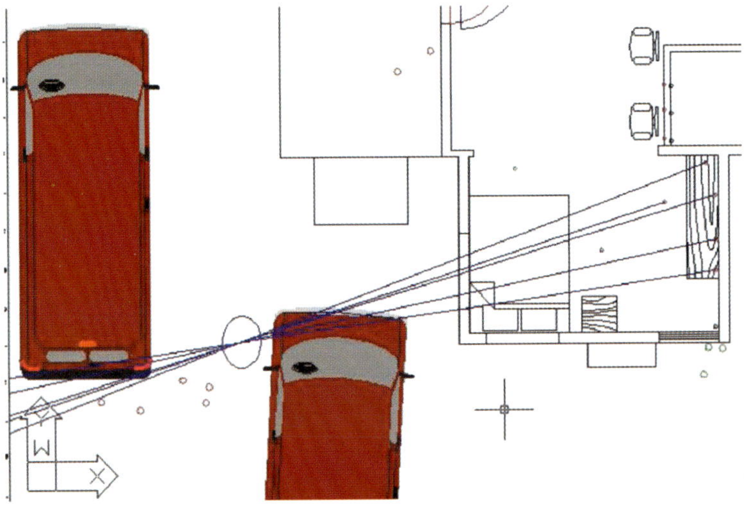

Figure 7.64

The two-dimensional drawing software used here allows the "drilling down" through the drawing to display the measurements both locating each bullet hole and describing its properties. The remaining five shots fired by Dom went through the window (Figure 7.64). The details of these trajectories and their path inside the building are represented in the drawing in Figure 7.65.

Officer Zwank's Shooting Positions and Trajectories

Zwank took a position by the other bedroom window. Only shots from this position struck and killed Whitehorse: 11 hits, all at the same angle and all roughly from the same position. The bloodstains, as well as the gunshot wounds themselves, tell us that Whitehorse was lying prone, face down on his stomach, possibly with his hands behind his head (wrist-perforating gunshot injury). This suggests that Whitehorse may have been complying with

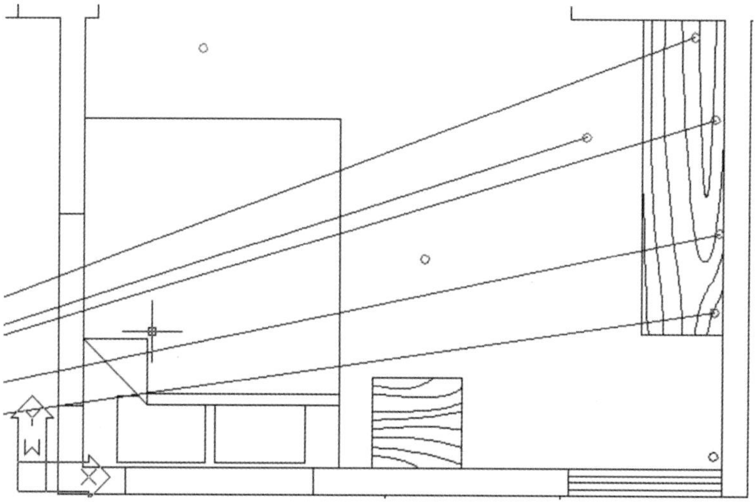

Figure 7.65

the officer's instructions to "get to the ground, hands behind head." He had to be in this position when Zwank shot him.

For example, the decedent's gun lay under his body 4 feet from the east wall (D) of the bedroom, and 6 feet 8 inches from the south wall (E). Knowing from the diagram called "Outside scene of 1948 Vidalia Road." that wall (D) is 13 feet 6 inches long; we can place the gun in reference to the northeast bedroom window (B), the position occupied by Zwank when he fired at the decedent. Using the measurements provided, adding our measurement of the window itself, and noting that the inside measurements of the baseline D will be shorter than 13 feet 6 inches, we can locate both the gun and the victim in relation to the northeast window (Figure 7.66).

I make a quick calculation at the scene and then develop the data as required for a report or for a possible court appearance. Using basic geometry, we calculate that the gun was between 5 feet and 6 feet 10 inches from the northeast window. Since Whitehorse lay on top of the gun, presumably with the gun near his right hand, this would also place the decedent's head and upper torso between 5 feet and 7 feet from the window. The decedent was about 6 feet tall, so, at most, the decedent's feet were 11 feet to 13 feet from the window, assuming he was flat on his stomach with his legs stretched out straight. This position fits consistently with the autopsy findings and provides added explanatory significance to the autopsy findings. Remember that his right wrist had a perforating gunshot wound.

The bullet impacts in the apartment and Zwank's approximate position, given his height and arm length, allow the approximation of these trajectories (Figure 7.67).* This measurement process can be quite time consuming, depending upon the ease and the ability to locate the appropriate targets, primary, secondary, tertiary, and so on, struck by each bullet. Several of the more mystifying bullet positions become clearer as the work progresses.

* Chapter 1 describes the use of a trajectory envelope. Here, the window basically forms the trajectory envelope, with the muzzle moved in that area to accommodate each shot from muzzle location to impact point. From this data, some clear concept of an error rate can be developed and calculated for the measurements involved. The method will be the same as the method applied to develop any metrological error rate.

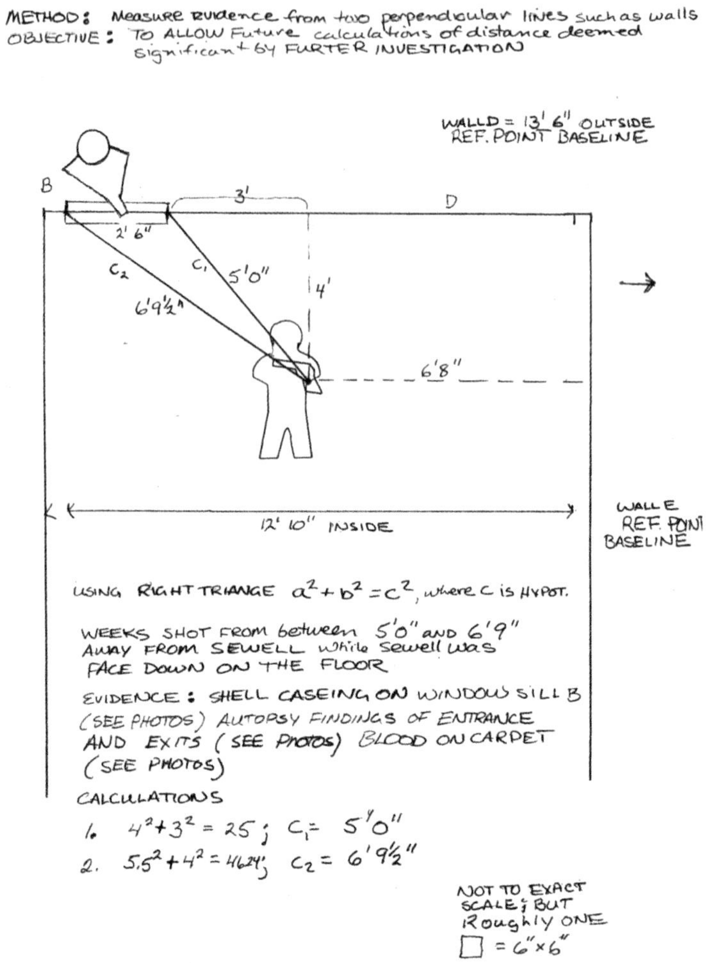

Figure 7.66

The bullet impacts on the doors, for example, seen in the earlier photos, now become more understandable. Consider the drawing in Figure 7.68 of the trajectories involved. The two shots were too high to strike Whitehorse. Scientists try to resist the temptation to over-interpret such data by supplying more subjective criteria to the explanation. For example, it may be tempting to suggest that shooting high is typical of an inexperienced shooter or that shooting 11 times at a prone suspect represents a type of shooting frenzy. I suspect that this is true, but that sort of inference, along with its defense in the courtroom, is best left to a psychiatrist or an appropriate administrative agency.

Additional documented shots fired by Zwank are drawn in Figure 7.69. We do know that Whitehorse was struck 11 times. Depending upon the quality and availability of the data, we may not be able to account for every shot fired in every shooting reconstruction. This becomes more obvious when faced with attempting to reconstruct the equivalent of a firefight among rival gang members.

This fact certainly applies to any scientifically based crime scene reconstruction or event reconstruction, regardless of type. One might begin to worry when the "storytellers" have answers for everything but data for nothing. With no reasoning to support the

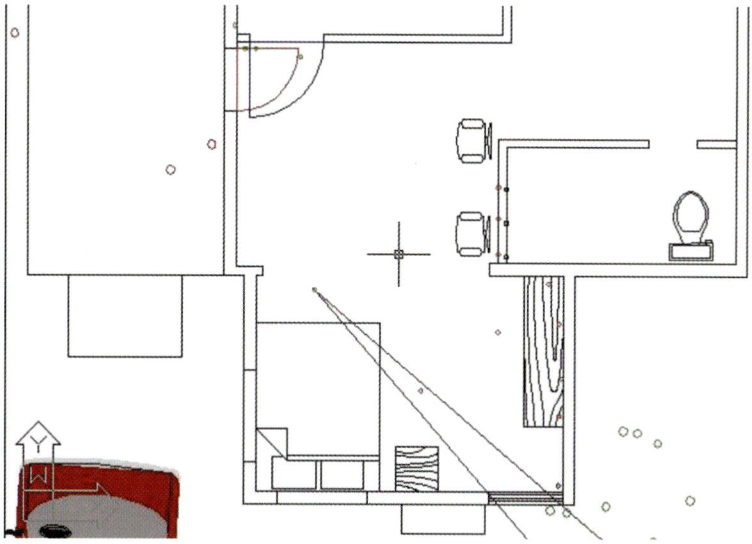

Figure 7.67

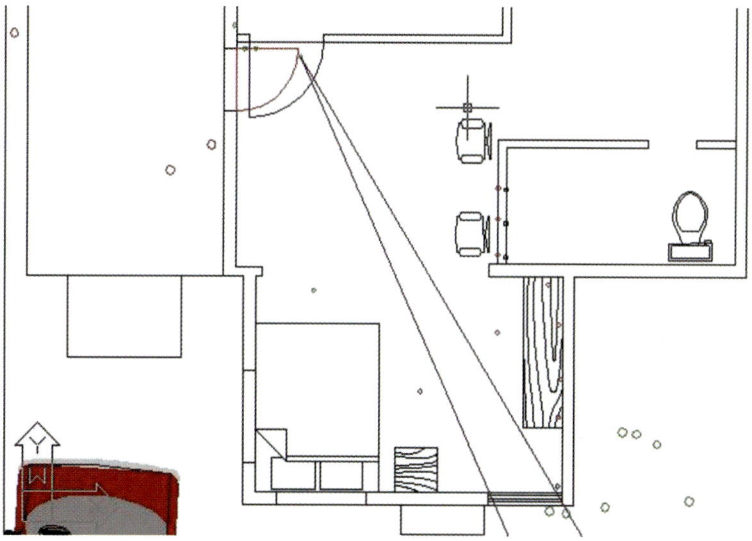

Figure 7.68

story they tell, the experience of the storytellers becomes scientifically irrelevant. Science involves reasons: measurements, testing, experiments, as well as critical observations. Science is data driven. Storytelling is merely entertaining (or not).

Even appropriately trained scientists may offer reconstructions that are mere storytelling. The temptation to state conclusions without performing the tasks necessary to support them raises the same questions as the student who writes a complex answer on a physics test without showing how that answer was derived. For all the teacher knows, the answer, correct or not, was copied from another student. It is therefore unacceptable.

Whatever the answers are, they are not scientific answers unless they are demonstrated products from appropriate applications of relevant scientific methods. Experience alone, the many years of working crime scenes, counts for nothing when it is offered as the only

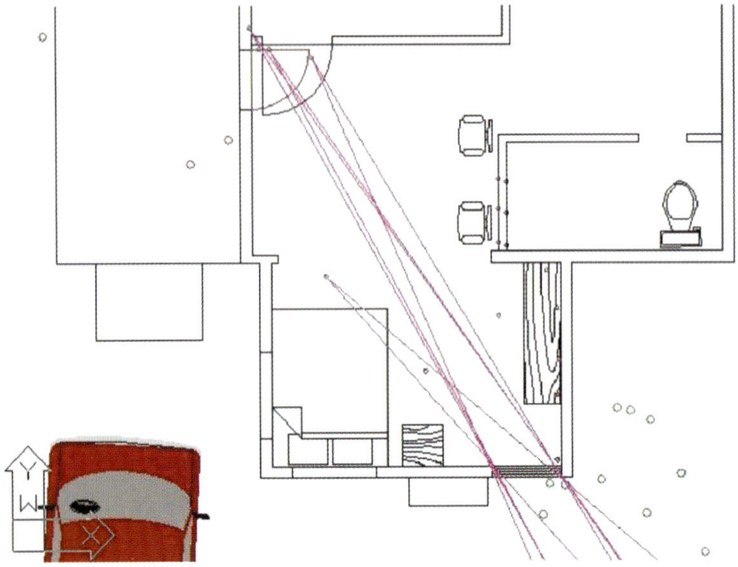

Figure 7.69

justification for an answer. Merely spending time at an activity does not ensure success. I heard a college sophomore complain about a failing grade point average: "Gee, what a waste of the last 5 years." And many years spent doing things the wrong way does not help much either. This Del Mar case reflects the reality of that problem and its ultimate cost to the criminal justice system.

II Nonshootings

A Flash in the Pan 8

Wrong rules the land and waiting justice sleeps.

Josiah Gilbert Holland, "God, Give Us Men!"

Introduction

Five kids waited in the car, and not patiently. They screamed, fought, and fussed while their mother locked the front door of their two-story house. Armed with a picnic basket filled with eats for a summer outing, Mom slid into the fray with basket in hand. The old Dodge wagon chugged off toward the freeway and the carefree summer day of fun began. Or so Mrs. Slovotski believed.

When the Slovotski clan got home, or to what was left of it anyway, they were met by hoses, fire trucks, squad cars, and even an ambulance. All blocked their way. Their house had black accents above broken windows and gray smoke drifting from a hole in the composition shingles. Water trickled from the threshold below the open front door, eventually joining a growing puddle on the concrete landing.

Adrian Slovotski had a good figure for a mother of five—an unusually good figure. An attractive face and an attentive manner came with the package. She had spent most of her time at work fending off the middle-aged lechers seeking to augment their paltry lives with some outside excitement. She made enemies that way. Her boss, in particular, had done everything in his power to make her life as miserable as his, particularly so after her vociferous rejection of his persistent sexual advances.

Times had been tough enough already. Her husband of 7 years had finally succumbed to the cancer that continued eating away at both his lungs and their savings account. After she declined her boss's propositions, she was dismissed from her secretarial position when he claimed she took off too much time to care for her ailing husband and her five kids. But at least she owned the house. Or she had, until now.

Overwhelmed, she felt she had only one way to turn. She had always called on her father for help. Now it was time she called him again.

When her father arrived, he saw a van labeled "arson unit." He immediately thought of her boss and the trouble he had caused her at work. Things would be all right. The extended Slovotski family joined forces: daughter and grandkids moved in with her parents. Time passed—first a week, then a month. That is how time works.

The homeowner's insurance paid off. Adrian cashed the check and got the required bids to rebuild the damaged structure. But before contractors could even submit the bids, a police officer arrived at the crowded home of Adrian Slovotski and her extended family. Just when she thought things could not get worse, they did. She was informed of her rights and arrested. The charge was arson. Her babies cried. She cried. Her parents cried.

Nevertheless, she found herself alone in jail, awaiting trial for arson. Time passed—now more slowly than she ever thought possible.

Basis for Arson Determination

The fire investigators' original finding was accidental fire due to electrical problems, but the lead detective working the case was not satisfied with that judgment.

At some point later in time, the detective claimed to have a confession, reported by a jailhouse informant, from Adrian Slovotski. The informant allegedly claimed that Adrian said she had put Crisco in a frying pan, put the pan on her stove's back-right burner, and turned the stove burner on high. Adrian had then removed the burner's stove knob, the informant said, and reinstalled it on the control rod misleadingly showing that the burner was off. After Adrian and her children left the house, the burner apparently had heated the Crisco, started the fire in the pan, and consequently burned down the house. Adrian's motive, the informant claimed, was to collect the insurance money, which she needed since she had been fired from her job.

Fire officials revised their charges based on claims made by the lead detective regarding the hearsay confession. They changed their determination from accidental fire to fire by arson. Apparently, no one questioned the detective's motives, even after it became known that Adrian had worked as his secretary before he fired her. The gung-ho deputy prosecutor also enthusiastically bought in to the revised charges.

With little, if any, additional investigation, fire officials simply agreed to go along with the detective. Using the same photos, the same investigation, and the same data, they now declared as arson what they had named an accidental electrical fire. They already had photos of the fire scene; they returned to collect various parts of the stove, as well as the pan that Adrian had allegedly filled with Crisco to start the fire.

The case came to the Final Analysis Forensics laboratory through two attorneys' efforts to provide a scientifically based reconstruction of the fire's mechanisms, as indicated by the physical evidence. They sought to understand the fire's origins and progress through the house, one way (accident) or another (arson). My analyses in this case considered the physical evidence from the fire and my own independent testing of its officially proposed origin, allegedly pointing toward arson. My conclusions were based upon these evidence examinations and my independent testing; all within the realm of the natural sciences.

On this case I consulted with a fire investigator, who later became the Alaska state fire marshal, but I am not, nor have I ever been, a "fire investigator." But my case experience, of necessity, includes fire-related deaths: those in burning vehicles, burning buildings, burning boats, and simply mysterious burning clothing.* That does not count cases involving bodies burned in attempts to dispose of any physical evidence. However, some fire cases involve issues that are based upon relatively simple applications of physics and chemistry, which do not require the expertise of a fire investigator to interpret and to analyze. Adrian Slovotski's is just such a case.

I have presented lectures and workshops to members of the International Association of Arson Investigators (IAAI) and have closer associations with some respected fire

* Through work at two medical examiners' offices and with the National Disaster Medical Service, one faces a variety of deaths about which one must at least learn enough to understand and explain.

investigators. I have also learned that, as in any field, some so-called fire investigators know less about basic physics and chemistry than is necessary for them to do their jobs in a reliable scientific manner. This unfortunate lapse can greatly hamper the scientific basis for any reconstruction of the cause and origin of a specific fire. Whatever the investigative field, a scientific basis of examination is a necessity.

Analysis of the Data

The photograph in Figure 8.1 was taken shortly after the fire had been extinguished. The photos show the point of origin, that is, the point where the fire started. The fire started on or near the stove. Fires burn up and out from the source. This pattern can be seen in the charring of the cabinets and damage above the stove. The exertion of positive pressure from the fire on the glass windows will push window glass out, and the broken glass will often be outside the house. When we find windows broken in and not out due to the fire, then investigators must seek the presence of accelerants. In such cases, due to accelerants, the much hotter fire will suck oxygen from outside into the room and literally implode the windows. The photographs do not provide that type of evidence for accelerants in this fire.

Window glass provides an excellent source to test for the presence of accelerants. The residues from the fire will collect on the window glass, providing a sample for gas chromatography. In this case, no use of accelerants was suspected, and no window glass was provided for testing. The glass appears to be broken outward, as one would expect from a fire without accelerants.

The vent hood over the stove has fallen onto the cooking surface (Figure 8.2). I keep looking at the white area to the right of the stove, which would be adjacent to the right rear burner. Between the fan hood and the large white patch on the wall, an outlet with charred wires that lead toward that white area seems to be the point of the V-shaped burn pattern in the kitchen (Figure 8.3).

Figure 8.1 The kitchen with the burn pattern.

Figure 8.2 The stove and counters.

Figure 8.3 The stove with the cabinet contents moved.

The first few photos show damage to the stove. The drywall behind and beside the stove is not burned through. It takes about 45 minutes to 1 hour to burn through drywall, unless a bromine compound is added to the drywall as a "fire retardant" during its manufacture. The stove looked like an interesting item among the debris: one worth a further look. If the state's theory were true, then the pan, the stove, and the burner heating the pan should provide evidence supporting the claim that this was the fire's origin.

Damage in the kitchen and throughout the house seems to focus in the area around the stove (Figure 8.4 to Figure 8.6). Long cracks with sharp edges appear in photographs of the window. They do not suggest, nor does anything else in the photograph, the presence of accelerants, which might point obviously toward arson. But arson does not always depend upon signs of accelerants.

Figure 8.4 The left-side counter beside the stove.

Figure 8.5 Another view of the left-side counter.

The state's theory claimed that a pan filled with Crisco, a type of shortening or fat used in cooking and baking, was used as a means to start the fire. This theory focused the analysis toward an examination of the stove and the pans used on its burners. If Crisco was used on the stovetop to start the fire, there should be some data to collect and to examine in order to test the claim. First, the data must be collected. Then, once we have a basis for testing, experiments should follow. How does Crisco behave when heated by a calrod burner?

The Stove

The kitchen stove proved to be an interesting item. It was a Whirlpool model with four calrod burners. Recall that the state's theory held that Mrs. Slovotski had removed the temperature selector dial from the right rear burner, reinstalling it to read off when in fact

Figure 8.6 Pan covered with debris.

the burner was actually set to high. The first problem I faced involved the stove's design: the stove in her kitchen did not have removable temperature selector dials. This model was equipped with a digital control board, located on the stove's upright back panel. No dials; not even buttons. Just a flat LED-backlit screen with icons to push.

This problem certainly raised questions about the thoroughness of the investigation and the collection of relevant data. The problems for the state's theory were just beginning.

Figure 8.7 shows the right-rear burner. This burner at least appears to be near the source of the fire given the typical V burn patterns. Note the greenish color on the burner.

All that remains of the stove's rear burner switches, located on the stove's rear LED panel above the burners, appears in Figure 8.8. The remains of the plastic panel have melted together. The switches for the right-rear burner cannot be identified. The switch control

Figure 8.7 Calrod burner.

Figure 8.8 All that remained of the stove's controls.

Figure 8.9 Transfer to the burner from the pan.

panel was destroyed without revealing any information about the nature of the switch, for example, on or off, or high or low. How did the investigators decide that the switch was put on high with the indicator altered to appear as if off was selected?

The arrow in Figure 8.9 points toward the greenish discoloration apparently associated with the lard-filled frying pan alleged to have started the fire, as discussed later. The transfer from the pan to the burner indicates that this burner was cooler than the pan: a Fourier series describing heat transfer explains the material transfer to the calrod element. This observation casts further doubt on the state's theory. Testing became very important in this case.

The foil-covered burner catch pan from under the right-rear burner appears in Figures 8.10 and 8.11. No surprise that its condition indicates an exposure to heat; it is part of a stove. Figure 8.11 shows the bottom of the same burner catch pan.

The Pan Containing the Crisco

The pan found on the right-rear burner shows a transfer between the calrod element and the pan's coating, a green enamel-based shell. The pan itself is a standard aluminum frying pan (Figure 8.12). The properties of this pan eventually pointed toward the correct explanation of the fire.

The design of a frying pan should be obvious. Its bottom is designed to distribute heat evenly throughout the pan, cooking the food evenly across its bottom. If such a pan is left

Figure 8.10 Foil on the catch pan.

Figure 8.11 The catch pan.

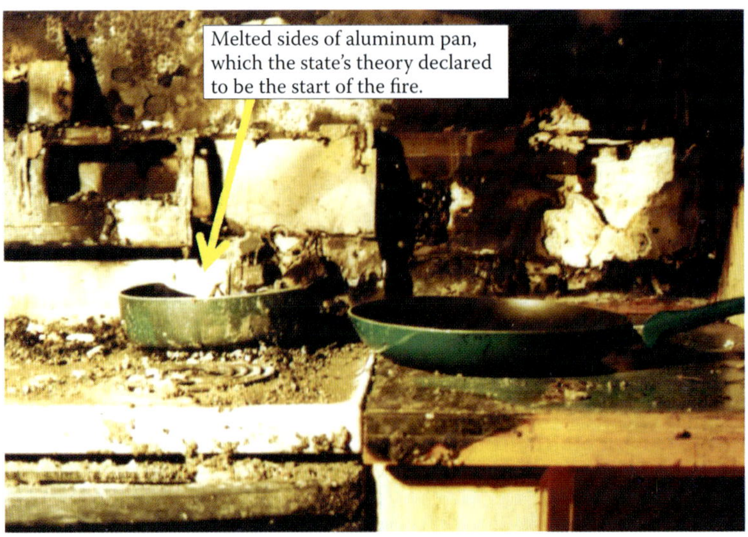

Figure 8.12 A green pan on the rear burner with another green pan on the right-side counter.

on a hot calrod burner and if the burner becomes hot enough to melt the pan, then the entire pan would melt, at least somewhat evenly. But this pan melted only on one side.

The area beside the right-rear burner, further to the right of the stove, appears to be the origin of the "V" pattern, indicating the point of origin. If this fire began and burned hotly to the right of the aluminum pan, then the side of the pan closest to the heat source could melt without distributing the heat evenly. The heat source was not in contact with the pan's bottom, but with the pan's side. This would explain the fact that only one side of the pan was deformed by the heat source. It would also explain the transfer of green enamel from the pan's bottom to the surface of the calrod burner.

The pan supposedly containing the Crisco and placed on the right-rear burner appears in Figure 8.13. Note the alteration of the green enamel on the bottom and the apparent pattern from the calrod burner left on its surface. The transfer of this material requires explanation.

The green frying pan's condition (Figure 8.14) raises a question that a routine investigation should attempt to answer: Why does the pan melt only on one side and from above

Figure 8.13 The bottom of the green pan with the ridge surface melted and a color change in the circle formed by the owner's contact with its bottom.

Figure 8.14 Inside the top of the green pan.

Figure 8.15 A side view of the pan's melted side.

Figure 8.16 A bottom view of the melted pan.

the designed heat source, the calrod burner? Several possibilities could explain this phenomenon, but none are addressed in the fire investigation.

Consider the pan pictured in Figure 8.15. This indicates that a source of extreme heat was generated above the burner, to the right of the pan, when facing the stove. That corresponds with the outlet beside the stove just above the right countertop and below the cabinets. The melting on the pan's topside is also visible from the view in Figure 8.16 of the pan's bottom surface.

Both Figure 8.13 and Figure 8.17 show the discoloration of the pan, and the enamel materials that were transferred to the burner element. This is a gold mine of data.

The breaker box seen in Figure 8.18 provided one more bit of data. Looking closely, one can see that the breaker for the kitchen outlets, including the outlet beside the stove, has been tripped (Figure 8.19). Figure 8.20 of the kitchen outlet shows that electrical problems obviously might in some way have contributed to the fire. The outlet servicing the Mr. Coffee appliance seen in Figure 8.20 was among those governed by the breaker.

The coffeemaker made by Mr. Coffee had been recalled for an electrical fault, which could, company officials said, cause a fire even if the coffeemaker was not being used. The recall covered several models of the popular appliance. The coffeemaker purchased by Adrian Slovotski was one of those subject to this recall. She apparently remained unaware of this problem with her appliance.

The only remaining evidence to either support or refute an electrical malfunction comes from the electrical outlet to the right of the stove into which the Mr. Coffee appliance was plugged (Figure 8.20).

Figure 8.17 Color change and transfer of pan materials found on the calrod burner.

Figure 8.18 The kitchen breaker is among the fused circuits in this electrical panel.

Figure 8.19 Circuit breakers can often reveal valuable information at a fire scene—their positions should always be examined.

Figure 8.20 The wires leading into this outlet once belonged to the power cord for a Mr. Coffee coffeemaker.

Mr. Coffee Coffeepot

Figure 8.21 indicates the location of the coffeemaker beside the outlet seen in Figure 8.20. The plug for the appliance is to the left in the photo, beside the green frying pan on the right-rear burner. Figure 8.22 indicates that the coffeemaker disintegrated during the fire. Note the broken glass on the pad upon which the appliance rested.

"V" Burn Patterns

The typical "V" pattern that fire investigators seek when looking for a fire's origin appears at the right above the outlet beside the stove (Figure 8.23). The origin of the fire is closer to the coffeemaker than to the stove (Figure 8.24).

Figure 8.21 Area used for the coffeemaker to the right of the outlet.

Figure 8.22 Broken glass from the Mr. Coffee glass coffeepot.

Figure 8.23 Overall view of the fire's origin.

Figure 8.24 The right rear burner is adjacent to the outlet for the coffeemaker and the coffeemaker itself—the pan on that burner was only melted on one side.

The patterns depicted in Figure 8.25 again strongly support that the point of origin is farther to the stove's right (facing the stove) than the state's theory proposes. The difference is in learning that there is another ignition source to the stove's right, in front of the "white" spot behind the counter on the wall. This is where Adrian Slovotski habitually plugged in her Mr. Coffee appliance. Indeed, the Crisco in the pan located on the stove's right-rear burner might have provided another easy fuel source for the fire, if it were there. There is no evidence to support the claim that Crisco was in the pan. Any additional burn patterns from the stove may have mistakenly been interpreted as the fire's source, absent other known ignition sources. The defective appliance seems to be the most compelling source of the fire.

Figure 8.26 shows a typical "V" pattern. The fire started near the stove but not necessarily on the stove. Fires burn up and out from their source. This pattern can be seen in the

Figure 8.25 Developing the "V" pattern.

Figure 8.26 The bottom of the "V" points toward the fire's point of origin.

Figure 8.27 The cabinet above the stove.

charring of the cabinets and damage above the stove. The conceptual jump from a fire to a stove is reasonable, since any stove is a source of tremendous heat in a kitchen. If investigators actually had taken the time to investigate, they would have discovered the other more likely source of the fire: electrical faults with a coffeemaker, which had been recalled by the manufacturer for defects known to start fires.

In Figure 8.27, note the crazing of wood on the closed cabinet door above the collapsed range hood, which provided additional information about the wood's proximity to the fire.

The Oil Lamp "Evidence"

The lamp pictured in Figure 8.28, wick in place, shows no evidence of being burned.

Figure 8.28 The broken oil lamp.

Figure 8.29 Items relocated by directing water from a fire hose in their direction.

Much of what appears on the counters in the scene photos depicts the displaced items associated with the firefighting process (Figure 8.29). Therefore, we may not be sure of the original locations occupied by such items before the firefighting process began.

This obviously applies to the oil lamp. Absent evidence of thermal damage and absent any positive test results for accelerants associated with lamp oil, no conclusions can be reached regarding the lamp's possible involvement in the fire. We have no scientific way to tell, for example, if the lamp even had oil in it. The lamp was not burned or charred, and in no way suggests that it was involved in the fire.

Fire Basics

Knowing a few basic points about the behavior of fire can help one assess some of the evidence in this case, as well as develop a limited reconstruction based upon scientific principles. Most of us have to look them up when specific questions arise.* Wherever data are offered as facts attributed to knowledgeable sources, conducting experiments or tests remains scientifically important.

1. Fire requires fuel, oxygen, and heat, which combine to form a chemical chain reaction.
2. Fuels, such as wood in a house or a fireplace, must be converted to vapors before they can burn.

* No scientist can maintain a brain filled with all sorts of facts, figures, formulas, and definitions. More important is knowing what one must research by knowing what is relevant in the first place. The key remains to be a logical method: knowing how to critically think through an issue, how to sort relevant from irrelevant data, and how to develop and assess alternative explanations. That ability forms the heart of critical intelligence so essential to the applications of scientific methods.

3. Solids and liquids do not burn. Only vapors burn.
4. There are four methods of heat transfer useful in determining a fire's origin:
    - Direct flame—Contact
    - Convection—Heated vapors vaporize other materials
    - Conduction—Heat transfers through an intimate contact, such as a steel beam
    - Radiation—Heat transfers through a wave motion
5. As part of the chemical chain reaction explaining fire, oxygen must attach to the fuel vapors in order for the fuel to burn.
6. A *flash point* is the lowest temperature that a substance can reach allowing it to vaporize. When combined with oxygen and an ignition source, it will burst into flames at this temperature.
7. *Ignition temperature* (also referred to as *autoignition temperature*) is the temperature at which a substance will ignite without a flame source.
8. Materials are called *flammable* if their flash point is below 100°F. (*Inflammable* literally means "not flammable." However, given the misuse of language, which literally defines the meanings attached to words with their popular use or abuse, inflammable seems to mean the same as flammable. If the tanker truck says inflammable it pays to stay away—do not assume that it is hauling milk.)
9. Materials are called *combustible* if their flash point is at or above 100°F.
10. Flame color approximates the fire's temperature (to the experienced observer):
    - Red colored flame = 1450°F and below
    - Salmon pink flame = 1550°F to 1650°F
    - Orange flame = 1700°F
    - Yellow flame = 1800°F to 1900°F
    - Pale yellow and white = 1900°F to 2000°F
    - White flame = 2000°F to 2300°F
11. Gasoline burns with a white flame at 2400°F, so whenever we see pale yellow to white flames, we also have reason to see a suspicious house fire. That is, unless a clear explanation for a gasoline source presents itself.
12. Dry chemical accelerants, such as thermite, burn at 3600°F. Anything in this range is not normal for a house fire and immediately becomes suspicious.
13. Clues at a fire that help establish the presence of accelerants include metallic objects that can provide a measure of heat in a house fire:
    - Copper pipes melt at 1981°F
    - Aluminum (e.g., door sills/frames) melts at 1100°F
    - Steel melts at 2550°F
    - Cast iron melts at 2780°F
    - Ceramics melt at 3400°F
    - Chrome melts at 3600°F
14. Photos in this case do not show abnormally high temperatures for a house fire; the flames themselves would not provide sufficient grounds to suspect accelerants or arson. None of the materials in the photographs indicate anything out of the ordinary.
15. In stove-caused fires, burn patterns develop directly above the stovetop. In this case, we see that the point of origin beside the stove does not quite present this same pattern.

Melting Points of Common Substances

The melting point of a material is the lowest temperature at which a substance changes state from solid to liquid. Aluminum turns from a solid to a liquid at 1100°F (1200°F pure form/alloy +10°F). Other substances have melting points that are listed in #13 in the previous list.

The aluminum pan seen in the earlier photos had to have been exposed to at least 1100°F on its side in order to melt, deforming as it did. Clearly, the bottom did not melt or deform. The enamel transferred from the pan to the calrod burner. Therefore, the bottom of the pan was not exposed to 1100°F temperatures, but the pan was hotter than the calrod burner.* That supports the conclusion that the source of the fire was not the burner but was above and to the right of the stove's right-rear burner. That location points directly at the outlet beside the stove powering the recalled Mr. Coffee coffeemaker.

Flash Points and Autoignition Temperatures

The lowest temperature at which vegetable oil or vegetable fat vaporizes remains between 450°F and 500°F. Crisco,† as a vegetable fat, will then vaporize at this temperature. If it does not reach its ignition temperature, then it will vaporize but not burn or burst into flame. Flash points for vapors require some external ignition source of suitable temperature. Autoignition temperatures, however, do not.

Lard has a large percentage of saturated fat, whereas vegetable fats such as Crisco have a low percent of these fatty acids. The autoignition temperature of most animal fats falls in the 550°F to 600°F range, compared with the autoignition temperature of most vegetable fats, which begins to appear at 685°F and usually ranges toward even higher temperatures. A calrod burner on the stove in question barely reached 475°F on its surface. It kept cycling on and off, even when the indicator was set to high. Apparently most modern stoves have built-in sensors designed to prevent overheating the cooking surfaces. At that temperature, Crisco should vaporize but could not reach the autoignition temperature of at least 685°F, the temperature required for Crisco to burst into flames without an independent ignition source. Nor would it reach the flash point of over 500°F required for combustion with a high-temperature ignition source. The burner, at barely 475°F, proves to be inadequate for both these tasks.

* Refer to the discussion of heat transfer principles developed by Fourier and modeled by the so-called Fourier series equations.

† Crisco, as the brand name for a hydrogenated cottonseed oil product originally intended as a material suitable to make candles, was reportedly derived from "CRYStalized Cottonseed Oil." Crisco was introduced to the public in 1911 by Procter & Gamble as a replacement for animal fat—lard—used in cooking and baking. Procter & Gamble sold both Crisco and Jif peanut butter to the J. M. Smuckers Company in 2002. The new owners reformulated Crisco in 2007, eliminating the trans-fat thought to pose a significant health risk to consumers.

Figure 8.30 Jean Baptiste Joseph Fourier.

J.B.J. Fourier, French Mathematician and Physicist: Heat Transfer

Jean Baptiste Joseph Fourier (March 21, 1768–May 16, 1830) (Figure 8.30) was a French mathematician and physicist famous for the Fourier series applications to problems of heat transfer. The Fourier transform equations developed through Fourier's law continue to immortalize his name.

Fourier's law describes, among other things, the phenomenon of heat conduction. As simply as possible,[*] the law states that heat transfers (moves) through any material toward another material at a rate proportional[†] to the area of the material involved in the transfer and the "negative gradient temperature," a vector[‡] in the direction that the slowest temperature increase moves.[§] Basically and more simply put, heat moves in the direction from highest heat to lowest heat at determinable rates and measurable magnitudes. One might add to that "with predictable effects."

Understanding the meaning of this law in its technical form requires at least an elementary knowledge of vector calculus. Studying partial differential equations provides a better foundation. With this understanding in mind, Fourier's law can be stated in two equivalent forms depending upon the purpose at hand. One form is called the *integral* form, when the objective is to look at the amount of thermal energy flowing into or out of a material as a whole. Another from is called the *differential* form, when the objective is to look at the flow rates or fluxes[¶] of energy locally.

[*] That is, while trying to include most of the concepts central to the mathematical effort involved in its application to physical systems. Conduction heat transfer is described by Fourier's law. In basic terms, it is the equation that we can solve for the heat flux vector (q) at a given temperature profile (T) and thermal conductivity (k). The minus sign ensures that heat flows down the temperature gradient as discussed and described.

[†] Two variables, x and y, are proportional if the ratio x/y is constant. Another way to put the relationship is to state that one of the quantities is proportional to the other quantity. The mathematical symbol $\propto$ is used to indicate that two values are proportional. For example, $A \propto B$.

[‡] A *vector* is a geometric term referring to anything having both a measurable length and a definite direction. In physics, Euclidean vectors represent physical quantities with both magnitude and direction. Other physical quantities, such as mass, are called *scalar* quantities since mass has no direction.

[§] That is measured perpendicular to the area through which the heat is flowing.

[¶] The term *flux* in physics refers to a flow of some volume per some unit of area; *flow* is the movement of some quantity per time. Flux, in physics, is a vector.

The issue to be addressed by Fourier's insights on heat conduction remains firmly practical: here, it is the transfer from the aluminum pan to the calrod burner in Slovotski's kitchen. Heat conduction between two solid bodies in contact remains the entire focus of the state's scientific evidence. The data require development and the results require explanation.

These areas from physics clearly fall within the domain and expertise of the forensic scientist. They do not fall within the typical practice of arson investigators or even the routine of more general fire investigators. While both relevant and important, there are other considerations when presenting a scientific reconstruction in a court of law. If any forensic scientist decided to lead a judge, the jury, both attorneys, and all courtroom spectators toward head-bobbing upright slumber, then discussing the Fourier series equations from the stand would be a preferred method. I do not recommend it.

The skill of presenting technical materials both simply and accurately must be part of any robust scientific reconstruction. Otherwise, very good scientific work may be accomplished, but no one who needs to learn about it will be any the wiser. One finds that presenting complex materials from physics or chemistry with these constraints requires a deeper understanding of the material and its application than can be developed through a simple classroom exposure. The skill can always be improved. It is difficult and frustrating, but worth the effort.

In Adrian Slovotski's kitchen, we can model a temperature drop at the interface between the pan and the calrod burner: the two metallic surfaces in contact. This brings us back to the green enamel on the calrod burner, which transferred from the pan's bottom to the burner.

Properties of Heat Studied by Fourier

Fourier studied and helped others develop a better understanding of various physical properties related to heat. The following list summarizes some of the areas covered:

- Conduction—Heat through contact and transfer
- Alloying—Nonchemical combination resulting in uniform mixture
- Plating—Mechanical surface bonding of metals
- Coating—Mechanical surface bonding of nonmetals or alloys
- Joining—Mechanical union of similar or dissimilar materials due to heat

The phenomenon of heat transfer covers the behavior of the calrod burner, the aluminum pan, and the enamel coating on the pan. Studying conduction, coating, and joining helps us understand the observation of a green enamel coating on the burner. For this to occur, according to the Fourier series and the work developed on the properties of heat, the temperature of the pan had to be greater than the temperature of the burner. Recall that the heat gradient vector moves from highest to lowest temperature. This move results in changing the enamel's state from solid to fluid. During the process, some of the enamel is transferred to the calrod burner, essentially coating the burner, depositing the melted enamel from the pan onto the burner.

The significance should be clear: The burner was cooler than the pan. Therefore, there has to be an explanation for the increased temperature of the pan that remains independent of the burner's temperature. The Crisco, even if it were present, would not account for this temperature increase (1100°F for aluminum). Neither would any of the materials in the kitchen. The best remaining explanation becomes an electrical fire resulting from some problem with the coffeemaker. The state's so-called scientific basis for an arson fire has collapsed like the side of an 1100°F aluminum pan.

Testing and Experiments

One way to test the state's hypothesis that the stove's right-rear burner heated an aluminum pan filled with Crisco and started the fire is to do just that. To confirm the theoretical work and to have explicit testing to present, we acquired a stove of the same make and model as the Slovotskis' stove and an aluminum pan of the same make, size, and enamel finish as the green pan found at the kitchen fire scene. The stove was installed in a controllable warehouse environment with non-water-based fire extinguishers available. The pan was filled with Crisco, and the right-rear burner, with the pan centered on its surface, was set to high. After 45 minutes, the warehouse was thick with smoke. The pan was blackened. There was no fire. The Crisco was gone. The pan uniformly distributed the heat and neither melted nor transferred its enamel surface to the calrod burner. Although this event would have made for a great deal of smoke damage, it would not have started a fire—certainly not the fire that burned the Slovotskis' home. The theoretical work had been confirmed by a simple follow-up test.*

Aftermath: Trial

The prosecutor was both persistent and vigorous in her attacks on the "evidence." The old *ad hominum* argument used to attack a person's claims by attacking the person remains a common logical fallacy. The prosecutor kept me on the stand for over an hour. Her attack ranged from "You are not an arson investigator, are you?" to "You no longer teach at the university, do you?" Regardless of the questions' relevancy, all one can do is to answer each question as clearly, respectfully, and truthfully as possible, hoping that the jury sees through the Sophistic ruse. It is always best to stick to the science, focus on evidence, and avoid the personalities. When forced, it remains best to take the high road and refuse to take the bait.

The state's case seemed to impress the jury. The state's "arson investigator," a responding firefighter, and, of course, the detective, testified for the prosecutor that the pan was the source of the fire, that the stove's burner was left on high with the knob in the off position, and that the stove's right-rear burner was the fire's point of origin. After only 2 hours of deliberation, the jury voted to convict Adrian Slovotski of arson in the first degree. When polled, the jury members said that they "ignored all the science stuff," adding that "the defense said one thing, the state something else," so "the two cancelled out and we disregarded all of it." They said that they convicted her because "she cashed her insurance check."

The attractive mother of five is now growing older in a women's prison. Her five kids are being raised by their grandparents.

* For a control, I repeated the test with another such pan but putting nothing inside. The pan got hot, smoked, and smelled, but did not change state from solid to liquid. Recall that the burner reached between 450°F and 475°F, and cycled on and off automatically.

The Party's Over

9

"Circumstantial evidence is," Mason told him, "as a matter of fact, about the strongest evidence we have. The big trouble with circumstantial evidence lies in its interpretation."

***Perry Mason Solves the Case of the Queenly Contestant*, Erle Stanley Gardner, p. 130, paperback edition, 1968**

Introduction

Drinking was not just a habit, it was a way of celebrating life. The young couple, Stephen and Shirley Vaughn, had their ups and downs, as all couples do, but both loved their kids, and as far as that goes, they loved each other, too. Both also loved to party. A party always involved alcohol. Today was no exception. The kids were visiting Grandma and Grandpa (Shirley's parents). They loved visiting as much as her parents loved their visits. It looked like a good day for the whole family.

The drinking started at noon, beginning with a beer each for Stephen and Shirley at home before the trip to their favorite bar, The Green Dragon. Before 6 P.M. the well drinks at the *Dragon* were under two bucks, and the beers were a buck apiece. No one seemed interested in the free snacks that the bar put out during their traditional happy hour celebration. The drinks remained the focus of entertainment.

Before the day's end, Stephen and Shirley were both quite intoxicated. Shirley felt a bit better after she vomited once on her way to the bar's bathroom. Stephen, however, had become a bit more sullen and belligerent. For some reason, after about 8 hours of drinking steadily, a conflict between the two always developed and escalated. Once awake the morning after, neither could ever remember what the conflict had been about. It was usually just a chemical thing, an attitude fueled by ethanol, a molehill turned into a Mount Everest.

Shirley and Stephen both ended up just behind their pickup truck in a parking lot, having what one of my friends calls a "discussion." That usually entailed yelling and struggling between invocations to the Lord and references to ordinarily unspeakable carnal acts. Somehow during the discussion a dagger ended up in Stephen's pericardium. Stephen lay with his back on the asphalt, sprawled over a history of crankcase leaks, just behind their Chevrolet Silverado. Shirley called 911 on her cell. The cops arrived and took Shirley into custody. She remained in custody and was eventually convicted of second-degree murder.

Just how these events transpired, how people and things got where they were, and how Stephen came to have a Tomahawk dagger penetrate his chest became issues in the case that I was asked to help sort out. This turned into a protracted task covering several years of work and multiple litigations. Most of my time involved trying to develop data that in most ordinary cases should have been easy to find. When any element of scene work, documentation, photography, and memory falls short, often there is not much that one can do. All a scientist can do is to develop whatever the available data and its analyses allow. The limitations imposed by the data must be at the forefront. That is what I attempted to do in this case. When that is done, often no one is happy, few are satisfied, and most are actively unhappy. That may become one unfortunate measure of scientific success. The place to begin: follow the blood.

Bloodstain Pattern Analysis and Scientific Method

Bloodstain pattern analysis involves the scientific study of the static consequences that result from dynamic blood-shedding events. The study involves analyzing the size, shape, distribution, and pattern of the stains, as well as the nature of their target surfaces. The purpose of the study is to help identify the nature of those blood-shedding events that produce the patterns under investigation. A bloodstain pattern analysis, when reasoning beyond specific blood-shedding patterns and their properties, becomes a specific form of crime scene or event analysis and reconstruction. It applies the tools of an event reconstruction but focuses specifically upon explaining blood-shedding events at the places and times in question. Evidence includes the number, character, and relationships among the bloodstains that are present at the scene, on the victim and the victim's clothing, on the suspect and the suspect's clothing, and on any weapons, vehicles, or other surfaces capable of sustaining bloodstains.

A bloodstain pattern analysis may include the identification of available stains as *human* bloodstains, as blood of a certain human blood type, or even as blood from specific individuals through a DNA analysis of relevant samples.* Qualified serologists and DNA technicians at forensic laboratories usually provide such analyses. Blood testing has developed great sophistication over the years, with DNA becoming the gold standard for individual identification.

Conclusions and Their Basis

Whenever one works on any case, the conclusions developed are limited by the available data. If that data changes or there are additional scientific advances applicable over the course of work, then conclusions must be reevaluated in light of those developments. My conclusions are always based upon and limited to the information available for review. In this case, my initial work was based upon the data collected by the police department, at the autopsy, and through the analyses of both the suspect's and the decedent's clothing. All the analyses are performed at the Final Analysis Forensics laboratory. Of course, one must list the data from which any conclusions develop. This list documents the data used for the work. Any additional data that influences further developments of this work are made clear. I listed the following data for the initial work on this case:

Evidence Provided

- Photograph of the scene (one) and autopsy photographs

Item Name

- Suspect blue jeans
- Suspect bra

* The genetic materials essential for DNA analysis are found in the white blood cells, never in the red cells. Red blood cells lack a nucleus and therefore lack the genetic materials required for DNA analysis.

- Suspect red shirt
- Suspect socks, one pair
- Suspect panties
- Victim watch
- Victim green pants
- Victim shirt
- Victim white socks, pair
- Victim pair tan boots
- Black-handled knife from vehicle
- Set of rolled friction ridge patterns (fingerprints) from Shirley R. Vaughn

My work developed the following conclusions based upon that data:

1. The decedent's fatal injury occurred during a struggle with the defendant in a parking lot.
2. The Tomahawk double-edged dagger penetrated the decedent's chest in a single, smooth motion.
3. The decedent fell onto his knees after sustaining his injury but before his death.
4. The decedent then fell or rolled onto his left side; he bled from this position.
5. The decedent bled onto the pavement and onto his shirt while lying on his left side.
6. After bleeding, the decedent was dragged on his back and left shoulder across some crankcase drips/grease on the parking lot pavement.
7. The defendant's right blouse seam and right cuff were both torn.
8. The defendant's left blouse cuff contacted both blood and vomit/stomach contents.
9. The defendant's blouse also has vomit/stomach contents on the left sleeve.
10. The defendant's pants have vomit/stomach contents on the left leg.
11. No blood appears on the defendant's pants.
12. The defendant had a blood ethanol level of 0.11 gram% when tested after the event.
13. The decedent had a blood ethanol level of 0.211 gram% when tested postmortem.
14. The defendant had blood on her fingers. She stated that she removed the knife from the decedent's chest.

Documentation in Reconstruction Reports

The documentation of a conclusion's development can be presented in various ways. I think that it is better to overdocument rather than underreport the scientific work on a case. In science, more is not necessarily less. Often, that "more" is what separates the ability to understand the development of and support for a conclusion from the mere pronouncement of a result. Why should anyone believe some listed conclusions? I have never found attributing a conclusion to some authority, any authority,* to be persuasive—certainly not

* An exception would be God's statement that he loves us and promises us eternal life. We believe it because God says so. Its truth is seemingly based upon the source of the statement. At least this is an exception to refutations of "arguments from authority" for a person of faith. In light of this belief based upon faith, evidence to support the statement can unfold, but like many a rational arguments of the same form, this one becomes circular: The evidence to support the statement is only evidence if the statement is true; the truth of the statement depends upon the evidence. Off to the intellectual races we go.

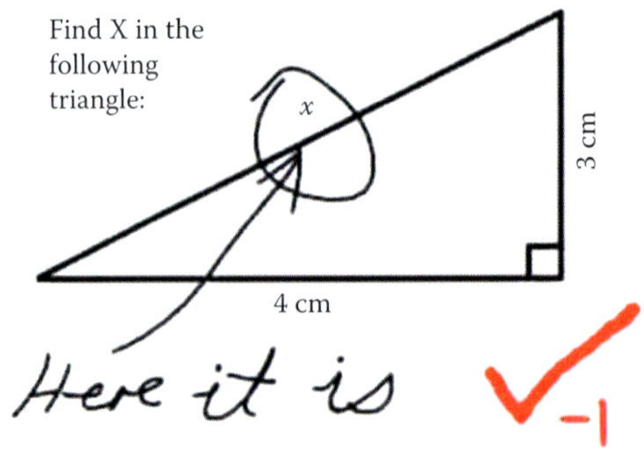

Figure 9.1 Author's correction on student exam.

from the scientific point of view. Yet, many reports simply state conclusions without documenting their derivations.

Crime laboratories have been criticized for unscientific practices in the National Research Council (NRC) report.* The evidence occasionally cited by critics includes reference to the one- or two-line "reports" offered by some crime lab personnel in complex cases. This does not necessarily represent the individual employee's choice; it may be the laboratory's written standard operating procedure (SOP).

Listing only conclusions does not advance the scientific enterprise. The comment made on a student's failing laboratory science exam applies here (Figure 9.1): It is important to show your work. Otherwise, for all the teacher knows, it may just be copied and has no significance as a measure of one's knowledge. *Derive* the answer, don't just *state* the answer. In fact, this failing among the forensic sciences has become so prevalent that it has spilled into both official and popular press reports about the forensic sciences as well.

The NRC published a report questioning the methods applied to a number of the forensic science disciplines (Figure 9.2). One criticism focused on drawing conclusions extending well beyond the given data. That sort of incaution still has various proponents scattered among the disciplines lumped together as the forensic sciences. They go way beyond the evidence. They eschew scientific methods or provide only lip service to their role. They are often great storytellers. Since they claim to have no science, they show only their conclusions. Jurors love it. Scientists hate it. And there we are.

A simple effort to show the work as clearly as possible helps others to evaluate the scientific basis of the opinion and to understand the processes used to develop the conclusions. This remains the basis for progress in natural science. However, things get complicated when science appears in court. Often the thread of advocacy becomes twisted within the efforts of science. This must be untwisted by the forensic scientist to the best of his ability.

The arguments justifying the one- or two-line report amount to, first, "Lawyers won't beat me up if I don't give them too much to beat me up with" or, second, "I have such a heavy caseload that I don't have time to write long reports." Both excuses fail to justify the lack of effort involved in taking this documentation path seriously. Both excuses invite even more criticism from lawyers that can harm otherwise good, reliable, scientific work products.

The first excuse might invite an attorney to criticize the work even more specifically: "Well then, what are you hiding that might merit criticism?" Perhaps, if you in fact do have

* The National Research Council report, *Strengthening Forensic Science in the United States: A Path Forward*, National Academies Press, Washington DC, 2009.

Figure 9.2 Discussions about the scientific practices thought to involve the forensic sciences have been published and widely discussed.

a heavy caseload, then you may not remember on the stand what you did on that case in the laboratory or precisely how you did it. A detailed report could help you remember.

Second, if you do have a heavy caseload, then the attorney will invariably attempt to confuse you with some conflation of the various cases that you worked around the time that you worked on the case at issue, running them together in order to invite confusion and misstatements. Again, a complete report can help avoid these two unnecessary traps set to confuse rather than further reliable science. It pays to remember that advocates have a job to do too, but scientists have a different job to do. All we can do is our scientific job the best way that we know how. When supporting documentation is absent, scientific suspicions must be raised and addressed. It is better to do it in a report while seated in front of your computer than it is to do it in the courtroom on the stand, flying by the seat of your pants.

Data and Interpretations

Some brief assessments of the data can also be helpful in presenting work on a case. In some cases, like this one, the dearth of photographs of the scene became an issue.

Scene of the fatal injury—The decedent was not photographed at the scene. Only one photograph was supplied of the area where the decedent fell onto the parking lot.

Description—The parking lot location shows some blood as well as crankcase drips from previously parked vehicles.

Injuries—The fatal injury is described at autopsy as an irregular 3 × 1.5 cm stab wound into the upper central right chest near the front midline. The single sharp-force injury measures about 12 cm from the entrance on the anterior chest wall to the terminus at the right atrium. The wound is from the front to back.

Other relevant injuries are described as two 1-cm moist purple-red abrasions at the front of the left knee with avulsed skin directed from medial to inferior. On the distal medial right lower leg is a diffused purple-gray contusion about 4 cm in diameter.

Toxicology reports indicate that the decedent's blood had 0.211 gram% ethanol present.

Overall conclusions—The fatal injury appears to have occurred during some type of struggle in the aforementioned parking lot. The struggle involved a double-edged dagger. At some point, the decedent may have fallen onto his knees, sustaining the afore-described nonfatal injuries.

Medical Examination: Defendant's Injuries

When the defendant was taken into custody and examined, she reportedly presented no bleeding injuries. Therefore, she was not the source of any of the blood found at the crime scene or on the items of evidence. In my own assessments of the clothing evidence, I used presumptive blood tests only (O-toluene and phenolphalein) on selected stains. Without further serological analysis, I am unable independently to confirm that the stains were human blood or, specifically, that the stains were human blood from the victim.

Bloodstain Pattern Analysis: Examination of Bloodstains and Trace Evidence

Drawing or roughly sketching the evidence can be helpful for developing and organizing observations and measurements. Make it clear that the drawings are not to scale and serve only to organize and visualize your approaches to the items on the table.

Decedent's Clothing

Overall analysis—The decedent lay on his left side after falling to his knees in the parking lot. He bled almost entirely to the left side of his T-shirt (soaking and wicking stains). At some point, he was dragged on his back and left shoulder across the parking lot, given the grease/oil stains and voids on the back of his shirt as well as near his waist.

Victim's green pants—Bloodstains appear on front left cuff, thigh and fly area both left and right; only grease/oil stains appear on the back surfaces of the pants (Figures 9.3 and 9.4).

Victim's cut T-shirt—Soaking bloodstains appear on the left front with minor transfers on the left rear shoulder and apparent grease oil stains on the back with drag voids present (Figures 9.5 and 9.6).

Victim's pair of white socks and pair of tan boots—The socks present no bloodstains; the boots have only dirt, possibly associated with the parking lot, on the soles and heels (Figures 9.7 and 9.8).

Victim's watch—No relevant evidence appears on this item (Figure 9.9).

In a report documenting a scientific reconstruction, drawings become a very useful tool. A drawing or sketch is not fancy, nor is it particularly compelling demonstrative evidence. Photographs accomplish that goal far better and with much less effort. However, the

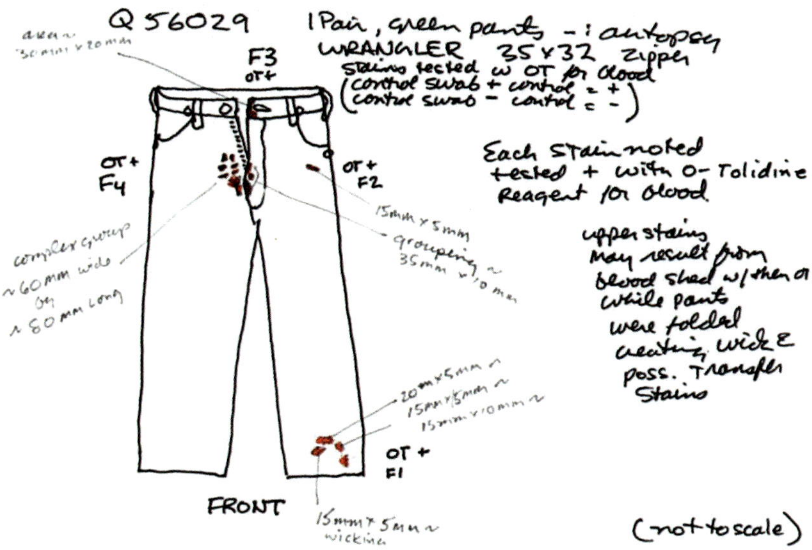

Figure 9.3 Sketch with notes—front of the victim's pants.

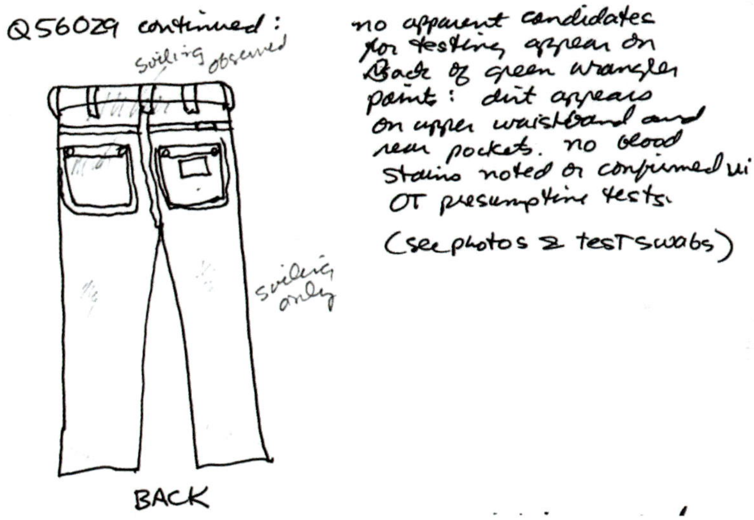

Figure 9.4 Sketch with notes—back of the victim's pants.

purpose of making drawings, in my view, is to develop one's ability to make observations: to see things as they are and to remember them and their significance.

Often I draw an item of evidence in order to help clarify just how the item should be processed: Which nondestructive test method should be applied first? What will that test tell me? What about the second step? How do they fit together? Why should it be done? What is that spot, dent, scratch, and so forth, on the left side, and what made it? Drawings become a visual inventory of the evidence made on your own through seeing it with your own eyes and capturing it with your own hand. That is better than any camera. I find that it is more helpful than the photos when I am on the stand, even if no one else ever sees the drawing.

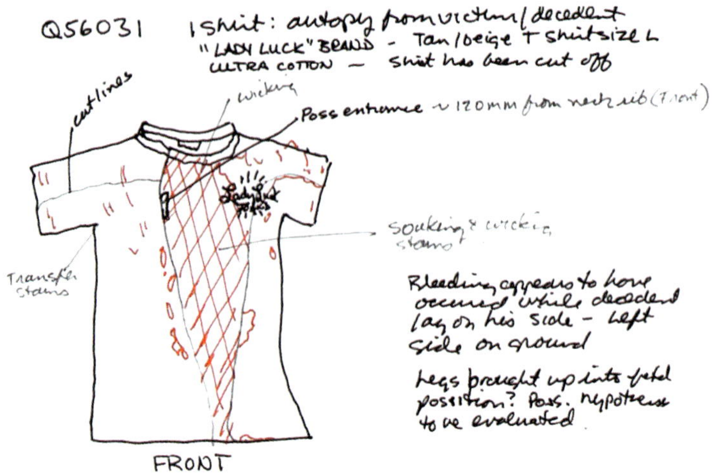

Figure 9.5 Sketch with notes—front of the victim's shirt.

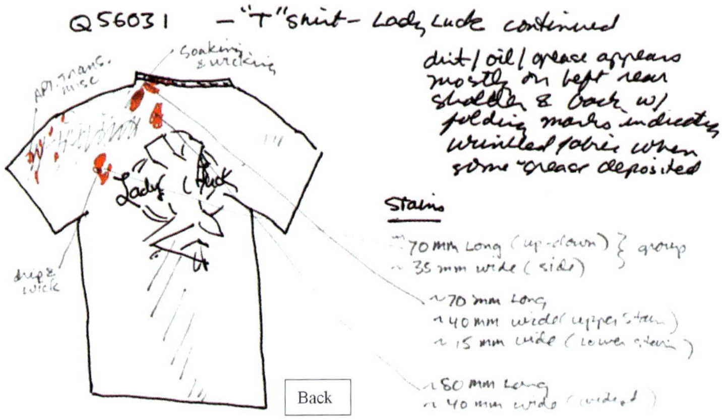

Figure 9.6 Sketch with notes—back of the victim's shirt.

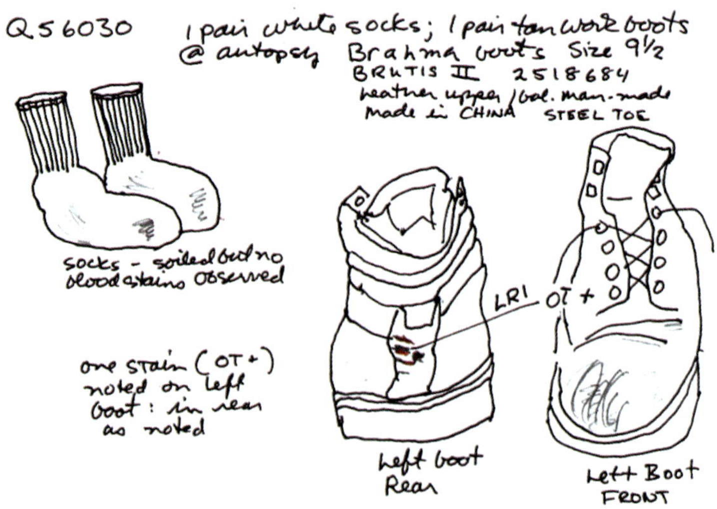

Figure 9.7 Sketches with notes—the victim's socks and left boot.

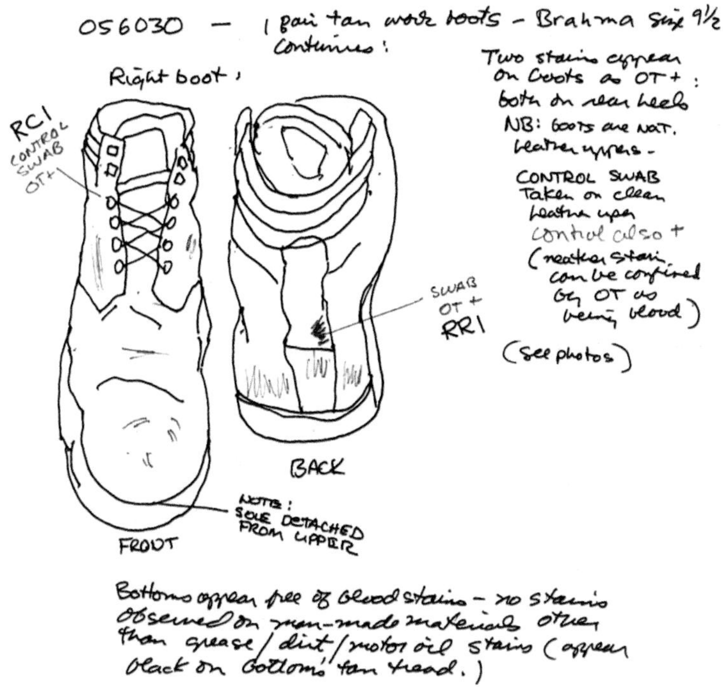

Figure 9.8 Sketches with notes—the victim's right boot.

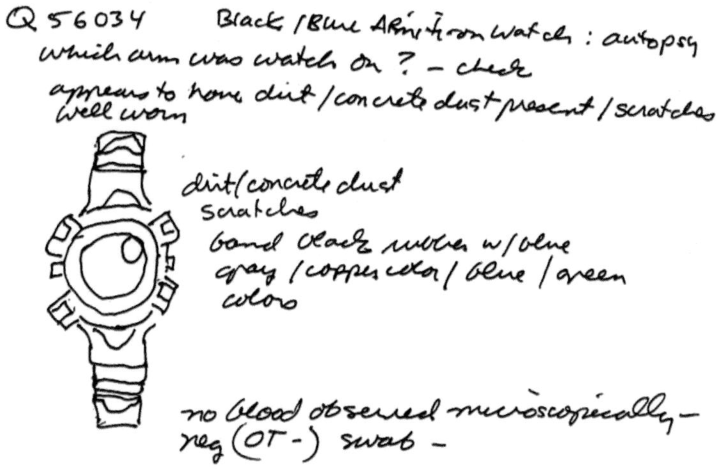

Figure 9.9 Sketch with notes—the victim's watch.

Defendant's Clothing

Item #3— Bra: No data of value can be established (Figure 9.10)
Item #5—Panties: No data of value can be established (Figure 9.10)
Items #7/#8—Socks: No blood appears on sock; grease and oil do appear on the sole (Figure 9.10)

Drawing the evidence also permits one to think about and reevaluate the significance of the item. Data may arrive later that changes the significance, one way or another. It does

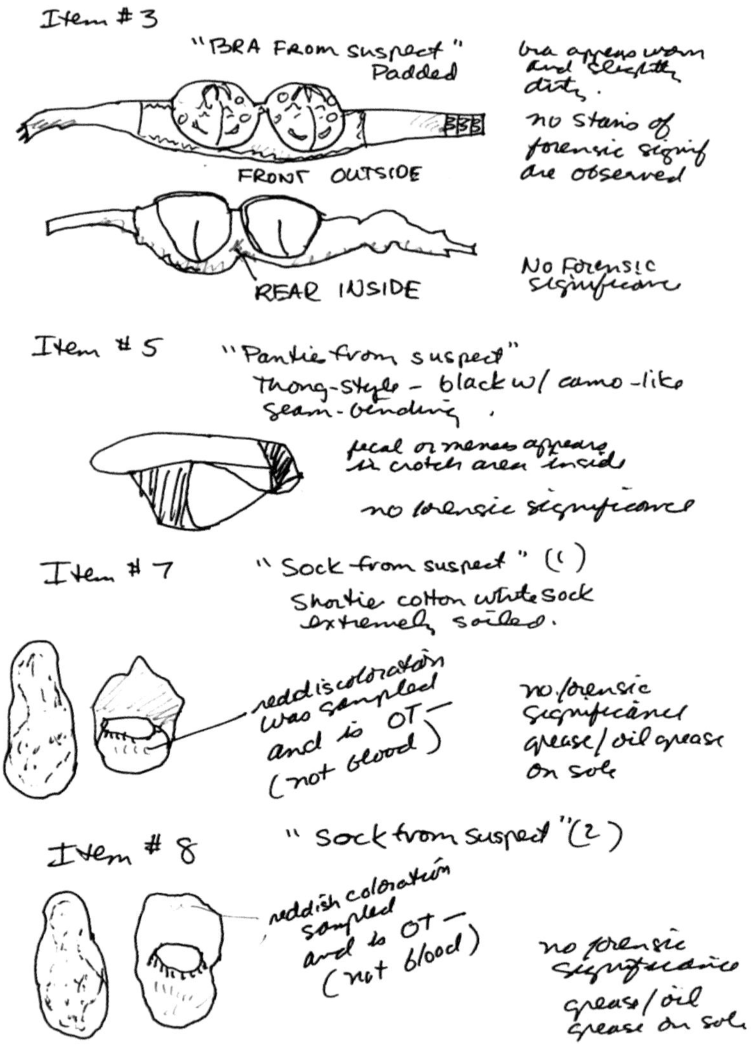

Figure 9.10 Sketches with notes of defendant's clothing—Items #3, #5, #7, and #8.

not hurt to consider an item that may have no value now but may be important later. The same principle applies, of course, to documenting a crime scene.

At the outset, we often do not have clear ideas about what may be or may become important evidence. The principle of "informed overdocumentation" applies here, too. Draw, measure, photograph without scales, photograph with scales, and take video of a walk-through to help visualize and understand the overall scene. Remember to turn off the voice recorder; some scene comments are best left undocumented.

The suspect's clothing, in this instance the undergarments, appear to lack any scientific value—at least at this point in the development of evidence. However, since I certainly do not have Edger Cayce's ability to see the future, documentation may be useful for later developments in the data.

Of course, there are nonscientific considerations in documenting evidence as well. It remains easier to document a piece of evidence having unknown probative value than it is to explain in court why you did not document it.

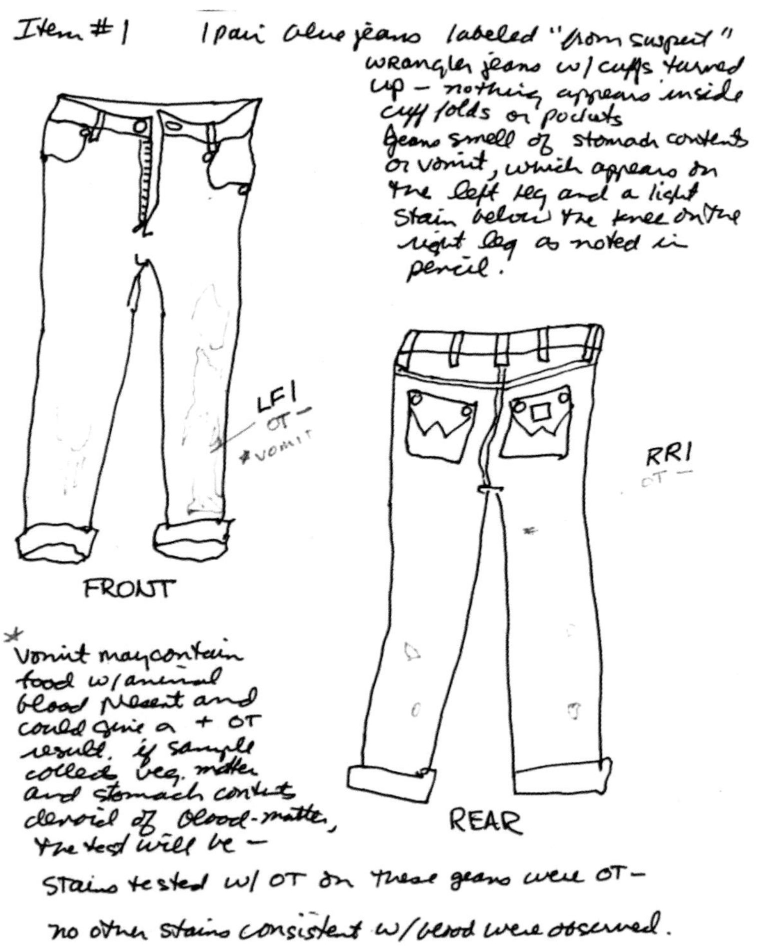

Figure 9.11 Sketches with notes of the defendant's jeans—Item #1.

Item #1—Blue jeans: Possible stains were tested for blood with negative results. Vomit/stomach contents appear on the left front leg and on the lower right knee (Figure 9.11)

Item #2—Red shirt: A seam on the right side is torn, as is the right cuff; blood (O-Toluene/phenolphalein +) appears on the left cuff along with vomit/stomach contents (Figure 9.12)

Photographs can often accompany the drawings, depending upon the nature of the evidence and what the photos are intended to show. For documentation, separating photos and drawings seems to work well since they usually serve different purposes.

The Dagger

Although a dagger is a weapon, the word *weapon* can be misleading. In fact, a pencil can become a weapon. I find that the term "weapon" is too general to be useful in a scientific report. If it is a knife, best to say knife; if it is a dagger say dagger. A gun can be a revolver,

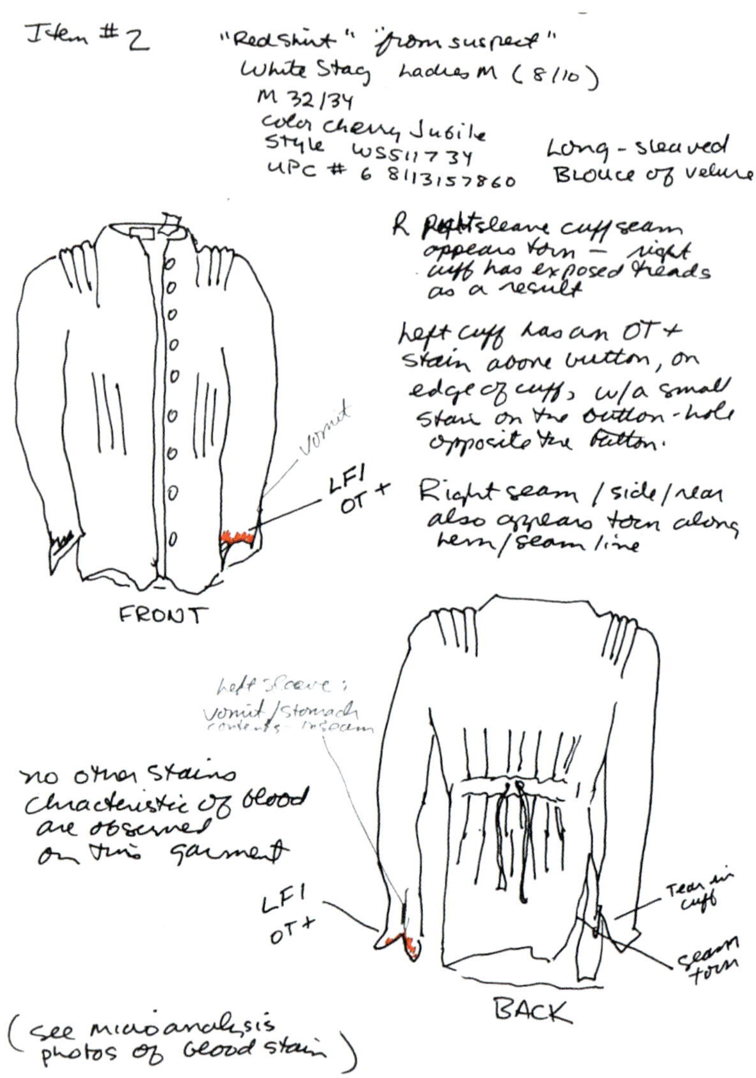

Figure 9.12 Sketches with notes of the defendant's shirt—Item #2.

a pistol, a rifle, or a shotgun, with multiple varieties under each classification. It remains a virtue to be as specific as the data warrants. If it is unknown, then it is acceptable to call something an unknown weapon. One needs to be general in that case, since we do not have specific data to point in any other direction. Blunt-force trauma from a 4 × 4 looks very similar to blunt force trauma from a 2 × 4 if the same 1½-inch edge strikes the victim inflicting the same number of blows. It is best to remain within the limits set by the data.

In this case, a black-handled dagger plunged into his heart led to Stephen Vaughn's death. A dagger has a single point with two edges, both of which are not usually sharpened. A knife, on the other hand, has a sharpened edge on one side and a blunt edge on the other. It has a point but not a sharpened point like many daggers. The injuries produced by a dagger appear unlike those produced by a knife. Figure 9.13 from a different sharp-force injury case helps to demonstrate some of the differences. The margins of the injury help us to determine which weapon was used. Here, both ends appear to come to a point,

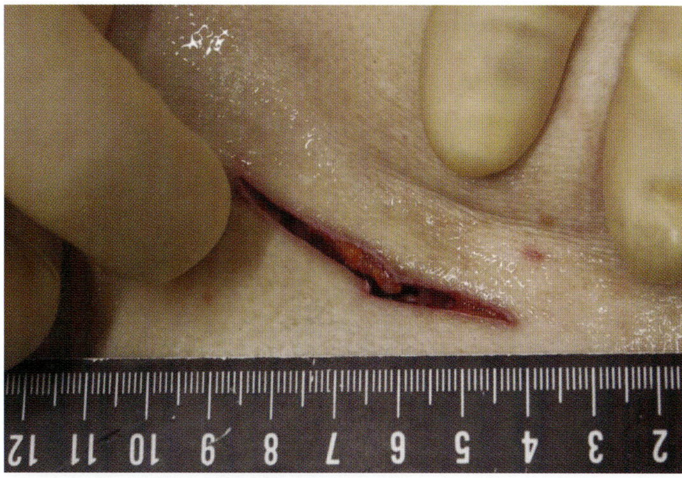

Figure 9.13 An articulated sharp-face injury.

suggesting that the weapon was a dagger. But an examination of the wound track and its margins when articulated show the classic boat shape of a single-edged knife. In this case, although unrelated to the Vaughn case discussed in this chapter, a single-edged kitchen knife produced the sharp-force injury in Figure 9.13.

In another unrelated case involving a dagger, the penetration was limited, yet the injury exhibited the typical double-edge pattern common for this type of knife. If the dagger moves while inserted in the body, other injuries can result. Note the ragged edge in the middle of the injury in Figure 9.13. If this appears bilaterally and the wound track is straight, similar patterns can suggest the type and design of the dagger used in the death.

The injury in Figure 9.14, again from a different case, was also produced by a single-edged knife. Note that the end of the perforation comes to a point bilaterally, suggesting a dagger was used to inflict the injury. But if other injuries appear on the body and if they leave perforations with one end that comes to such a point with the other end flat and straight,

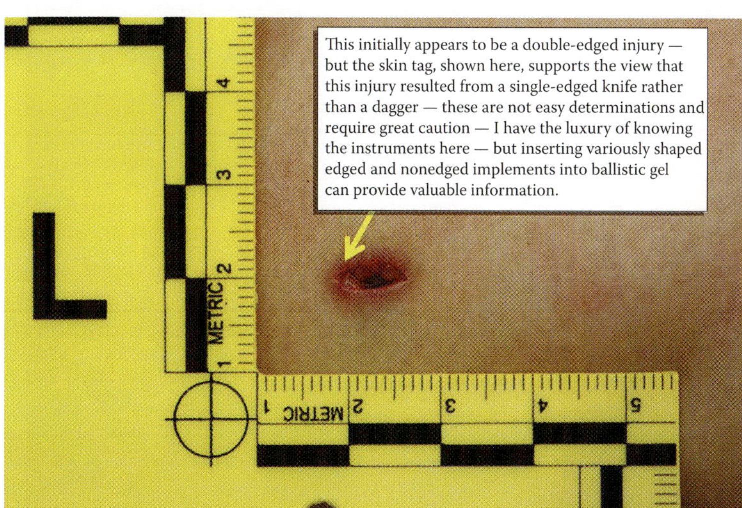

Figure 9.14 A single- or double-edged instrument?

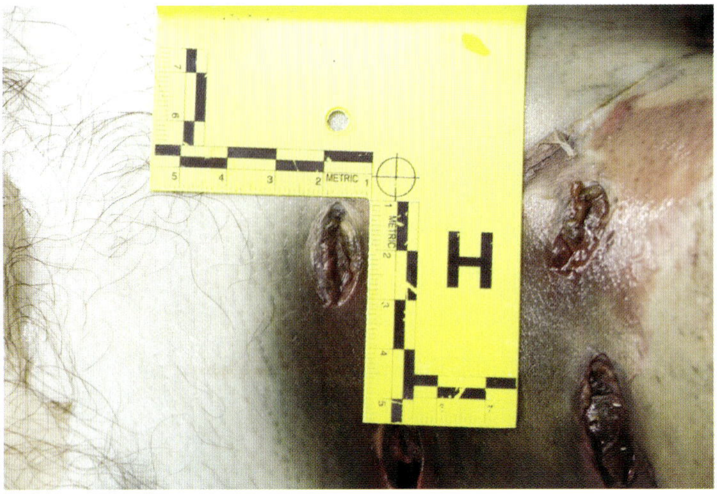

Figure 9.15 Sharp-force injuries from a very different case.

then a single-bladed knife was involved in the homicide. Sometimes the skin tag is difficult to see at first, but when the margins are closed together, one can make a clearer observation.

Decomposition presents additional challenges for interpreting knife wounds. In Figure 9.15, the weapon used was a kitchen boning knife—seemingly the kitchen knife of choice for sharp-force homicides. The margins, when articulated, show the classic boat-shaped entrance. One might be led astray if the interpretation develops faster than the observations. In cases with multiple sharp-force injuries (again unrelated to the Vaughn case), one must examine each injury to ensure that the same type of knife was used to inflict all the injuries.

The sharp-force injury in Figure 9.16 was produced by the knife pictured in Figure 9.17. In this case, the knife also pursued a straight course into the body. However, the photograph

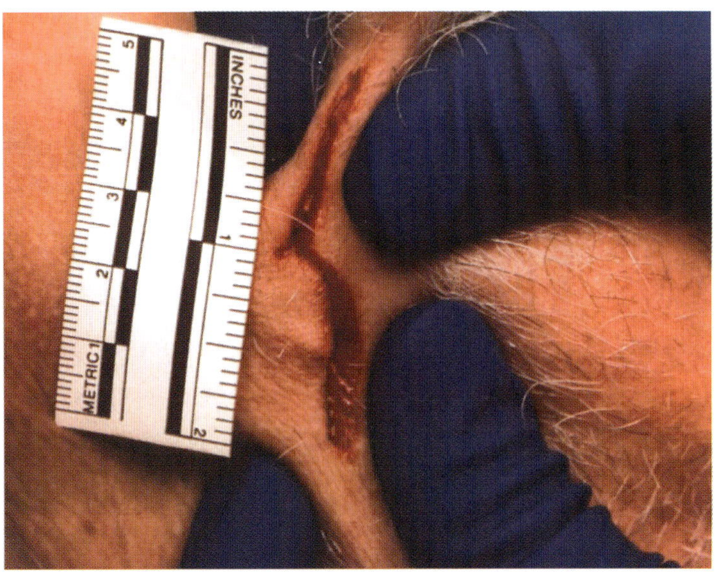

Figure 9.16 Yet another case shows the articulation of a very different sharp-force injury.

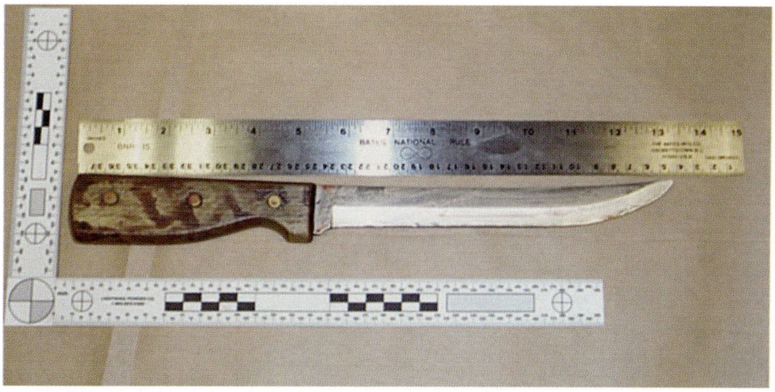

Figure 9.17 The knife producing the injury, pictured in Figure 9.16.

shows two such injuries next to each other. Depending upon the wound track, this may result from either one insertion, with the knife moving upward (making the lower injury in the photo the first given the location of the knife's sharp edge) or two insertions closely spaced. The autopsy may be able to help decide which event occurred, but that is not always the case.

The same double point appears as in the dagger injury in this case as seen in Figure 9.18. Daggers usually have sharp points with dull edges. Evidence of dull edges can be seen on the skin as well as in the wound track, usually in the form of lacerations (tears) along its course into the body.

Description

This dagger is a Tomahawk brand double-edge dagger about 23 cm from the tip to the end of the handle. It has an accompanying sheath (Figure 9.19). In documenting its condition, I found it to have a sharp point, average dagger edge, with no oxidation.

Bloodstains Present on the Dagger

Apparent bloodstains (O-Toluene/phenolphalein[*] +) appear on the Tomahawk side of the blade as well as on the same side of the handle farthest from the blade. Apparent bloodstains (O-Toluene/phenolphalein +) also appear on the blade side but only near the hilt. No blood appears in the junction between the blade and the handle. Blood stains appear on one side of the hilt.

Further Blood Evidence on the Handle

A set of "plastic impressions" (patterns in blood transferred from the contact of a bloody object or objects to another object receiving the impression) appear on the handle's side (the side closest to the "T" in Tomahawk and closest to the "8" in "XL 238.") These impressions were photographed and studied for possible identification of the friction ridge details.

[*] $C_{20}H_{14}O_4$ is the chemical formula for this presumptive blood test reagent.

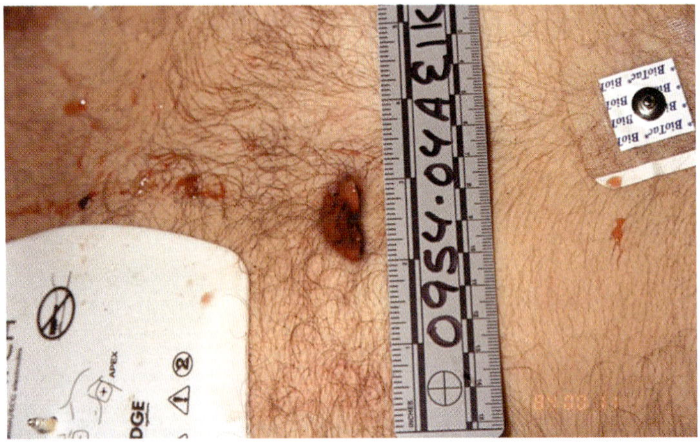

Figure 9.18 A sharp-force injury from this case.

Figure 9.19 A dagger and sheath.

One cropped photomicrograph[*] of the impression appears in Figure 9.20. The figure shows what is called a plastic impression on the dagger. A plastic impression generally has raised surfaces leaving a 3D effect. Analysis strongly suggests that the impression was produced by a bloodstained thumb and that the lines evident in the photograph represent friction ridge detail from a fingerprint. Exemplar prints from the defendant, Shirley R. Vaughn, were obtained and comparisons between the known prints and the unknown pattern on the dagger handle were attempted. Although there are obvious similarities between both left and right base line friction ridges on the ten-print card and on the dagger, no conclusive identifications can be made.[†]

[*] A photomicrograph is a photograph taken through the lens of a microscope. A microphotograph is just a very tiny picture! (Best never to confuse the two as suggested earlier.)

[†] I was assisted by my colleagues, Latent Print Examiners Scott Verbonus, Michele Tripplet, and Steve Everest, of the King County Sheriff's Office Latent Print Lab in Seattle, Washington; and Dr. Glen Langenburg, of the Minnesota Bureau of Criminal Apprehension, in Minneapolis, Minnesota.

Figure 9.20 Friction-ridge impression from the dagger.

I attempted to increase the detail using various wavelengths of light from an ALS,* a micro Luminol† mist, and a Blue Star‡ mist, all without success. The clearest impressions remain those captured in the photomicrograph reproduced in Figure 9.20. Only general characterizations can be made from an analysis of the stain itself. The impression lacks sufficient individualizing elements.

Regardless of the identity of its origin, the bloodstain on the handle does show that a previously bloodstained object left the impression while pulling on the handle in a direction away from the blade. This data still remained important to consider. As additional materials became available to me in the laboratory, the data began to develop and connections among the evidence began to appear.

Locard's exchange principle remains useful: every contact leaves a trace. But not every contact leaves a *useful* trace, and not every trace remains *available* indefinitely (Figure 9.21).

Figure 9.22 to Figure 9.39 document various attempts to visualize the friction ridge detail on the dagger's handle. Currently, we might use one of two vacuum metal deposition (VMD) devices to develop such potential prints on various items.§

* ALS refers to an alternate light source—a powerful full-spectrum light with multiple filters, selectable to produce specific wavelengths ranging from the ultraviolet to the infrared.

† Luminol is a chemical that luminesces in contact with the heme in dried bloodstains. Heme is the deep red, nonprotein ferrous component of hemoglobin, $C_{34}H_{32}FeN_4O_4$. ("Heme" is short for hematin.). Luminol is very sensitive but not specific for blood. Luminescence is a chemical reaction during which electrons are released in the form of light energy. This is distinguished from fluorescence, which is a property of substances that reflect light energy of a certain wavelength when light is supplied by a separate light source. Luminol does not require a separate light source. Other blood detection chemicals, such as fluorescein, do require a separate light source. Fluorescein is a synthetic organic compound soluble in water and alcohol. It is also used as a fluorescent tracer for many medical applications.

‡ Blue Star is a proprietary form of Luminol developed by a colleague in France. Blue Star has certain advantages over traditional Luminol and some analysts find it simpler and more convenient to apply.

§ Go Evidence, established by Brian Orr and Scott Verbonus, works as a Final Analysis Forensics affiliate on cases requiring fingerprint analyses. We are one of the only laboratories in the world with vacuum metal deposition devices for items as large as a car door and as small as a tack, a nail, a cartridge case, or a bullet.

Figure 9.21 The bird struck the window leaving its impression on the glass. But like each of us, it will not last forever.

Each image is taken with a different light and filter combination. Altering the angle of both the camera and the lighting can sometimes help to visualize data otherwise difficult to view. As the above photos illustrate, options are few. Even if one can obtain images of friction ridge detail, the data may not provide any basis for identification. Friction ridges are common—it is the individual characteristics in a print that examiners search for and document. None of these individuating characteristics appear on these images.

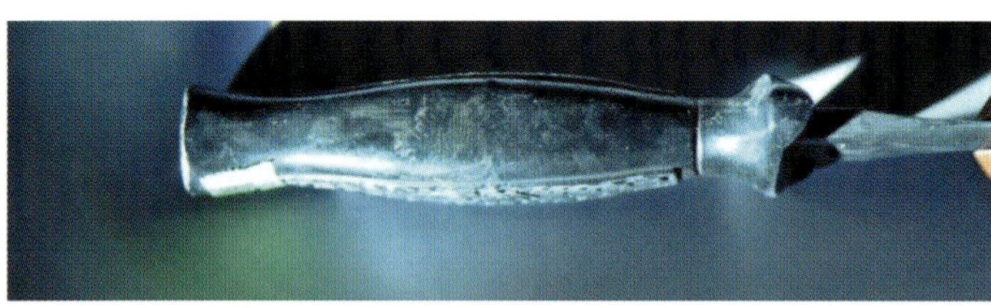

Figure 9.22

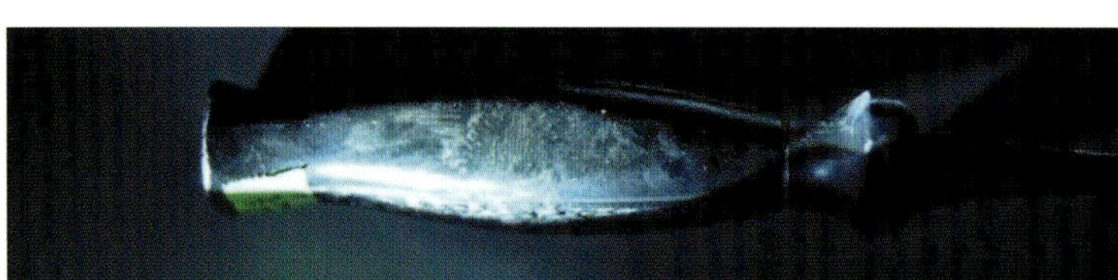

Figure 9.23

Figure 9.24

Figure 9.25

Figure 9.26

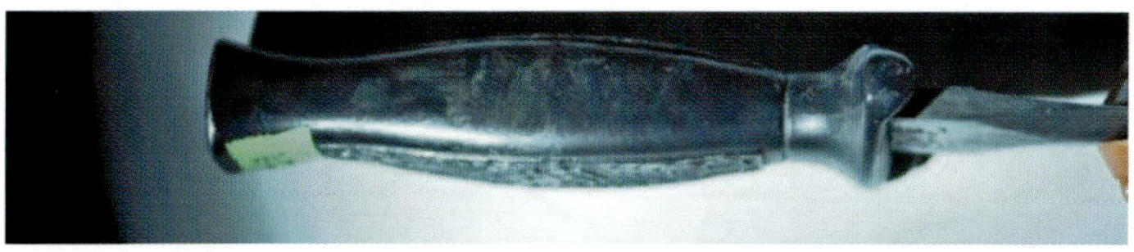

Figure 9.27

Figure 9.28

Figure 9.29

Figure 9.30 Photomicrographs of the dagger show traces of blood and the properties of the blade, hilt, and handle.

Figure 9.31

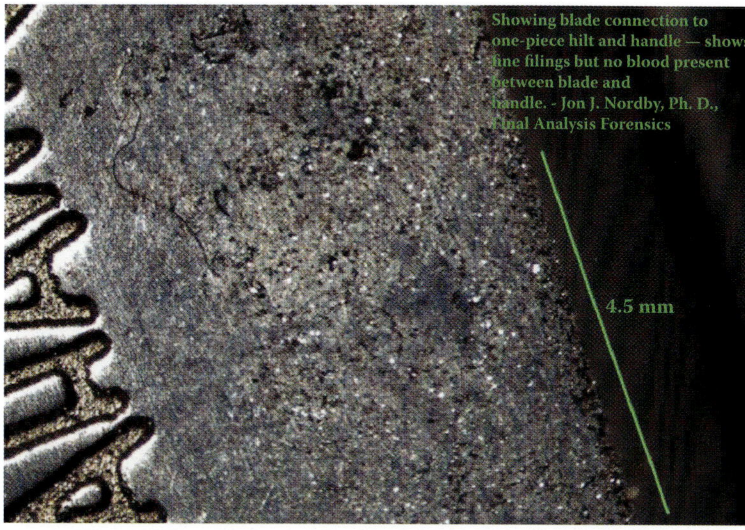

Figure 9.32

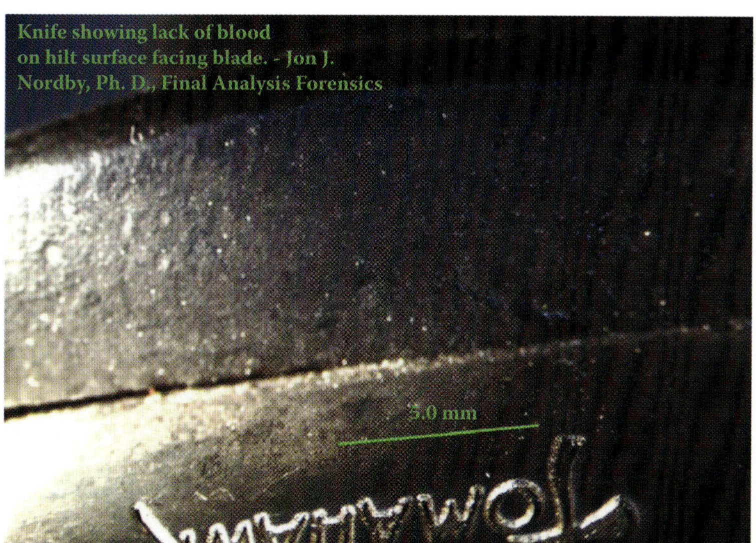

Figure 9.33

Figure 9.34

Figure 9.35

Figure 9.36

Figure 9.37

Figure 9.38

Figure 9.39

Drawings of the Dagger

See Figure 9.40 to Figure 9.42 for my notes about the dagger.

Observations of the injury include recognition of any hilt marks, injuries such as contusions or lacerations produced by the material separating the blade from the handle or from the base of the handle itself. No such injuries appeared on Mr. Vaughn. That observation provides information about the force and the depth of insertion.

Another question involves what to call the sharp-force injury suffered by the decedent. When an injury is called a *stab wound* the connotation carried with the term suggests that a knife was used in a certain way, usually that a knife or a dagger delivered the injury after travelling through an arc of some description.

This injury was not such a stabbing injury. The medical examiner describes an injury with a very limited wound track—the dagger went in at 90 degrees and did not rock, move, or twist. It appears to have come out in exactly the same manner. That makes this sharp-force injury an atypical stab wound.

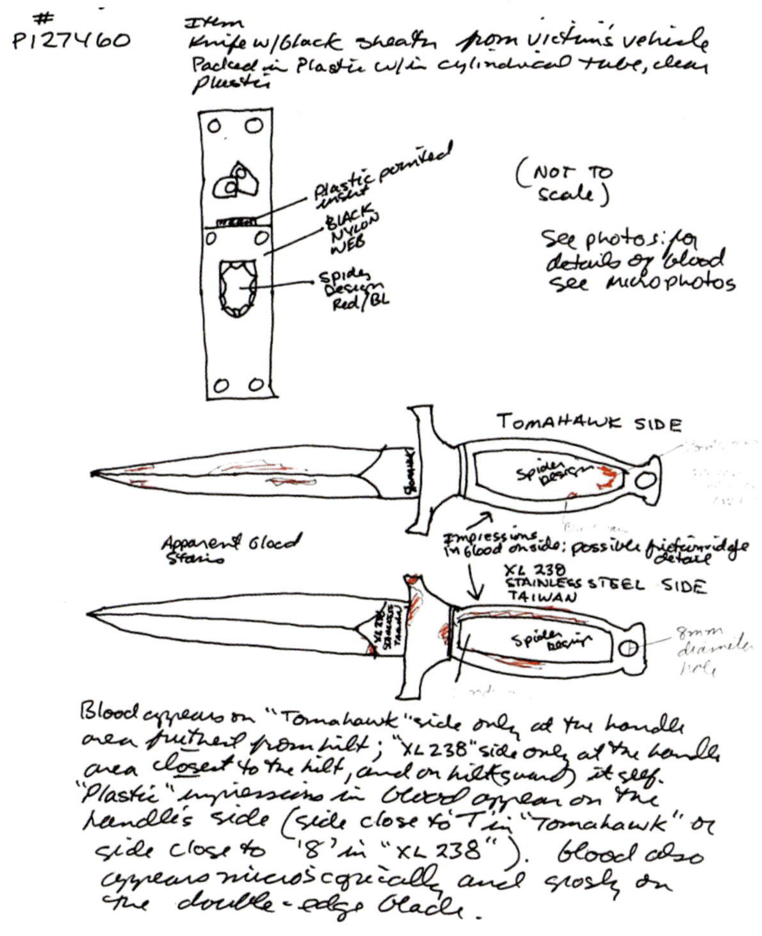

Figure 9.40 Notes and drawings—preliminary work with the evidence can be developed with added observations from other tools of observation and testing.

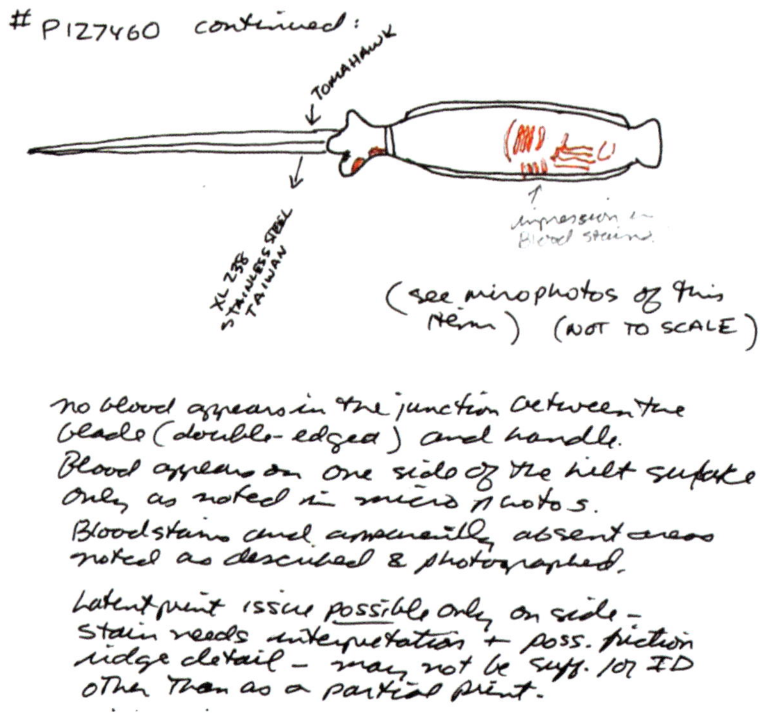

Figure 9.41 Presumptive blood testing must be followed up with tests that confirm the presence of human versus nonhuman blood.

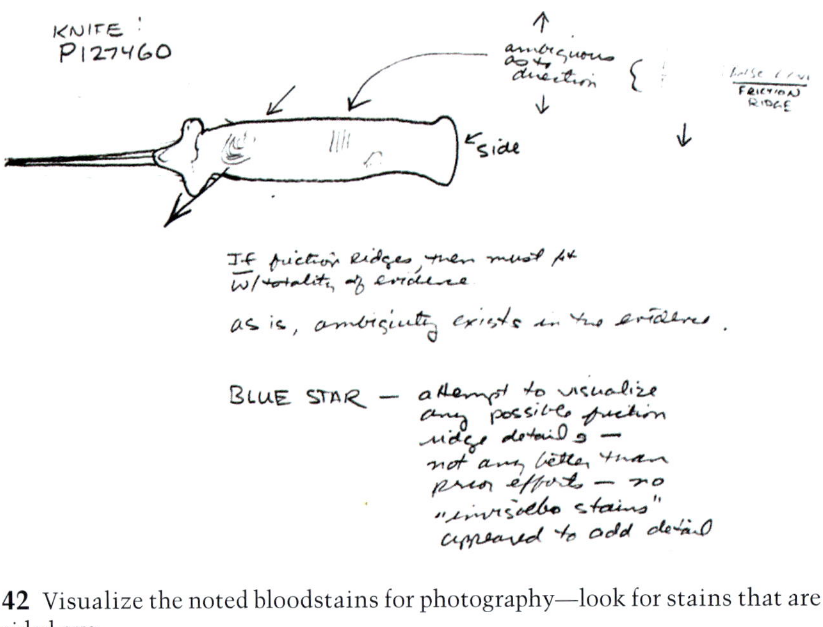

Figure 9.42 Visualize the noted bloodstains for photography—look for stains that are invisible to the unaided eye.

Event Reconstruction Based upon the Data Supplied

Important caveat: This sequence of events is not meant to represent a fixed time line, but only a rough structure of the events that could occur within a few minutes of each other or that could occur with a significant time lapse between items. Some of these events may have occurred simultaneously or in an alternate order. Such a development of the events leads to, and eventually must support the conclusions in the numbered list in the section at the beginning of this chapter—"Conclusions and Their Basis."

1. The decedent's fatal injury occurred during some type of a struggle in a parking lot.
2. The Tomahawk double-edged dagger penetrated the decedent's chest in a single, straight, smooth motion.
3. The decedent fell onto his knees after sustaining his injury but before his death.
4. The decedent then fell or rolled onto his left side; he bled in this position.
5. The decedent bled onto the pavement and onto his shirt while lying on his left side.
6. After bleeding, the decedent was dragged on his back and left shoulder across some crankcase drips/grease on the parking lot pavement.
7. The defendant had blood on her fingers when she was taken into custody.
8. The defendant claimed that she removed the knife from her husband and put it on the pavement.

This list of conclusions requires supporting data developed through observations, testing, and data analysis. The reconstruction presented does not impress jurors who have come to expect a work product just like those developed on popular television programs such as *CSI*. The development and support of the reconstruction is also much less dramatic and much more time consuming than television scripts allow. Nonetheless, a report on a death case should develop and present that support. Whether the attorneys trying the case ask for that reasoning in court is well outside the forensic scientist's control. All one can do is to present the work and the reasoning in the report. Provide it accurately when asked.

Photographs Documenting the Evidence

Scene photographs are obviously part of the evidence. Figure 9.43 shows the bloodstains on the parking lot where the death apparently occurred. Note that there appears to be crankcase deposits from previously parked vehicles present on the paved surface. This information, if accurate, supports the interpretation from analysis of the decedent's T-shirt that he was at some point dragged a short distance across the parking lot's surface while on his back.

In addition to the scene photographs, evidence photos obviously help document the items examined in the laboratory. Some items are readily captured in photographs and their interpretation remains unproblematic. For example, Figure 9.44 to Figure 9.58, document my examination of the dagger. Figure 9.59 to Figure 9.61 show the observations and measurements. Not always, but sometimes, photographs alone have the potential to be misleading if they are not examined with a practiced and trained eye.

Figure 9.43 Scene photo of the parking lot.

Figure 9.44 The dagger removed from Stephen Vaughn and the sheath found in the truck's cab. The movement of the dagger from the victim to the truck remains unexplained.

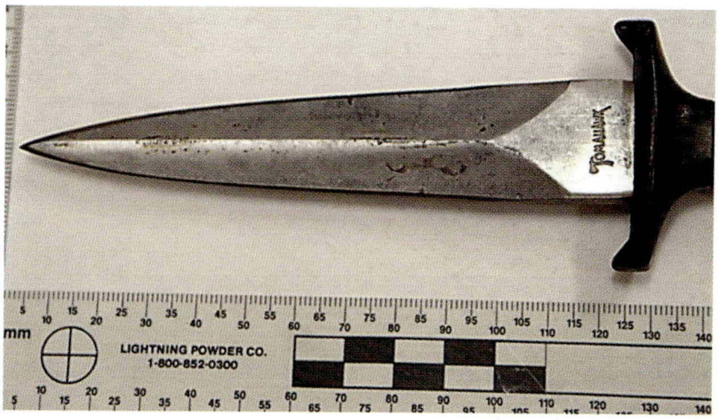

Figure 9.45 Various views of the dagger (see also Figure 9.46 to Figure 9.58).

Figure 9.46

Figure 9.47

Figure 9.48

Figure 9.49

Figure 9.50

Figure 9.51

Figure 9.52

On dark materials, if it is difficult to see the bloodstains, using infrared (IR) light proves useful. Figures 9.53 and 9.54, taken by my colleague Jim Pex, illustrate the use of IR photos to help visualize bloodstains on dark material. Other basic darkroom techniques can also help visualize trace evidence.

If an IR camera or film is unavailable, sometimes altering the contrast or the gamma values in Photoshop or some other digital image program provides the necessary clarity to see the relevant components of the image more clearly. In Photoshop, the photo's history is preserved so that the steps taken to produce the image can be repeated by anyone evaluating the work.

In Figure 9.64 to Figure 9.73, I adjust the contrast in their gamma values to make the blood data more visible and the morphology more straightforward to document. It is a good practice to keep both images. At Final Analysis Forensics, we make every effort to store our images as TIFF files or raw images. They usually have the best resolution and do not degrade when the files are repeatedly opened, copied, or otherwise manipulated. It is easy to create small JPEG files to use for smaller report files, e-mails, and so forth. Issues involved in clarifying photos offer another reason to draw the items showing the stains and measurements rather than to rely only on the photographs.

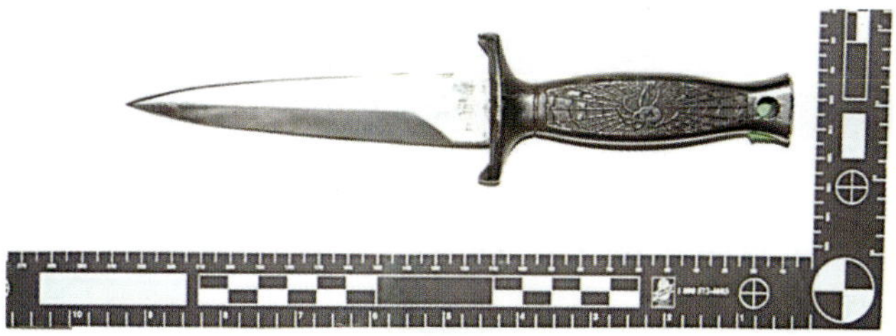

Figure 9.53

Figure 9.54

Figure 9.55

Figure 9.56

Figure 9.57

Figure 9.58

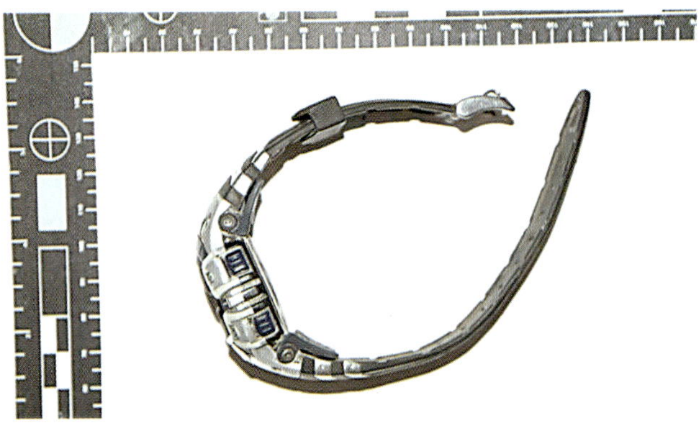

Figure 9.59 The watch removed from the victim before the autopsy appears at this time to have little useful information relevant to any effort to understand his death.

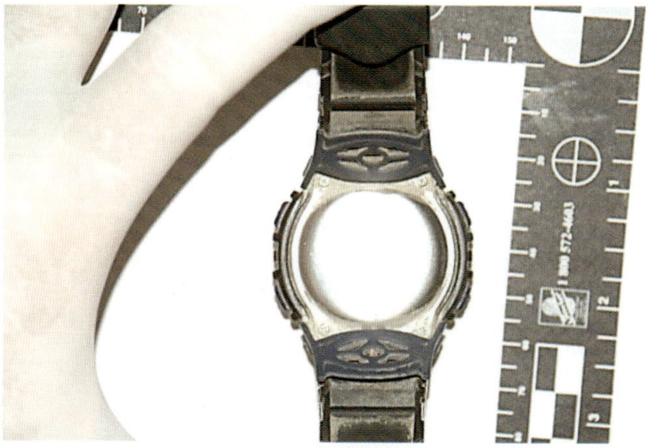

Figure 9.60 Back of the victim's watch.

Figure 9.61 Like most items, one must first look and determine their contribution to any data set used to explain events. In this case, the watch does not add anything of value.

Figure 9.62 Example from a different case: Shows an ordinary photo taken under full-spectrum light with no filter on the lens. Compare this photo to the photo in Figure 9.63.

Figure 9.63 This shows an IR photo of the same item from a different case. IR photography requires either IR film, or a suitable IR digital camera. I use a digital SLR. This photo was taken using IR film.

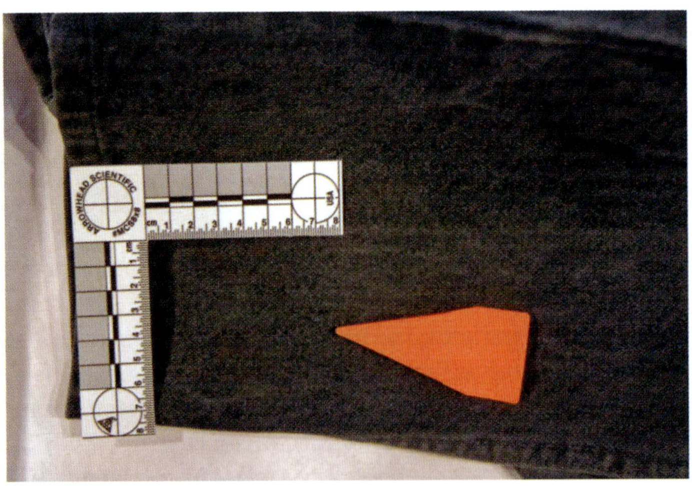

Figure 9.64 The victim's pants with bloodstains are noted and documented (see also Figures 9.65 and 9.66).

Figure 9.65

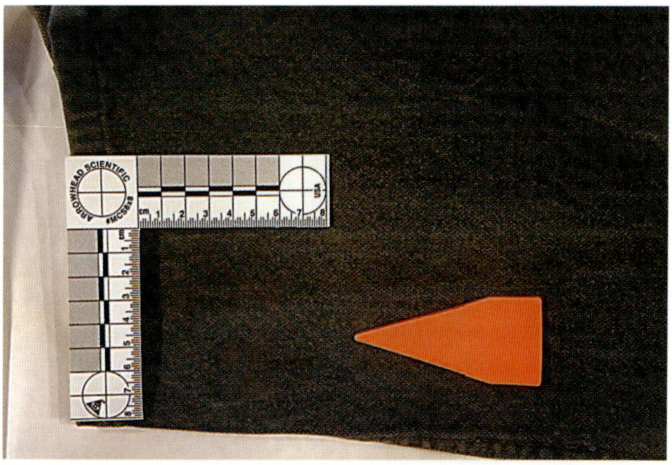

Figure 9.66

Figure 9.67 The blood is difficult to see in this photo. (Used with permission from the photographer—Jim Pex.)

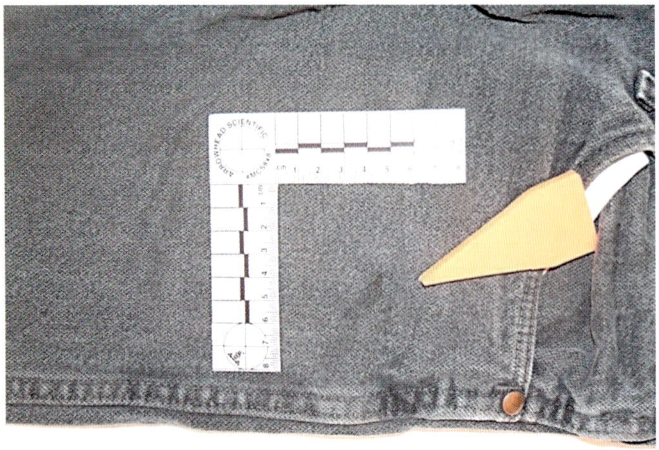

Figure 9.68 The blood is easier to see by changing the photo's gamma value. (Used with permission from the photographer—Jim Pex.)

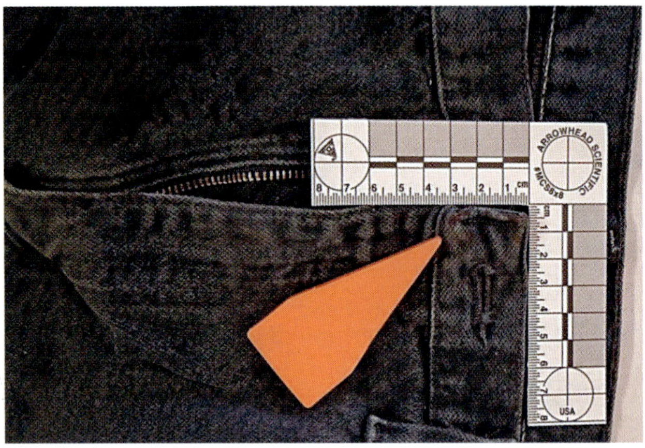

Figure 9.69 An ordinary full-spectrum light appears to show stains.

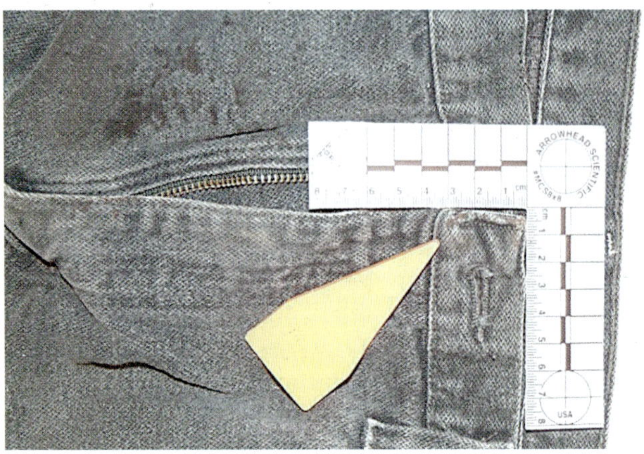

Figure 9.70 Adjusting the gamma reveals that some of the "stains" are simple wear and shadows, and some, present around the ends and by the button are apparent bloodstains.

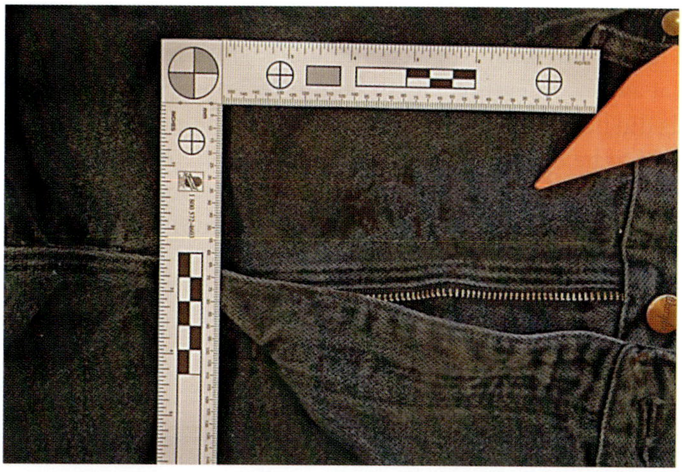

Figure 9.71 Stains can be seen, but photo do not always reveal or capture what the analyst sees.

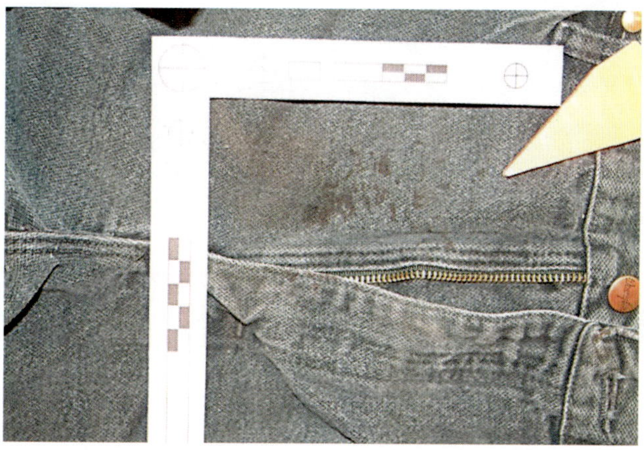

Figure 9.72 The photo depicts the stains seen in Figure 9.71, with an adjusted gamma value. The difference in photos provides another reason to draw the evidence—not simply rely on photographs.

Figure 9.73 The backside of the decedent's pants.

My examination of the decedent's T-shirt is documented in Figure 9.74 to Figure 9.84.

The socks (Figure 9.85) had no detectable bloodstains and had nothing of value to offer. But the boots that appear in Figure 9.86 to Figure 9.97 do have some valuable data present. Both of the boots did have oil and directional drag marks on the upper heel. Some spots on the boots appeared to be blood and presumptive tests were made. Using a control showed, as expected, that anywhere a sample was taken on the boots, the test remained positive. Some leathers are noted for this property. Any useful presumptive test for the presence of blood must also test negative for the unstained leather, otherwise more advanced methods are called for.

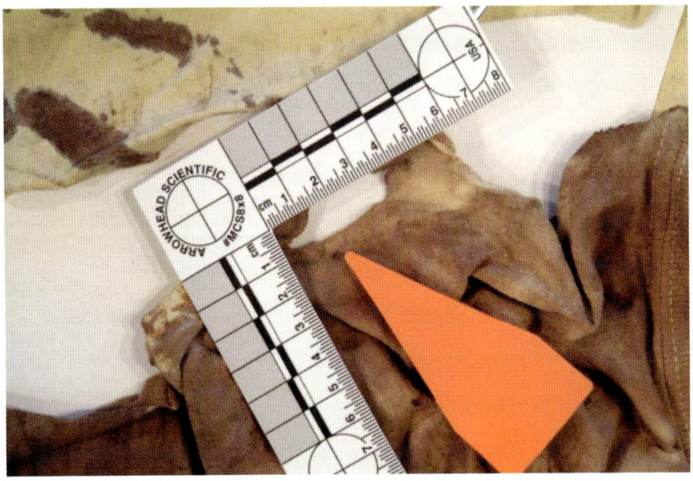

Figure 9.74 The overlapping orientation of the shirt with the dagger blade injury may be difficult to determine; the shirt may have moved during any struggle. The difficulty is compounded when the garment was cut off by EMTs and the blood hardens, stiffening the shirt. (See also Figure 9.75 to Figure 9.79.)

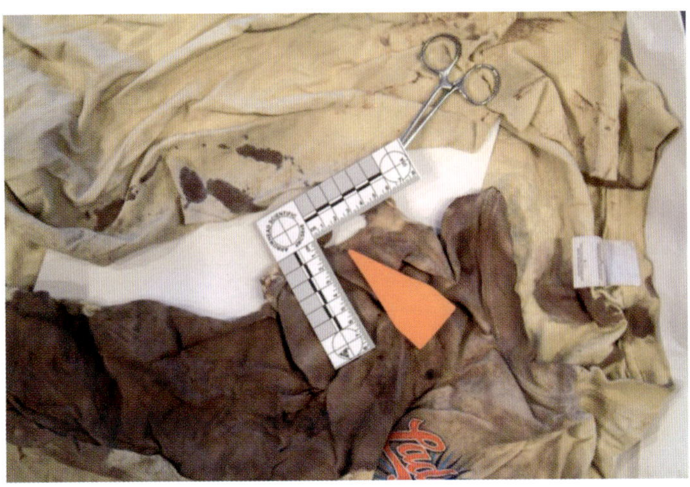

Figure 9.75

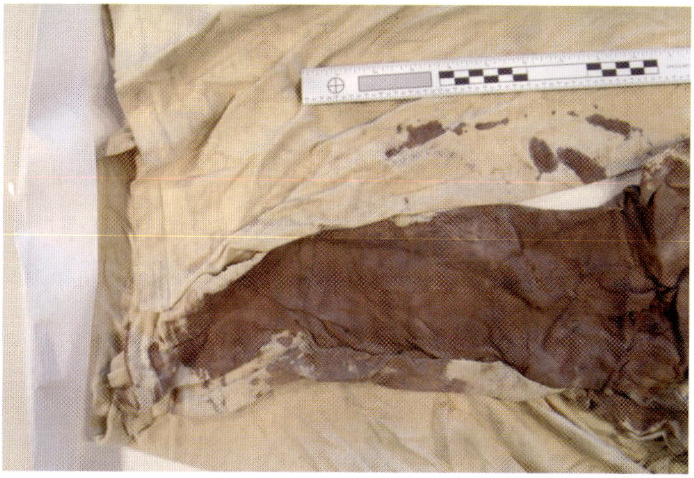

Figure 9.76

Figure 9.77

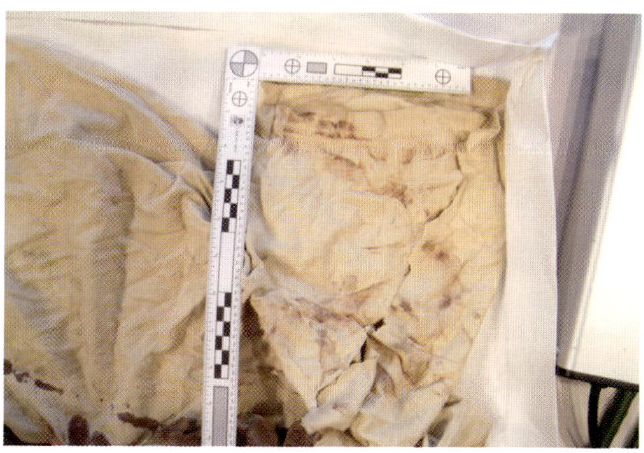

Figure 9.78

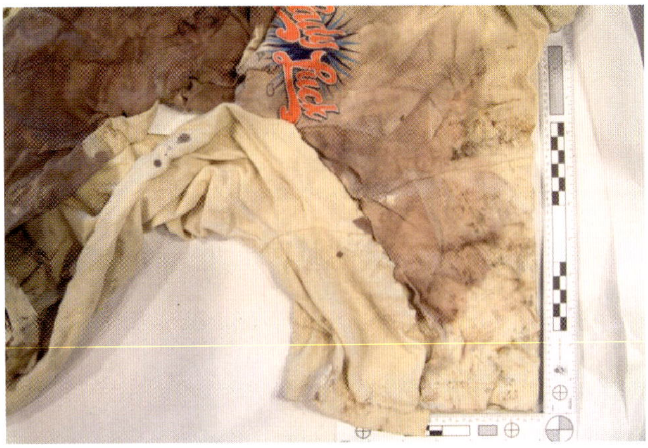

Figure 9.79

Figure 9.80 The oil stains (confirmed by FTIR) from the parking lot appear on the back of the T-shirt. The marks indicate a direction: movement from top to bottom, from the right shoulder to the bottom of the shirt.

Figure 9.81 Oil and bloodstains.

Figure 9.82 When the fabric is a light hue and the stains or marks are darker, sometimes the contrast/gamma altering functions in Photoshop can make the images more clear. Here, the images can speak for themselves without darkroom enhancements or tricks.

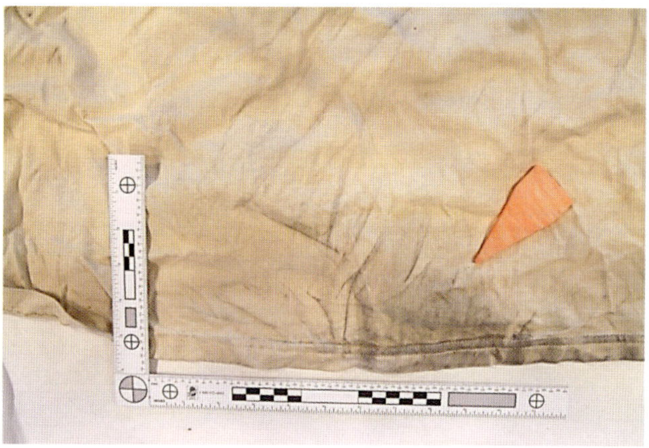

Figure 9.83 Another view of the stains.

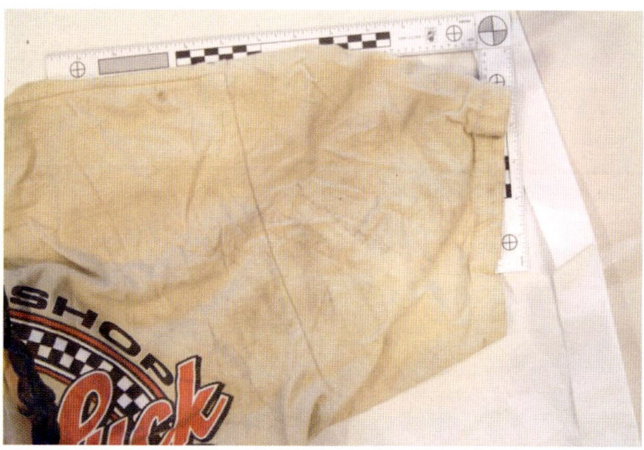

Figure 9.84 Another view of the stains.

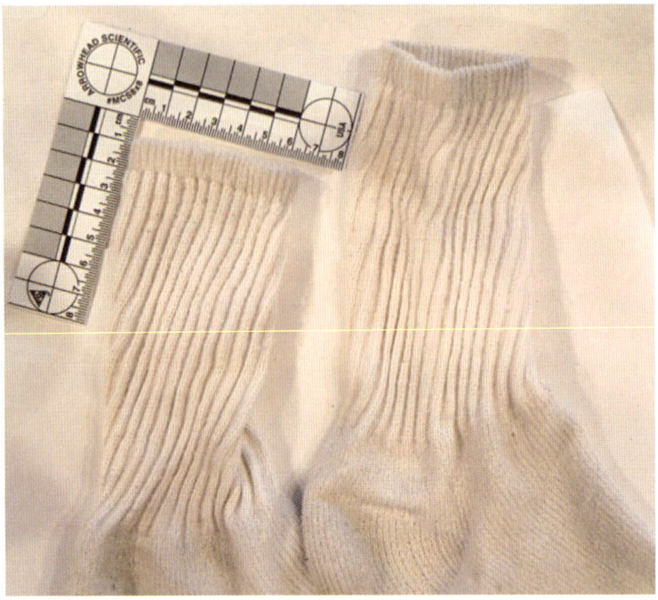

Figure 9.85 The socks were soiled without any other stains.

Figure 9.86 Different views of the stains on the victim's boots. (See also Figure 9.87 to Figure 9.97.)

Figure 9.87

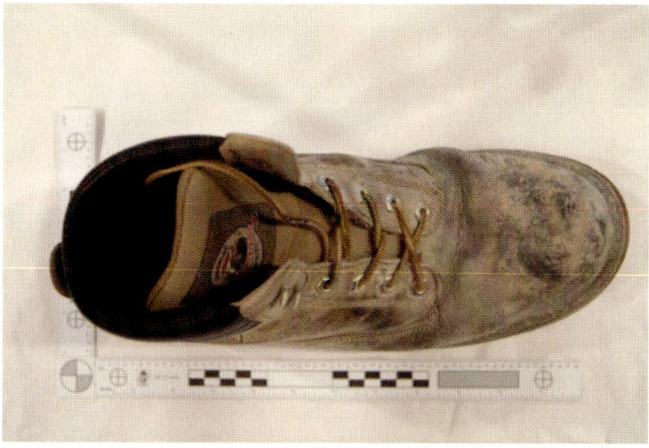

Figure 9.88

Figure 9.89

Figure 9.90

Figure 9.91

Figure 9.92

Figure 9.93

Figure 9.94

Figure 9.95

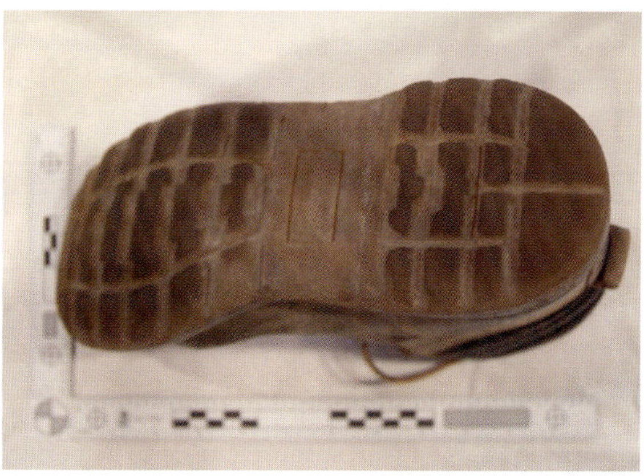

Figure 9.96

Figure 9.97

The defendant's clothing appears in Figure 9.98 to Figure 9.120. The presence of bloodstains can help provide data to answer questions about her position and her actions related to the bleeding victim. I also noted the presence of emesis on her clothing.

Vomit appears on the left leg of her jeans; blood appears on her right knee (Figure 9.101). The examination of most bloody clothing brought into the lab is done under a filtered fume hood to safeguard against infectious disease, but also to ensure that no cross contamination or ambient contamination from other laboratory sources occurs. As a negative control, we put a prepared SEM stub in the examination area to ensure that what we find on the evidence item does not originate from the laboratory.

Often, a digital microscope is useful, easy, and a quick way to visualize various items of potential value on clothing. Some digital scopes come with built-in ALS capabilities, making the examination process even easier at the microscopic level. Once large areas have been examined, then the relevant areas can be studied, samples can be taken, and the areas can be mapped in supplemental drawings or photographs.

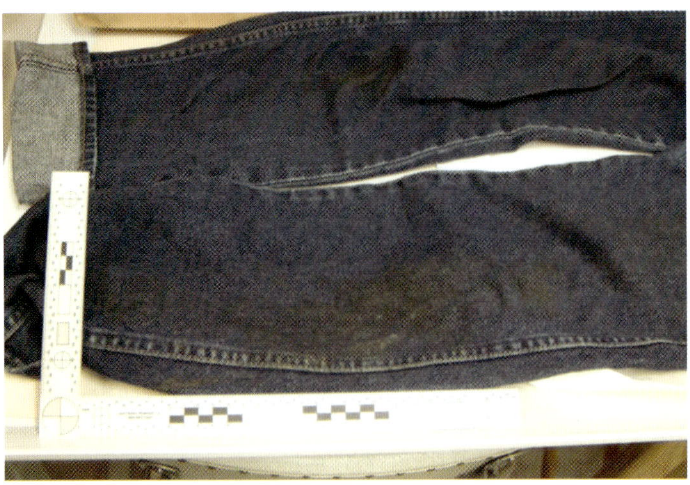

Figure 9.98 Apparent stains are shown—presumptive tests to be performed. (See also Figures 9.99 and 9.100.)

Figure 9.99

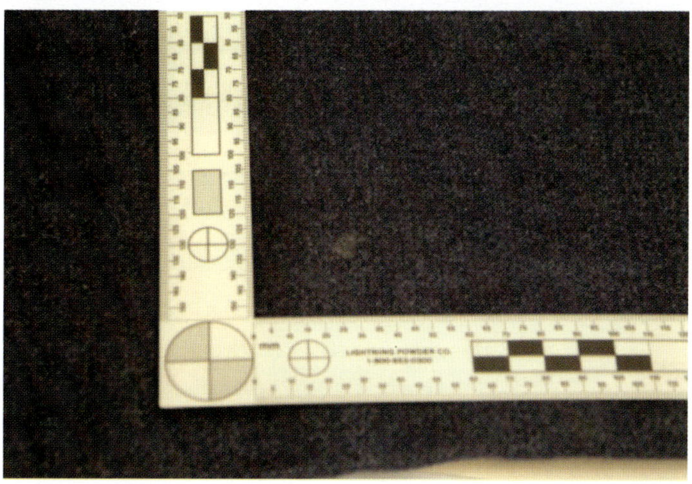

Figure 9.100

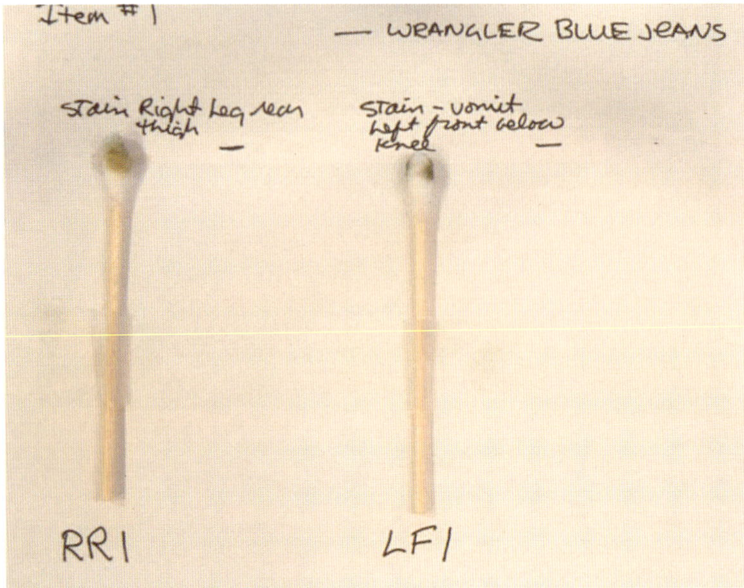

Figure 9.101 Testing the stains on the defendant's jeans.

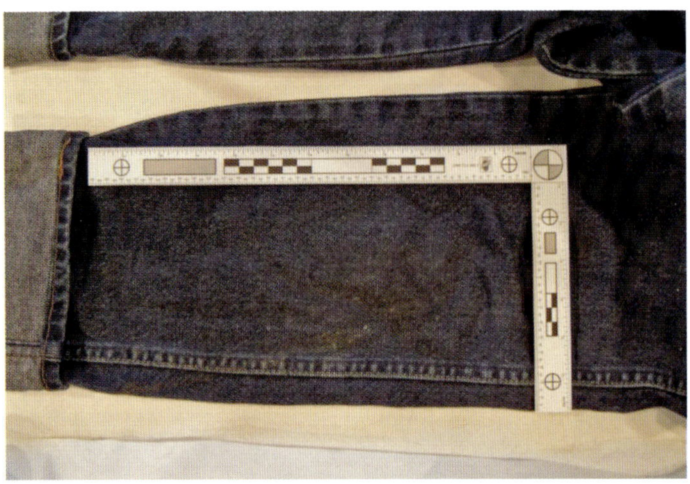

Figure 9.102 Stains on the defendant's jeans (see also Figure 9.103 to Figure 9.107).

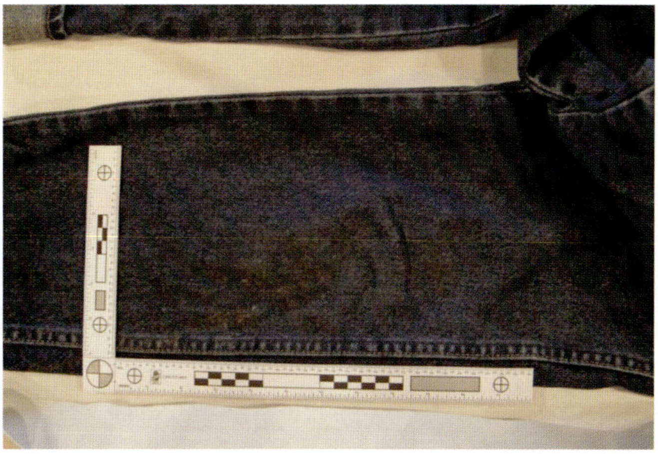

Figure 9.103

Figure 9.104

Figure 9.105

Figure 9.106

Figure 9.107

A fume hood must have specific filtration capabilities in order to capture any airborne hazards including any pathogens that might be present with the evidence. Different filters are designed for different purposes—from capturing hazardous chemicals to trapping biological agents and explosive vapors (see Figure 9.108 to Figure 9.110).

All too often, we are asked to find blood on red or otherwise dark materials (see Figure 9.111 to Figure 9.120). The weave and fabric itself often provides its own challenges. IR photographs, as stated earlier, are one solution; another, as explained previously, is to vary the contrast and gamma values. I have found that digital microscopy also helps determine the morphology of a potential stain as well as to help identify the nature of the stain itself (Figure 9.130 to Figure 9.132).

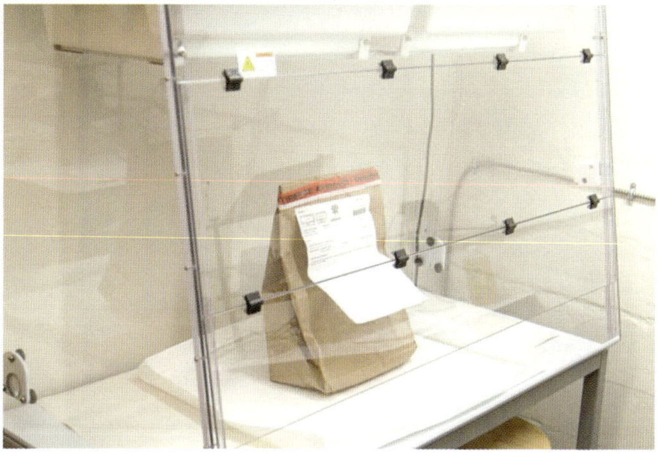

Figure 9.108 All potential chemicals or biohazards are examined under a fume hood with the appropriate filters in place.

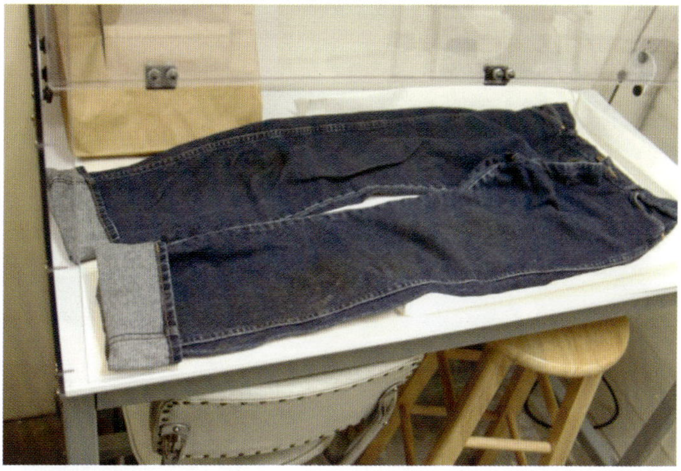

Figure 9.109 The suspects jeans under a fume hood.

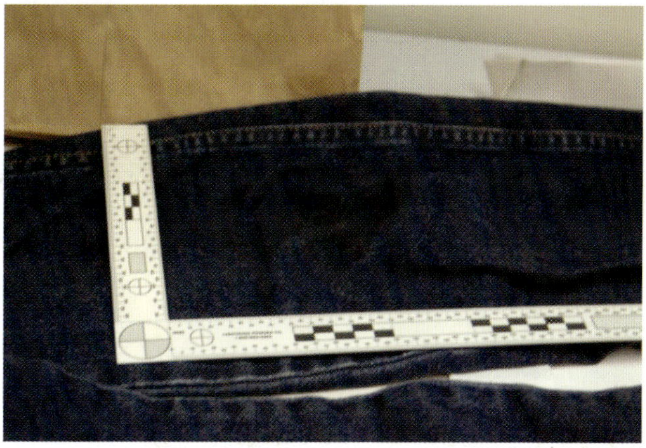

Figure 9.110 Documenting the jeans to accompany the drawings.

Figure 9.111 Stains on the defendent's shirt (see also Figure 9.112 to Figure 9.120).

Figure 9.112

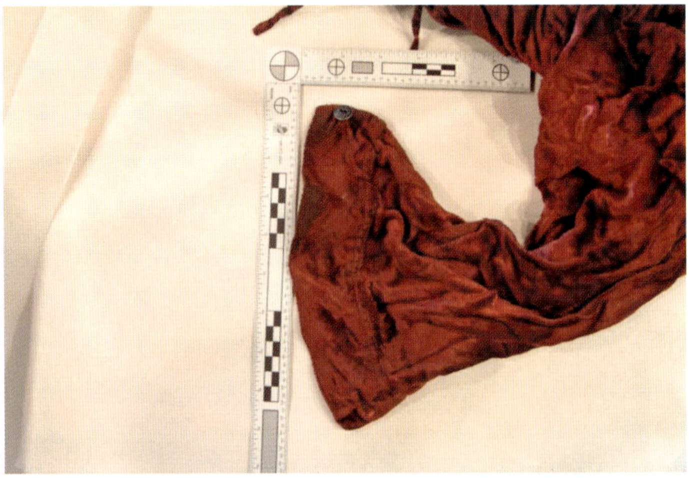

Figure 9.113

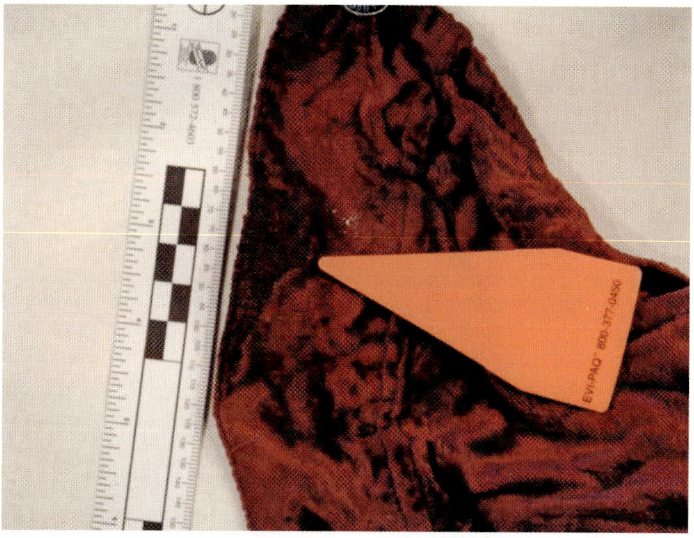

Figure 9.114

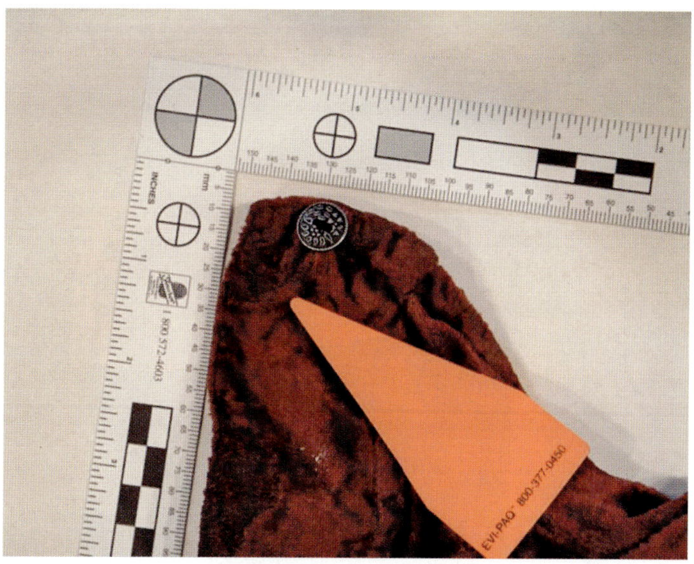

Figure 9.115 The blood on the cuff and its pattern suggests that this stain originated when the cuff came in contact with a blood source. No impact spatter appears, just soaking and wicking stains were observed on the garment.

Figure 9.116

Figure 9.117

Figure 9.118

Figure 9.119

Figure 9.120

Figure 9.121

The type of stain can show the mechanism of deposition. Microscopy of the blouse is shown in Figure 1.121 to Figure 9.129.

The undergarments (Figure 9.133 to Figure 9.135) also lacked any value for data development in this case. Regardless, all items must be examined and documented despite any preanalytical opinions about their relevance. If we do not look, we may never know if those opinions were correct or incorrect. It is often easier to look at an item than it is to explain on the stand why you did not bother to examine that item. In other cases, however, the explanation of an item's irrelevance can be very compelling. Each path requires a considered scientific judgment. Some laboratory SOPs do compel the examination of every item. Sometimes, with limited time and limited budgets, hard decisions must be made and justified by scientific as well as economic reasoning.

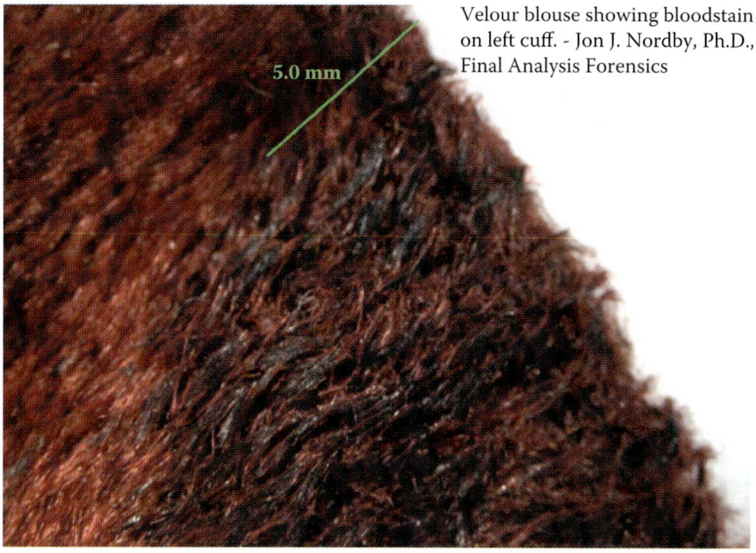

Velour blouse showing bloodstain on left cuff. - Jon J. Nordby, Ph.D., Final Analysis Forensics

Figure 9.122

Blouse showing bloodstains on outside of left cuff. - Jon J. Nordby, Ph.D., Final Analysis Forensics

Figure 9.123

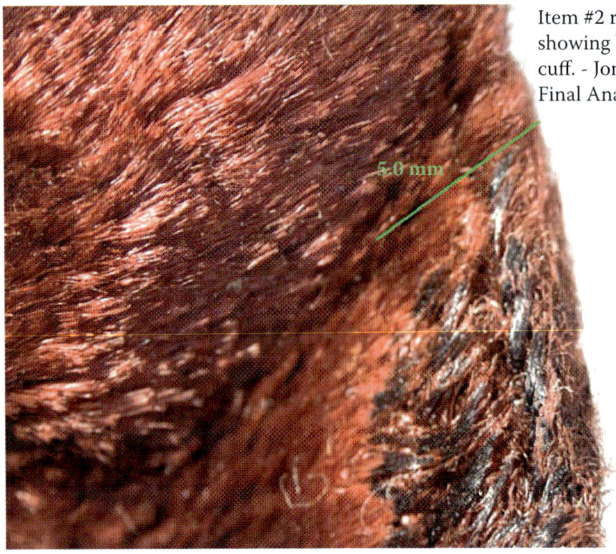

Item #2 red velour blouse showing bloodstain on left cuff. - Jon J. Nordby, Ph.D., Final Analysis Forensics

Figure 9.124

Blouse showing bloodstain on left cuff by button hole. - Jon J. Nordby, Ph.D., Final Analysis Forensics

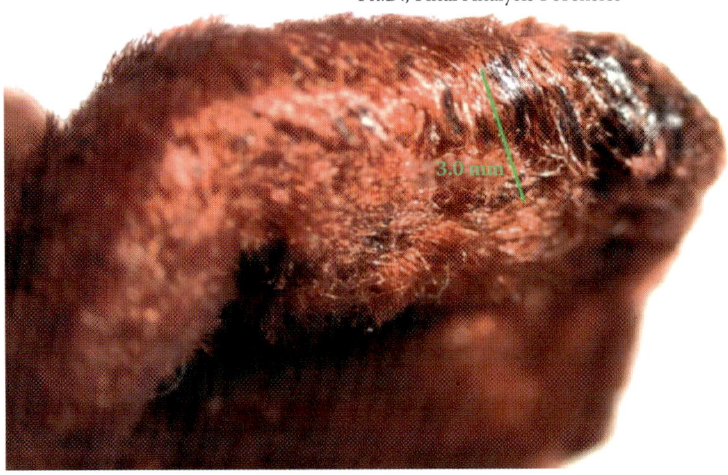

Figure 9.125

Item #2 red velour blouse showing bloodstain on inside left cuff. - Jon J. Nordby, Ph.D., Final Analysis Forensics

Figure 9.126

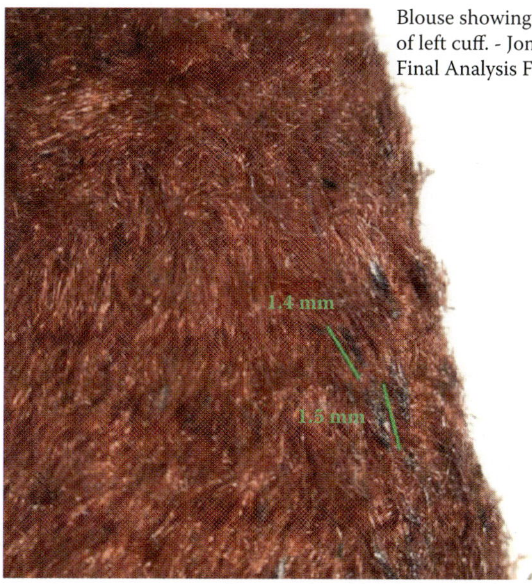

Blouse showing bloodstain on inside of left cuff. - Jon J. Nordby, Ph.D., Final Analysis Forensics

Figure 9.127

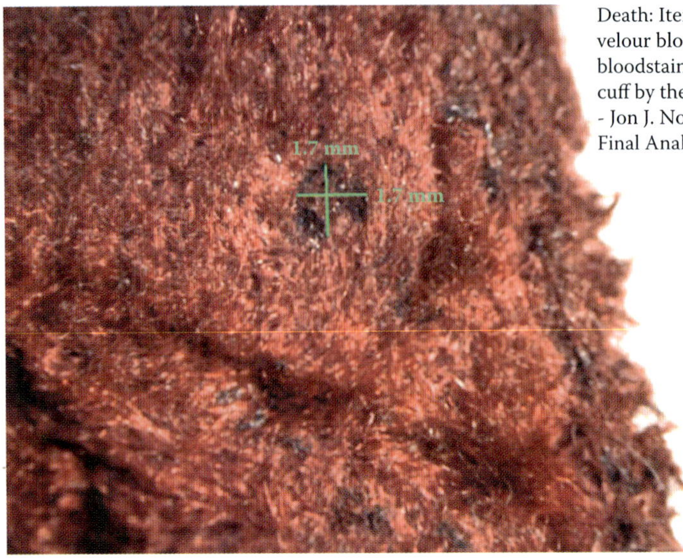

Death: Item #2 red velour blouse showing bloodstains on inside left cuff by the button.
- Jon J. Nordby, Ph.D., Final Analysis Forensics

Figure 9.128

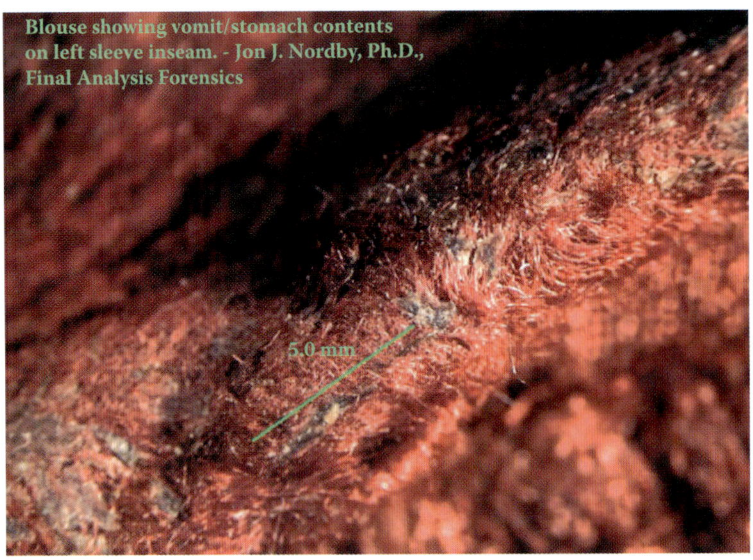

Figure 9.129

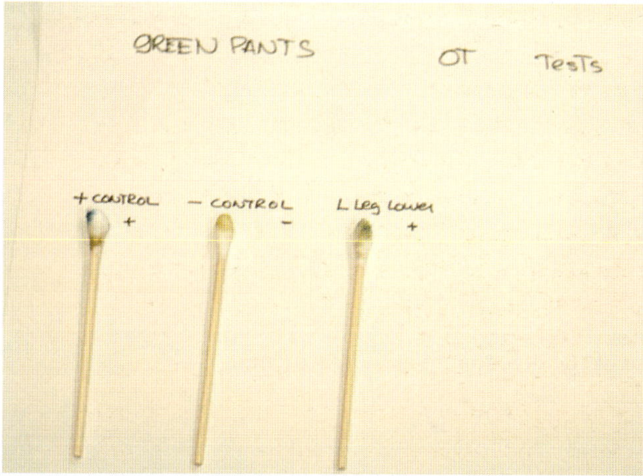

Figure 9.130 Presumptive blood tests are recorded in notes—the time element is critical, so the time elapsed between the test and photograph can be recorded in the photo log.

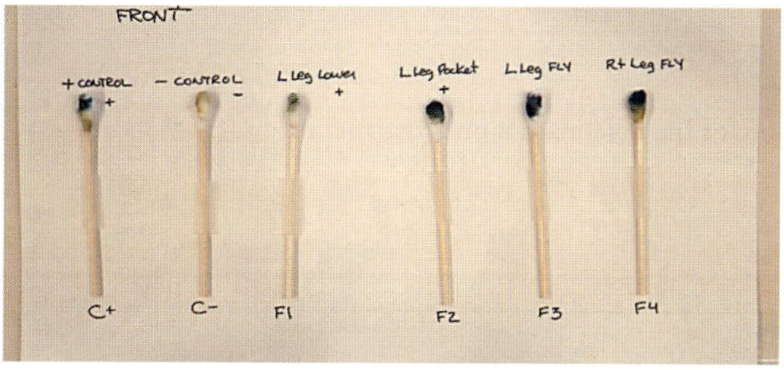

Figure 9.131 Presumptive blood tests of the pants with controls and areas tested.

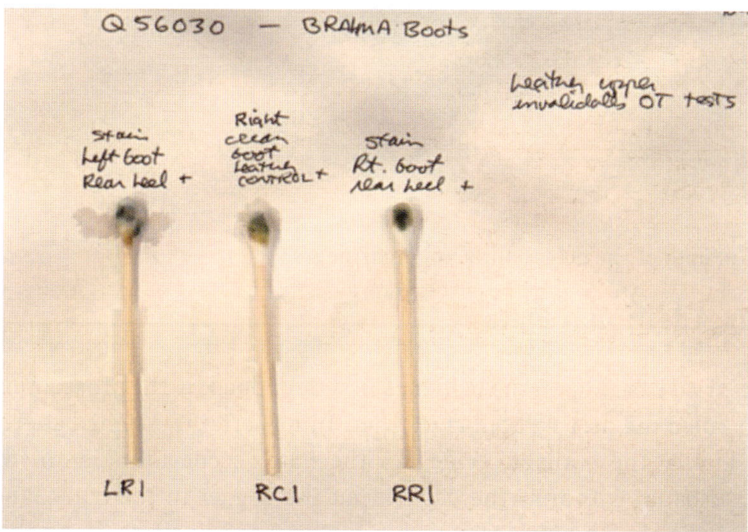

Figure 9.132 Boot presumptive; each were positive.

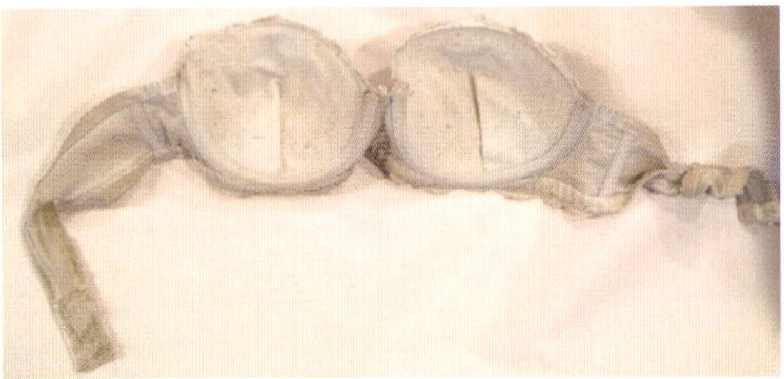

Figure 9.133 Photo of the defendant's undergarments.

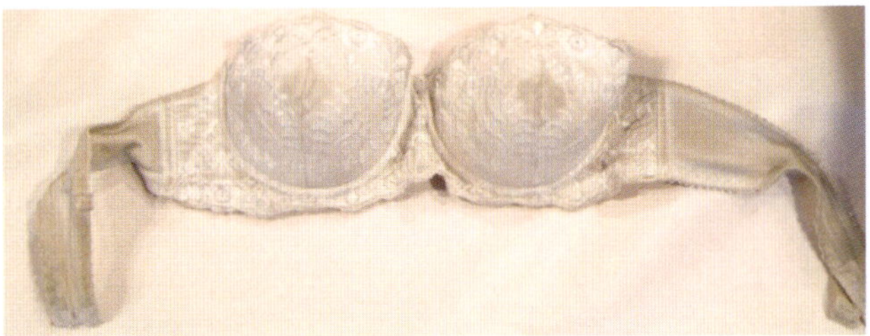

Figure 9.134 Photo of the defendant's undergarments.

Figure 9.135 Photo of the defendant's undergarments.

Developing Additional Data

This case was tried based upon the evidence first developed by the prosecution. The medical examiner ruled the death a homicide. That ruling certainly seemed appropriate based upon a first look at the available evidence. The state's scenario maintained that Shirley stabbed her husband, withdrew the dagger, put the dagger in the truck, and called 911. However, studying the physical relationship of the two principles doing their dance-of-the-drunken-tales in the parking lot developed more questions than answers. The recognition, documentation, and collection of physical evidence also left much to be desired. Basic questions that scene work could have answered through proper data collection simply had no answers. No scenario offered could rely on all the physical evidence: there did not seem to be much physical evidence to rely upon.

For one, no one could recall or say how the dagger traveled from the parking lot into the cab of the Vaughn's Silverado. The removal of the dagger from Stephen's chest also remained a mystery. Shirley had called 911 on her cell, and when police arrived, the dagger was not in his chest. That means that someone at the scene removed the dagger. Shirley seemed to be the best candidate. However, she did not recall doing so. Even if she said she did, some corroboration should at least be attempted. Some responding officers stated that the dagger lay near the body in the parking lot when they arrived. Others could not recall its location. Clearly, any more work depended upon receiving and developing additional data. Fortunately, such data had been discovered by the police and photographs that were not previously available, appeared. More data arrived.

Scientific reports ideally detail each of the assumptions, analyses, comparisons, and associations, as well as the foundations, for any interpretations developed from the available data. Additional data, when provided, can lead to additional developments in the assessment of conclusions derived from the original data set. With additional information, which was not available initially, I was asked (if possible) to answer the following questions:

1. What is the most likely *manner* of Stephen Vaughn's death?
2. What is the most likely *mechanism* by which the dagger was removed from his body?
3. What is the most likely *mechanism* by which the dagger traveled to the pickup and ended up under the passenger seat?

Any answers to the questions must be developed reliably from the available data to merit any evidential consideration. The problem of limited evidence and the problem of contrary conclusions having the same probability led me to conclude that Stephen's manner of death should be classified as undetermined. This essentially means that the evidence is equivocal among suicide, homicide, and accident as alternative manners of death. Anything more specific in this case went beyond the data and became mere speculation—speculation that does not meet any accepted standard for evidence, either scientific or legal. As such, this speculation violates the cannons for the scientific foundations of crime scene reconstruction as I have outlined them in this book. Appeals to medical or scientific authority are simply not relevant here from the methodological point of view.

To address the second and, specifically, the third question, I considered the additional evidence discovered by police and I requested the following items:

1. The original source of the scene digital photographs with time and date stamp, or associated software information attached to all such photographs to ensure that one could determine the order in which photos were taken. (This would provide at least some basis for assessing when items such as the dagger were in specific locations in some specific order.) A photo log would have helped in this regard, but one could not be produced because no photo log was taken.
2. The original source of the silver-emulsion film photographs; either the negatives (preferable) or a contact print of the negatives, again providing date, time, and the order they were taken. (Ordering the photos would provide some account of the items' movement in time as well as in space.)

3. The original photo log, typically kept for all crime scene photographs, listing the camera used for each series of pictures was not available. But identifying the cameras used can help assess the data captured in the various images that now became available. All that we had were more photographs.

I found that some of the new photos were scanned,* which leads to additional questions about their evidential value—questions that can be answered through access to the above data. If a mixture of film and digital cameras were used, then this scanning can be easily explained and the order of the photos can be established through the above documentation. Documentation of any statements made by the investigating officers would be important.

Statements about processing this crime scene remained important data, specifically the items of evidence including the parking lot itself, the vehicles present at the time the body was initially attended to by first responders, and the Vaughn's Silverado and how it was handled as evidence. Any recollections about the dagger and its chain of custody from the scene to the laboratory were needed. With this, any explanation of how the dagger traveled from the body or the parking lot's surface to the vehicle and then to the laboratory seemed so basic and so important that I hoped some memories could be restored. They were not.

This data would help to address the questions asked of me by the court. But when writing a report under these circumstances, one should make it clear from the outset that such information to be developed from the additional photographic evidence may not help to answer any of these questions.

I applied the photographic data to these questions by constructing a virtual model of the victim's body based upon the available physical evidence, including the autopsy data, the clothing, and the dagger itself. The model can be used to investigate a body's movements in the parking lot based upon the bloodstains, the grease stains, and drag marks, as well as on the statements made by the investigating officers. The results from this type of work are usually presented as images in a report. The images, in order to be reliable, must depict the necessary conditions for the observed data. Using a virtual model requires the identification of both the assumptions made and the purposes illustrated: possible, probable, or to help visualize alternative scenarios, which may then be evaluated by appeal to the physical evidence.

Another approach although more labor intensive, is to scan the clothing on a representative mannequin and stitch the images together using a 3D program as described in Chapter 1. The model, with the actual clothing in place of a rendering, can then be moved around in virtual space in order to investigate trajectories and positions. Figure 9.136 to 9.143 represent still pictures of the program's functions in the shotgun shooting death of two men†; a case unrelated to this one but that shows the potential usefulness of such an approach to the evidence.

* A Kodak 1640 film scanner is listed for some key photos, whereas a Kodak DX4530 Digital Zoom Camera is listed for others. Neither provides the necessary data to document the evidential standing of the photos in terms of time, date, or sequence when related to other photos or scans.

† The scanning and imaging work presented here was done with my partner Eugene Liscio, AI2 Forensic Animation, Imaging & Illustration.

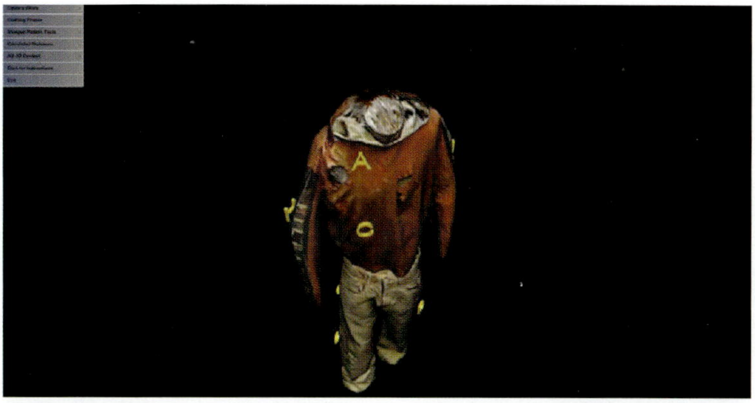

Figure 9.136 The clothing as worn is scanned. The letters designate the injuries suffered by the victim. This different case developed through an analysis of the body and the clothing and lining them up to match. We still do not know the actual positions of the body when the multiple shots occurred, but some shots are not possible from certain positions and a process of elimination can help begin to sort the data.

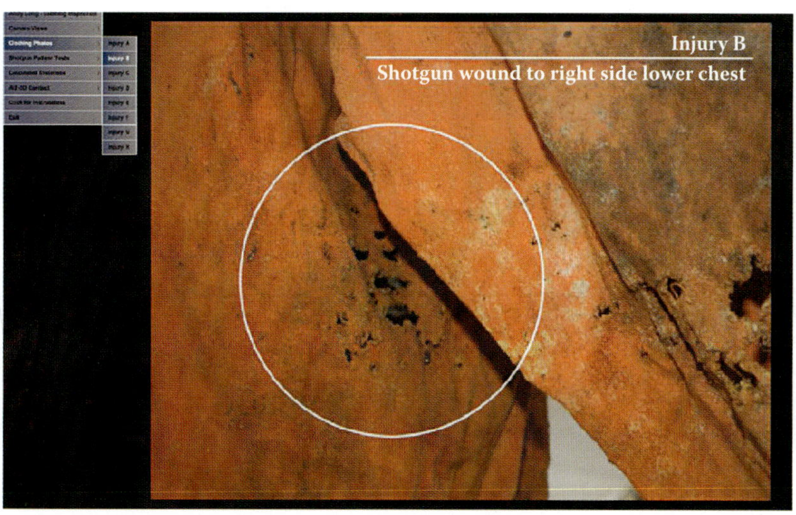

Figure 9.137 The menu to the left allows the user to choose a close-up of the injury in question from the autopsy and the clothing. The clothing defect is shown.

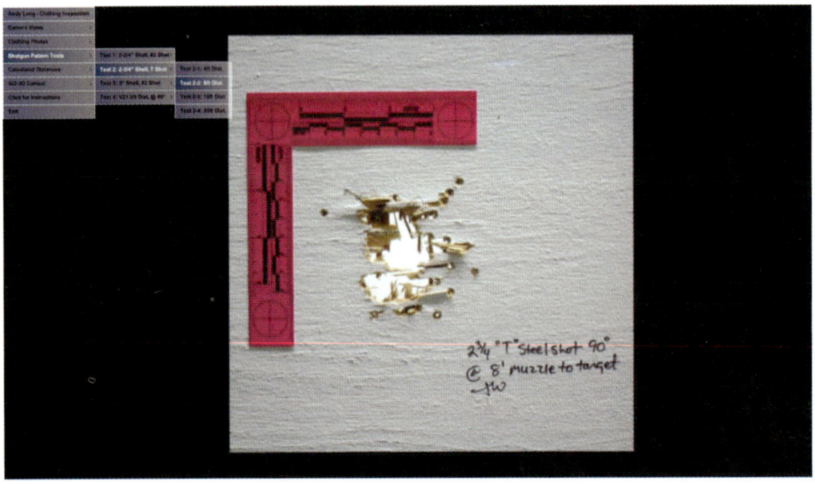

Figure 9.138 The menu also presents the muzzle-to-target test data showing the effects on targets from various distances. This allows a reasoned approximation of the distance based upon the pellet spread, the injury, and the direction of the shot.

Figure 9.139 The menu then permits the user to click the resulting trajectory on or off. The trajectory as a tube includes the trajectory envelope and error rate as discussed in Chapter 1. This is useful for nonshooting cases as well and also provides a powerful tool to use in court.

Figure 9.140 This image shows the trajectories of shots striking the upper body. There are two cameras to select: one for the upper body and one for the lower body. This makes it easy to examine the autopsy data, the clothing analyses, and the calculated trajectory data together with the raw data from which they are developed.

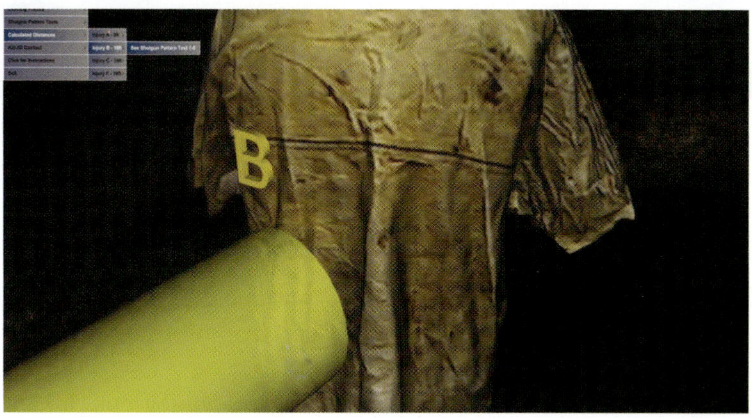

Figure 9.141 The shot to the rear of the second victim's back visualizing the muzzle-to-target data. The menu allows the user to view the autopsy results documenting each injury and its measurements, the clothing damage and measurements, as well as the shot-spread test targets. The same principles work equally well with blunt-force or sharp-force injuries.

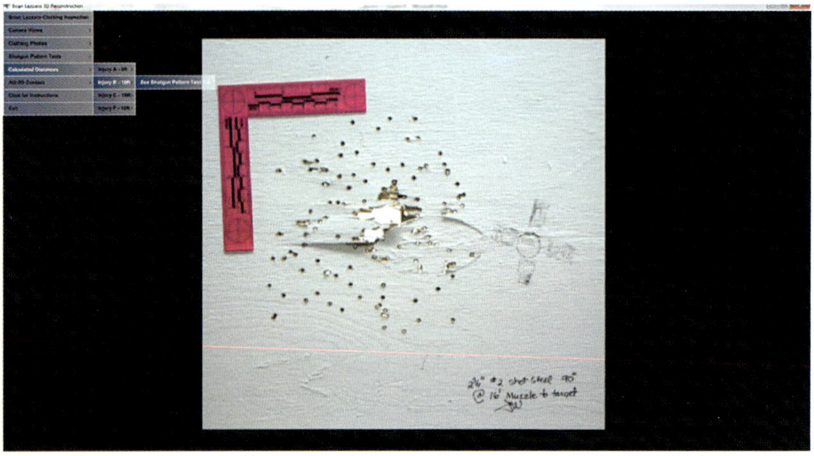

Figure 9.142 The data developed for the spread of #2 shot corresponding with the injury and the clothing damage can be evaluated and viewed in the program. (Note the shot cup impact on the target.) The same damage appeared at autopsy on the injury.

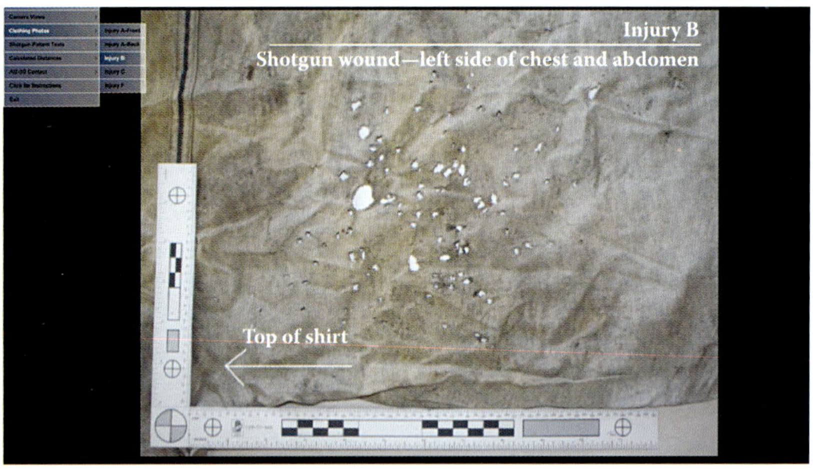

Figure 9.143 The shirt damage and the autopsy images corresponding with the trajectory letter and the test target can be compared and printed if necessary.

Brief Illustrated Reconstruction

Figure 9.144 to Figure 9.152 show an illustrated reconstruction of the sharp-force case presented here. Short of the movable 3D photos discussed above from another case, 3D still images can also provide useful illustrations.

The additional disclosed photographs included original photographs taken of the dagger in various locations as well as photographs of the decedent taken by the medical examiner as part of the autopsy. Although caution must always be attached to using photographs, they are often all that remain of scenes, bodies, and even physical evidence. Some experience with photoanalysis and photogrammetry should be a required skill for producing scientifically based reconstructions. Knowledge about depth-of-field limitations, lens distortions, and lens-to-object distance determinations form only some of the points requiring attention.

Figure 9.144 Google Maps street view of the scene as it appears looking southeast.

Figure 9.145 Google Maps street view of the scene as it appears looking southwest.

Figure 9.146 Google Maps street view of the scene as it appears looking northwest.

As with any physical evidence, images and the work derived from their analysis present limitations that must be recognized by the analyst and stated in a report. Although television shows depict exotic abilities to zoom in and grab clear details, such as faces captured in surveillance videos of moving cars, that is not how digital pixels work. All one gets, if anything, is a blur of essentially nothing at all.

Figure 9.147 Three-dimensional rendering that demonstrates the most probable positions of the decedent's left hand and the dagger, after the dagger penetrated his chest. Recall that we do not know how the dagger entered his body.

Figure 9.148 Three-dimensional rendering that demonstrates the most probable positions of the decedent's left hand and the dagger after the dagger penetrated his chest and he fell to his knees.

Figure 9.149 Three-dimensional rendering that demonstrates the most probable position of the decedent as he lay on his left side.

Figure 9.150 Three-dimensional rendering that demonstrates the most probable position of the decedent's left arm and the dagger as he was pulled by his right arm.

Figure 9.151 Three-dimensional rendering that demonstrates the most probable position of the decedent's left arm and the dagger as seen in Figure 9.150 as he was pulled by his right arm when seen from the opposite direction from Figure 9.150. No dagger is clearly visible from this perspective.

Figure 9.152 The only police photo showing the area behind the truck where the decedent lay on his left side. Apparent bloodstains appear along with oil stains typical for any parking lot. The caveat warning about relying solely upon photographs to develop data is widely chanted, as almost a mantra, for crime scene reconstruction. The reality, however, is that photographs often are the only remaining evidence of the scene; photographs are literally all that we can work with. While caution is crucial, photoanalysis and photogrammetry supply indispensable tools that can contribute reliability to the process of depending upon photographs.

Dagger, Decedent's Left Hand Injury and Bloodstained Clothing

Figure 9.153 to Figure 9.162 show photographs of the dagger, the decedent's left hand injury, and bloodstained clothing.

Figure 9.153 shows the dagger and sheath, which were not initially provided for my evaluation. Note what appears to be a linear transfer bloodstain on the file folder used as a surface for the photograph. Tests confirmed that it was blood.

Areas of staining on the left hand correspond with areas of staining visible on both sides of the dagger. The bloodstains visible on the decedent's left hand do not originate from any visible or described injury to that hand. All the blood originated from the decedent's chest wound, which was a direct push–pull type entrance, described at autopsy as having no wiggle or turning of the blade once inside the chest cavity. The blood on the decedent's left hand and, therefore, the blood on the dagger, originated from the decedent's sharp-force injury to the chest.

Figure 9.153 Photo of the dagger and sheath.

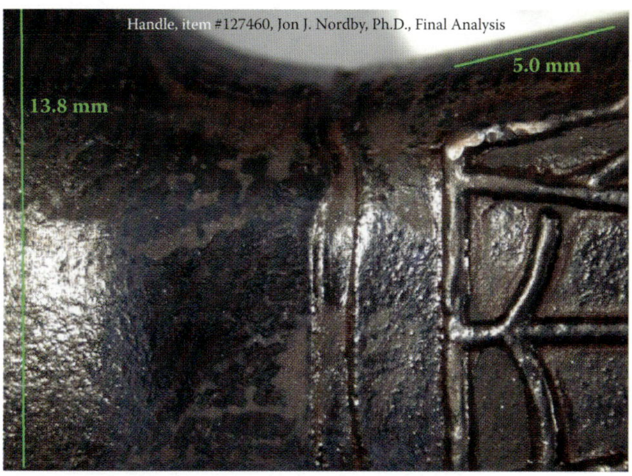

Figure 9.154 Photomicrograph of the dagger showing blood deposits as confirmed by tests.

When initially examined, the dagger revealed contact transfer, smear, and smudge stains. The location of the stains documented on photographs taken of the dagger during that examination can now be compared with photographs taken of the decedent's left hand before the autopsy, which were not initially available. In addition, photos of the dagger as seen in these photos were not available during my initial laboratory work. This is one good reason to state that one's conclusions are based upon the available data and that if more data surfaces, the conclusions are subject to change.

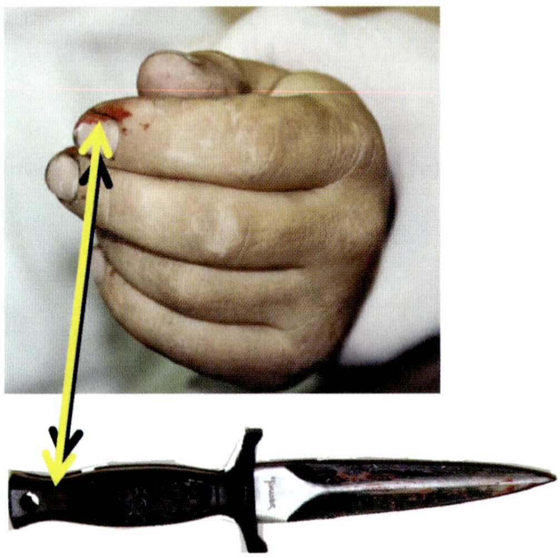

Figure 9.155 This shows the relationship of blood on the left hand with blood on the dagger.

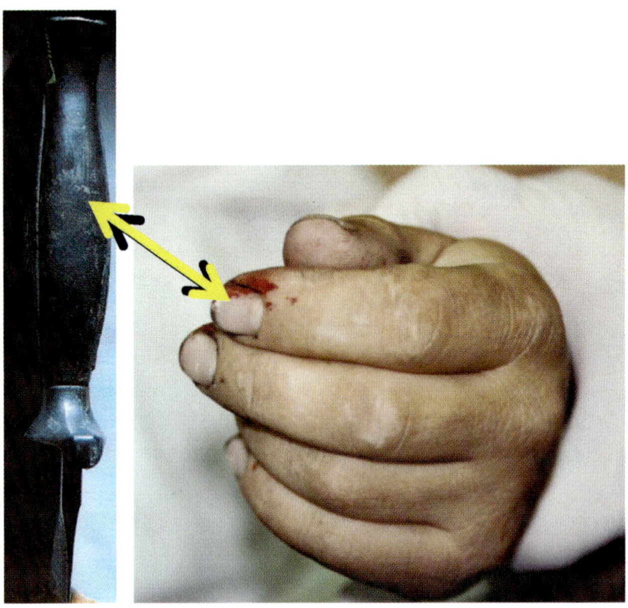

Figure 9.156 This shows the relationship of blood on the left hand with blood on the dagger.

The examination of both sets of photographs—my original evidence photos and the subsequent photos of the hands and dagger—reveals a correlation between the stains on the dagger and the stains on the decedent's left hand. These correlations involve plastic impressions of friction ridge detail found on the dagger (to be a plastic impression means that the impression is not only two dimensional but also three dimensional in nature). They also involve noting the significant presence of bloodstains in one location on the hand and an absence in another on the dagger.

The photographs are intended to show the areas of correlation and interest. Although some effort was made to scale the dagger size relative to the hand size, nothing of evidential value depends upon the accuracy of this scaling process. The significance involves the overall pattern of the stains found on both the dagger and the left hand.

Blood volumes adhere to the dagger blade, indicating that the dagger was wet with blood when removed from the decedent's chest.. The handle also presents flow patterns with adhering stains to the margins of the design on the handle.

During my initial examination of the dagger, I found sand and dirt typical of debris from a parking lot. The particles showed signs of an oil residue upon microscopic examination. Although no microscopic comparisons can be made since the evidence is not available from the decedent's hand, the photographs showing the decedent's left hand capture the presence of what appear to be typical dirt and sand from an area where cars are parked. The void on his palm area also means that his hand was not pressed completely flat against such a dirt-covered surface.

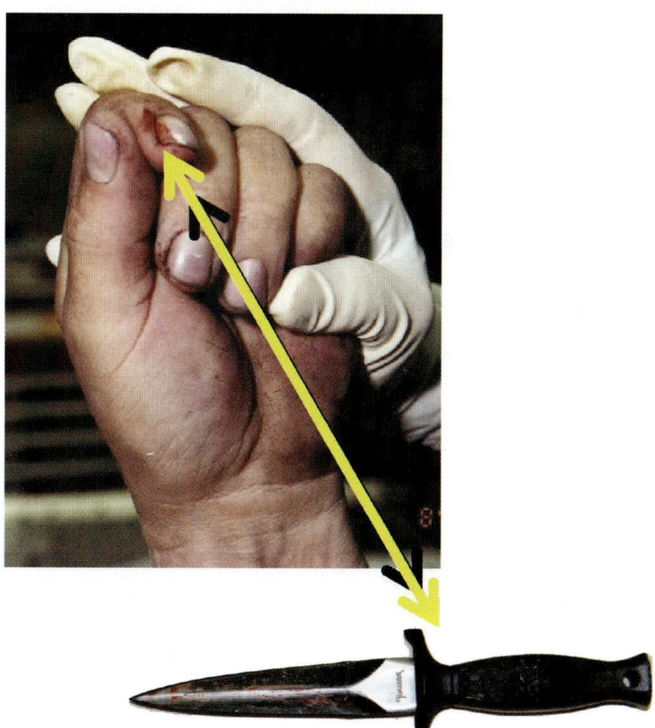

Figure 9.157 This shows the relationship of blood on the left hand with blood on the dagger.

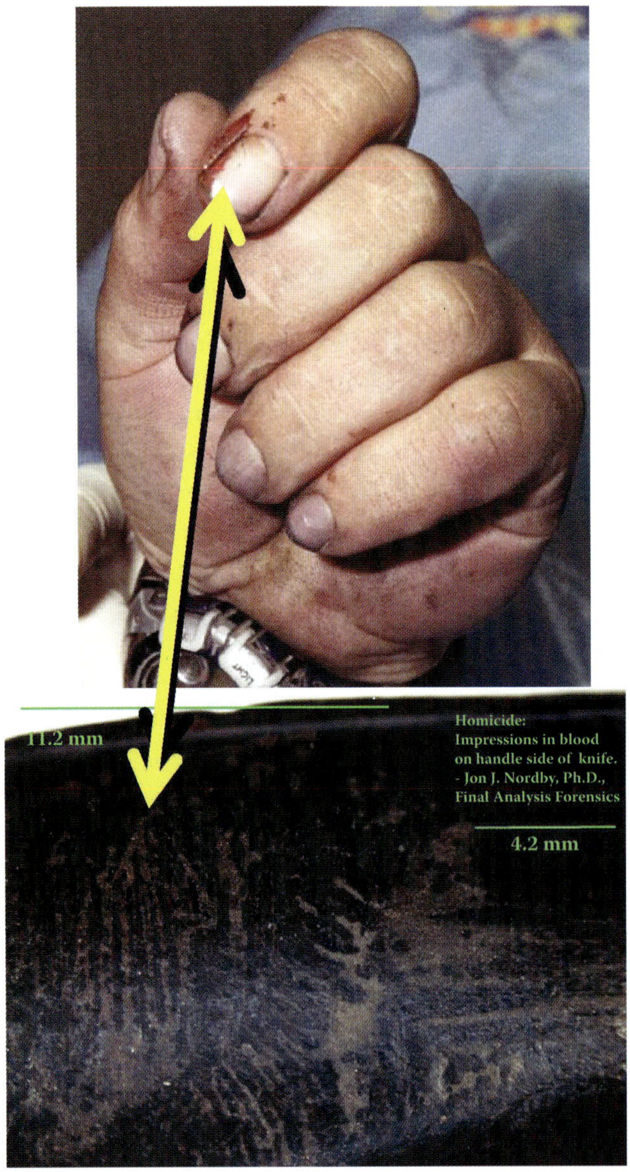

Figure 9.158 Photos showing the ridge detail. Again, the individual characteristics used for print identification do not appear.

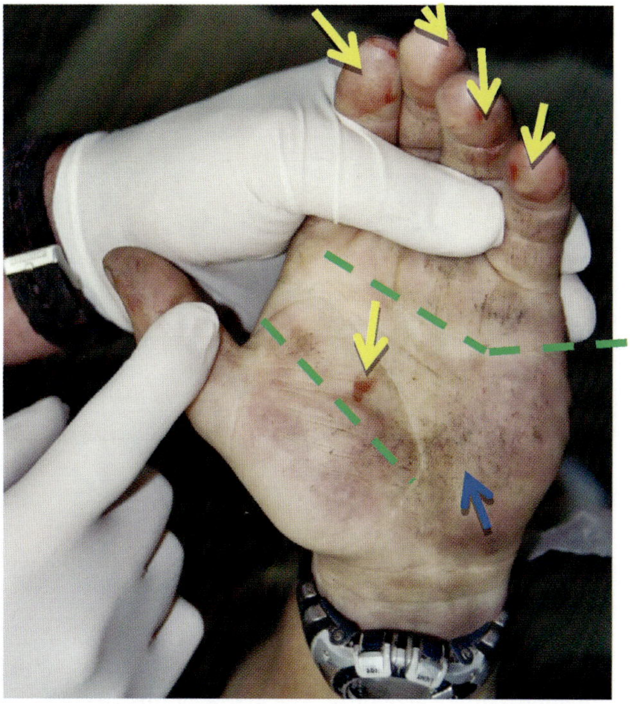

Figure 9.159 Photo showing the void area on the left palm.

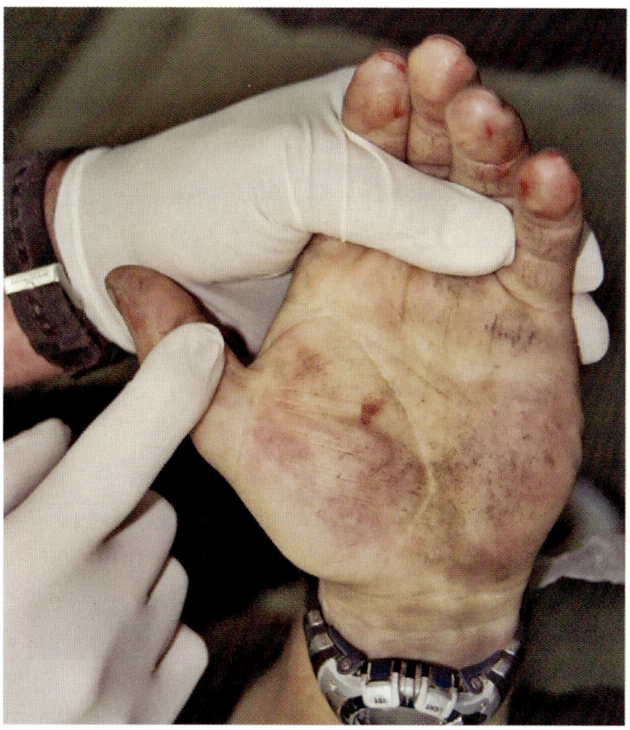

Figure 9.160 This is the original photo as received showing the palm surfaces of the decedent's (Mr. Stephen Vaughn's) left hand. Mr. Vaughn was reported to be left-handed.

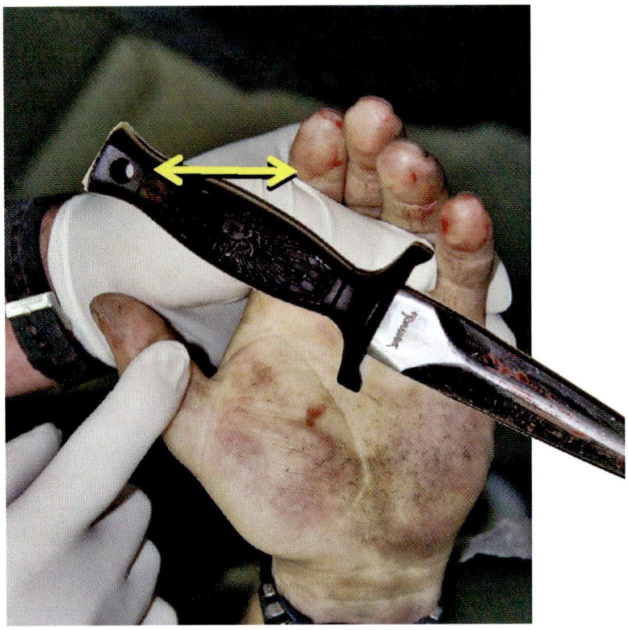

Figure 9.161 The dagger bloodstains related to the bloodstains on the decedent's left hand. The dagger is shown in approximate size only. The arrow points to corresponding bloodstains while subsequent photos show relevant voids as well as corresponding bloodstain details.

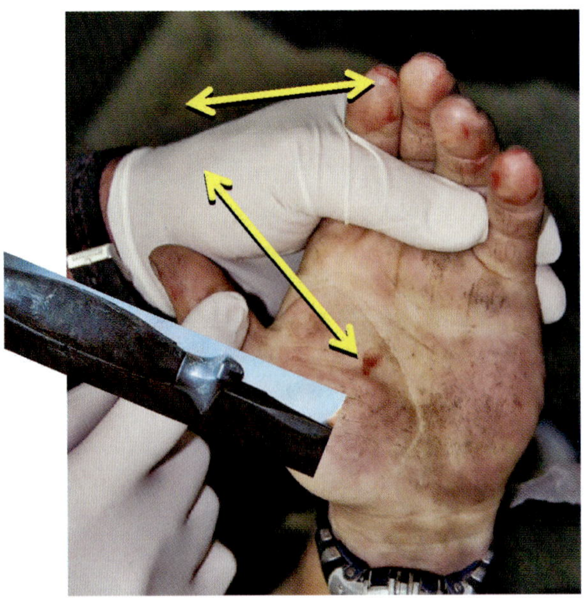

Figure 9.162 The dagger is shown in approximate size only; this shows the transfer sources for bloodstains found on the dagger.

Although identification is not possible, the ridge detail presents an impression in blood that shows that the digit involved slipped upward toward the upper end of the dagger's handle (away from the body) with the dagger inserted into the decedent's chest. The blood on the handle (prior photomicrographs) flowed on the handle and was therefore "wet" with sufficient volume to flow.

The void in the palm (Figure 9.159) shows that some object was present when blood was shed and when greasy dirt was in contact with the hand. Note that just below the little finger there is an apparent grease stain that does not transfer to other areas of the palm. Also, blood from the fingers does not transfer to the palm beyond the area noted on the palmer thumb area.

Blood on the decedent's palm on the dagger and on the first and second fingers indicates that the index or the second finger transferred blood to the palmer surface. The blood present on the fingers can be seen in the photos taken of the backside of his left fingers. The void is visible from the web of his palm (above the thumb) to the outside surface of his hand (area immediately below the little finger on the palm).

The decedent lay on his left side as indicated by the soaking and wicking bloodstains visible on his T-shirt (Figure 9.164) and as documented in my original analyses. The entrance is side-to-side on the shirt paralleling the side-to-side orientation on the body.

Greasy dirt on the T-shirt (Figures 9.166 and 9.167) shows that the decedent was dragged across the parking lot area by his right arm, with his left arm trailing. Note the dirt on the left and right shoulders on the front and back as the shirt bunched up for the turn and the initial movement. The force indicates that he was turned onto his back from a

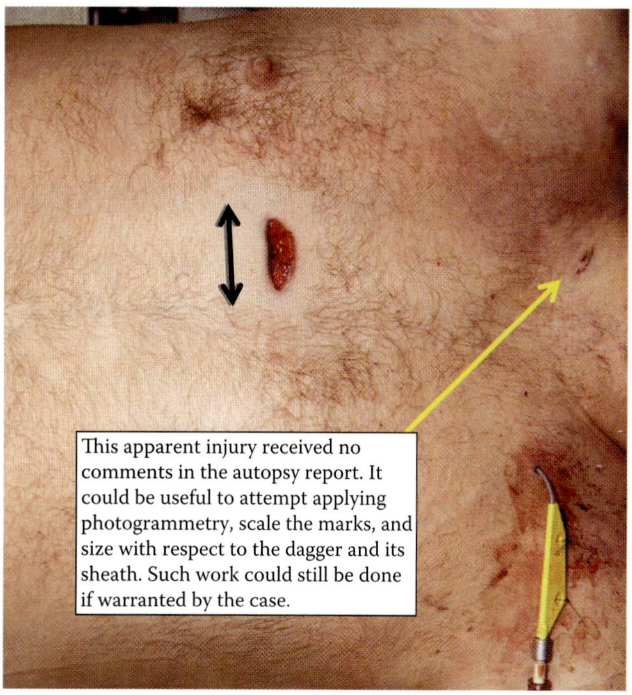

Figure 9.163 An autopsy photo showing the orientation of the dagger blade to the body; the dagger blade entered with beveled edges toward the lateral aspects of the torso.

Figure 9.164 This shows the soaking and wicking bloodstains on the T-shirt that were previously documented.

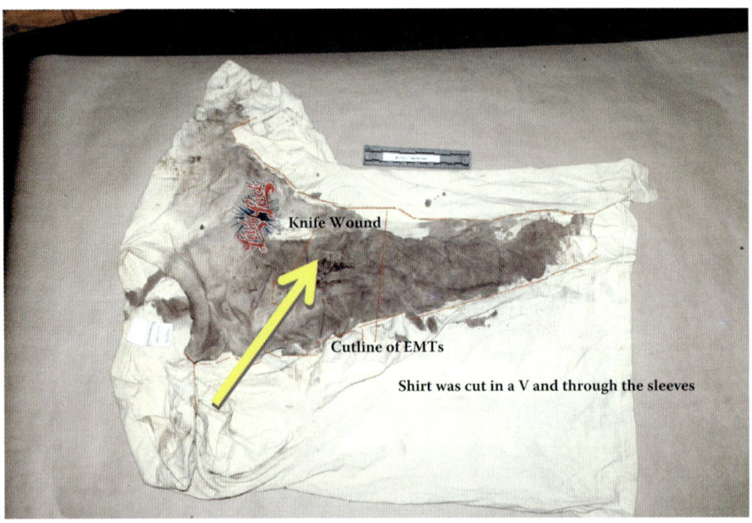

Figure 9.165 Another view of the T-shirt showing that Mr. Vaughn bled while lying on his left side.

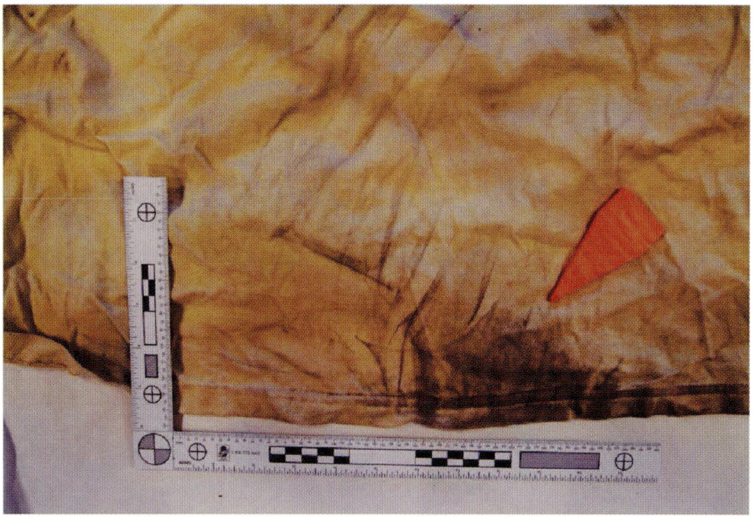

Figure 9.166 This shows the dirt and evidence of dragging along the oily surface of the parking lot.

Figure 9.167 This shows the dirt on the back left shoulder of the decedent's T-shirt.

position on his left side. This force would eventually have pulled the shirt toward his pants. Movement on the parking lot is documented through the dirt and bloodstains on the decedent's clothing. The stains and the folds leaving voids indicate the direction of movement.

Lab notes from prior analysis show the bloodstains and the dirt on the clothing (Figure 9.168 to Figure 9.170). The shirt and pants stains show his body position while bleeding with the dagger removed before Shirley Vaughn attempts to drag him by using his right arm. Blood from the decedent's hand and the dagger transferred to his pants as he lay on his left side. This conclusion was developed from an examination, which is documented in Figure 1.173.

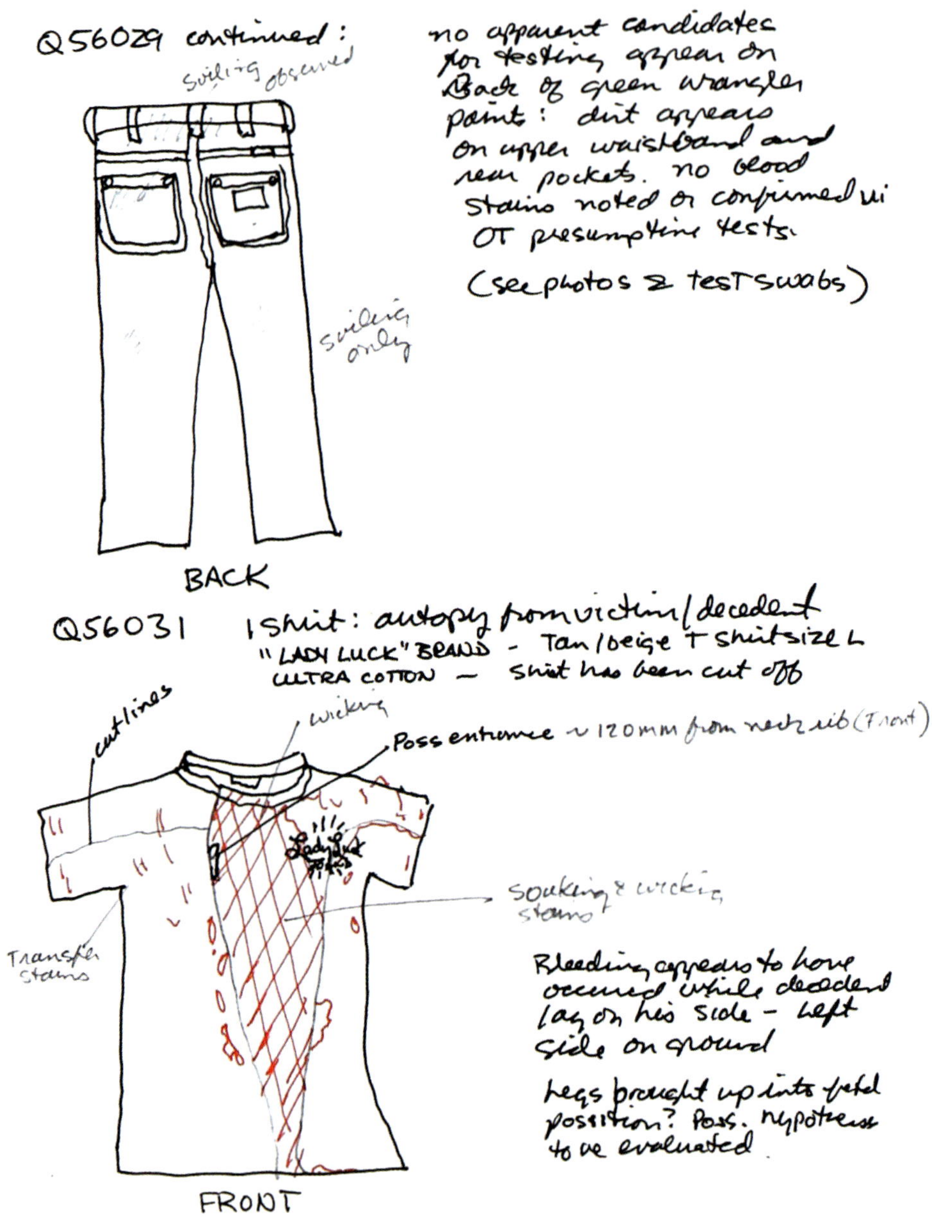

Figure 9.168 Lab notes reproduced from my initial report documenting the stains on the clothing.

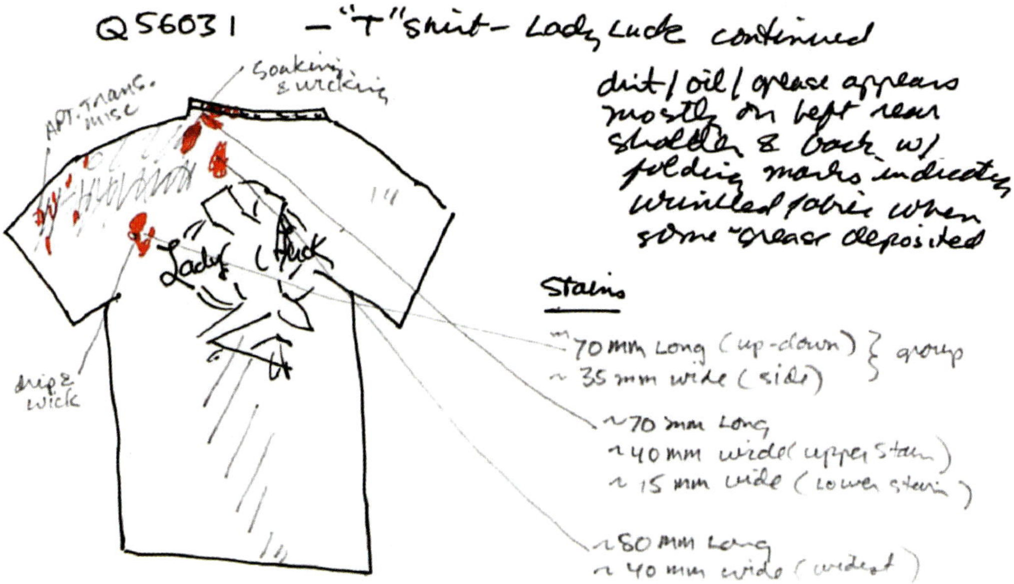

Figure 9.169 Lab notes showing the back of the decedent's T-shirt as documented.

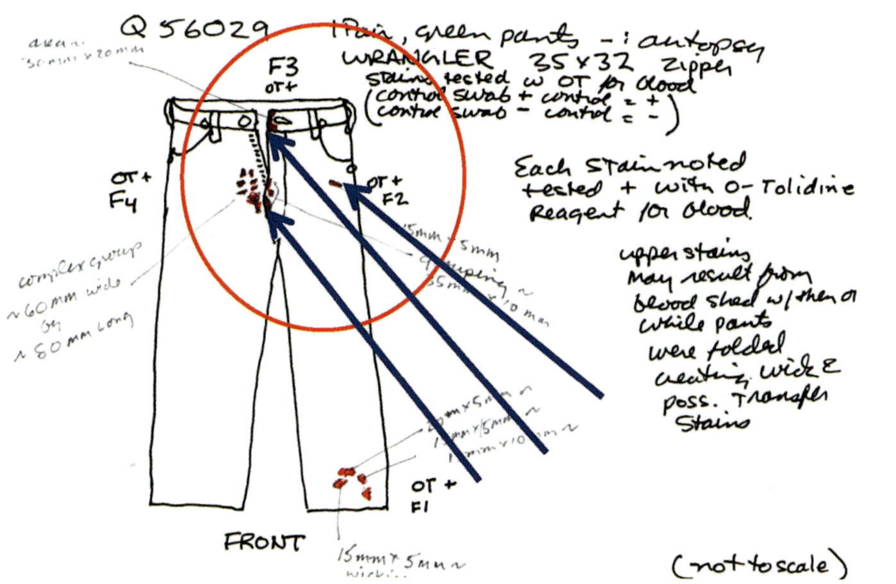

Figure 9.170 Lab notes from my prior report showing bloodstains, which now have added significance given the new data.

Blood appears as transfer stains in three identified locations (Figure 9.171 to Figure 9.173). The blood from his left hand and the dagger produced these transfer stains as his left arm, having withdrawn the dagger, moved toward the surface of the parking lot. He lay for some period of time on his left side. This position accounts for the bloodstains visible on the left side of his T-shirt.

Figure 9.171 This shows the decedent's jeans with significant bloodstains present on the button flap.

Figure 9.172 This shows the decedent's jeans with significant bloodstains present above the zipper.

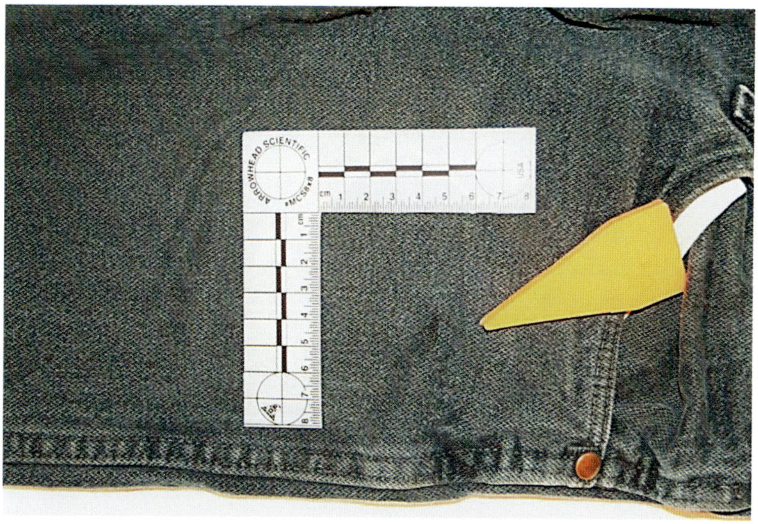

Figure 9.173 This shows the decedent's jeans with significant bloodstains present on the left pocket.

Bloodstains on the pants (as documented for prior analyses) confirm the blood transfers that one would expect if the dagger were in the decedent's left hand with the hand moving from the midline to the left lateral and ending on the ground.

The reconstruction data supports the following conclusions:

1. Stephen Vaughn had his left hand on the dagger when he was actively bleeding, with his hand palm-side up and the heel of his hand toward the chest, fingers closed around the handle.
2. Debris (dirt, grease, sand) and blood appear to be present on his left hand and also present on the dagger in corresponding locations typical of gripping the handle.
3. He was dragged on the parking lot surface, pulled by his right arm during which his body turned from lying on his left side backward and toward the truck onto his back.
4. He probably removed the dagger from his chest while still on his left side, as seen from the resulting bloodstains and dirt on both his hand and the dagger, with even more wet blood present on the dagger.
5. If Shirley Vaughn removed the dagger from her husband, then she would have had blood on her right hand (she was reported to me to be right-handed). She did not appear to have any blood on her right hand; therefore, she did not remove the dagger from her husband.
6. If there was no blood on Shirley Vaughn's right hand (with some on her left hand) and if the dagger was removed from his body before he lay on his back, then Stephen Vaughn removed the dagger (supporting bloodstains and trace evidence plus prior probabilities).
7. When Shirley Vaughn reportedly pulled her husband by pulling on his right arm, his body turned onto his back from lying on his left side. As he was turned onto his back and was pulled slightly toward their truck, the dagger could have been too close to his body for Shirley (intoxicated) to see (both prior probabilities and posterior probabilities).

Figure 9.174 The arrows show the back of Stephen Vaughn's heels. Avulsed and stained with dirt from the parking lot surface, the back of his boots shows an upward movement (toward his head) consistent with the attempt to drag him with his right arm.

8. Shirley was unable to move him very far by pulling on his right arm (with his left arm dragging beside his torso).
9. If the dagger was moved by this dragging movement, and if Stephen Vaughn removed the dagger, then Shirley Vaughn may not have seen the dagger (posterior probability).
10. If she did not see the dagger, then she could not have removed it, nor could she have placed it in the truck (since both actions logically entail seeing the dagger).
11. The stains on the decedent's hands and the reported lack of stains on the defendant's right hand decrease the prior probability of her removing the dagger from his chest. The posterior probability of her removing the dagger from his chest decreases the prior probability of her placing the dagger in the truck but remains dependent on additional unavailable data.
12. Given the stain patterns visible in the photographic evidence and given the lack of any documented bloodstains on the defendant's right hand, it is more probable (than not) that she did not touch the dagger. If she did not touch the dagger, then she did not move the dagger. If she did not move the dagger, then she did not place the dagger in the truck.

The Dagger and Its Location in the Truck

The lack of photo documentation helps obscure any answers to the questions of how the dagger got into the sheath and how the dagger got into the truck. Photographs provided do not help answer the questions; instead they simply add new questions.

The photographs of the dagger were presumably taken by crime scene investigators. In these images we see the truck floor and a seat area in the background (Figure 9.175 to Figure 9.178). Figure 9.175 shows the dagger in the sheath, however, Figure 9.176 shows the dagger in the sheath and under the seat with the addition of a string. This string is typical of

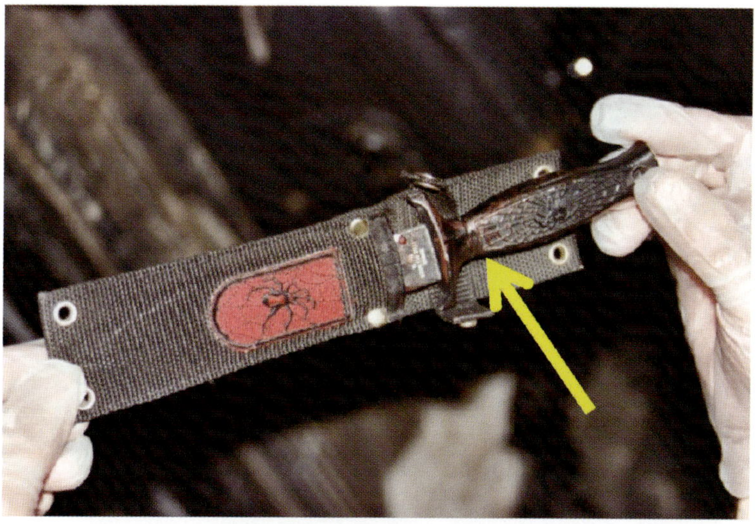

Figure 9.175 The dagger in the sheath. Its location is not clearly documented. I received this photograph after my initial review of the case. Note the heavy deposition of blood on the handle, as well as on the hilt of the dagger. Touching the handle with bare hands and putting the dagger into the sheath would be sufficient to transfer blood from the dagger to a hand or from a hand to the dagger.

Figure 9.176 The dagger both in its sheath and in the truck. It appears to have added a string to its handle.

an evidence tag used to help identify items of evidence. The presence of this string suggests that the dagger has been positioned under the seat. What was its original location? How did it get from the parking lot into the truck? Such questions remain unanswered.

Additional photographs simply confuse the issues even more. Note the inconsistent positions of the dagger among the photographs taken of its position in the truck. For a report, these must either be resolved or documented despite any limitations that they impose. In both Figures 9.176 and 9.177, reportedly taken of the dagger as it was found in the truck, what looks like a string appears. In Figure 9.178, no evidence tag or string

Figure 9.177 Another view of the dagger in its sheath with a string apparently tied around the handle.

Figure 9.178 Yet another view of the dagger in the truck. This time it may even appear in an entirely different location in the truck.

appears. It also shows the dagger in a totally different orientation to the seat and the transmission tunnel. Note that the prior photographs show the dagger handle with associated string against the transmission tunnel, whereas Figure 9.178 shows it in a completely different orientation. These inconsistencies support one conclusion: They do not support any statement about how the dagger got into the vehicle. (Again, note the blood on the dagger handle and hilt.)

In prior cases, when we found an item of evidence such as a skull with maxilla in one state (Washington) and the associating mandible for that skull in another state (Oregon), the question always arose: How did the skull move from one state to another? We had studied the decomposition* patterns of disarticulation and were learning at that time about "animal transport," a mechanism by which skeletal remains become scattered about in an outdoor crime scene. For the skull, we devised a new category: "maniacal transport." In the case of the skull, the perpetrator himself moved it. Maniacal transport became a category used to explain unexplained mechanisms of evidence movement. All I can say about the dagger's move to the truck is that it remains mysterious and therefore can be classified as "mysterious transport."

Brief Summary of Overall Conclusions

I was asked to consider three questions for this analysis and asked to include the new photographic data provided to me. They are repeated here as follows:

1. What is the most likely manner of Stephen Vaughn's death?
2. What is the most likely mechanism by which the dagger was removed from Stephen Vaughn's body?
3. What is the most likely mechanism by which the dagger ended up in the pickup truck under the passenger seat?

These answers must be based upon the available evidence to merit any evidential consideration. In my view, answers must, to be useful, be established to a high degree of medical or scientific certainty—again, as based upon the given physical evidence, not merely appeals to authority. If someone truly is an authority, then he or she ought to be able to provide scientifically sound *facts* to support the given conclusion or, in the absence of that, be able to explain why such facts are not forthcoming. Authority may get one's testimony admitted, but authority without clear factual support remains just another opinion.

What Is the Most Likely Manner of Stephen Vaughn's Death? Undetermined

The problem of limited evidence, combined with equal probabilities for a given alternative answer, led me to conclude in my first two reports that Stephen Vaughn's manner of death must be "undetermined." This finding essentially means that the evidence is equivocal among suicide, homicide, and accident as alternative manners of death. Anything more specific, in my scientific opinion, goes well beyond the data and becomes mere speculation—speculation that does not meet any accepted standard for evidence, either scientific or legal. As such, this speculation violates the cannons of the scientific foundations for crime scene reconstruction as I have explained them. Appeals to the medical or scientific authorities are simply not relevant here.

As far as I can tell, the only reason provided for concluding that Stephen Vaughn's death was a homicide is an appeal to authority: In essence, it is a homicide because the authority says so. This remains outside the province of science and rests within the subjective realm of mere opinion. For every such opinion, a contrary opinion can be found. This

* See "Is Forensic Taphonomy Scientific?" Jon J. Nordby, pp. 31–42, Chapter 2 in *Forensic Taphonomy: Volume II*, W. D. Haglund, and M. H. Sorg, editors, CRC Press, 2002.

is a problem when no reasons support either opinion. Reasons of a scientific sort, however, are required here.

The question about the manner of death becomes: How did the dagger come to enter Stephen Vaughn's chest? Despite the evidence of blood on his left hand, despite his being left-handed, and despite the transfer stains and trace evidence fit between his left hand and the dagger handle, we still have no reasons to decide that the dagger entered his chest through self-infliction rather than through, say, a mutual struggle that resulted in the dagger being pushed inadvertently into his chest. Nor can we rule out that he intentionally positioned the dagger against himself in a self-destructive taunting threat, and that Shirley Vaughn, in an act of defiance, pushed the dagger inward. Nor can we rule out the possibility that somehow, in a struggle, the dagger was inadvertently pushed into his chest through the uncontrolled movements of struggling drunks. The point to understand and to emphasize here is that all the physical evidence would look the same in any of these mutually exclusive alternatives. To pick one over any other, in the absence of any fact that tips the balance toward homicide over accident, or suicide over homicide, is irrational. The rational choice, then, remains undetermined. We simply do not know how the dagger came to be inserted into Mr. Vaughn's chest. We do, however, know why we do not know.

What Is the Most Likely Mechanism by Which the Dagger Was Removed from Stephen Vaughn's Body? Mr. Vaughn Removed the Dagger

The evidence provided in the photos of the decedent's left hand and of the dagger support the contact of his hand with the dagger handle after the injury occurred. We simply do not have sufficient information to decide if his hand forced the dagger into his body. The bloodstain and trace evidence do support the claim that he removed the dagger from his body. With the dagger removed, he lay for some moments on his left side, bleeding. With the dagger removed, his left arm and the dagger lay next to his torso. When Shirley Vaughn grabbed his right arm, he was turned onto his back, with his right arm over his head, and his left arm and the dagger toward his body.

(A certainty factor and error rate discussion appears in Appendix A.)

What Is the Most Likely Mechanism by Which the Dagger Ended Up in the Pickup Truck under the Passenger Seat? We Do Not Know

Given the ambiguous evidence of the dagger's position in the truck—showing it in at least two different places and under at least two different sets of factual circumstances (with the string and without the string)—we can only say that the dagger came to be in the truck through maniacal transport. In other words, it got there, we just do not know how. We do know that a human agency was involved.

Therefore, the evidence supports the following statements as fact: First, Stephen Vaughn's manner of death is undetermined; second, Stephen Vaughn removed the dagger from his own chest; third, we do not know how, when, or how many times the dagger came to be placed into the sheath and inside the truck.

The photograph of the decedent's left hand again presents the possibility that the blood print on the dagger originated from the decedent. The photographs of the suspect's hand (Figures 9.179 and 9.180) do not show blood in the relevant locations. But again, even if sufficient ridge detail were present on the dagger to confirm this possibility, nothing of

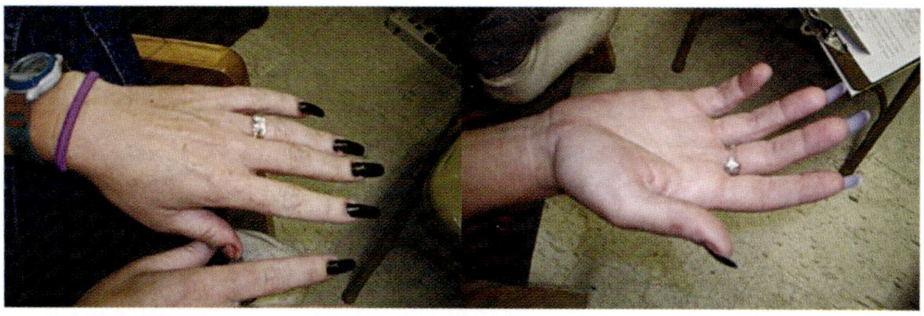

Figure 9.179 The dorsal surfaces and palm surfaces of the suspect's hands.

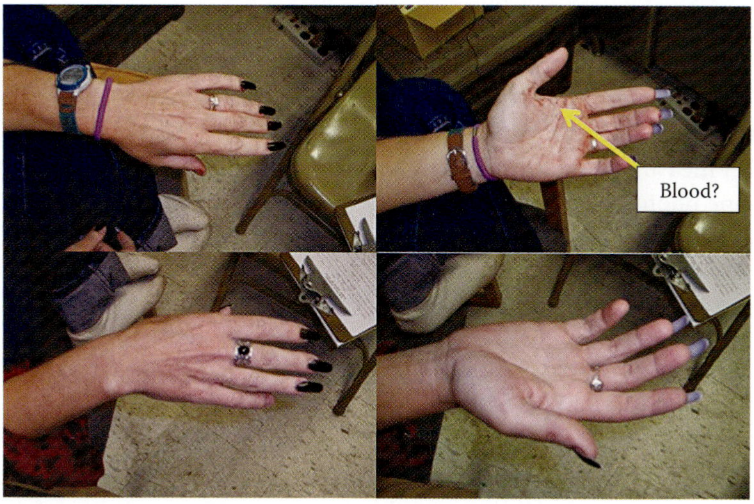

Figure 9.180 The dorsal surfaces and palm surfaces of the suspect's hands.

relevance regarding the act of inserting the dagger into his torso follows. The dagger had to produce blood by being inserted before blood from that injury could be present on the decedent's hand. Therefore, the contact with the dagger, which could be documented by a positive friction ridge match, does not provide evidence about the actions surrounding the insertion of the dagger into the decedent.

Photographs of Shirley's hands taken by detectives immediately after she was taken into custody provided additional data to consider (Figures 9.179 and 9.180). If the results developed thus far are correct, then the additional photos ought to fit consistently with these results. If they do not, then science dictates that we revise and reexamine all the data. If they do, then we have additional supporting data, or neutral data, depending upon the case.

In the photos, Shirley's left hand appears to have bloodstains, but her right hand does not appear to be bloodstained. It is reasonable to ask about the source of blood on her left hand and equally reasonable to ask why no blood appears on her right hand. According to testimony by Detective Larry Kidd, he did not bag Shirley's hands because he did not see blood on her right hand and he believed that the blood on Shirley's left hand resulted from applying pressure to the decedent's chest as instructed by the EMTs while she was on her cell phone.

According to information supplied by her mother, Shirley is right-handed. The inference then is that she would use her right hand to remove the dagger from the decedent's chest, carry it to the truck in her right hand, and put it in the truck again using her right hand. However, if she used her right hand to hold her cell phone, she could have used her left to remove the dagger. Assuming that these inferences are correct, two questions arise: What explains the apparent lack of blood on her right hand?* What explains the apparent blood on her left hand between the thumb and first finger?

From the scientific perspective two problems immediately arise. The first problem is that we do not rely on testimony relating secondhand observations alone to determine the presence or absence of blood on any surface, including a suspect's hands. Much in bloodstain pattern analysis depends on testing to confirm or disconfirm relevant observations. Without such testing, it is difficult to evaluate the testimony of witnesses, other than to accept their statements as true while at the same time being cautious about the scientific weight that they can be used to bear.

The second problem is that we cannot identify the presence of blood through photos alone. No reagents can be applied to photographs to confirm or disconfirm that the photographed surfaces react to presumptive tests for blood. We must even move cautiously when doing bloodstain pattern analysis from photographs, especially photographs that are not close-ups with measuring devices present and with appropriate lighting and contrast. For example, in one photo of the right palm, I inserted an arrow pointing toward an area that I would evaluate with reagents and closer visual inspection if the evidence had been available (Figure 9.180).

There are other issues that arise with respect to these inferences about the lack of blood on Shirley's right palm combined with her right-handedness. The first is that there is not a great deal of blood on the dagger to be transferred to another surface, such as a hand. Depending upon the time interval between the injury and the removal of the dagger, the small amount of blood that I documented on the dagger could have become dry enough not to produce obvious transfer stains through contact. Without the ability to test the right palm for the presence of trace deposits of blood, no scientific conclusions follow from this apparent lack of evidence. The timeworn saying is "the absence of evidence is not the evidence of absence." But be careful not to turn it around: the correct statement remains that the "evidence of absence is *not* absence of evidence."

Another issue concerning these inferences about the lack of blood on Shirley's right palm concerns testimony about the cell phone contact with EMTs. According to reports, Shirley had the cell phone in her right hand while talking with EMTs. If she kept the cell phone to her ear, and if she is right-handed as reported, she would have used her left hand to accomplish the tasks undertaken while she was on the phone. If she handled the dagger in some manner while on the phone, then there is no reason to suppose that she did not continue to use her left hand. Therefore, the lack of blood on Shirley's right palm has no testable scientific significance.

The remaining consideration to be covered in a report of the work on this case concerns the elements of the physical evidence related to Mr. Vaughn's manner of death.

Manner of Death Issues

The case data present what is often described as an equivocal death. I stated this opinion on the stand during my testimony.

* This must be stated as "apparent" since I know of no tests performed to identify blood on the defendant's hands other than a simple visual inspection.

Q. *Attorney:* Forensically, and the evidence you've looked at and everything you've seen, can you say how the knife got into the decedent's chest?
A. *Dr. Nordby:* No.
Q. *Attorney:* In your opinion, based on this forensic evidence, could anybody determine that?
A. *Dr. Nordby:* Not based on an analysis of the physical evidence. The evidence just doesn't speak to that question in this kind of situation. It's not atypical. Just can't get that from here. I can't as a scientist, get there from here.

This opinion is based solely on the physical evidence available to me—the province of the natural sciences. It means, simply, that from the physical evidence alone, no single manner of death immediately overrules any other manner of death (except natural in this case, given the injury). The obvious leading candidates for manner of death in this case are homicide and accident. Within 'accident' one would include an accidental self-infliction or an accidental two-party infliction. A suicidal self-infliction would be extremely difficult to support based upon the physical evidence. Although it is accurate to say that from the physical evidence alone, one cannot rule out self-infliction, the blood evidence and the alcohol factors of the two principals involved in the case better support an accidental or homicidal manner of death. (The description by the medical examiner of the character and physical features of the sharp-force injury suffered by the decedent supports this ambiguity.)

Here homicide of course does not mean murder; it simply means death at the hands of another person. However, from the physical evidence, other than to rule out a natural death, there appears to be nothing to support any one manner—accident, suicide, or homicide—over any other manner. This is what is called an "equivocal death." The official manner could logically be ruled undetermined in this type of case in the absence of compelling evidence to the contrary (such as witnesses, however, even witness observations could remain ambiguous on the relevant issues). Any problems attempting to prove that the injury was suicidal are problems that also apply to proving that the injury was homicidal or accidental. We are without scientific grounds for supporting one manner of death over another.

Given the scene, the autopsy, and the physical evidence along with the circumstantial evidence, the actions of the two intoxicated participants in this case appear to be more an instance underscoring the dangers of struggling with sharp and therefore dangerous objects rather than an instance of deliberate murder. However, it is sufficient to point out that the difficulty of supporting this interpretation of the physical data simply mirrors the difficulty of supporting any other interpretation of that data. That is essentially the meaning of an equivocal death given the ambiguous nature of the physical evidence. That is why I answered no to both questions from the attorney.

When authors write mystery stories, they seldom seem to choose real mysteries. Perhaps that is because real mysteries require long, tedious, and difficult hours of work. That image of long, boring hours just does not play well as visual or transcribed entertainment. If one wants a real mystery, well, here it is. Is this case, a homicide or an accident? All the scientist can do is dig out and present facts and their reliable interpretation. When the interpretations are based upon very little, we have to say so and make it clear. But in that sort of case we often seem to run into the self-assured storyteller, certain of conclusions and oozing with experience. That type of professional "expert" does not need scientific methods that just get in the way of a good story.

The Picture of a Crime

10

> By undue profundity we perplex and enfeeble thought; and it is possible to make even Venus herself vanish from the firmament by a scrutiny too sustained, too concentrated, or too direct.
>
> **Edgar Allan Poe, *The Murders in the Rue Morgue***

Introduction

Chadwick's PM Mart sold everything from salami to comic books. Most customers shopped after 5 P.M., spending their money, whatever its source, on beer, wine, and Thunderbird. The traffic generally slowed after 2 A.M. Chester Chadwick was thinking of closing his store from 3 A.M. to 6 A.M. due to the lack of business. But his clerk's minimum wage could be offset by one or two sales during that shift. So far, he managed to cover the costs of remaining open. But one customer at 4 A.M. on a Sunday morning in October changed his mind.

Clerk Reed Wenzel looked up from behind his counter to see a short, skinny customer enter and walk straight toward the refrigerators in the back of the store. Wenzel could see his movements in the convex mirror mounted in the corner between the refrigerators on the back wall and a bank of freezers on the sidewall. No other customers perused the Mart's merchandise. The small man, dressed in dark clothing, then walked up the next aisle toward Wenzel's counter, his hand in his coat's right pocket.

The knife was not big, as knives go, but Reed Wenzel did not take chances. When the little man walked behind the counter, the knife aimed in his direction, he immediately gave up the cash register's contents and watched the man leave the store. The police arrived in minutes. Fortunately, a surveillance video camera with a fish-eye lens, hidden in the wall behind the cash register, recorded the entire robbery. That the robbery was caught on film was a bonus for Chadwick, since he had installed the camera primarily to watch over the clerks' interactions with the cash register (Figures 10.1 and 10.2).

The problem with surveillance videos remains to be their notoriously poor quality (Figure 10.3). This video was no different. Nonetheless, within 3 days, police had arrested the robber. At least that is what they said. Wenzel identified Richard Barry in a lineup, the prosecutor proffered charges, and a jury of Barry's peers convicted him of first-degree armed robbery. As a habitual petty crook, he occupied a small cell in the county jail, sentenced to a stretch of 5-to-10. But this offense upped the ante. Hearings were underway to decide if this was his third strike. If so, he would be sentenced to life without parole.

His public defender called Final Analysis Forensics. He asked me to examine the video to see if it showed anything even remotely relevant to the case. I remained skeptical given the traditionally poor quality of convenience store surveillance videos. But if one does not look, one does not learn. The natural sciences, after all, are based upon observations.

Figure 10.1 The camera lens, hidden in the wall.

Figure 10.2 This view is from behind the counter. The video camera is on the wall to the right.

Figure 10.3 The video camera builds distortion into the lens in order to capture a wider field of view.

Issue and Focus

Not all reconstructions have the same form; nor do they exactly apply the same methods. The quality of the data often helps dictate the method (see Figure 10.4). Sometimes the methods must be developed and applied for a specific, limited purpose. Not all such reconstructions involve a death or a gun. And not all reconstructions involve crimes actually using a knife or a dagger to draw blood from a victim. This case, as it turned out, centered on the stature of the perpetrator pictured in the surveillance video taken during the robbery of Chadwick's PM Mart (Figures 10.5 and 10.6). Given the perpetrator's apparently diminutive stature, a question logically arises concerning the actual innocence or guilt of Richard Barry. In fact, in contrast with the perpetrator pictured in the video, Barry lacks a similar diminutive stature.

Discovering the uniqueness of a tool, such as a gun's barrel, for example, helps to identify any bullets fired from that specific gun. The identification is strongest when the uniqueness is dramatic. These so-called *accidental characteristics* involve the marks, chips, scoring, and accumulated damage that an object acquires over time, use, and abuse. They are distinguished from so-called *class characteristics* shared by all objects of a given type. For example, all Smith & Wesson 44 cal Model 29 revolvers share the class characteristic of right-twist barrel rifling of the lands and grooves. However, not all Smith & Wesson 44 cal Model 29 revolvers share the marks made by little Tommy when he jabbed a rat-tail file into daddy's revolver and got it stuck.

Unmistakable accidental characteristics prove to be one useful feature of an object that can establish uniqueness. Human stature, while a class characteristic and not an accidental characteristic, does function to eliminate some groups of humans in favor of some other

Figure 10.4 This image came from the same video camera. I asked the clerk to stand in his usual position when he was working on the night the store was robbed.

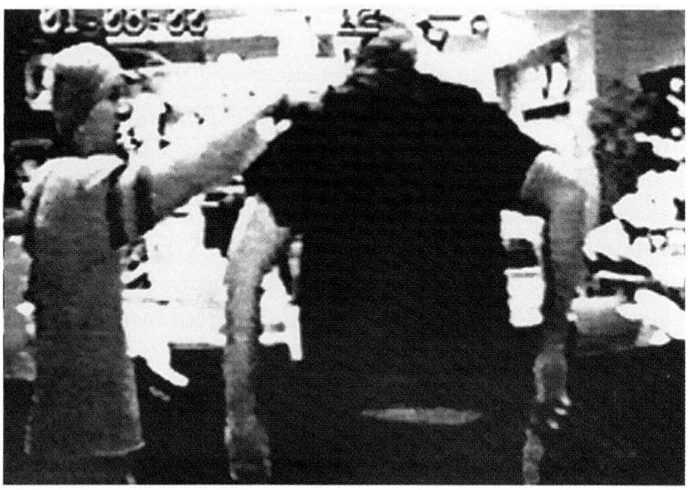

Figure 10.5 The perpetrator approached the clerk to his left with a knife in his right hand. The images are poor, to say the least.

Figure 10.6 The clerk turns toward the robber and attempts to keep the knife off his neck.

groups. If the suspect and the perpetrator can be shown to belong to different classifications of humans according to height, then we have sufficient data to eliminate that suspect as the perpetrator. That is similar to discovering that the bullet killing little Tommy, for example, was fired from a 1911-A Colt 45 cal pistol and not from a Smith & Wesson 44 cal revolver. That is strong evidence; in fact, stronger than the potential quibbles generated by the details of a comparison among members of the same class.

Consequently, the focus of my scientific efforts became gleaning as much objective data as possible about the height of the perpetrator from the videotape of the robbery, then comparing that data with known measurements of Barry's height. On the surface, this would seem to be a relatively simple matter. In the CBS television series *CSI*, agents would just zoom in on the video, import some sort of techno ruler, and simply record the data.* However, the fantasy of television technology has not yet made its way into real-life forensic practice. The images on the video of the robbery do not contain sufficient pixels to zoom anywhere without turning the already grainy image into a blurry maze. Nor is there sufficient data to make such measurements easily. However, within a suitably refined scientific context, with specific limitations in mind, they can supply useful information.

The present challenge, then, involves the attempt to validly compare the stature of the robber with the stature of the suspect (see Figure 10.6). Identification becomes a much more straightforward task to answer the obvious, refined question: Does the height of the robber match the height of the convict? If there is a dramatic difference in height, the answer can be supported. If they are similar, or even identical, then all bets are off—an identification of one individual of height x as the individual in an image of height x does not work because the members of the class are just too numerous. That demands a precision and an accuracy that the data does not support.

Scientific Precision and Scientific Accuracy

Precision, in scientific terms, concerns how much a result varies from one result to the next: the smaller the variation, the greater the precision. On the other hand, in the natural

* They might just as well create a new tool to extract DNA from the video as well.

sciences, *accuracy* concerns how closely the results come to some known "true" result: the closer the results are to a known true result, the greater the accuracy.

The Videotape Data

Surveillance cameras of the type installed at Chadwick's PM Mart have tiny apertures with a spherical lens in order to capture the widest possible field of view through a pinhole-size lens. However, in optics, the price paid for a small, inconspicuous lens with a wide field of view is "depth of field." In practical terms, this means that distortions of size and even shape occur in objects appearing away from the optical center of the lens. If I knew the exact properties of the lens, I would be able to calculate the distortions involved. However, I was unable to gain access to the camera or to any data describing its properties. Chadwick refused to allow the public defender access to his camera. Appeals through motions to the court also failed: the judge ruled that the defense once had the opportunity to examine such data in the trial. Now it was too late. Chadwick could do as he pleased.

Without knowing the exact properties of the surveillance camera lens, it is impossible to derive mathematically accurate and precise calculations of the relative heights or sizes of objects seen in the video (see Figure 10.7). However, the following observations remain helpful nonetheless:

1. In this wide-angle video lens (fish-eye), distortion appears most severely on both the far left and far right sides of the visual field, as well as on the top and bottom. This can be appreciated by imagining half of a clear glass ball, with the center portion of the ball closest to the eye with the ball curving away from the eye toward the circumference. Images seen through the center will exhibit less distortion than images seen further away from the center, given their greater distance from the "flat" properties incorporated into the lens.

Figure 10.7 The clerk drops his arm, still facing the robber as he prepares to hand over the cash.

2. Unlike our example, the surveillance lens is not ball shaped but designed so that the field of view is wider than it is high, therefore limiting distortion from top to bottom while exhibiting distortion from side to side.
3. With this type of lens, objects appearing on the far left or the far right appear to be both farther back and smaller than they actually are in reality. Top and bottom images appear less severely distorted, but this is due to the size of the lens aperture as noted in #2.
4. Objects captured nearest the center of the field are depicted more accurately sized than objects toward the periphery of the lens.

Without such optical and mathematical data, mathematical scientific precision cannot be achieved. However, the issues in this case, focusing on the relative sizes of the perpetrator compared with the convicted defendant, are not best resolved by using precise measurements. Instead, we require *reliably accurate relative comparisons* sufficient to answer the question posed to the data. With the conditions for the analysis suitably stated, precision can be developed. In other words, with precision developed in the analysis, different analysts using the same data will derive the same conclusions using the same method.

In the natural sciences, accuracy is often achieved without precision. Precision carries with it a demand for repeated *exactness of measurement* sometimes to the second or third place of the decimal. Lay people often confuse accuracy with precision. This often leads to serious confusion and occasionally leads incautious scientists to import accuracy to their work unwarranted by the question posed to the data. It all depends on the nature of the question posed to the data, and the nature of the data itself.

Suppose that I ask a scientist "how far is it from Seattle to Tacoma?" with the purpose of my question being the need to decide whether to walk, ride a bicycle, take a bus, or charter an airplane. If the answer offered is "about 30 miles, depending" then that answer avoids the problems of making extraneous assumptions about beginning and ending points, and has the accuracy required for my decision concerning transportation. The data offered was inexact, and the answer appropriately reflects that fact, yet the answer remains reliably accurate for my purpose.

If the scientist replies, "it is 34.567 miles," I should rightly question his assumptions. Although his answer appears to be extremely accurate and "scientific," I immediately doubt its accuracy. After all, how does the scientist get such an exact figure? I never specified an exact beginning point or an exact ending point. With this distinction in mind, a method for approaching the videotape of the robbery required to address the question of stature began to take form.

Method: Photogrammetry

Recall from Chapter 1 that photogrammetry involves deriving reliable measurements from photographs.* Since a photo image from this video camera takes objects from the 3D world and projects them on a flat 2D image plane, we lose the depth information, as well as distance perspective. This loss is exacerbated by the distortion built into the fish-eye lens

* *Photogram* is a photograph and *metry* refers to the science of measuring, from the word *metrology*.

Figure 10.8 Video cameras in stores typically operate with wide-angle lenses in order to capture events in a wide field of view—these views often require some corrections to account for distortions among images.

discussed earlier (Figure 10.8).* However, photogrammetry involves the reverse. By knowing the exact position and orientation of the camera in 3D space along with the focal length of the lens, we can regain some 3D information. But we do not have access to the camera and therefore do not have the required information on its focal length.

The choice of photogrammetric technique can depend upon the type of data to be developed. Sometimes, an existing photograph permits the extraction of an important piece of measurement data. Without the lens data, a reliable method must be developed and tested based upon this possibility.

The basis for this method is "relative approximation."† Our concern here is not to compare the exact sizes of the perpetrator and the convicted suspect, Barry, but rather to focuses on their relative sizes. By "relative sizes" we mean their sizes in comparing one with the other. We are not able to derive the perpetrator's exact size from the video, but we are able to derive his or her size relative to Barry's size (Figure 10.9). Although our data lacks exactness and precludes our ability to measure each individual to within a fraction of an inch, it does allow us to compare their statures and to accurately determine their relative sizes. If their sizes differ significantly, the results should be conclusive. If not, then the data do not support an answer to the question.

To this end, the most basic technique becomes to capture images with known sizes by measuring the heights of several subjects, including Barry. These heights are then compared with known heights captured and measured in the center of the surveillance video's field of view. The center of the lens provides the most reliable image with the least distortion.

* The distortion around the margins is the price paid for the wide field of vision from a small lens.

† See my article and book chapter (in a physics treatise) about the nature and use of approximations in the natural sciences: "How Approximations Take Us Away from Theory and Toward the Truth," by Nancy D. Cartwright and Jon Nordby, *Pacific Philosophical Quarterly*, July 1983; as well as *How the Laws of Physics Lie*, written by Nancy Cartwright, with Jon Nordby coauthoring Essay #6, "For Phenomenological Laws," pp. 100–127, Clarendon Press, Oxford University Press, New York, 1983.

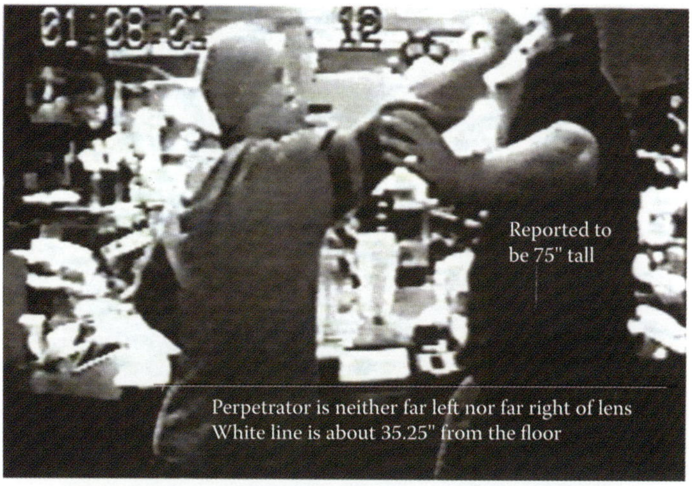

Figure 10.9 The images here are centered in the field-of-view of the lens—known heights are useful but must be carefully considered.

Approximate scaling is then applied to introduce visually similar images into the video environment. Relative heights may then be compared accurately by visual inspection.

When I mentioned relative heights, I was not referring to my nephew, but it turns out that he was about the right size for comparison purposes. I attempted to develop a close-to-life-size image of the robber's stature, again with serious limitations noted. I then posed my nephew in a similar position beside a counter of the same 35¼-inch height (Figure 10.10).

When I discovered that my nephew possessed the qualities that adequately represented the relevant qualities of the robber, I began to make further comparisons. Of course, these comparisons are subject to error, and the error must be considered and reported as it develops. The errors can compound as well. On occasion, this compounding process presents a challenging bit of mathematics when calculating error rates or rates of acceptable variance.

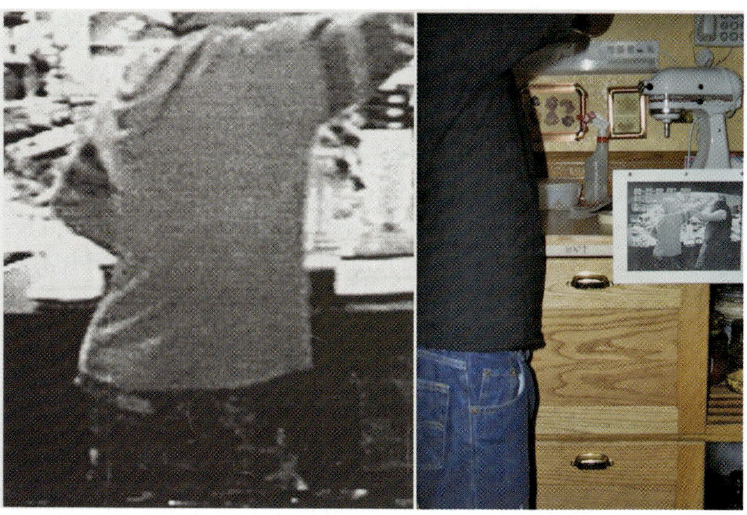

Figure 10.10 I attempted to develop a close-to-life-size image of the robber's stature. I posed my nephew in a similar position beside a counter.

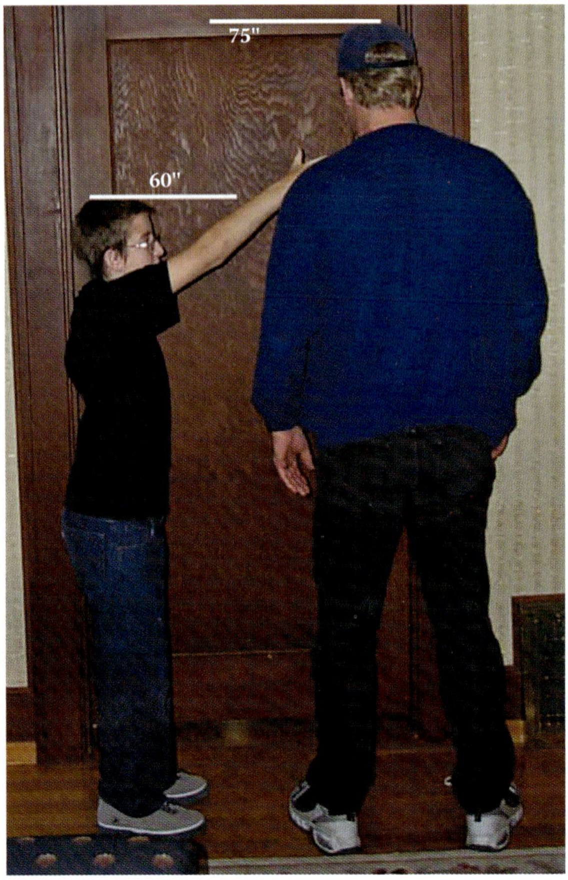

Figure 10.11 The errantly convicted inmate standing next to me for comparison—establishing two known height measurements.

When the error rates reach a certain point, however, the errors are no longer "acceptable variance." At this point the error rates become "unacceptable variance." At that point, a new approach becomes warranted.

For planning processes, I estimate the focal length of the camera capturing the image, and stood next to my nephew, bent slightly forward like the clerk in the video (Figure 10.11). (Perhaps just like the robber, my nephew was also ready to escape.) In spite of this, I captured several images to help judge the accuracy of my work.

Using known heights of these two suspicious characters, I was able to test my calculation methods to help assess the relative heights of the objects pictured in the grainy video.

Error Rates

Error rates (meaning the number of times relative heights vary from known heights) of course apply if we were producing actual numerical height measurements.* Every measurement made in science introduces the concept of error. However, at this simple level, the work product concerns simply a visual comparison of a known with an unknown. It

* Refer to Appendix A for a discussion of error rates, their interpretation, and their development.

is meant to depict a relative comparison of objects in the video environment; in this case, relative comparisons of the perpetrator's and Barry's heights. By applying this method, accuracy is built in by using comparisons of a known height with the unknown height, thereby approximating the relative status of the unknown by visual comparison. In this case, seeing truly is believing only if a scientifically reliable method produces the comparison that we see.

Error, of course, does enter into the picture, literally. To overcome any error and reach a result, the difference between the known and the unknown heights must be greater than any induced measurement errors. In other words, the taller the known height and the shorter the unknown height, the more reliable the comparison becomes. The error rate, then, can be overcome to accommodate the ±5% range that can be calculated for the comparison project.

Photographic Comparisons

Interpretations approximating relative sizes are not impossible, even given the serious mathematical limitations and the number of facts remaining unknown. The photographs presented here show known-sized objects as well as comparisons of known sizes with objects in various shots taken from both the test video and the surveillance video depicting the robbery (see Figure 10.12).

Their purpose is to show various attempts to "fit" known-sized objects scaled into the surveillance video by using measurements of known-sized objects depicted in the videos. These efforts must be taken for what they are: at best, approximations of relative sizes based upon incomplete scaling data. For reasons discussed earlier, exact sizes cannot be determined here. Missing data includes the exact properties of the fish-eye lens used in capturing the video images as well as the role of aperture size. Yet even if such data were available, the method of relative approximation developed and applied in this case is the most conservative, reliable approach: it makes the fewest assumptions and reaches much less specific conclusions regarding measurements involving units of length.

The Best Method Possible versus This Method of Relative Approximation

Of course, the best avenue to settle this matter once and for all would be an "exact conditions test." Such a test would involve taking Barry to Chadwick's PM Mart, and photographing him in as close to the same positions that the perpetrator occupies in the video of the crime, using the identical camera with the identical lens and the same type of videotape. In the absence of that, we must be satisfied with relative size determinations imperfectly captured by other photographs introduced into the video's photo environment. This effort becomes long and arduous, reaching very modest conclusions when compared with the fanciful abilities of the television crime dramas' video analysts. However, a limited conclusion based upon all of these scaling efforts can be derived. The goal of a scientific method applied to case data is not some sense of exactness but some confidence in its reliability.

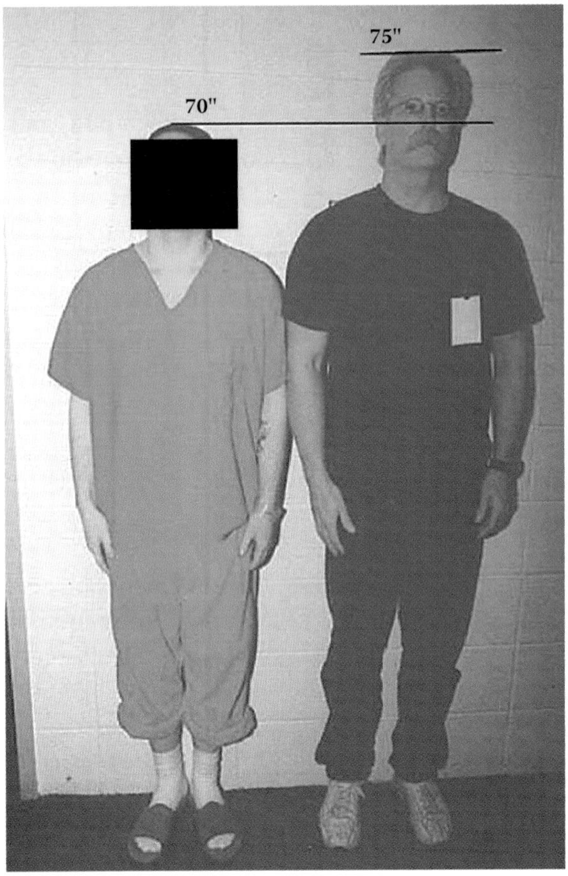

Figure 10.12 The man convicted of the robbery stands in jail to the left. The author stands to the right.

Figure 10.13 With great difficulty and much trepidation, I sized my nephew and put his image in the still of the robber taken with the surveillance camera. He feared that he might be arrested for the crime. I assured him that he was safe. Probably.

Conclusions

As developed through these various measurements and scaling efforts, the perpetrator pictured in the surveillance video is relatively smaller in stature than Barry. The perpetrator is closer to 5 feet than to 6 feet tall, whereas Barry is closer to 6 feet than to 5 feet tall. Therefore, Barry is not pictured in the surveillance video that recorded the perpetrator at the scene. The basis for this conclusion also includes anthropometric calculations, first introduced into forensic anthropology by Alfonse Bertillon. Although Bertillon's work did not represent hard science in his time, it does represent statistical inferences based upon documented average human proportions observed among many known human physiognomies and their principles of development. The conclusion from this additional effort also helps confirm that Barry is significantly taller than the perpetrator pictured in the surveillance video showing the robbery of Chadwick's PM Mart.

Caveats

The photo documents may contain the potential for misunderstanding and any incautious interpretations must be avoided. (For example, I would not like to see my nephew, pictured as an individual closely approximating the perpetrator's height for comparison purposes, arrested for this crime based solely on these experiments!) Suitably understood, when viewed for the purposes they were created—as relative size indicators—they do retain a measurement value but only must be viewed as a relative comparison without units of measurement.

These images are not subject to the abuse of science, which would occur if a transparency of Barry's face were used as an overlay on top of the extremely vague visage of the perpetrator as captured in the video. This could never even remotely be construed as any type of identification. We could take a transparency of my grandmother's face and make it fit the vague visage found on the images of the perpetrator's face. That becomes an extreme example of precision at the expense of accuracy; fantasy in place of science.

Addendum: Bertillon's Anthropometric Calculations for Identification

Bertillon's method involved measuring the size of the face, ears, nose, shoulders, trunk, and other key parts of the body in the effort to develop a single number unique to that individual, much in the way that such a number can be developed from fingerprint examinations. The application of the relative approximation method, however, remains distant from this ambitious effort.

Instead, as a simple check, we can use the statistics developed by Bertillon regarding the relationship among various measurements representing percentages of human height. The distance in males measured from the floor to just above the iliac crest is taken as representing 55% to 60% of an individual's height. (Again, these percentages represent historically interesting statistical claims rather than hard science. One must not put too much onto this data for predicting specific individual heights. This becomes simply another approach to the problem posed by this case.) Like the method of relative approximation, this statistical method also makes no claims for individual precision, yet it may have value for purposes of a comparison between a known and an unknown height.

Basic Calculations

The countertop, visible in the video, actually measures 2 feet 11¼ inches from the floor. The perpetrator's iliac crest, as seen in the video, is approximately even with the countertop's height. If 60% of a male's height (from just above the iliac crest) is measured to the countertop at 2 feet 11¼ inches, then the following approximations can be developed:

$$.60x = 2'11\tfrac{1}{4}''$$

$$.60x = 35.25''$$

$$x = 58.75''$$

$$x = 4'8\tfrac{3}{4}''$$

If we use the statistical derivation of 55% as the percentage of a male's height (from just above the iliac crest), and this height appears to be even with the countertop measured at 2 feet 11¼ inches, then the following approximations can be developed:

$$.55x = 2'11\tfrac{1}{4}''$$

$$.55x = 35.25''$$

$$x = 64.09''$$

$$x = 5'4''$$

Based upon these approximations, the perpetrator's height is closer to 5 feet than to 6 feet.

Based upon the measurements of Barry's height, his is closer to 6 feet than to 5 feet.

Based upon the difference in height using either anthropological statistic, the perpetrator cannot be Barry.

Again, such updated anthropometric analyses as developed by Bertillon are useful only if we acknowledge a great individual variation among these percentages and abandon the grander scheme of deriving an individual number for each person on earth. Nonetheless, the percentages of stature estimations can be useful in providing ranges of values for comparison purposes among homogeneous members of any given population.

The Data Applied to This Analysis

- One black-and-white videotape of the robbery taken by a security camera, provided by Chadwick's PM Mart
- One enhanced version of the security camera video
- Booking photos of Richard Barry
- Four still photos taken from the enhanced tape
- Scene measurements and five photos of Chadwick's PM Mart
- Test videotape taken by the security camera at Chadwick's PM Mart

Conclusions

The difference in height between the perpetrator and the convicted suspect became obvious to the court, the prosecution, and the defense. Peer review of my report to the court supported my modest conclusions. The accused received a pardon. The robber remains free. My nephew remains nervous.

Cold Case Crackpots*

11

A fanatic is one who is sure the Almighty would agree with him,
If only the Almighty had all the facts.

Unknown

Once to every man and nation comes the moment to decide
In the strife of Truth with falsehood, the good or evil side.

James Russell Lowell, "The Present Crisis"

Introduction: "Cold Cases" and Forensic Science

This case involves the stabbing homicide of 16-year-old Amelia Wentworth-Hobbs on November 14, 1973. She lived the life of a typical Midwestern teen, both popular and athletic with an increasingly active social life. Her parents were rightfully proud of her academic accomplishments and her moral character. Whenever she went out with friends, she always called home if she missed her school-night curfew of 9 P.M. When 9 o'clock came and went without a phone call, her parents became alarmed. They called all her friends without result. By the early hours of the next morning, they were frantic and called the police. When the officers of the small, rural enclave responded, a county deputy made the chilling discovery. He found her lying behind a large rock between the milking barn and the cow pasture. It looked like she had been dead for at least 4 or 5 hours. Nothing would ever be the same for the Wentworth-Hobbs family. The search for her killer was on.

Hers is considered to be a "cold case" because it remains unsolved and unclosed since her tragic death occurred about four decades ago. Her family lived with the horror of her death and with the knowledge that her murderer remained free. This perversion of justice must never become officially tolerated. Too often, it has been accepted as one of the difficult realities of policing. But things eventually change, and hopefully for the better—better means catching the actual killer, not just putting someone in jail and calling it good enough. That effort takes time.

Such cold cases have recently regained some of their original heat: developments in the natural sciences in general, and in the forensic sciences in particular, have helped extract solutions from once-insolvable cases. New technology, such as DNA typing, digital photographic methods, computers, photonics, and many additional tools in the trace evidence arsenal have contributed to the solution of cold crimes when such tools are applied

* A "crackpot" in this sense is an eccentric person, especially one with bizarre ideas: a person whose ideas are foolish, harebrained, or otherwise devoid of reason and rational support. Most conspiracy theorists exhibit these characteristics, whereas others embrace reasoning by committing every sort of documented logical fallacy from Aristotle to the present.

appropriately to properly preserved evidence. Contemporary television dramas also lead the public to believe that such cold cases are solvable with a little leg work combined with luck and dedicated investigators. Other crime dramas uncover the guilty and exonerate the innocent in large part due to some advance in the natural sciences. On television, evidence provides the proof of guilt beyond a reasonable doubt satisfying all legal requirements. It makes a good story.

Real cold cases demand that the best and most careful science possible be applied to all the available data. However, in this homicide case from 1973, all of the physical evidence except the scene photos, autopsy photographs, and a knife was destroyed. Therefore, none of the newly developed, more advanced scientific testing is possible. No DNA work can be performed on any of the bloodstains recovered on clothing, or on any of the apparent bloodstains collected from the wood, rocks, or dirt.

In fact, from photographs alone, we have no *scientific* reason to conclude that these visible stains are, in fact, blood, let alone human blood from a particular individual.* Perhaps the murderer cut himself and dripped blood at the scene. Or perhaps a drip trail suggests some location for the initial sharp-force injury. If we had the evidence, it could be tested. Perhaps DNA profiles could be developed. However, this is not possible in the absence of such data. The physical evidence from the homicide case of 16-year-old Amelia Wentworth-Hobbs was destroyed by Napier Valley Police about 30 years after the crime.

The destruction of the physical evidence obviously hampers the application of all the forensic sciences—there are simply no data to work with. We have only the scene and autopsy photographs and, hopefully, some tissue samples retained in perpetuity by the medical examiner. What remains for scientific consideration, then, are the photographs taken of the scene and at the autopsy, as well as the reports generated by investigators and the medical examiner. The development of evidence then appears to turn on any contemporary developments in photoanalysis techniques that help us reconstruct the homicide data sufficiently to identify the perpetrator or at least to lead us in the proper direction. This crime scene reconstruction must then be scientifically robust enough to provide proof in a criminal case and honest enough to note its significant shortcomings. This requires applying the best science to the remaining available photographic data.

Nothing can derail a challenging cold case faster than trying to solve it by turning a pseudoscience into a science or, as in this case, by turning a potentially useful technique for the discovery of evidence into a method for its interpretation and analysis. The waste of misplaced time and misapplied effort destroys any future opportunity for a productive shot at solving the case and convicting the perpetrator with scientifically directed efforts that withstand judicial as well as forensic scrutiny. In the court of public opinion, giving pseudoscience an official stamp of approval leads to the mistaken belief that the hocus-pocus machinations of a pseudoscientist represents the work done by legitimate crime laboratories and reputable independent forensic scientists who actually work in scientific fields.

Tracking, as a discipline of any sort, has supplied no current or past developments in photoanalysis, photogrammetry, or footwear impression evidence. In short, it is not scientific as it is applied in this case. Nor does the report titled "Forensic Photo Analysis"

* People have a hard time with this point, but we cannot test for human blood or test stains that appear on photographs for DNA. It is just a picture, not actual blood. The inference may be sound, with highly probable premises, valid form, and even true conclusions, but it must remain an inference. No physical tests can confirm the truth of the conclusion.

apply any principles of analysis. The report simply lists the author's conclusions, essentially based solely upon his authority and experience. No reasoning is forthcoming. No principles and no basis for the conclusions are developed; no work is shown. It remains, simply, nonscience. Recall that this does not logically entail that all the statements or all the conclusions are false: It means that true or false, the conclusions have no support developed as reasons establishing their truth. It is hard to reason with a position that rejects reason at the outset.*

Science versus Nonscience

The basic properties of scientific work, described in the "Introduction" at the beginning of this book, include the methodological assurance of reliability. A tracker's conclusions, reached from examining decade-old photographs, even if they turned out to be true conclusions, reveal no reliable method that licensed them. The "Introduction" mentioned the noted psychic and clairvoyant Edgar Cayce who made many startling predictions while in a sleep-induced trance. Many, if not most, turned out to be true. We may marvel at such an ability to "see things others fail to see," but this approach cannot be characterized as 'scientific' regardless of the truth Cayce's predictions display. Not all such predictions based upon unique skills at "seeing" turn out to be true. Consider the often-told historical tale in physics of the rise and fall of N-rays allegedly discovered by French physicist René Blondlot (Figure 11.1).

Blondlot worked as professor of physics at the University of Nancy in France. He did not appear to be an obvious crank. He received scientific awards and enjoyed membership in the Academy of Sciences. X-rays, recently discovered by Wilhelm Conrad Röntgen, a German physicist, who both detected and produced electromagnetic radiation in a wavelength that we now call X-rays. They were a popular subject for further investigation by

Figure 11.1 Professor René Blondlot (left) and Wilhelm Conrad Röntgen (right).

* Of course it does not take long to note that such a position is inconsistent or contradictory at the outset. Presumably they offer some reason to support their rejection of reason, in some form or other, that is, at best circular, at worst contradictory. The same argument forms apply to positions such as moral relativism and other positions that trivialize or marginalize logic and human reason.

many physicists of the time. Blondlot joined in that investigation. He tried to polarize X-rays and as a result, in 1903, he announced the discovery of a new invisible radiation, which he called N-rays.

To visualize these N-rays, he developed aluminum prisms and lenses, which he claimed "focused the N-rays." When moving a fluorescent thread across these focused rays, he said that he detected and displayed their spectrum.

The American spectroscopist Robert Wood* attempted to perform Blondlot's N-ray experiments but could neither detect nor display any spectra. In response, Blondlot performed his experiments while Wood observed. In one experiment, Blondlot directed his N-rays onto an electric spark. He claimed that he saw the spark's brightness increase as a result. When Wood stated that he could not see any increase in brightness, he was told "that his eyes had not been properly sensitized."

Blondlot showed Wood photographic plates that he claimed captured the increased brightness of the spark, but Wood remained unable to see the distinction. Blondlot then demonstrated the N-ray's refraction by the aluminum prism through a 3 mm cardboard slit. Blondlot claimed that he detected a 0.1 mm change in intensity at the N-ray's focal plane. Skeptical of this, Wood secretly modified the test apparatus: he removed Blondlot's aluminum prism from the spectrometer. Despite the missing prism, Blondlot continued to describe the N-ray spectrum.† His special ability to see N-rays, an ability unique to Blondlot, is a form of expectation-laden observation: seeing what one expects to see instead of seeing what is there. The experimentally discredited N-rays soon vanished from physics altogether.

The use of tracking from decades-old photographs to reconstruct the movement of victim and perpetrator at a cold-case crime scene is said to require the same type of seeing purportedly possessed by Blondlot. This tracker, the court was told, can see things that ordinary folks like scientists or lawyers (including judges) cannot see. Mind you, the tracker had not just the ability to interpret things that others fail to interpret, but actually to see things that others will never be able to see. Quite an auspicious claim.

The public defender asked for a review of the tracker's report and requested that I divine, if possible, the secrets of this Casey-like ability to see details, motives, and moods anew from photographs of an ancient crime scene. This variation on traditional tracking obviously is a marketable skill that has become in tremendous demand by both sheriffs' offices and prosecutors' offices. I guess they have never heard of Blondlot and N-rays. Maybe they would be in the market for an N-ray generator. I could build one in no time at all.

Case Review

This homicide case applies the enterprise of "tracking," also called by its practitioner (depending upon the purposes at hand) "forensic photo examination," since photos remain the only evidence in this homicide case from 1973.

* Robert Wood wrote the classic text *Physical Optics*. He discovered atomic fluorescence as a property of all atoms.

† Wood published the results in *Nature* (volume 70, p. 530, September 29, 1904).

Specifically, I was asked by the court to accomplish the following tasks: (1) to discuss the scientific standing of this tracking work as forensic photoanalysis presented as a form of crime scene reconstruction; (2) to discuss any scientific contributions that bloodstain pattern analysis might provide to this case; and (3) to identify the potential contributions of contemporary developments in the forensic sciences to the physical evidence in this case if that physical evidence from 1973 were somehow available today.

Summary of Basic Conclusions

The forensic photoanalysis provided through the tracker report had no scientific standing and no cognitive status above unsupported, speculative opinion. The descriptions of movement and motive claimed to be visible in the photographs are simply fanciful storytelling. These pronouncements have the cognitive status of any claim advanced as "true because I say so, and because I am smarter, and because I have unique abilities to see things that no one else can see."

I also concluded in my report that despite the destruction of the physical evidence, some potentially useful scientific work may yet be performed on the photographic evidence. However, I pointed out that such work has not yet been produced—certainly this work was not produced by the tracker and certainly not by applying tracking as any sort of disciplined approach to photoanalysis. Because the physical evidence was destroyed, any bloodstain pattern analysis has extremely limited value since we cannot identify photographed stains as blood belonging to a specific individual, as human blood, or even as blood of any type. Nor can we scientifically determine that any blood associates with any particular event in time. Worse, none of the stains were photographed at 90 degrees and none had any useful measurement scales in the photos; scales would allow image rectifying.

Shoe Print Evidence

The tracker's methods remain unstated in his report. They supposedly involve just "seeing." However, tracking or reading "signs" in order to follow a trail in nature is a reasonably well understood and certainly a very useful activity.* The addition of canine support in the form of cadaver dogs and other search animals plays a central role in many death investigations in which searches are conducted.† The objective remains the discovery of victims, tracking suspects, or finding potential evidence. It does not involve the analysis or interpretation of that evidence once it is discovered. The value of tracking is measured by the success in finding the object of the search. Once it is found, tracking ceases.

Inferences made by reading signs along the trail are tested and confirmed, or tested and refuted, by their success or failure in reaching the next sign, and so on. Yet, the world of

* Tracking, when practiced according to its principles, of course remains a worthy effort with which I find no fault. The difficulty arises when the principles of tracking are applied to areas clearly requiring other skills that also are outside the area covered by the practice.

† Cadaver dogs were used to help locate human remains at the World Trade Center site in 2001. They provided delightful company and received a lot of love from all of us working in the autopsy area.

pine needles and rocks differs from the world of silver emulsion and pixels. Confirmations of observations made in the wild do not occur through observing photographs alone. Nor do trackers commonly follow photographic signs to the exclusion of signs visible in the natural environment. Although undoubtedly useful for the observation and recognition of potential evidence, tracking itself is not a discipline equipped to analyze or interpret such observations by applying the scientific method. Thus, tracking as a practice of potential discovery does not qualify as a branch of evidence interpretation or photoanalysis.

The tracker's report for this 1973 murder investigation presented statements that appear to constitute some type of a tracking crime scene reconstruction developed exclusively by studying photographs, questioning witnesses, and reading reports. In that respect, the subject matter appears similar to the data that are relevant for any crime scene reconstruction project. But the data relevant to tracking is found in the wild, not in photographs. The statements made in the tracker's report fall far short of any scientific standards for crime scene reconstruction established as a point of law through various court decisions: by Federal Rule 702,* by *Daubert*,† by *Kumho Tire*,‡ or even by the older *Frye*§ decision. As such, no notion of method is addressed even when considering tracking as a potentially relevant practice and even when considering tracking as a nonscientific practice that falls under the *Kumho Tire* ruling.

These unsupported statements assume at least some apparent expertise in footwear impression evidence and in photoanalysis as a form of crime scene reconstruction. While my experience includes several areas among the forensic sciences and forensic medicine, producing a scientifically reliable crime scene reconstruction also demands some expertise in photography, photoanalysis, and photogrammetry. Producing such a reconstruction relies (often exclusively) on photographic data to initiate and define the project. To present this issue here, consider more general scientific approaches to impression evidence provided by photographs. This consideration takes us to an assessment of the tracker's report forensic photoanalysis sections, and a distinction between his project and the task of developing a reliable scientific event reconstruction.

* Rule 702 covers the admissibility of scientific testimony in courts of law. "If scientific, technical, or other specialized knowledge will assist the trier of fact to understand the evidence or to determine a fact in issue, a witness qualified as an expert by knowledge, skill, experience, training, or education, may testify thereto in the form of an opinion or otherwise, if (1) the testimony is based upon sufficient facts or data, (2) the testimony is the product of reliable principles and methods, and (3) the witness has applied the principles and methods reliably to the facts of the case." (As amended April 17, 2000, effective December 1, 2000.)

† The *Daubert v. Merrell Dow Pharmaceuticals*, 509 U.S. 579 (1993), ruling establishes the admissibility standard for scientific "expert" testimony. Applying Rule 702, the decision ruled that the judge is the gatekeeper for determining what counts as scientific knowledge, for determining relevance and reliability, and for deciding what counts as "scientific knowledge." Scientific knowledge is defined as any statement that (1) is falsifiable, (2) subject to peer review, (3) has a known or potential error rate, (4) maintains known standards and controls in testing and experiment, and (5) remains generally accepted by the relevant scientific communities.

‡ *Kumho Tire Co. v. Carmichael*, 526 U.S. 137 (1999), is a United States Supreme Court case that applied the Daubert standard to expert testimony from nonscientists.

§ To meet the *Frye* standard, any scientific evidence presented to the court must be "generally accepted" by the relevant practitioners in the scientific community. This criterion of general acceptance applies to the techniques, principles, or procedures used to develop evidence for the case.

Necessary Preliminaries for Photoanalysis

Tracking from photographs recalls an episode from *The Andy Griffith Show* involving Deputy Barney Fife and "Old Blue," the wonder dog who has joined the Mayberry Sheriff's Department as Barney's new tracking dog. During the hunt for an escaped fugitive, Barney shows Old Blue a wanted poster with a picture of the wanted escapee; the dog looks, sniffs, and then Barney turns the dog loose to track the suspect. That is funny in a television episode. But it is not funny in a murder case. The present case remains relevantly similar: we recognize that sticks versus photos of sticks, dirt versus photos of dirt, suspects versus photos of suspects contrast things that differ in kind.* Tracking, it seems at the very least, requires the actual sticks and the real dirt, not merely their pictures. "Reading sign" in nature requires different skills than reading images in photographs. Of course, reading images in photographs also requires different skills than reading sign in nature.

The tracker does not discuss anything about his method in the report that he presents to the court. This suggests a limited, nonscientific approach. It matters little if the report originates from a state crime laboratory or from an independent "expert."† He appears not to follow any scientifically accepted protocols for either analogue or digital photographic evidence. Such protocols have been developed by the Scientific Working Group on Image Technology (SWGIT) and the Scientific Working Group on Digital Evidence (SWGDE) as applied to traditional darkroom or to digital enhancement techniques. Such techniques could be performed upon the photographic evidence in order to make the images from 1973 clearer, especially the images alleged to show footwear impressions.‡

After viewing scene photos with improved resolution, the tracker states in his follow-up report that the improved image, visible in the new photos, "affords different lighting." This is not, of course, technically correct. The comment reveals an unfamiliarity with photo processing and photoanalysis that appears throughout his various reports. It also betrays him by demonstrating his total ignorance of physics and optics.§ Perhaps the contrast and the brightness of the image are different from the images in the previously developed photos. But there is no different lighting. Whatever lighting and camera setting was actually used to capture the photos at the time they were taken remains unchanged.

One way to improve the existing images even more would be to have the film negatives "drum scanned" to develop a very large digital file. These photos were redone in 2008 and could have been drum scanned, but perhaps no one working on the case was aware of this possibility. The result would bring out much more detail, and the digitized images could

* Attributed to an unnamed source, a child on a tour of the FBI laboratory in Washington DC saw some wanted posters. When their purposes were explained, the child wisely replied, "Well why didn't you just keep them when you took their pictures?" Why indeed.

† Those "practitioners" who call themselves "expert witnesses" apparently do not understand that one is an expert only if the court in which one is to testify qualifies one as such. Otherwise, there is no such thing as an expert witness outside of a specific courtroom in a specific case.

‡ Briefly, digital techniques are acceptable provided that they function within the boundaries of the old-time darkroom techniques for film processing used to improve an image by manipulating features such as contrast, brightness, size, and so on.

§ The classic text is *Principles of Optics: Electromagnetic Theory of Propagation, Interference and Diffraction of Light* (7th ed.), Max Born and Emil Wolf, Cambridge University Press, 1999. This book is a classic 20th century work and undoubtedly the most influential book in optics published in the last 40 years.

have been processed with any modern photo-editing software, such as Photoshop. The tracker did not do this enhancement.

Producing enhanced images* from scanned copies of original crime scene photos does not create data since it functions like a traditional darkroom technique. For any work on these photographs, I would have used image-enhancement techniques with traditional darkroom counterparts. These are limited to and include adjustments of brightness and contrast, color balance, cropping, dodging, and burning. According to SWGIT, "documentation of enhancement steps is not necessary when using traditional darkroom techniques."

Despite the fact that such documentation of the traditional darkroom techniques remains unnecessary according to SWGIT, our protocols at Final Analysis Forensics include listing all traditional darkroom techniques applied to these images to foster the aforementioned features. The steps that could be taken to improve these images by using traditional darkroom techniques and photo-editing software are limited to:

- Copying a TIFF file; saving AS a copy in a JPG format file
- Cropping
- Adjustments of brightness and contrast
- Adjustments of color balance

However, this is baseline photoanalysis for such images containing potential pattern evidence. Other techniques are available in the photoanalysis tool bag. None appear in the tracker's forensic photoanalysis. Photoshop now keeps a history of how an image is processed, allowing scientists checking the work product to repeat the same steps taken to develop the images.

Photogrammetry Again (Measurements from Photos)

To produce measurements from photos, (certainly a relevant issue for doing a forensic photoanalysis of footwear impression evidence at a crime scene), certain data must be gathered. The following questions must be asked and answered to derive traditional measurements:

1. What kind of camera took this photo?
2. What kind of lens was used on this camera?
3. What is the focal length of the camera lens for each respective photo?
4. Does there appear to be any lens distortion in the images?
5. How does this lens distortion affect one's measurements?
6. How far is the camera from the object in the photo?
7. Since the ground is uneven in several places in the photos, how does one account for the distortion resulting from this variation in perspective?
8. Has the effect of perspective in the images been corrected?
9. How was the perspective corrected without the use of a right-angle scale or other reference grid?

* SWGIT provides a helpful discussion of image improvement techniques and their application to photographic evidence.

Figure 11.2 Hermann Rorschach, circa 1921.

The tracker neither raises nor answers any such questions in his reports. The other option would be to explain why such questions may not be answerable given the data. He does not do that either. Nor does he present any discussion of these points when asked about them in his courtroom testimony. For this reason, his impressions involving the relative size or placement of purported impression evidence in the photographs remain subjective and unsupported. Many impressions of footprints reported to be over other footprints in fact remain invisible even to practiced eyes.* Similar limitations appear with his efforts at pattern matching, which he bases upon mere subjective assertion rather than upon accepted measurement techniques, 1-to-1 functions, or some other isomorphism developed through the proper manipulation of the photographic data. The tracker's reports read like the ramblings of a psychologist's patient interpreting inkblots while taking the Rorschach test.† That is the way the testimony sounded as well. At one point while on the stand during the *Frye* hearing, the tracker was describing some invisible shape hiding in a photograph and said, "You see that, don't you judge?" The judge replied, "No, I don't see anything." It reminded me of a college psychology course: specifically Rorschach's inkblots and the stories they generated among his patients (Figure 11.2).

Of course, pattern matching and the interpretation of footwear impressions differ in kind from subjective reactions to shapes based upon needs, goals, and fears. But the tracker never offers any such account. In fact, he offers no account at all. Whatever one lists as the inkblot approach's intellectual defects, at least Rorschach was correct to claim that such interpretations reveal a person's thinking processes. The tracker's thinking process, revealed by his interpretations of shapes on photographic paper, proved to be a very non-scientific process at that, but were nonetheless quite revealing.

* Bill Bodziak and I both worked on this case. Bodziak's expertise, among other things, lies in footwear impression evidence. The tracker consistently referred to footprints in his testimony. Footprints are very different from footwear impressions: the former require bare feet, the latter require shoes of some sort. Presumably precision in terminology does not constitute any part of the tracking discipline as practiced by this witness.

† Hermann Rorschach's book *Psychodiagnostik* was published in 1927. He developed his so-called inkblot test, which was used for any number of diagnostic and treatment applications. It even found use as a tool for studying imagination and consciousness.

Pattern Matching and Analysis

In order to do pattern matching from photos (certainly a relevant issue for doing a forensic photoanalysis of footwear impression evidence at a crime scene), certain data must be gathered. The following questions must at least be asked and, hopefully, answered:

1. Do you have any traces or outlines of the boot patterns in the photos for physical comparison?
2. Have you matched the traces in the soil or on the grass to the boot patterns?
3. How was this matching of traces in the soil or on the grass accomplished?
4. Did you measure the spacing between the treads in the prints?
5. Did you measure the curvature of the tread near the outer edges of the sole?
6. How do these compare to the photo of the known boots belonging to the suspect?

In the scene photos, the police officers appear to be standing right next to or even over the impression evidence. How can the tracker distinguish between the prints made by one flat sole from those made by some other flat sole when they appear only in a photograph? His efforts at pattern matching are again mere subjective assertion rather than well-accepted measurement, 1-to-1 functions, or an isomorphism developed through proper photoanalysis.

Just how these shoe/boot prints are measured or correlated to a suspect's boots is of obvious interest (Figure 11.3). Even with the digital data as it is presented on the improved resolution CDs, there is a way to get some quantitative measurements. To do so, for example, trace the impressions using photo-editing software and then try to match to the photo as traced with the boots as photographed. This does require scaling, however.

Alternatively, I would use a computer-aided design (CAD) package to measure some of the tread angles and patterns and then do the same with the photo of the boots. I first applied the process to various types of injuries. But I found that it works well for tire tread impression evidence as well as for developing footwear impression data.

Figure 11.3 These do not appear to be compelling photographs documenting the search for footwear impressions; the photographer appears to be stepping all over the evidence.

Figure 11.4 This shows the placements of "targets" when photos are taken to be used in 3D image production. (See Chapter 1.)

At the scene, after camera calibration, take photographs using standard photogrammetry targets placed on each side of the tire impression print (Figure 11.4).

The resulting image, after processing with a 3D software package, appears in Figure 11.5. The development of an appropriate image of a suspect tire and the image processing steps taken in the software appear in Figure 11.6.

An image of the known tread can be developed and compared with the tread impression (Figure 11.7). First, class characteristics are determined for both; then any potential individual impressions can be investigated. The comparison can be developed from photographs only, but this is difficult.

Measurements are critical in establishing relationships among individual and class characteristics. Some accounts of natural science claim that if there are no measurements made among the data, then it is doubtful that any scientific work is being done. This notion that mathematics is the language of natural science developed along with the intellectual revolutions accompanying the Renaissance. In the 20th century, logic as the language of mathematics took its proper place in the natural sciences. As shown in Figure 11.8,

Figure 11.5 This is the impression that results.

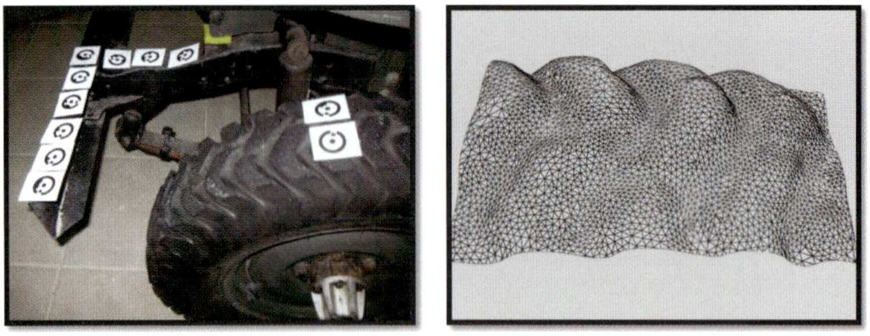

Figure 11.6 Another 3D image developing a "point cloud."

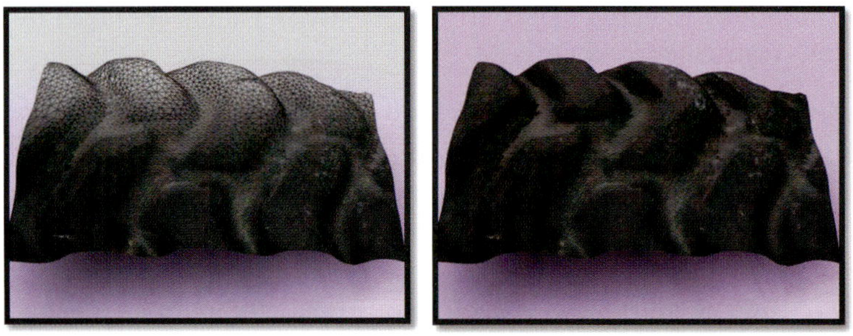

Figure 11.7 The development process reaches 3D results, which can be moved in the software to view the image from different positions.

Figure 11.8 The author developing overlays.

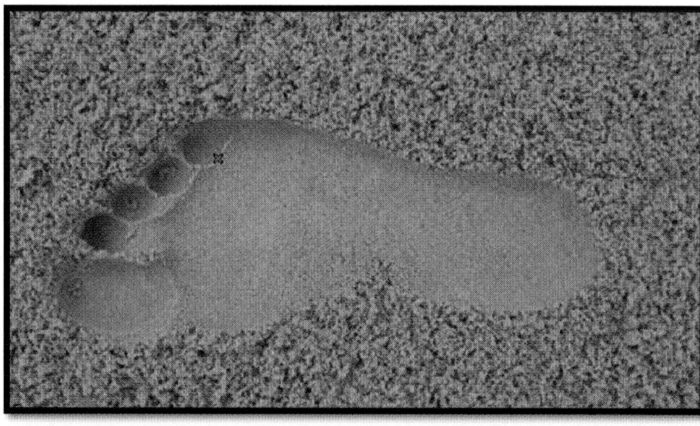

Figure 11.9 A footprint (as distinct from a "footwear") impression.

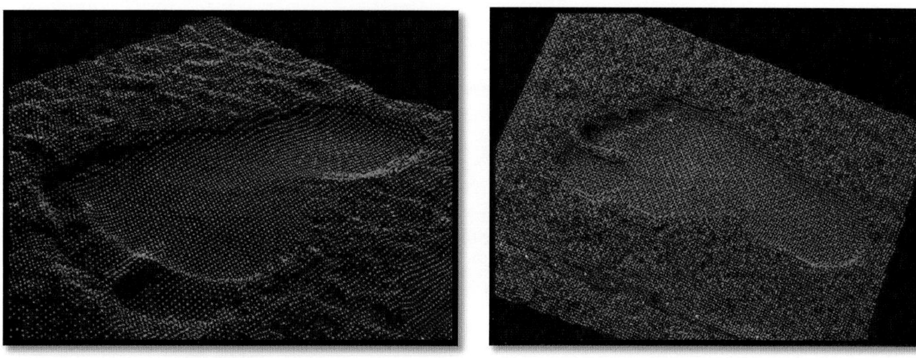

Figure 11.10 Point cloud data.

old-school overlays are often used when we have some known footwear to compare with the unknown footwear impression. Sometimes that process is all that is required to address the facts at issue.

Both overlays and photogrammetry can be used for comparisons between footprint impressions and known feet (Figure 11.9).

From the point cloud, the image is stiched together developing a 3D model useful for measurement and comparison purposes (Figure 11.10). The real benefit of 3D modeling involves the ability to make an infinite amount of measurements from the scale image.

Some police agency crime scene units were sold fish-eye camera systems that can take 360-degree images (well, almost) of the crime scene and allow the user to manipulate the photographs by making observations from the camera's point of view throughout the image. It is just like Disneyland. Although this has some obvious value for recording images and documenting the scene, it has no value for making measurements or developing data. I find these cameras to be expensive toys with limited scientific value when the intent is to preserve the scene so that one can make measurements or manipulate the data to answer questions that invariably arise after the scene is long gone. That is the virtue of scanning the scene rather than simply photographing it with an admittedly cool but scientifically useless tool.

Depending on the software, the 3D image can be developed in several ways; but now we have an image that allows for measurements, much like the overlay in Figure 11.11. Additional steps can develop additional points by filling in the gaps, generating an image.

Figure 11.11 Point cloud data is used to develop measurements.

Figure 11.12 Digital or analog calipers are traditionally valuable tools. All such tools require validation.

The points make measurements possible and to scale. The total station also works on a similar principle.

Regardless of technique, the method remains the same: develop and compare measurements, note physical properties composing class characteristics, and search for those exhibiting individual characteristics (see Figures 11.12 and 11.13). Whether using a scanner or acetate, the principles are identical.

The 1973 crime scene photograph in Figure 11.14 remains to be the clearest image involving footwear impression evidence on the ground. However, the problem remains to be the need to find a 90-degree angle from the lens to the ground—a task that would be

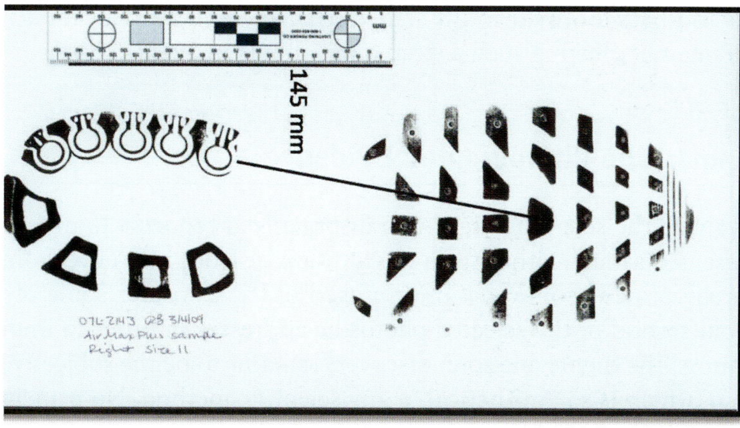

Figure 11.13 Measurements of various spatial relationships: the virtue of the digital approach is the ability to account for the curvature of shoe prints in the dirt and the ability to flatten the images.

Figure 11.14 A footwear impression in the dirt at the scene.

possible by having a known circle in the image. Since the 1980s, the standard American Board of Forensic Odontology's (ABFO)* right-angle scale has been used to perform this function, among others. Alternatively, I would use a CAD package to measure some of the tread angles and patterns, and then do the same with the photo of any suspect's boots. If the relative angles and spacing of the treads were the same, it could give an indication that they may be related. But these are class characteristics and do not function to positively identify the footwear leaving the impressions seen.

The tracker did none of this, making his assertions simple conjectures without scientific support. Virtually nothing in his report is quantitative: no systematic comparisons of shoe sizes among visitors to the crime scene and no effort to document the population of

* The late Thomas Krause, DDS, was instrumental in its development.

so-called stars-and-bars footwear in the state during 1973. (How relevant is it to this case if I said that I remember having such a tread on my boots in 1973?)

Tracker Report: The Photographic Evidence

One way to evaluate the scientific merit of a disorganized report is to dissect the various statements presented as fact, either from observation or from inference from some set of data,* and thereby provide some sort of improved order. Consider some of the tracker's statements about several of the selected photos he addresses in order to underscore their conjectural nature. The theme one soon discovers remains to be the subjective, "only I can see it" approach, which is so antithetical to any scientific method. No principle of charity can fix that defect. It is substantive and pervasive.

From a photo, he states that there was no precipitation the night before, which may be true, but there was precipitation at the time that the photographs were taken, according to the scene log provided by officers. And the weather records report intermittent rain and snow. Comments follow after quotes from his narrative.

"07:33 A light snow began falling."
"07:45 Detective and Deputy began taking pictures of the scene."

Also, how long did it take the deputies to take all of their photos? It could have been half an hour later when some photos were taken. Is it possible, then, that the ground was wet again when the photos would have been taken?

- The tracker states that the "photos show dry sandy soil clinging to the side of the sock and shoe of the right foot. We see clearly many small rocks in this photo that have dry surfaces."

But are some dry and some wet? What is the difference? Was this area protected by the blanket covering the victim? Time is an issue. Any consistent inference from this observation must consider the moisture and its absence: Were the small rocks disturbed after 7:33 A.M.?

- "We see a significant amount of unidentifiable footprints [sic] alongside the body, in the area of the feet of the body, and the footprint trail toward the paved driveway."

So, he sees footprints,† but they are not identifiable. Does this potentially mean they could include "footprints" from the suspect and the victim? As far as the scientific data goes, all footwear impressions discussed remain unidentifiable: they do not point toward any individual, any shoe, or any particular time frame. This fact compromises any probative value attaching to these observations.

* I apply what is known as the principle of charity to any such work. Science depends upon peer reviews for progress. The principle states that one must develop the material by using the clearest, most logically consistent interpretation possible to assess its contribution to our knowledge. This enables the reviewer to focus on the substance of the reasoning, not just noting minor errors that have no affect on the results.

† Obviously, the tracker means shoe prints, since footprints is a term reserved for bare feet. If he were a student of footwear impressions, he would know that. Obviously he is not.

- "In this photo we see several stars-and-bars footprints. These prints tell us that the wearer was not merely walking through the area, but was in fact actively moving his/her feet in this small area."

The question remains: How does he determine this without taking any measurements? What quantitative factors might assist in this interpretation? Through photogrammetry we may be able to develop some supporting data such as stride length or even the depth of various impressions in soil, sand, grass or other surfaces. No such data appears to have been developed.

- "These prints are unaffected by moisture, they are clear and definite in detail and are not affected by any significant period of dry weather."

This is puzzling because the reports state that it is snowing and that the weather appears moist. It is unclear how the tracker distinguishes dry from wet areas on various surfaces given the limitations of simply looking at photographs without applying any enhancement techniques. Even with such techniques, identifying moisture in photographs is difficult at best. Reasons to support the interpretation are, of course, absent. All one can say is "it looks dry." This is apparently the basis of his claim despite facts to the contrary. (Water drops are visible on the outer surface of the plastic bags covering the decedent's hands/arms. Crime scene protocols, even in 1973, recommend using paper bags over a decedent's hands to better preserve the evidence and to prevent condensation from developing inside the bags. We simply do not know when the drops on the outside of the bags originated.)

- "We see the investigators arm and hand holding the measuring tape and a scrape on the ground that appears to be caused by the ball of the foot area of the soft-sole shoe pushing violently sideways against the ground."

Here the term "violently" is hyperbolae, exaggeration, and just downright misleading. Did he do any testing to determine how much displacement or force it takes to make a similar impression in that type of soil? What if the soil was simply soft? One of the impressions appearing near the 2 × 4 frame was relatively deep. This depth reveals that the impression appears in a patch of soft soil, not that the impression reflects violence. Violence seems to suggest that soil is kicked up and thrown. Was there any evidence of this? Again, this illustrates the nonquantitative and subjective nature of his impressions.

- "The light snow fall moisture is apparent on the grass blades in this photo and we note that the footprint sign mentioned is prior to this moisture."

So the ground is, in fact, wet? These claims again fail to cohere without some explanation of the differences: some ground is wet, some ground is not wet, the effects of the trees, and so on. However, no such factual information appears in the reports. In an autopsy photo showing the decedent lying on her back with plastic bags over her arms water appears on the plastic bags. (Paper bags should be used to preserve such evidence including fingernail scrapings and trace evidence from the decedent's hands.)

- "We do see evidence of time frame appropriate ground surface disturbance."

This claim appears to be without any clear meaning: What does he mean? Can he cluster impression evidence and both date and age footwear impressions from the photographs? Without identifying the various possible sources of flat footwear impressions, it is unclear how such a position could be supported. This claim exemplifies statements to be eschewed by the natural sciences. In this sense, the prints (even in the best-case scenario where they are fully individuated) become like fingerprints—they may exist, but we have no way to show when they first appeared.

- "What is significant is that all of the photos that we have examined of this scene show a relatively small area."

This might be true, but this should not imply that relevant footprints were contained or limited to that small area. Did anyone check the suspect's trail from the house to the neighbor's house? Were there any more stars-and-bars pattern prints anywhere else on the property? Were any found on the neighbors' property? Such data are at the very least relevant to these issues. The fact that plaster casts were destroyed along with the rest of the evidence severely limits further progress.

True to the shifting logic of pseudoscience, the prosecutor asked me during my testimony, "Dr. Nordby, you are not a tracker, are you?" Of course, my answer was no. Then came the invariable statement that I knew would follow: "Well, then, the tracker *sees things* that you *can't see*." My reply was easy, but I'm afraid it went over her head: "No. Anyone should be able to *see* it if it is there. What anyone may not be able to *do* when *seeing a sign in nature* is to interpret it." I added my own example. "In medicine, everyone should be able to *see the injury*, but of course not everyone has the expertise to *treat it*. It's not a matter of seeing: it's a matter of interpreting."

For example, in Figure 11.15 everyone can see the image to the left, but not everyone may know how to interpret the shapes. If I provide some hint in the form of the photo on the right, one might make some reasonable guesses. However, we should each be able to see the same thing on the left. If one speaks Russian and English, the image in Figure 11.16 should help.

I next waited for the prosecutor to compare the tracker with Galileo, another favorite move of pseudoscientists. The idea is that their theories are rejected or held in contempt just like Galileo's theories were rejected and held in contempt in his day. But they fail to note that the scientists in Galileo's day, then called natural philosophers, took his theory very seriously. They had reasons provided by Galileo to evaluate and to ponder. In fact, it

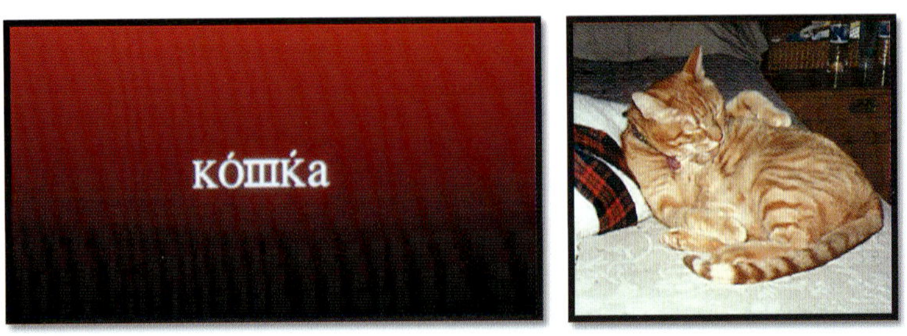

Figure 11.15 The puzzle on the left; the hint on the right.

Figure 11.16 The solution.

was the Catholic Church that rejected Galileo's theory out of hand. But no one in natural philosophy thought that Galileo was a kook. They may have thought that he was mistaken, but he was not some nut advancing alien views. This is not true of the tracker's ability to see things in photographs that no one else can see. N-rays will always be with us in one form or another.

The Photos in General

It is difficult, especially in the absence of any photo enhancements, to see "footprints on top of footprints" in most cases (recall the need for data to be observable, not necessarily interpretable, to anyone regardless of training). The original negatives and extensive enhancement and comparison work may help clarify and identify actual photographic data. However, nothing in the tracker's work moves beyond subjective speculation. He has not shown through photos that "both shoe prints (stars-and-bars and soft-soled flat) in a very small area one stepping upon the other indicating both subjects were there at the same time and physically engaged." Multiple sources of flat shoes other than those belonging to the victim, the lack of ability to individuate impressions, the lack of any method to identify a time frame for making prints over prints, and assumptions about what it means to be physically engaged renders these claims to be simple, unsupported speculation.

Another shockingly blatant speculation appears regarding these time frames. The tracker states that "these photos and the footprint impressions are of the time frame as to have been made after 5:00 P.M. in the evening of November 14th and prior to the time the body was discovered at 6:00 A.M." One simply wonders how in the scientific world he could support such a claim, even if for other reasons yet to be discovered, that the claim is coincidentally true. He also claims to be able to read the emotional state of any subject leaving a footprint—happy, sad, or mad. Remarkable, just not scientific.

Although the tracker may have identified some class characteristics of footwear, he admits that he cannot individuate any of the footprints that he claims to identify. If this is true, then he cannot distinguish one flat sole from any other or one stars-and-bars print from any other. He makes no progress toward developing a scientific crime scene reconstruction based upon his forensic photoanalysis. In fact, no reliable contribution to the case is provided by the tracker's photographic tracking (Figure 11.17).

Figure 11.17 This digitally sharpened photo provides one example of the tracker's marvelous ability to decipher movements and motives from marks left in the dirt. The arrows represent "footprints" that he is able to interpret.

Bloodstain Pattern Evidence

Any bloodstain pattern analysis in this case is, of necessity, based upon the provided photographic evidence and perhaps a visit to the scene some decades after the fact. The other important source of information for the bloodstain pattern analysis is the autopsy report and supporting photographs.

The jacket and sweater worn by the decedent would have essentially soaked up and blocked much of any arterial spurting, with the blood wicking through the material. Any such arterial spurting or gushing sufficient to project blood onto the rockery would not occur unless the knife were somehow moved or removed from the injury and at the same time there was a relatively clear path through her clothing for the blood to "spurt" from any compromised artery. The sharp-force injury seen at autopsy does not appear to represent multiple insertions and removals of the knife. In fact, any movement of the knife seems to have been from anatomically upward and inward pressure on the weapon's handle to leave the injuries described. Some pressure against the inserted knife moved the knife handle upward toward the victim's head leaving the injuries, visible upon internal examination, of the pleural cavity. Such movement may have allowed this blood volume to reach the rocks as seen in Figures 11.18 and 11.19.

An "arterial spurting" event may explain some of the apparent bloodstains seen on one of the rocks. Stains also appear on the other rocks and on the wood around the rockery

Figure 11.18 Blood volume.

Figure 11.19 Blood on the rocks.

area. No effort was made to measure these stains to determine their pattern relationships or to determine their point of origin. (This effort may be possible given photogrammetry and items of known size in the photos but would be extremely difficult to do from these photos alone.) Although some of the smaller stains (not easily seen) could appear typical of expirated blood, the autopsy report reveals potential sources of bleeding in addition to the stab wound in the decedent's back. She also had multiple scratches on her face and a broken nose. These injuries, as well as the stab wound, would also put blood in the decedent's mouth, potentially producing bloodstains through dripping, atomizing, or even through being cast off by sufficient movement.

The autopsy surgeon describes a ¼-inch laceration or tear on the inner aspect of the decedent's right lip that, although not a serious injury, could account for bloodstains at the scene, including blood from exhalations or breathing as well as blood dripping while she walked or moved. This bleeding obviously occurred before her fatal sharp-force injury. She also has avulsed areas or abrasions on her face. These could be the result of contact with the rocks or even a blow of some type. Such events could produce many different bloodstain patterns depending upon the actual mechanism(s) involved and their locations. Without physical scene evidence or more complete scene documentation, this is difficult to determine. Such matters must remain to be questions.

Without the evidence to test, we also cannot exclude another person as the source of the drip stains observed on the wood or of other stains typical of cast-off (such as a larger volume drop mixed with saliva from the lip injury released during a forceful head movement of some sort). Nothing scientifically precludes bleeding from the assailant, who may have suffered an injury while handling the knife while inflicting the fatal injury.

Decedent's Position(s) and Movements Based on Bloodstain Pattern Analysis and Autopsy Data

The bloodstains present on the decedent's green sweater, on her jeans, and on her jacket, as well as the debris on her pants legs and the blood on the rockery suggest a semiseated position, perhaps leaning against the rockery. The stains appear to originate from active bleeding of various sorts, not excluding a breached artery, and blood soaking through the clothing perhaps onto the rockery through contact and movement of a blood source against the rocks. Unfortunately, adequate photos to develop such data of the bloodstains on the rocks were not taken. Multiple mechanisms, combined with the drips and possible expirations noted earlier, may account for these complex stain patterns. This accounts for observations recorded at the scene, photos of the clothing, and photos accompanying the descriptions of her injuries.

The bloodstains on her clothing, especially the soaking stains over her right hip and buttocks onto the jeans, suggest that she may have bled while in a semiseated or back-leaning position, perhaps with pressure against the knife exacerbating the interior wound track through movement of the protruding knife handle anatomically upward, in turn resulting in the associated downward internal damage done by the blade as described by the autopsy surgeon. The stains on her clothing, together with her position when first observed by detectives, suggests that some of these clothing stains may not result from the heart pumping but may be due to gravity when the decedent's position was changed from one of leaning on the rock to one of lying on her back.

This change in position could have resulted from unknown factors, including a possible inadvertent, unconscious, and completely understandable repositioning by her father after she was first discovered by the deputy. We can only be clear that, despite any efforts on her part (demonstrated by the ground debris appearing in her clenched fists), she was not successful at moving herself into this prone position as discovered by the detectives. This fact can be inferred by noting the postmortem lividity described next and by remembering the obvious point that deceased persons seldom successfully move themselves very far.

Postmortem lividity provides some of this additional information on her position. Fixed lividity takes approximately 8 hours to develop, depending upon the circumstances. Fixed lividity appears on the decedent's left back, above the hip with an area of blanching appearing between the left shoulder and the left back, which appears larger than the area receiving pressure from the bra. Blanching also appears on her left hip area, with associated lividity described on the medial right thigh and the lateral left thigh. This blanching suggests pressure to the tissues preventing the deposition of hemolytic byproducts in the associated dependent blanched areas. Lividity also appears on her left upper shoulder and left upper arm. This suggests that she may have been leaning against an object providing irregular pressure, with her left arm and left shoulder lower than her right arm and right shoulder. No lividity appears on her anterior legs.

The presence and appearance of this lividity also suggests that she remained in this semiseated, semileft-leaning position for at least 8 hours depending upon circumstances, many of which remain unknown. Given the rockery and her possible positioning to accommodate these postmortem data, she may have appeared with her face down, leaning forward at the hips, bent slightly leftward, with the knife protruding at a slightly upward angle from her back. However, this cannot be concluded with any certainty.

Limited Crime Scene Reconstruction: Postmortem Position Only

Important caveat: This is limited to inferences supported by the bloodstain pattern and autopsy evidence as seen in an inadequate set of photographs in the absence of any physical evidence. This covers only a partial account of the decedent's positions and movement after her death. The three drawings are not meant to depict significant detail, only to depict relative positions.

The photo in Figure 11.20 shows the decedent's position after she was moved away from the rock. Her initial position, to account for the bloodstains on the clothing and the lividity as well as the stains on the rock, can be approximated very roughly in the three drawings that follow under #2 below.

Figure 11.20

Figure 11.21 Illustration.

Figure 11.22 Illustration from another viewpoint.

Approximate relative position, eventually with the left shoulder lower than the right (Figures 11.21 and 11.22). (These images are for illustrative purposes only and are only to show a relative position given the considerations of postmortem lividity as described in the autopsy report.) (Note: Drawings by E. Liscio are not to scale.)

1. See Figure 11.21 for general position.
2. See Figure 11.22 for shoulder angle.
3. The ground debris in her hands suggests that she attempted to use her hands to move from something like this position (Figure 11.23) after her injury.

   Again, the drawings are presented to demonstrate relative positions only. Her left leg and her pants legs both show a change in position (Figure 11.24).

Figure 11.23 Looking at the illustration from above. This 3D render differs greatly from an ordinary "drawing" or an "animation" (with moving images).

Figure 11.24 Another view from the above illustration—one virtue of a 3D render is the ability to look at the scene from multiple points of view. Another virtue is the ability to develop measurements from each render. Neither is possible from a single drawing.

4. Her positions must account for observable bloodstains. The time sequence or even the nature of each position remain unknown given the lack of data (from Figure 11.23 to the photo after movement in Figure 11.24).

   Bloodstains show that her head had to be down and positioned to allow the flow to the right while the blood was still flowing (Figure 11.25).

   The bloodstain patterns help account for the various positions she occupied and tell us about some of her movements after the fatal injury.
5. No available data support inferences about her intermediate positions, even the position necessary to produce the bloodstains shown in the photos (see Figure 11.25 to Figure 11.34).

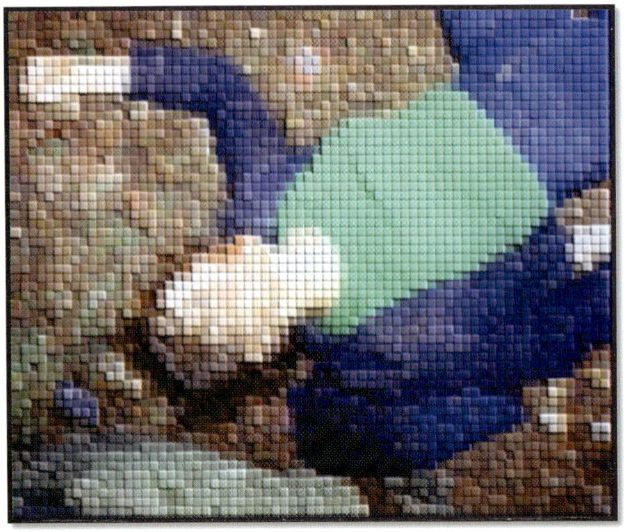

Figure 11.25 The image is blurred to honor the victim's privacy. It is used to represent the position of the victim when she was first discovered.

Figure 11.26 The bloodstains on the rocks are visible here.

6. The decedent rested against these rocks in some manner as suggested earlier (Figure 11.26). She leaned against the rocks in this location for at least 8 hours in order to develop the lividity observed in the photographs and the stains observed on her clothing.
7. Apparent bloodstains appear on the rock shown in Figure 11.27. It also appears that the rock has moist or wet patches visible. Note the smooth-soled boot present in the lower-right corner of the photograph. This fact alone renders any footwear impression evidence in the area suspect at best, contaminated at worst, but unreliable in any case.

Figure 11.27 Note the photographer's foot in the picture making footwear impressions questionable at best.

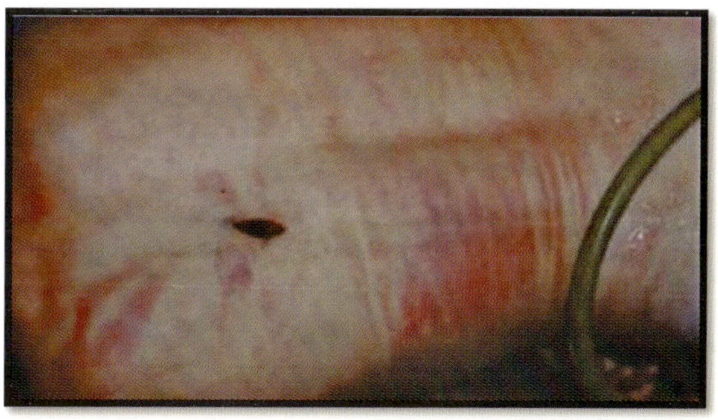

Figure 11.28 Decedent's back showing the injury and lividity.

8. Bruising appears on the lower margin of the knife wound from the handle (the guard faces the sharp edge) (Figure 11.28). The wound's margins show (Figure 11.29) the blade's movement, which could, under suitable circumstances, account for the blood deposited on the rocks. The wound supplied an opening, allowing considerable blood leakage depending upon the body's position(s).
9. Blood patterns on the clothing correspond to the body position indicated by lividity (Figure 11.30).

   The knife blade guard is anatomically downward; the knife's angle shows the handle against the upper margin of the wound.
10. The clothing shows extensive bleeding with soaking and apparent wicking (Figure 11.31). The arrow denotes an area with an apparent flow pattern from the injury to the head. She was in a position with her head down and the wound up for this pattern to occur. The bloodstains indicate that she was either moved or moved herself after sustaining the injury.
11. The bloodstains on the clothing and the jeans show active bleeding as well as potential limited passive leaking from movement after death (Figure 11.32).

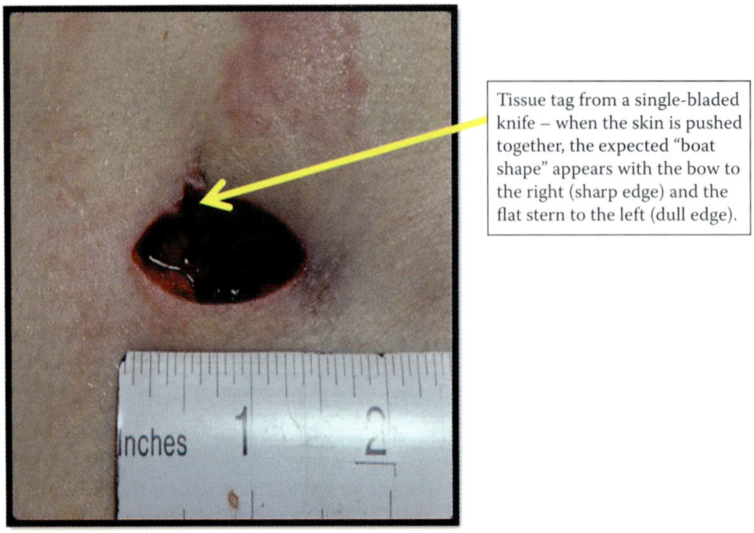

Figure 11.29 Sharp-force injury.

Figure 11.30 Blood flow here is from various mechanisms.

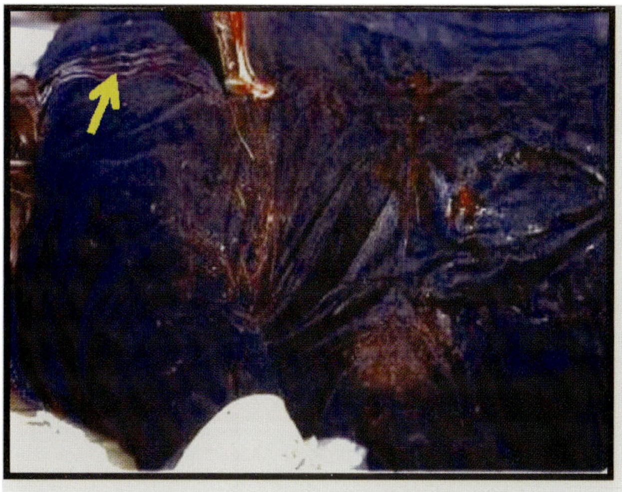

Figure 11.31 The bloodstains visible here show that blood flowed from the wound toward the neck—so at some point, the head was lower than the bleeding injury.

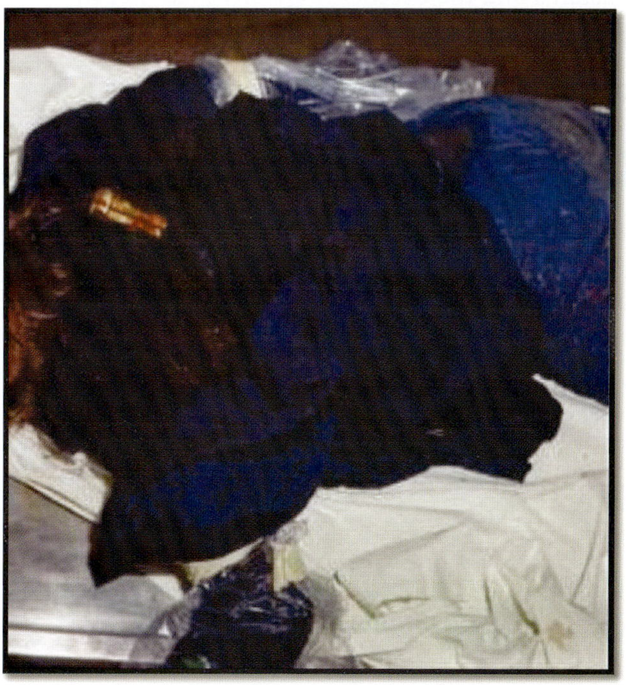

Figure 11.32 Soaking stain on right side.

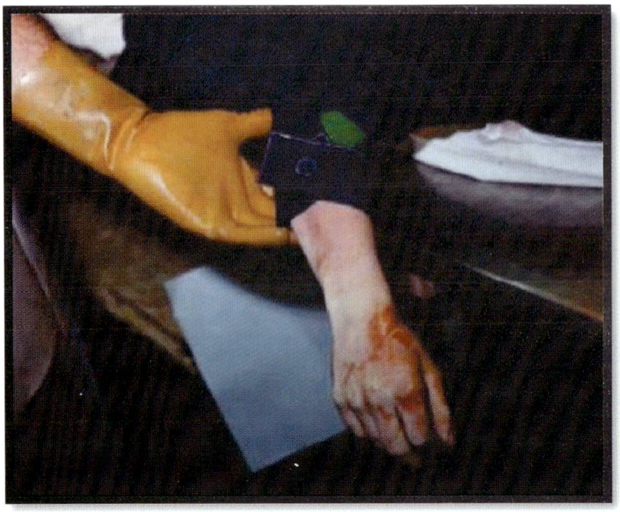

Figure 11.33 Bloodstains on the decedent's right arm and hand.

12. At least three potential sources of blood appear from her mouth and nose: internal bleeding due to the stab wound, the lip laceration, and the facial abrasions.
13. Blood on her hands appears to be contact transfer staining (Figure 11.33).
14. Complex bloodstains (more than one mechanism) appear on the clothing, the rocks, the face, and the hands. The blood in Figure 11.34 results from active bleeding. The bloodstains and the decedent's lividity place her in some position above the upper blood volume and then near the lower volume seated on the ground

Figure 11.34 Photo taken at the scene showing bloodstains on the rocks—some dark areas are simply wet with ambient moisture.

and leaning on the rock to the left. She appears to have been seated on the ground between the above rocks leaning slightly left. She remained in this position long enough for the dependent lividity to become fixed.

Sciences That Could Have Been Applied in 1973, in 1986, or Now

DNA analysis as we know it today had not yet been developed in any practical sense in 1973. Based upon the state of the serological sciences in 1973, the testing that could (and should) have been performed and followed up in this unsolved homicide include:

1. The fingernail scrapings, including the possibility of finding exculpatory evidence, especially given the photos of the minor scratches on the defendant's arm.
2. This data could have been applied to include or exclude suspects based upon tests available in 1973 using the materials collected from under the decedent's fingernails.
3. The blood evidence found on the one suspect's jeans could have been subjected to blood-typing tests that could have tied this other suspect directly to the victim or could have been used to exclude that connection.

Negative Effects of Evidence Destruction

The destruction of physical evidence in this case eliminates the development of data both from the tests available in 1986 (at the time the evidence was destroyed and the case file pages were marked "settled"), as well as eliminating the potential of developing evidence through the application of scientific advances in areas such as DNA typing today. If evidence had been preserved, contemporary DNA analysis may have been applied to identify the killer.

For example, even in 1986, a test of the victim's clothing for blood evidence from the assailant would have been possible. Or the blood evidence on any suspect's clothing could have been tested for origin and type. That includes bloodstains observed in the scene photos of the rocks, wood, grass, or dirt. The blood possibly could have been distinguished from the decedent even through ABO typing. Or, though no sexual assault was evident, a rape kit test of bodily fluids processed with today's science might have revealed additional information of potential value.

What appear to be wet patches are visible in Figure 11.35. Note the boot present in the lower right corner of the photograph. This fact alone renders any footwear impression evidence in the area suspect at best, contaminated at worst, but unreliable in either case.

Conclusions

A reconstruction, corrected and presented to the court, considers only the scope of the available physical evidence, together with its quality, appropriate collection, proper storage, and valid interpretation. It is important to remember that a scientifically based sharp-force injury event reconstruction, a critical review of alternative reconstructions, or suitably

Figure 11.35 Rock showing both moisture and bloodstains.

corrected scientifically based conclusions do not present the entire story of a crime. They do not tell us who did it, but they do tell us something about how it was done.

In this case, the jury had much more information that remained beyond the scope of the natural sciences. It is certainly an error to conclude that the results of a crime scene reconstruction reveal the guilt or innocence of every accused person. Its value and its scope depend upon the significance of the physical evidence and the application of the reconstruction itself to the questions before the court. Sometimes the reconstruction can only tell us some limited things about the commission of the crime. Anyone capable of accomplishing these events may become a suspect. Therefore, any relationship of that data to the identification of the perpetrator may depend upon factors well beyond the evidence that is suitable for scientific analyses.

It was some 30 years later that detectives developed a suspect, arrested him, and prosecutors brought the case to trial. The jury voted to convict. When polled, jurors said that they had ignored the tracker and his spurious speculations. The scientific evidence could show how Amelia Wentworth-Hobbs died but not who had killed her. Their verdict had depended upon other factors. The jury was insightful enough to recognize and decide the case based upon this evidence. A correct verdict must fit with all the evidence. The basis becomes similar to the coherence theory of truth—if something is true, then the world must be consistent with that truth. If not, then we have a question requiring some answers. In this case the jury, with their verdict, finally answered a family's painful questions.

Bombs Away

12

Blind respect for authority is the greatest enemy of truth.

Albert Einstein

Introduction

A terrorist bombing has less-than-subtle applications. Suicide bombers blow up innocent citizens along with themselves. A ton of fertilizer mix destroys a federal building and the lives of the people in it. Roadside bombs called *improvised explosive devices* or IEDs kill our soldiers on routine patrols. Bombs need have no political motivations. In Chicago in the 1930s, many gangsters were vaporized along with their automobiles. Explosive devices appear in many forms and for many purposes throughout history. The apparent purpose of the device in this case involved ending an unhappy marriage. Divorce would have been cheaper and ultimately, more effective.

Enter one Charles Pertzborn III. As a Certified Public Accountant (CPA), his life was filled with ledgers and numbers, bills and receipts, charge slips and purchase orders. By all accounts, his was about as far from an exciting life as a corpse in a funeral home. When he did not bring his work home with him, he spent his spare time looking over his stamp collection—or rather, the stamp collection that his grandfather gave to him as a 16th birthday present. That had been at least 40 years ago by now. Over all these years, Charles did not add a single stamp to the pages of his album. He just looked. This practice irritated Bernice Pertzborn more than a thistle in a baby's diaper or a ... well, you get the idea.

Bernice always had some task for Charles to accomplish. These projects kept him away from his stamps. Most such activities involved outdoor projects, such as painting the house, scraping and painting the trim, mowing the lawn, washing Bernice's new Geo Tracker, or sweeping the walks and driveway. Regardless of the work, Bernice always began her part in each project by scrutinizing Charles from the porch, standing with her fists resting on her hips, elbows splayed out to each side. At some point, she would start shaking her head back and forth. Then, predictably, she shouted advice and instructions over the sound of any equipment that Charles was using in these projects. Invariably, the advice turned to criticism and the criticism soon morphed into a string of obscenities that would leave a death-row inmate cringing in fear and embarrassment. This pattern invariably repeated itself each time for each task throughout 30 years of wedded bliss.

The stout Mrs. Pertzborn also happened to love her new Geo Tracker with more passion than she could ever muster for the diminutive Mr. Pertzborn. According to his wife at least, Charles not only failed as a husband; he also fell way short as a driver. His incompetence was surpassed only by his temerity. She never let him forget it. His life with that Geo consisted of washing and waxing; her life with the machine was filled with pleasure drives, scenic tours, and camping trips. Why, Charles even had to take the bus to the hospital for his hernia surgery. She could not take him because she was on one of her overnight camping trips—always taken alone. At least that is what she told Charles.

Charles caught wind of the other man sharing camping trips with his wife when he intercepted a racy note from a "Stevie-weevie" to a "Bunny-honey." While he was dutifully vacuuming the Geo's carpeting, he found the note next to an empty condom wrapper caught between the car's seats. This, Charles later told detectives, was the insult that plunged him from humiliation well past anger and deep into the heart of murderous rage. He would take no more. And neither would Bunny-honey.

That is how Charles came to put aside his stamp album to concentrate on bomb-making procedures found on various Internet sites. He became a student of stealth. In his mind, he had developed into a top-rated and greatly feared assassin, courting danger at every turn, punishing evildoers and defending the oppressed. In reality, he had just traded in his stamps for a Walter Mitty–style fantasy life. Yes, it truly was fraught with danger all right, but only for Charles Bartholomew Pertzborn III.

Once Charles methodically settled on a design for his bomb, he spent several nights alone in the garage with the Geo, crafting the device while Bernice slept. The design he chose used the Geo's brake-light circuit to initiate the ignition process. With the bomb's proximity to the car's always-full gas tank, the chances for complete conflagration looked pretty good. After the final night's work, he felt somewhat sad that an adoring public would never appreciate the genius of his brilliant design.

After he went to bed, his troubles began in earnest. Bernice discovered the charge slips that Charles' double-entry mind had saved when he purchased the wire, the 50 cal cartridges, and the .300 Wetherby magnum rounds called for in the bomb's plans. That was bad enough, but like any well-organized secret agent, he saved these receipts, along with the plans that he had drawn, and clipped them both to the Geo's owners' manual. He had used the book's diagrams to locate the vehicle wiring points used to initiate the explosion. There was an explosion all right, and Bernice was at its center. But it was not the type of explosion that Charles had planned.

Bernice pummeled Charles, first as he lay defenseless in the marital bed and then as he lay unconscious on the bloody sheets, before deputies arrived to haul him away in handcuffs. He sat in his cell and cried like any top-flight secret agent would—especially if that agent were married to Bernice Gwendolyn Pertzborn.

When they are discovered, explosive devices often are sent to the Bureau of Alcohol, Tobacco, Firearms and Explosives (ATF) for analysis. Investigators routinely issue reports on their nature, danger, and effectiveness. With very good reason their work is well respected, thorough, and applies the latest advances in the natural sciences. But it always remains important to check even the best scientific work. If that work contains basic errors and the work continues unchecked, the potential misapplications of science to crimes may repeat. When compounded, such strings of scientific error deliver blows from which the justice system may never recover.

The Geo, including Charles' modifying handiwork, was returned from the ATF along with the investigators' report. The prosecution accepted the report at face value: Charles had built a deadly device capable of killing, maiming and destroying anyone or anything nearby when the Geo's brakes were first applied.

The public defender's office took on Charles' case and attempted to assemble a credible defense. He was charged with multiple felonies, including building a "deadly device capable of inflicting grievous bodily harm on innocent human beings," or some such wording.

No one seemed to think that any review was necessary. After all, the ATF is a government agency second to none. The fact that the report describing the device in this attempted murder case was presented by the ATF seemed alone sufficient to end all questions about its conclusions. But even the best applications of natural science involve the potential for error, both methodological error (the error built into the methods themselves) and practitioner error (the mistakes humans make applying those methods). After all, the discovery of error is one reason why evidence is preserved: so that the evidence can be reexamined and reevaluated as a check on the work done by others, as well as to develop any results overlooked in the prior work.*

Despite this general acceptance of the ATF's conclusions by both the advocate for the state and the advocate for the defendant, I was asked by the court to evaluate the scientific work done by the ATF on the explosive device installed by Charles in his wife's car. The loaded Geo, as analyzed by the ATF, was shrink-wrapped and sent to the Final Analysis Forensics laboratory where I put it on one of our lifts, opened the package, and began my examination.

Each scientific report issued from Final Analysis Forensics receives a peer review as part of our SOP.† Another scientist is asked to review the work and uncover any errors of either omission or commission. Peer review has been one of the standard bases of natural science for a hundred years or more. It is difficult, if not impossible, to develop a peer review of many reports issued from state crime laboratories: a one- or two-line statement of conclusion does not make a scientific report. The ATF, on the contrary, issues a full report in most cases, at least in all the cases that I have seen. I began to work on this case as I would on any peer review of a colleague's scientific report.‡ That requires an independent review of the data collection and development along with a review of the report itself.

The Device and the ATF Review

My work focused exclusively on the device described by the Explosives Technology Branch of the ATF as "consistent with an improvised explosive device."§ The report further describes this device constructed by Charles and its functioning as follows:

* Sometimes the state's preservers of evidence take their important role too far. Not only do they preserve the evidence, but they oppose any independent analysis. When a proper, nondestructive scientific evaluation of the evidence is completed, then that scientific work must be checked. If the evidence is collected but not examined, then an independent examination should be performed to evaluate its scientific significance. After all, why do we preserve evidence in the first place?

† Standard operating procedure. Having written policies and procedures is one of the many requirements for ASCLAD (American Society of Crime Laboratory Directors) accreditation. A component of such policies and procedures include a statement of SOPs.

‡ Some would argue that the ATF personnel have no peers. That comment is meant by some to elevate the ATF at the expense of everyone else. I do not know anyone at the ATF who would subscribe to that view.

§ A scientist must keep in mind that science does not apply to all the evidence forming a case; certainly not all the data in a case involves the natural sciences. Seldom, then, as a natural scientist does one have the perspective to form an opinion about guilt or innocence—unless, of course, the entire case consists of physical evidence. Then it is necessary to avoid bias as much as possible and best to continue to present sound applications of natural science to the best of one's ability regardless of adversarial consequences.

The Device's Construction, According to the ATF

> This device was constructed using a 50 caliber cartridge case modified by having a small hole drilled through the spent primer. A .300 Wetherby Magnum cartridge case was modified in the same manner. The two cartridges were compressed together and contained a quantity of explosive powder identified by ATF laboratory as a mixture of smokeless powders. The device utilized and a [*sic*] length of insulated braided wire having a few of the strands running in the casings to act as a filament-type ignition fuse.

The Device's Function According to the ATF

> Applying electrical current to the wire would cause the wire strands to heat and ignite the explosive powders, causing the device to explode. This type of device is capable of causing property damage and/or injury or death to persons near the explosion.

The Device Errata

My overall conclusion considered only the properties of this device, and was based both upon an examination of this device documented by photographs and tests with a functionally identical test device that I built in the laboratory. The ATF's device description and functional analysis originates with an explosives enforcement officer, and was reviewed by the supervisory explosives enforcement officer, and approved by the then-chief of Explosives Technology Branch. I concluded, contrary to the conclusion supported by the Explosives Technology Branch of the ATF, that as actually constructed, applying electrical current through the wires present in this device would not cause the strands to heat, and would not ignite any explosive powders present. Further conclusions, supported by the reasoning I explained, hold that the agent conducting the analysis inaccurately describes the device in his report. At best, this actual device is an ill-conceived misfit, exhibiting its designer's (and analyst's) poor grasp of the basic principles of electricity or, more charitably, its analyst's inaccurate observations combined with a failure to conduct any specific scientific tests. The report appeared to me as a cut-and-paste timesaving template—almost a form letter—meant to apply to hundreds of generic bombs in hundreds of similar cases. If so, this approach to conducting a scientific investigation runs counter to every rational standard at the foundation of science and, indeed, of reason itself.

Description and Analysis of the Device

Description

The device arrived in three parts. They consisted of the following:

1. A 50 cal military cartridge case with a hole drilled through the primer with an 8-foot length of 24-gauge insulated 7-strand tinned wire leading to the case. This wire was then stripped of its insulation, and the bare wire was wrapped around the base of the case. A short length of insulated wire emerged from this bare wire around the base and led through the hole drilled through the spent primer cap. The wire and the hole have what appears to be wax applied, purportedly to seal

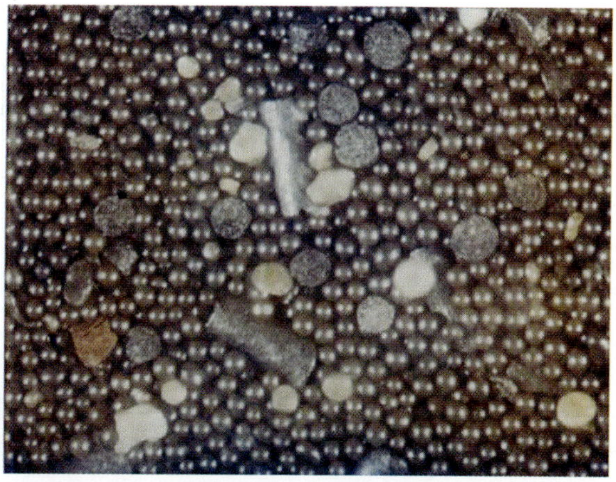

Figure 12.1 A mix of smokeless powders, containing some nonpowder fillers but no Taggents.

Figure 12.2 The 50 cal case, and the bare wire around the base with the insulated wire running through the hole in the cartridge case.

the wire's entry. Insulated through this hole, the 7-wire strands were then bared and clipped, leaving three short strands and one longer strand free inside the case (Figure 12.2 to Figure 12.4).

2. A .300 Wetherby Magnum cartridge case, also with a hole drilled through the primer, and prepared similarly to the 50 cal case. The same design principles were followed, with a 5-foot length of 7-strand tinned wire leading to the case, bared wire wrapped around the case, and 5 strands of bare wire protruding from the insulated section of wire through the hole, again apparently sealed with wax (Figure 12.5 to Figure 12.7).
3. A packet containing 14.66 grams, or 226.3 grains, of mixed smokeless powder, excluding any removed by previous investigators, was described as being loaded into both cases. The powder had no Taggents present (see Figure 12.1). Tangents, mixed with the explosive, help to identify its source through the information that they carry. The information leads investigators from the manufacturer to the distributor then to the seller, and through them, eventually to the buyer.

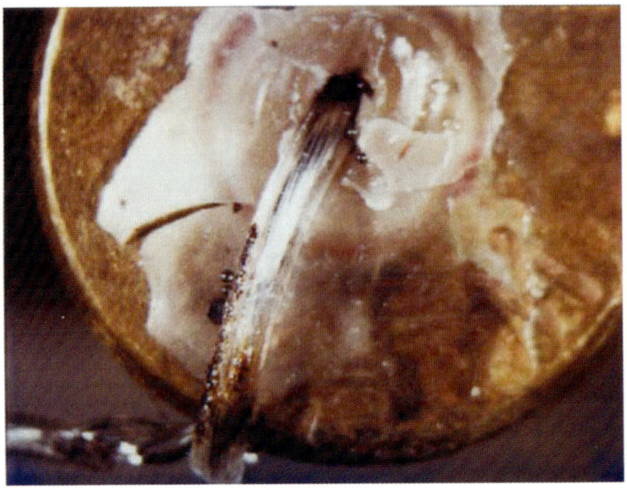

Figure 12.3 The 50 cal case and the insulated wire running into the hole drilled in the base. Note the wax around the hole.

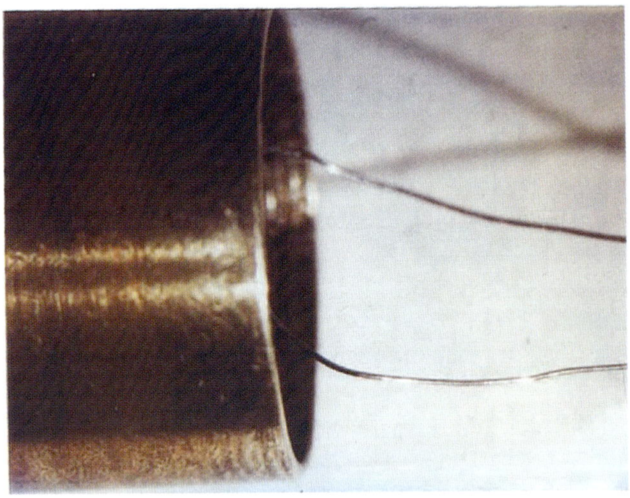

Figure 12.4 The 50 cal case and two of the strands of wire emerging from its interior.

The .300 Wetherby case was then reportedly pressed into the 50 cal case without benefit of insulation. The cases were reportedly then taped together with black electrical tape.

Analysis

For purposes of this conceptual analysis, the device means the aforementioned assembled components, with wires and powder in the cases and the two cases pressed together with black electrical tape around the union. Elementary electrical theory, based upon Ohm's law, dictates that applying current from a 12-volt car battery of any description or condition to the device as described, with one lead to positive and the other to negative, will fail to heat the wire strands inside the device. The same is true when the device is wired into an automobile's brake-light circuit.

Figure 12.5 The .300 Wetherby case with the bare wire wrapped around the base and the insulated section leading to the hole in the primer cap.

Figure 12.6 The .300 Wetherby case with the insulated section of wire entering the hole drilled through the primer cap. Again, note the appearance of wax.

The basic components of electrical theory are the volt, the volume or push of the current; the ohm, or the restriction to the push; and the amp, or the stream of current pushed past all restrictions. Electricity through wire is often described by an analogy to water through pipe, with water through a hose analogous to electrons through a wire. (Volts being the amount of water in a holding tank ready to go; ohms being the size of the faucet opening from the tank resisting the flow of water; and amps being the stream of water through the hose connected to the faucet.) Electrical current, like water, always follows the path of least resistance.

In this device, the path of least resistance for the current is actually through the cartridge cases. The wires connect directly to the cases, and current, following the path of least resistance, will flow through the large pipe provided by the conductive cases. It will, as a matter of natural law, bypass the strands of wire inside the cases, since these strands

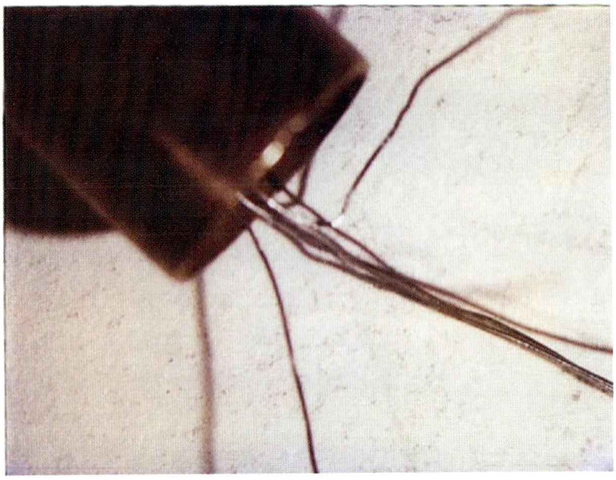

Figure 12.7 The .300 Wetherby case with five strands of bare wire protruding from the end. Note that shadows also appear and should not be confused with the actual wire.

offer the highest resistance to the current flow. (The size of the pipe is very small and current will flow instead through the bigger pipe provided by the cases, therefore bypassing the little pipes.) To claim that the wires inside the cases will heat when current is applied to the wires attached to a battery, one to "+" the other to "–" is to claim that for this device alone, Ohm's law is invalid. This is not a defensible commonsense position. It is certainly not good science. However, in the interest of sound science, the claim will be tested. Doing any technical review and developing conclusions about either a design or a device can benefit from testing the device or building a replica, if possible and practical. This device was simple enough to merit such a test.

Testing the Device

I built a similar device (Figure 12.8), omitting the gunpowder since the issue focuses on the behavior of the 3- and 5-wire strands inside the cartridge cases. In place of the .300 Wetherby case, a small brass plug was substituted to enable visual inspection of the wire strands inside the case(s). The device was constructed on a wooden board, its components labeled, and it remains available for independent examination and analysis, or for courtroom exhibit and demonstration. Its construction allows for many different scenarios involving the strands of wire inside the cases: touching the cases, not touching the cases, touching each other, not touching each other, and so on. A red jumper wire allows the plug and case to remain in contact, as the actual device was constructed, while permitting a view of the bare wire strands inside the case. The wiring of the device attempts to replicate the brake-light circuit of a 1995 Geo Tracker.

Similar 24-gauge, 7-strand tinned wire was used (see Figure 12.9). It was connected to a 12-volt battery in two ways: first, a direct connection of the wires from the device, one to the + and the other to the – terminal of the battery. Second, the device was connected as hypothetically envisioned, in parallel with the brake light of the 1995 Geo Tracker. The device behaved similarly under both test conditions: the wires inside the 50 cal case did not heat. Current simply passes through the union of the two cases described in the actual device.

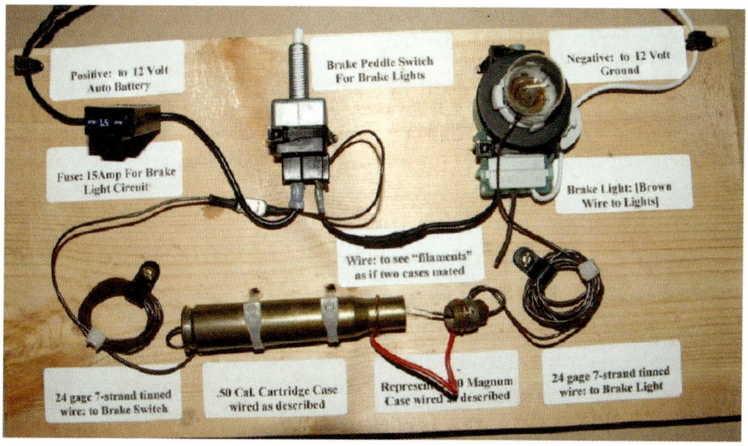

Figure 12.8 The device that I built exactly reproduces the circuitry of the original "bomb." I labeled items and mounted them on a board for easier viewing and demonstration.

Figure 12.9 The original 7-strand, 24-gauge tinned wire.

When connected directly to a 12-volt auto battery, the 24-gauge wire became hot only where directly connected to the battery. This represents a "dead short" where + and – are directly connected, with no intermediary device. Since the wire behaves like a dead short, the device draws no current load whatsoever. The path of greatest resistance is the 24-gauge wire itself: both the case(s) and the small strands of wire under test remained unaffected and cool.

When connected to the battery via the brake-light circuit of a 1995 Geo Tracker, results with the device were similar. However, the parallel connection of the 7-strand, 24-gauge wire, one to the brake switch and the other to the brake-light bulb, reduced the resistance of the wire by avoiding a dead short. When wired in parallel, current will also flow through the wire from the switch to the brake light, reducing current through the small wire to survivable levels. In no examined case does the wire inside the case(s) heat in any measurable manner.

Although specific measured values will vary with the condition and charge of the battery, the contamination of any connections in the circuit, or other circuits at issue, and the sensitivity of the digital multimeter (DMM) used, the following values are illustrative of the pattern ordained by Ohm's law.

When 11.6 volts from an old 12-volt battery are connected to the brake-light circuit and the switch is released, the following average (mean) measurements of resistance were obtained over a series of 20 tests:

At 200-ohm sensitivity, the wire from the light to the plug (.300 case stand-in) was 02.2 to 01.5, while the resistance across the cases was 0.03, or no measurable resistance within the error rate of a cheap pocket DMM.

At 200-ohm sensitivity, the wire from the brake switch to the 50 cal case was 03.5 to 02.6, while the resistance across the cases was 0.03, or no measurable resistance within the error rate of a cheap pocket DMM.

There was no increased resistance measured in the filaments of wire into either case, since in the actual circuit the current follows the path of least resistance that is across the cases (case and plug) and these filaments touch either the case, or each other, or obviously, the original wire.

When 11.6 volts from a very old 12-volt battery are connected to the brake-light circuit and the switch is released, the following average (mean) measurements of amperage were obtained over a series of 20 tests:

At 10 amps sensitivity, with the switch on, +.01 amps flow across the wire from the light switch to the 50 cal case; –.01 amp flows across the wire from the .300 cal (plug) to the light, and predictably 0.1 amp flows across the cases. (Here + and – indicate direction of current from positive to ground [or –]. Simply reversing the leads to the battery will change the values from + to –, and – to +.)

The further repetition of such tests remains unnecessary since Ohm's law explains why these results occur, that the current will follow across the cases and not through the strands. This always occurs regardless of the many variables that may alter the actual values. Here, only the emerging pattern is relevant.

Alternative Contexts and Their Effects on the Device

Since the device will not serve to ignite the powder, or anything else, no further examinations involving the powder or actual explosions of it under different circumstances were undertaken. Even if this device is suspended in a gasoline tank above the liquid gasoline level in order to add gasoline vapors to the potential power of an explosion, nothing will result. The device would have to be capable of working and would have to involve the vapors in the tank. (One gallon of gasoline has the explosive power of 83 pounds of trinitrotoluene, or TNT.)

Conclusions

This device would not cause the strands to heat, and would not ignite any explosive powders present. The ATF agent inaccurately describes the device in his report and misidentifies its

Figure 12.10 The author with a dangerous device of his own design and construction. His wife is smart enough to stay well away, ever ready to dial 911.

potential to injure or harm. Sometimes a device that could injure or harm is not built with that intent, despite the undeniable fact that it could (Figure 12.10). This alleged explosive device, on the other hand, exhibits its designer's poor grasp of the basic principles of electricity and would not cause an explosion.

The charge involving possessing and constructing a dangerous explosive device were dropped as the court ordered. Given my report, the ATF agents withdrew their report and adopted mine. In science, no one is perfect—we all make mistakes. It is only a mistake if we do not admit it and move to correct it. That is how science progresses.

As far as I know, the attempted murder charge still applied. Even if someone fails, or in this case, is too "stuck on stupid" to succeed, it should never be permissible even to attempt to blow up one's wife.

The Brothers Grimm 13

One week he was well, the next he was dead. (*Binnen acht Tagen war er gesund und tot.*)

Alfred Krupp, Eulogy, 1967

Introduction

It may have been a mistake to live in the same house. The trailer had been convenient when their mother was alive, but now she was gone and the two shared her house as a joint inheritance. The trailer remained convenient, parked beside the house, an extension cord running through the basement window and into the front sliding patio door. Nonetheless, they were both here. Together again, living in the same house where they grew up.

Evidence of their youthful battles remained on the walls of their childhood room. Divots in the walls, holes in the ceiling, and the favorite, a large dent in the wooden floor where Raymond's baseball bat struck after the glancing blow just missed his brother Bert's head. Yet, they played together and stuck together. When a neighborhood foe threatened one, the other sprung to his defense.

As adults they maintained an uneasy truce. But after their mother died, that truce began to falter, its force weakened by repeated infusions of alcohol. Both had retired and lived off the generosity of their mother. Their only expense for the last year had been the beer and wine that fueled their disputes.

Today was no different. By lunchtime, angry words reduced to a string of obscenities; the volume alone would offend the devil himself. Raymond could not remember who brought the 12-inch kitchen knife to the table. Alcohol strengthened the battles but weakened the memories. Somehow, that kitchen knife found its way into Bert's neck, entering the soft tissues above the sternal notch and exiting just above the curve of his right shoulder. Both brothers had been quite surprised by the blood, but for Bert, the surprise did not last very long. Raymond called 911. He walked out the back door toward the trailer and threw the cordless phone on the grass. The phone stuck antenna first into the dirt.

Scene Description

At the scene, the victim appeared face down on the kitchen floor. Blood seemed to coat every available kitchen surface. There were smears, wipes, swipes, arterial spurts, all sorts of stains—enough to fill an entire textbook on bloodstain pattern analysis with representative photos.

Arterial spurting indicated that the victim bled out at the scene leaving stains no higher than approximately 36 inches from the floor. His injuries consisted of one sharp-force perforation of the neck. This wound was a 7¾-inch front-to-back stab wound of the left

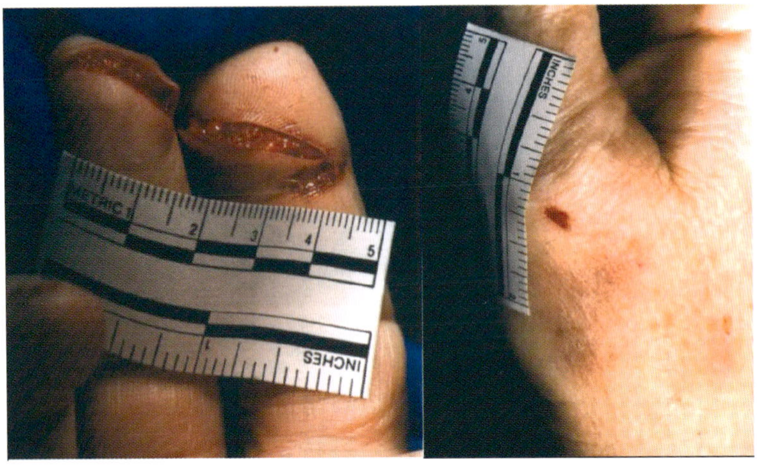

Figure 13.1 Incised wounds to the victim's right fingers and a nick on the right thumb.

anterior neck, transecting the jugular vein and carotid artery; the knife blade had exited from above the left shoulder blade. There were incised wounds to the left hand (Figure 13.1) and minor incised wounds to the anterior neck region and temple. The temptation is to call the hand injuries defensive wounds, but that introduces a conclusion into what should be merely initial observations: descriptions of the wounds without conclusions about their status or mechanism. Not yet.

From the bloodstains and his injuries, it appeared that all Bert's injuries occurred during a struggle with his brother in the kitchen involving a kitchen knife. (See Figures 13.2 and 13.3.) The paramount question soon became the mechanism for the fatal injury: Just how did the knife come to perforate Bert's neck? The question was much the same as the question raised in the Vaughn case (Chapter 9)—answering the "how" question became

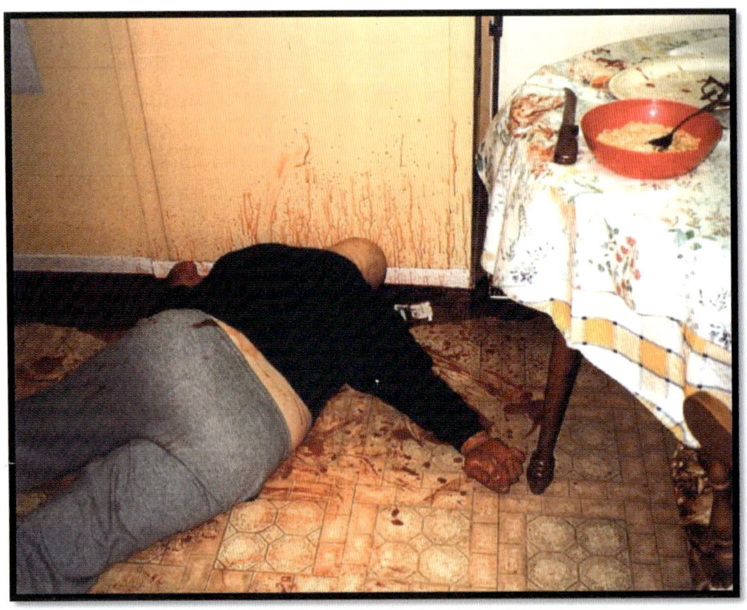

Figure 13.2 The victim's position after he slid down while being held by his brother.

Figure 13.3 Bloodstains on the floor provide a great deal of information—describing the stains must be the first step—then measurements, followed by interpretations.

the focus. With an answer to that question, perhaps an answer to the two other obvious questions would develop: In what manner and by whom?

Examination of the Suspect

When Raymond was taken into custody and examined, he presented no bleeding injuries of any sort. Therefore, he was not the source of any of the blood in the kitchen, unless the blood had come from some earlier misadventure. The crime laboratory response team sampled and tested selected stains on Raymond's clothing (Figure 13.4), showing first that they were blood and second that they were human blood probably from his brother Bert. They doubted that DNA testing would be necessary.

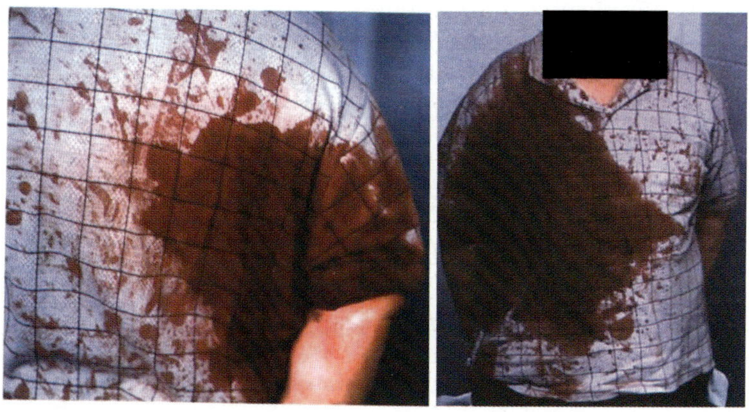

Figure 13.4 The suspect's bloody shirt. He said that he held his brother with his right arm. The bloodstains support his claim.

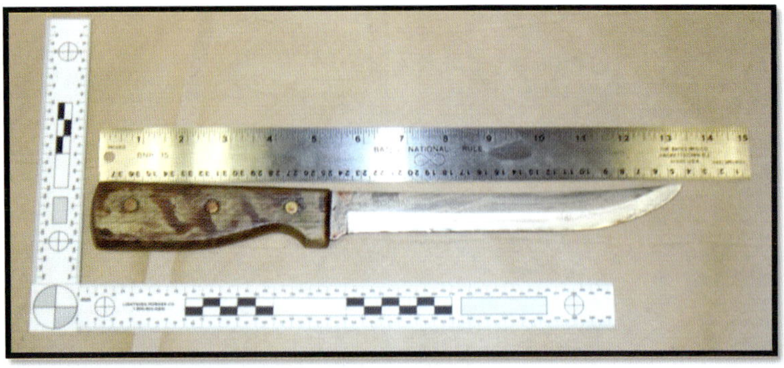

Figure 13.5 The boning knife that perforated Bert's neck.

Samples are saved, of course, because it is impossible to foresee what tests and techniques may apply in the future. An important element of scientific work with data from a crime scene involves its collection. We do not always, if ever, have insight into what data may become critical evidence, or what data might require analysis and examination in light of future, as yet unforeseen, developments. The collection process assumes primary importance at the scene. If we fail to collect data when the opportunity presents itself, we lose the opportunity that data affords. The opportunity may never arise again—it is a one-shot proposition.

Examination of the Knife

The kitchen knife protruding from Bert's neck was an Old Homestead boning knife, stainless steel, made in Japan (Figure 13.5). It measured 13.5 inches tip to handle. It had been well used throughout the years. Now it was bent, possibly from moving while inserted in Bert's neck. The edge was dull; the point, about average for a boning knife.

The Scenario

Verbal accounts of how things happened at crime scenes are important for this scientific work, as well as for any scientific work in physics or chemistry. The how, who, and the context encompassing the why may differ, but the principles of investigation remain the same. The given verbal account(s) functions like a given hypothesis. The testing and experiment, as well as the available data and its properties, become the means to assess the truth or falsity of the hypothesis.*

Raymond said that the two had been drinking all day; Bert preferred beer while Raymond drank wine. After drinking through breakfast, the argument increased in both volume and intensity. Bert sat down to his lunch, while Raymond prepared his. The knife

* Many times the testimonial evidence provided by witnesses or by self-interested participants in events leading to someone's death is irrelevant to the scientist. Their information simply does not carry scientific weight. The value of some accounts, however, is to provide some hypothesis to compare with the physical evidence. That hypothesis may bear the weight of the entire investigation. The other avenue, of course, is more common in the natural sciences. An examination of the physical data helps to generate the hypotheses for further testing and the development of an explanation for the events in question. I address the logic of this process in *Dead Reckoning: The Art of Forensic Detection*, Jon J. Nordby, CRC Press, 1999.

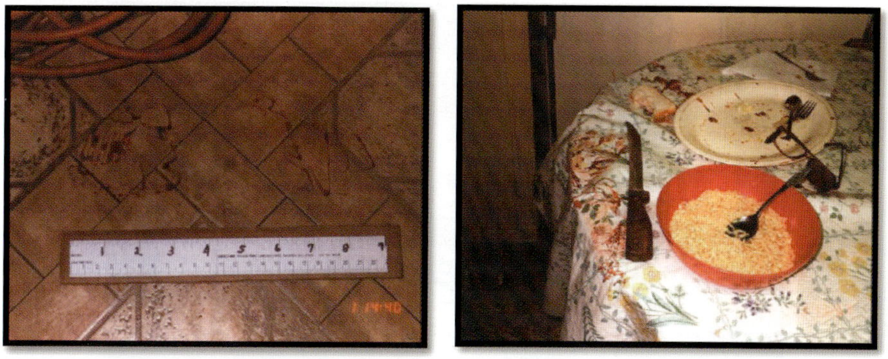

Figure 13.6 Bloodstains on movable items are difficult to use in locating any point of origin for the bloodletting events, but they do supply useful information.

cut through the meat and cheese composing the meal at hand. Raymond wielded the knife and prepared his lunch while drinking another beer (Figure 13.6).

Raymond said that, somehow, he ended up at the table, shoving his brother. His brother shoved back. Somehow Bert got cut pulling the knife away, the struggle escalated, and at some point during these unchoreographed movements, the knife ended up through Bert's neck. He said that it stunned them both. The blood kept coming, soaking them both. Raymond said that he bent down as Bert slumped to the vinyl floor and held him until he stopped breathing. Then he stopped bleeding. Raymond, calm at his interview yet with eyes beginning to exhibit the properties of shock, said that he grabbed the cordless phone, called 911, walked out the back door toward his trailer, and threw the phone onto the grass. He stood outside smoking a cigarette while waiting for whoever would come to address the mess defiling his mother's kitchen.

Testing of Force Required to Reproduce the Victim's Fatal Injury

In order to measure the force required to push a knife of the aforementioned brand, style, and condition into a human neck, tests were conducted using a pig. First, I built a device that would hold a knife while measuring the force in pounds required for that knife to penetrate and perforate an object (Figure 13.7). The device uses a digital scale registering in pounds. Pushing the knife will measure the force. Two readings result: one, the force required to penetrate the surface and, two, the force required to perforate the structure once penetrated.

I used a recently killed pig to approximate both human skin and the structures of the human neck. Six penetrations with six perforations the length of the blade through the neck were attempted (Figure 13.8) with the following results:

Attempt Number	Force to Penetrate	Force to Perforate (the Knife's Length)
1	8.6 lb	2.2 lb
2	12.0 lb	1.8 lb
3	13.0 lb	2.0 lb
4	10.0 lb	2.0 lb
5	9.0 lb	2.0 lb
6	7.0 lb	1.9 lb

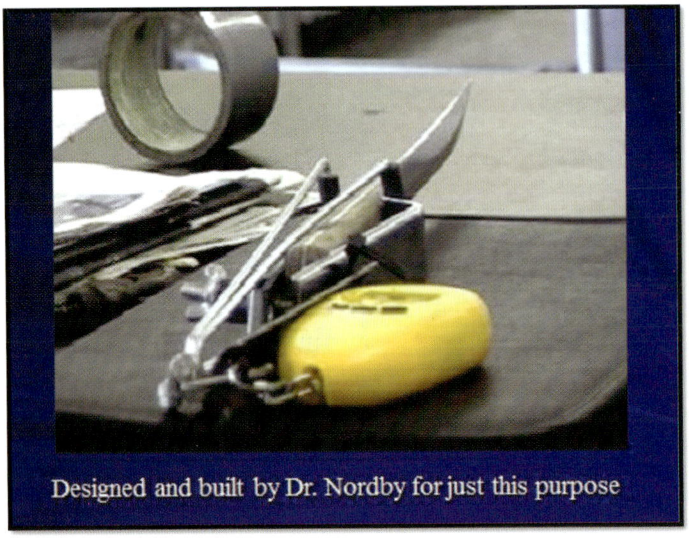

Figure 13.7 The "stab-o-meter" uses a digital scale attached to a clamp, which holds any sharp-force instrument when the sharp instrument meets resistance, the digital scale provides a measurement of its value in pounds.

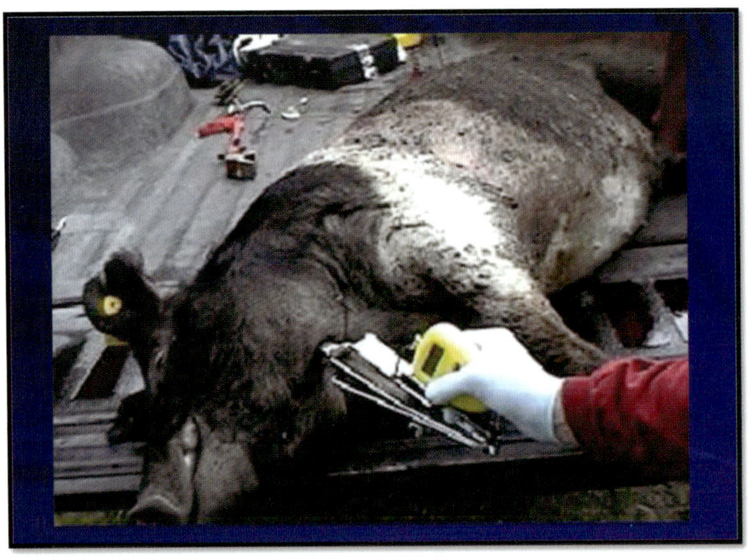

Figure 13.8 The stab-o-meter is used to measure the resistance of the knife, in this case, as it is pushed into the neck area of a freshly euthanized pig.

Examination of Bloodstain Patterns and Trace Evidence

The bloodstains on the clothing are first described and then documented. I use similar approaches, depending upon the type of case and its individuating circumstances. Describing the decedent's clothing came first (Figure 13.9), then the suspect's clothing (Figures 13.10 and 13.11).

Decedent's Clothing

Sweatpants front—Saturation stains on right and left upper legs; soaking stains crusted on left thigh area; dripped blood on right lower leg; impact staining mixed with drips on inner area of right leg and crotch

Sweatpants back—Right leg swipe stain (rear pocket area); contact transfer stains left side rear with dripped staining

Sweatshirt—Saturation stains front; crusty thick staining; not significant staining on the back side

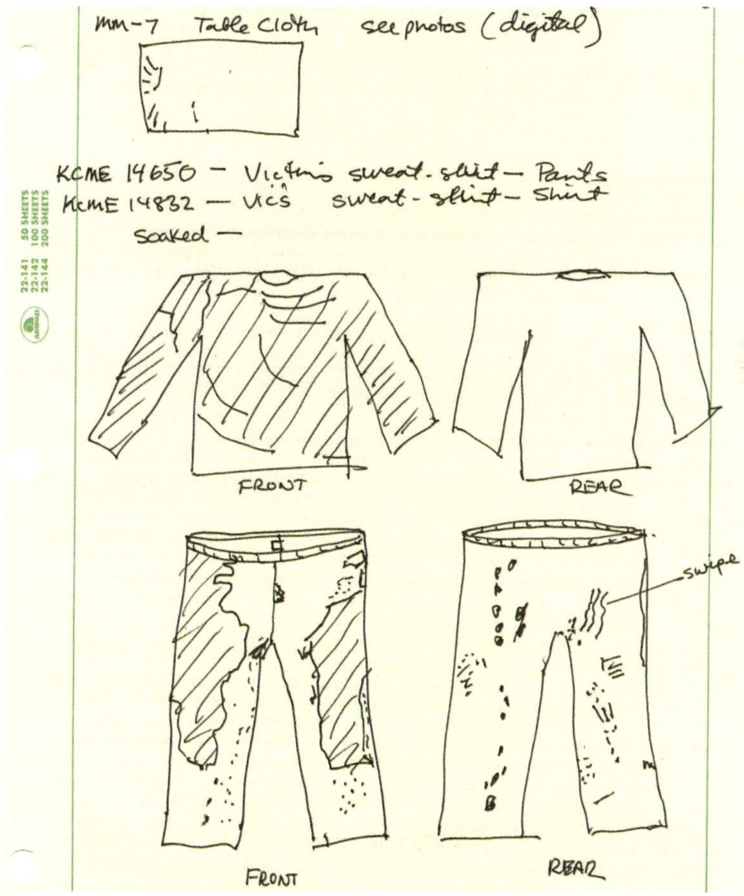

Figure 13.9 Sketches of the tablecloth and the decedent's clothing.

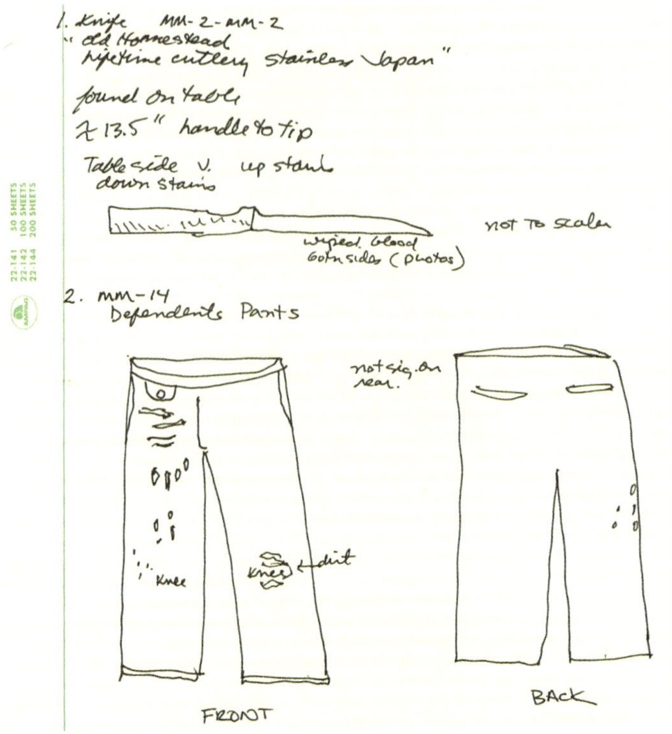

Figure 13.10 Sketches of the knife and the defendant's pants.

Socks—Dripped stains on front of socks exposed above slipper
Slippers—Both soles covered with stains; tops with dripped blood; soaked bloodstains

Defendant's Clothing

Shirt, front and rear—Saturation stain on right side soaked through to the back side; contact transfer stains, dripped blood; possible impact spatter left front and rear
Pants—Transfer stains, dripped, possible impact stains, not significant staining on back side

Rough sketches here serve to jog the memory more than to aid documentation.

Tablecloth and Plate

Movable items, such as objects on a table and tablecloths, require careful observation. The stains and voids allow some interpretations of placement during the blood-shedding events, but caution must attach to any such conclusions. Here, I consider the movable items and then consider the floor, walls, and ceiling.

For the movable objects, I documented projected stains on the tablecloth; directional stains on the top of the plate; a relatively large contact transfer stain with gravitational flow apparently from a bloody hand putting the knife on the table. Impact stains also appear on the plate showing directionality. A pair of eyeglasses also rests on the table. One lens is missing and bloodstains are present.

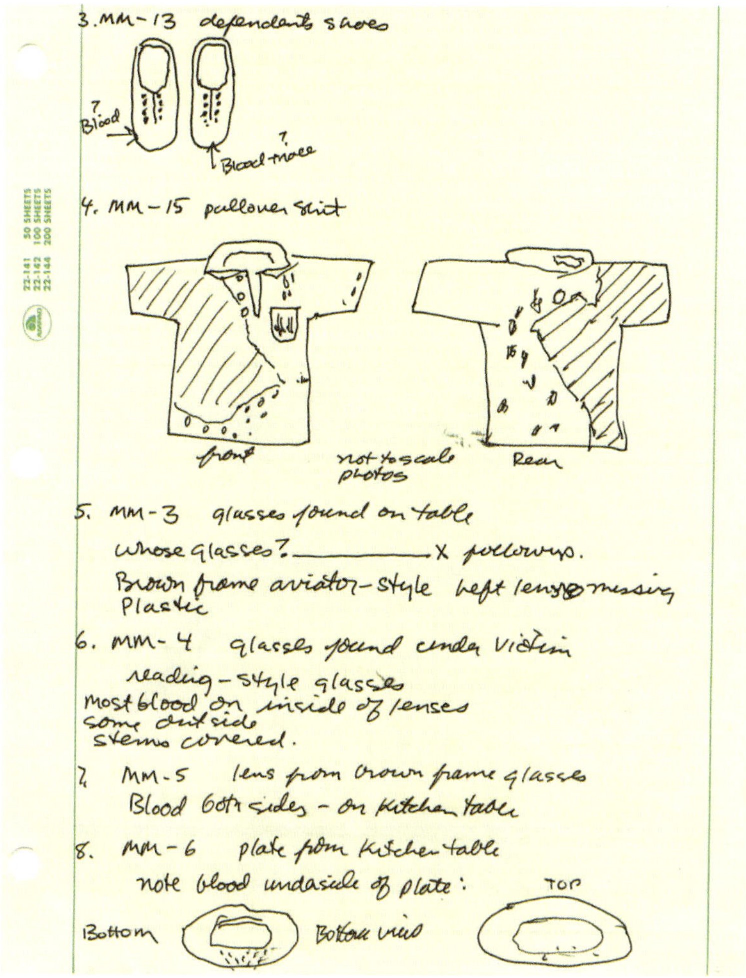

Figure 13.11 Sketches of the defendant's shoes and shirt, along with notes on other items found at the scene.

Crime Scene Analysis: One-Story Residence

Scene Conditions and Diagrams

When I revisited the scene some time later, the house had undergone extensive remodeling. The kitchen floor had been replaced, and the cabinets separating the kitchen from the dining room had been removed. The walls, ceiling, and cabinets had been painted an off-white color. The same refrigerator occupied the same space as it did the day of the brothers' struggle, but new light fixtures had been installed. I went to remeasure the kitchen walls and floors to double check the distances and size of the kitchen scene. The table, chairs, and the other items of evidence, of course, were long gone. Rechecking measurements not only helps to orient the photographs of the scene, but it also provides an opportunity to verify the data. In the course of my measurements, I discovered that previous investigators had incorrectly oriented the house on the north–south compass. I was able to confirm, by going through the data, that this error raised no problems for any data analyses and resulted in no further errors.

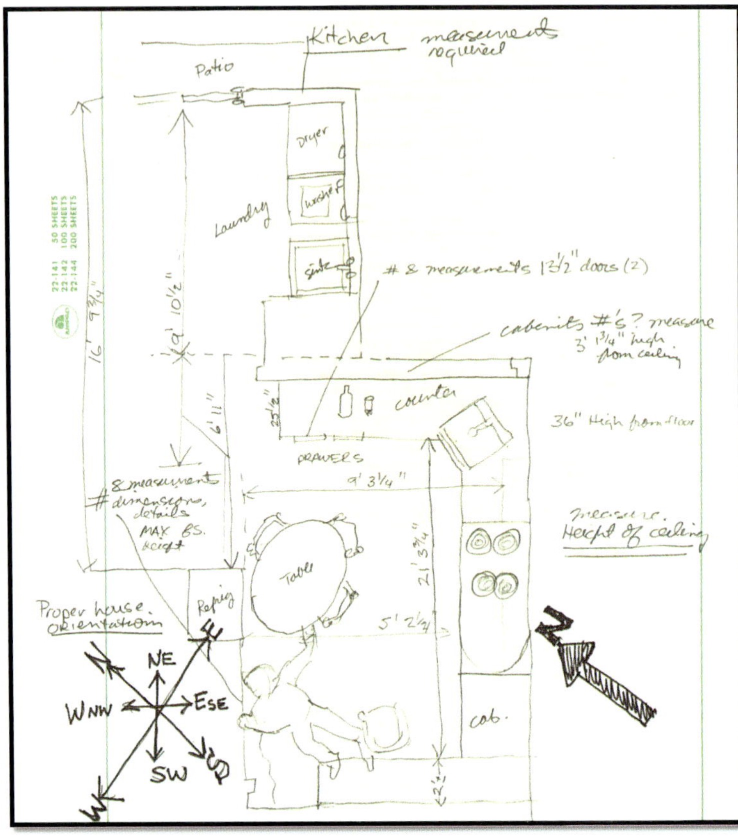

Figure 13.12 Notes for any later 3D modeling require measurements of the relevant objects along the Z axis, as well as along the X and Y axes.

The rough sketch is just that, very rough. As notes, sketches provide an orientation for the photographs as well as a check of the measurements (Figure 13.12). They can then be used when a final scene drawing captures the evidence and its locations in the house. I have used the old reliable 2D scene drawing and computer-aided design (CAD) programs for many years. But I now find that a 3D approach provides better documentation and much more useful data. With proper 3D documentation, reliable scene measurements can be made directly from the images long after the scene is gone. The orientations of bloodstains, bullet entries, and bullet exits in the rooms, as well as cast-off bloodstains on ceilings, are just some of the items that can be usefully documented using this approach. It remains critical to remember that events occur in three dimensions, not in just two dimensions. In the Cartesian approach to drawing, there is an X, a Y, and a Z axis.

The main deficiency with this crime scene's documentation concerned the absence of any specialized measuring devices in the photographs that the crime scene unit took of the bloodstains on the walls, table, and floor. The exception was the photos of the floor with an ordinary ruler beside what appear to be bloody footprints. Unlike ordinary rulers, the American Board of Forensic Odontology (ABFO) measuring devices become a necessary part of photos, documenting the physical evidence in cases of blood shed. These ABFO rulers use a circle to help us reorient photos to 90 degrees to determine the length and width of relevantly selected stains. These measurements, in turn, are used in calculations that help the analyst to interpret the stains. The rulers also have a small photo calibration section of gray and black that can be used the same way that brightness and contrast sheets have been used for silver emulsion film, as well as for digital images.

The crime scene drawings presented similar deficiencies that needed to be overcome. Ordinarily, some method is adopted to locate significant objects in the scene. For example, to locate a chair, we might pick a wall, and measure from the wall to a fixed point on the chair such as the right front leg. To locate that chair in the scene precisely, we might then choose another wall that is perpendicular to the first wall and measure the distance from that wall to the same right front leg. Thus, the item of evidence, in this example the chair, can be located for future reference, even if it is later moved. In this case, items of evidence were not located for any future reference by using any of the available methods. This can potentially present problems for scientists and investigators who must perform their work on the evidence, long after the evidence has been moved, relocated, or even destroyed.

Accurate and complete documentation of the crime scene is one of the most fundamental and important tasks in any criminal investigation. It is literally the data to be used in a scientific crime scene reconstruction. Checking the data whenever possible is an essential step before developing any scientific reconstruction. This checking does not consist of finding fault, or assessing blame, or addressing competence. The purpose has historical and substantive precedence. The work is only as good as the data. Any analysis based upon inaccurate, faulty, or otherwise corrupt data has the value of one sand grain in the middle of the Sahara Desert.

Additional problems facing any eventual analysis of this event focus on the fact that blood was shed on many surfaces that move just as humans involved in a struggle move, items such as plates, tables, tablecloths, and chairs. Since they are moveable, they are not as useful as fixed objects, such as floors and walls, for establishing the relative positions of participants in an active struggle.

However, despite the absence of standard measuring devices in the scene photographs, the photographs of the scene are of sufficient quality and quantity to overcome the absence of measuring devices. Instead, we use the photographic depiction of the type of bloodstain and the relative location of the stains to reconstruct the movements of the participants in the struggle resulting in this sharp-force injury death. Anytime a digital camera is used, we have a potentially valuable measurement tool. Scene photos can be used to reorient the pictured items and with photogrammetry as one approach, we can often overcome problems of omitted measurements and inadequate documentation.

We can also rely upon photographs to help us understand the relative positions of objects moved during the struggle, such as the table and chair, in lieu of more detailed scene drawings. Despite these difficulties, the preservation of the clothing evidence and the photographs of the scene provide ample documentation for scientific work in this particular case. In another case, however, the type and volume of data available here might be inadequate to accomplish the tasks demanded by other circumstances.

Bloodstain Patterns at the Scene

Walls—No stains higher than 36 inches reported, or observed in photographs of the walls, refrigerator, and table (Figure 13.13)
Ceilings—No stains reported, recorded, or observed
Furniture—Table, tablecloth, chairs, cushion with bloodstains (Figure 13.14)

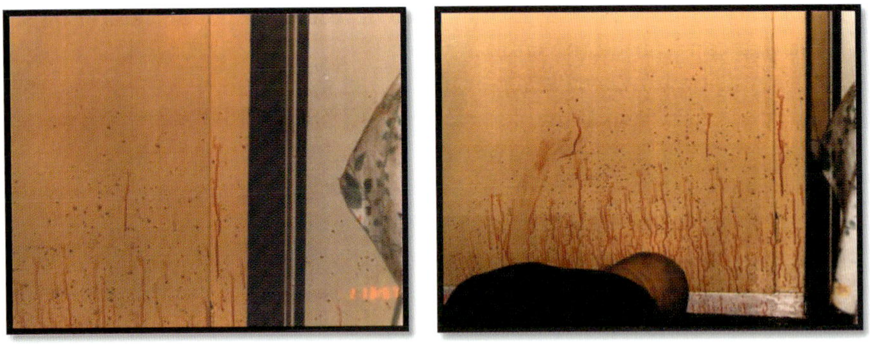

Figure 13.13 Wall stains.

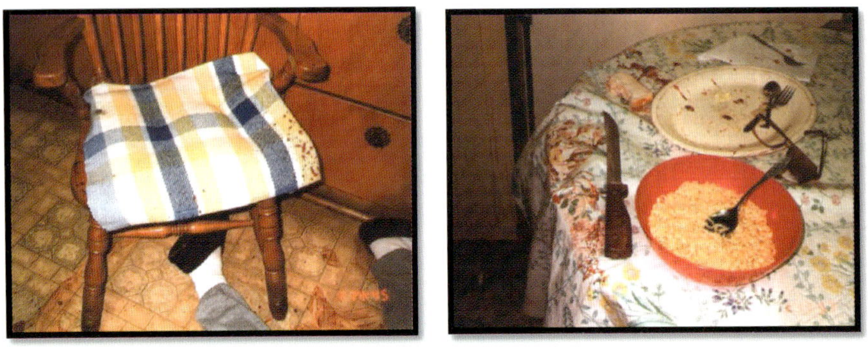

Figure 13.14 A chair, the floor, and the victim's feet on the left; his food with the tablecloth, plate, glasses, and the knife appear on the right.

The original police visit to the scene did not result in any collection of measurements. All references were nonspecific regarding the size and shape of bloodstains and their relative locations. Sketches with measurements were absent as well. None located items of evidence such as the plate, eyeglasses, eyeglass lens, table, and chairs, although a scale drawing of the overall scene did prove helpful. Developing measurements from photographs is time consuming and not always possible. It is always better to take measurements at the scene even if it is your belief at the time that they will never be used and that they amount to a lot of work to no apparent end. Again, we never actually know what may become important in the future. It is best to have the information rather than to explain later why you do not.

Crime Scene Reconstruction

Whenever a reconstruction's conclusions are stated and numbered in a report, it helps to reassess exactly what provides the basis for each conclusion listed and what kind of support those bases provide. Some statement summarizing the overall status of the project also assists those reading the report to appreciate the reasonable degree of scientific certainty that attaches to the work. Whenever the statements are too specific for the data or lack documented support, they do not warrant inclusion: best to dump those statements. A scientifically based reconstruction is not the same as a jolly good story. Unlike a good story, there will be gaps.

Figure 13.15 This photo shows bloodstains on the tablecloth—IR photos can often make the bloodstains visible on a blank surface to eliminate distracting design features of any target object.

As another exercise, after stating the conclusions, write down the supporting reasons. I use this to prepare for testimony as well as to confirm that the conclusion actually follows from or is supported by the data presented. If not, then more work is needed and that statement has to go.

This sequence of events could occur within a few minutes, once the struggle escalated, and some of or many of the events may have occurred simultaneously. It does not account for all the events, nor does it include subjective properties such as motive with the conclusions.

1. The decedent is sitting at the kitchen table but his legs are not under it and the suspect is standing; while in these relative positions, a struggle ensues.

   Void patterns on the seat cushion; dripping on the left side of the seat cushion; dripping and possible impact spatter on the crotch and areas between both legs of the decedent's sweatpants; dripped blood on top of the victim's socks and slippers.
2. The struggle escalates; a kitchen knife cuts the decedent, still seated at the table, producing minor cuts to the front of his neck and two major cuts to his right fingers.

   Bloodstains require a source; blood dripped onto the floor around the victim's chair; blood on the tablecloth wipe transfer or swipes (Figure 13.15); damaged eyeglasses found on the table.
3. The decedent remains seated for a few moments while bleeding as the struggle continues.

   Blood dripped beside the chair onto the floor; blood dripped onto the seat cushion, small bloodstains on the plate under the rim–impact spatter (Figure 13.16), blood dripped into blood, expirated blood.
4. The suspect moves between the decedent and the table, the table moves toward the refrigerator; the decedent's chair moves slightly during the continuing struggle.

   The tablecloth shifts, the table is pushed up against the refrigerator (floor marks); all major staining appears on the front or side of both the victim and the defendant.

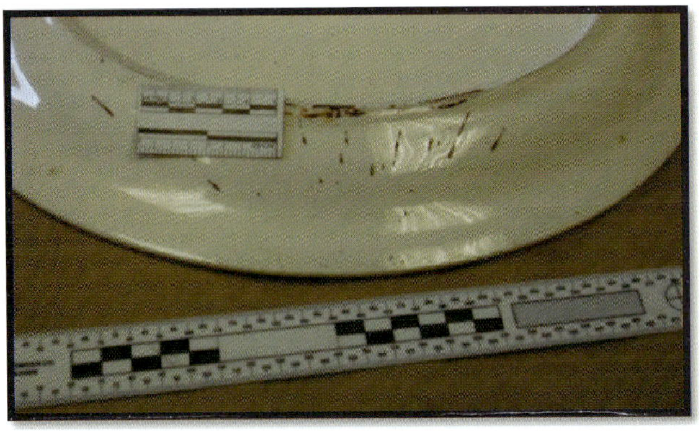

Figure 13.16 Apparent impact stains on the side of the plate.

5. The fatal injury occurs while the decedent is still seated or just as he is attempting to stand. The suspect faces the decedent or faces him slightly to the decedent's left.

   Arterial spurting is blocked by the defendant; no stains appear on the walls or the refrigerator higher than about 36 inches from the floor; projected blood appears on the wall to the side of the victim's chair and would have been higher if the victim was fully standing.
6. A force moves the neck while the knife is held in the neck, or a force moves the knife while the neck is held, or some combination of these movements occur while the knife remains in the decedent's neck.

   The knife blade is bent toward its hilt; the injury exhibits a butterfly-style appearance indicative of a knife recutting the entry wound when it is moved in a different direction (Figure 13.17).
7. The suspect is standing bent forward at the waist, slightly to the decedent's left and remains in close contact with the decedent.

   Blood soaked the defendant's right side; stains on the victim's clothing show soaking on the left side and contact transfers appear on both sets of pants

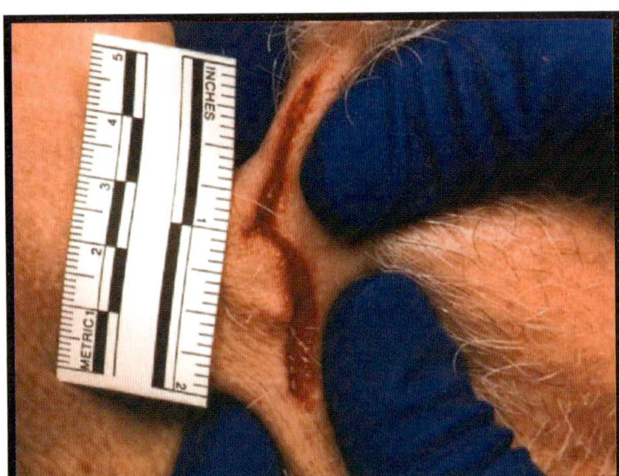

Figure 13.17 This photograph shows entrance defect in the victim's neck articulated at the autopsy. Note the movement of the knife through the tissue.

Figure 13.18 The victim's sweatpants on the left; the suspect's pants on the right. Note the knees on both.

(Figure 13.18); arterial spurting is directed low on the wall, bloodstains on the floor remain near the victim's chair no higher than 36 inches on the wall.

8. The decedent goes to his knees, either from a semistanding, or seated position; his feet slip through the bloodstains on the floor around his chair.

   Slippers show blood on the soles and a wiped stain appears near the base of the victim's chair; blood appears on the knees of the victim's sweatpants.
9. The suspect removes the knife, moves away from the decedent, and puts the knife on the kitchen table.

   A bloody hand puts the knife on the table; the decedent could not reach the table; he could not place the knife on the table; the knife entered the decedent's neck from the front and exited from the upper back.
10. While the decedent is on his knees, he slides forward, face to the floor.

    Low stains on the wall; wiped stain on the floor; final position of the victim.
11. The decedent moves his left arm, smearing blood on the wall with left hand.

    Smear on the wall; transfer swipe and wipe of the wall stains with a blood trail leading to his hand from this movement.
12. The decedent's left arm slips to the floor; he remains in this position when discovered by investigators.

    His arm rests on the floor after the swipe/wipe stain was made on the wall within reach of his left arm.
13. The suspect walks out of the kitchen with a black portable phone, calls 911, leaves the house through the back porch, and throws the phone into the grass.

    Footprints in blood were photographed and measured by police; the phone was found outside on the grass.
14. At some point, the suspect's left knee contacts ordinary yard dirt and grass.

    Dirt and blades of grass are found on his left knee over the bloodstains.
15. Suspect remains at the scene until investigators arrive.

    Suspect is arrested at the scene.

The bloodstains on the walls, floor, table, chair, and cushion with void areas, along with both brothers' clothing, provided most of the data in this case. During testimony, I was asked to consider and answer more speculative questions. The questions required the consideration of data from many other cases, as well as the data from this case. Such questions are much more speculative and require judgments based upon data that often remains unstated. That is one problem with such questions: no one can examine the data because the data remains invisible, locked inside one's case experience. I find it best to avoid that appeal to unstated data. When asked to address such issues, I try to limit the answers exclusively to data from the case at hand. This raises obvious questions about the status of questions requesting assessments of error rates (see Appendix A).

Further interpretive inferences from the evidence:

a. This scene does not show that the victim attacked the defendant (the victim remains seated until fatally injured) as the defense claimed.

   To gain the most basic mechanical advantage, the victim would stand up to attack his brother. The victim did not stand up and left no signs of attacking the defendant while eating at the table.

b. This scene does not support any claim that the defendant acted in self-defense, as the defense claimed.

   The victim could not have attacked the defendant while he remained seated and the defendant remained standing. Therefore, the defendant could not be acting simply to defend himself. The victim remained seated when cut by the knife.

Photographs from Final Analysis Forensics

See Figure 13.19 to Figure 13.28.

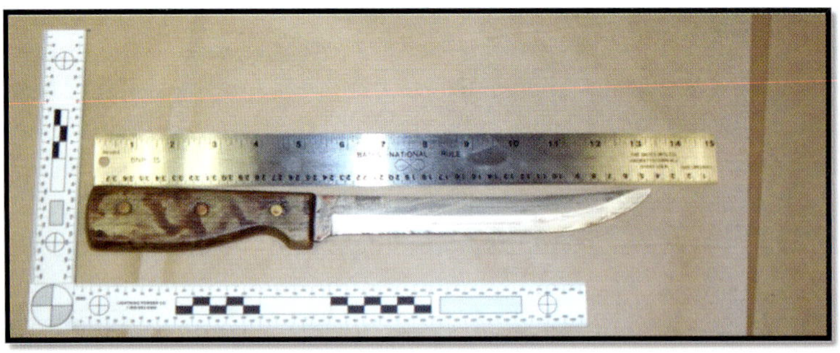

Figure 13.19 The knife used in the stabbing death of the victim.

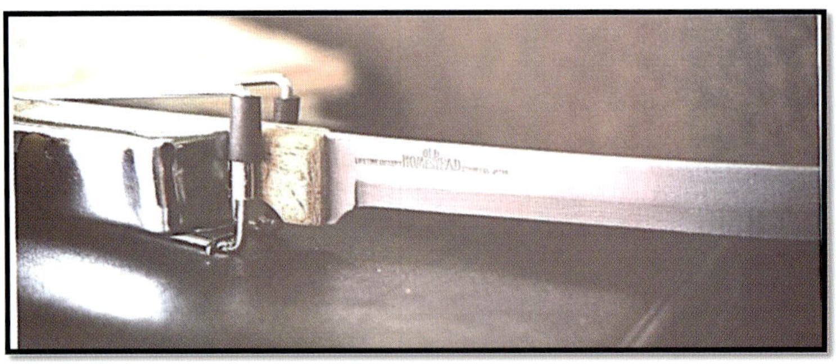

Figure 13.20 Another test using the same brand and model knife with the stab-o-meter.

Figure 13.21 Another angle of the knife used in the stabbing death of the victim.

Figure 13.22 A similar knife showing the use of the stab-o-meter to measure the force in either pounds or kilograms, required to push this knife through a pig's neck. The testing showed no differing values for both knives.

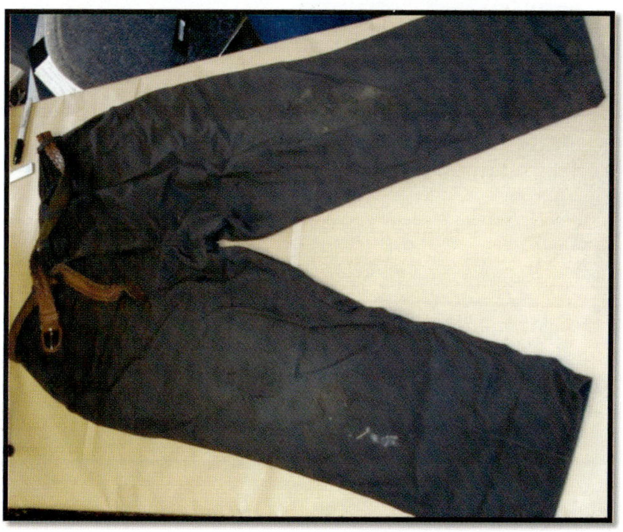

Figure 13.23 The defendant's pants showing bloodstains; again, note the knee areas.

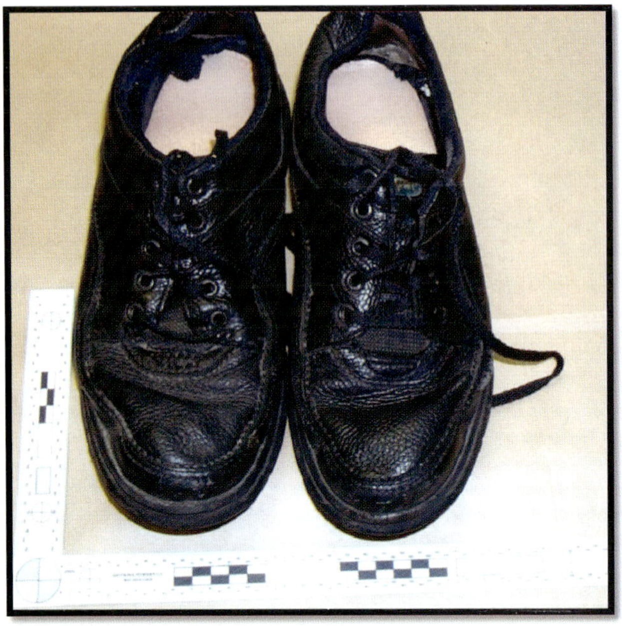

Figure 13.24 The defendant's shoes provided no significant information other than the presence of blood on the soles. (See the bloodstain patterns on the floor.)

Figure 13.25 The front of the defendant's shirt showing extensive soaking on the right side caused by arterial spurting from the victim.

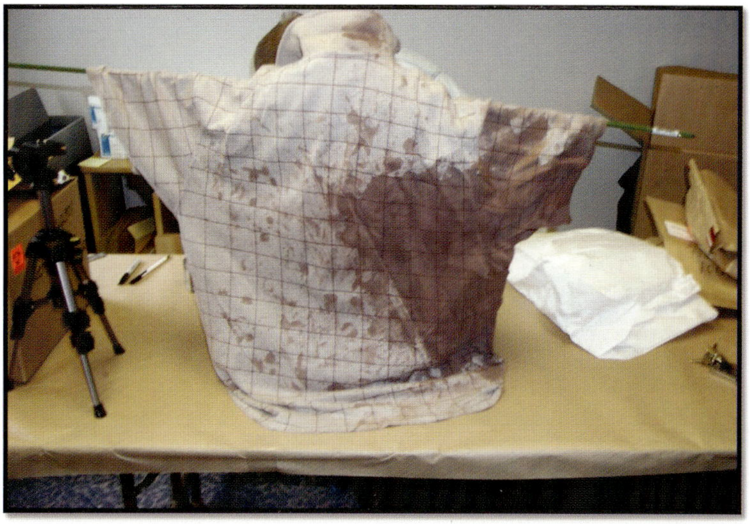

Figure 13.26 The back of the defendant's shirt showing extensive soaking on the right side caused by arterial spurting from the victim.

Figure 13.27 A pair of glasses found on the table with the left lens missing.

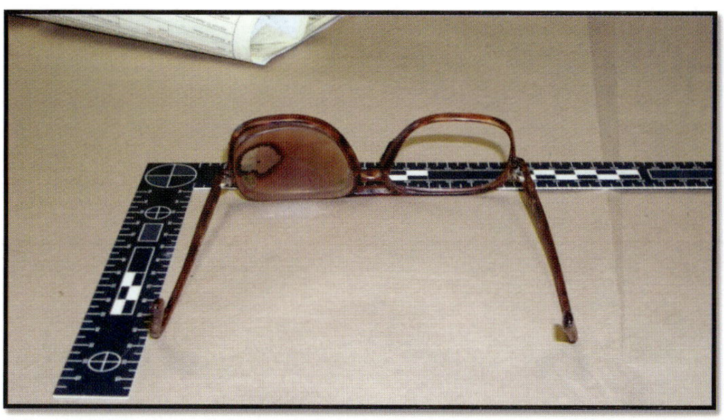

Figure 13.28 The back of the glasses found on the table. Bloodstains on the lenses may in some circumstances provide important data (see Chapter 2).

Figure 13.29 The tablecloth exhibits complex bloodstain patterns, meaning that the patterns result from differing mechanisms.

In addition, bloodstains on the tablecloth can be seen in Figure 13.29 to Figure 13.32. The cloth is made of plastic and does not absorb fluids into the coated fabric, allowing the bloodstains to flake off when dried. This type of bloodstain evidence should not be folded unnecessarily. The stains should be documented at the scene.

The decedent's sweatshirt (Figures 13.33 and 13.34) presents multiple bloodstain patterns reflecting his multiple positions after sustaining his fatal injury.

The decedent's sweatpants (Figure 13.35 to Figure 13.39) also provide valuable data in the form of bloodstain patterns. The multiple stains should be described and then classified by applying a definition. Most such definitions provide some insight into the mechanism producing the stains. The Scientific Working Group on Bloodstain Pattern Analysis (SWGSTAIN) has been charged with developing a consistent set of bloodstain pattern terms as well as documenting the scientific bases for the analysis of bloodstains. The photos of the decedent's pants show the various types of bloodstain patterns, including what are called cast-off stains, swipe stains, and transfer stains (see Appendix B).

Figure 13.30 Capturing these patterns can be challenging at a scene unless one has time to set up a digital IR camera and be prepared to test multiple filters.

Figure 13.31 Often varying the gamma and brightness values helps to improve the contrast between the blood and its target.

Figure 13.32 Photos such as this one can be adjusted to picture a right-angle view using the ABFO scale and warping.

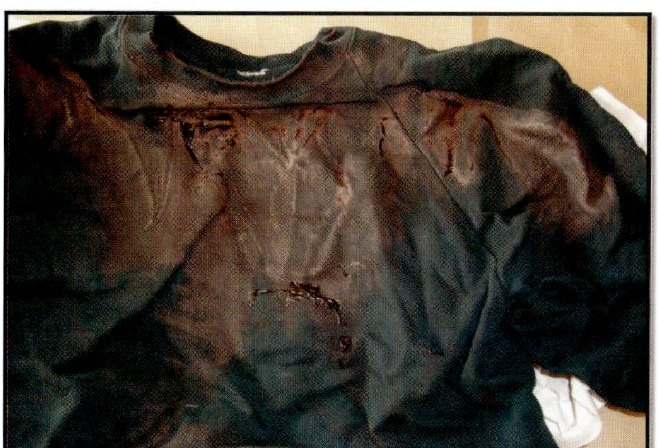

Figure 13.33 The decedent's sweatshirt.

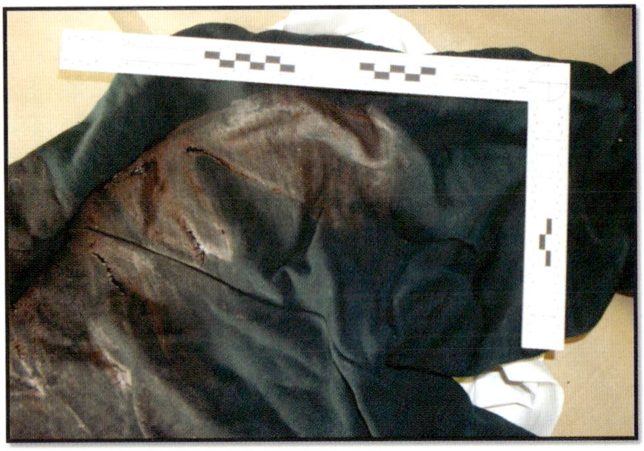

Figure 13.34 Bloodstains on the decedent's sweatshirt.

Figure 13.35 Closer view of the decedent's sweatpants.

Figure 13.36 Another view of the decedent's sweatpants.

Figure 13.37 Multiple stains appear on the garment from multiple mechanisms.

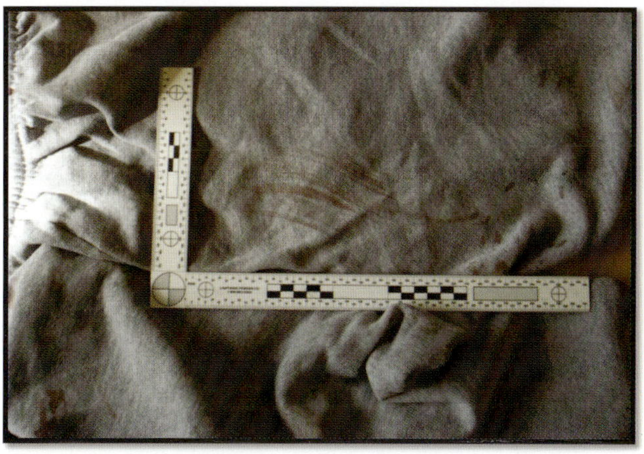

Figure 13.38 Bloodstains on the decedent's sweatpants.

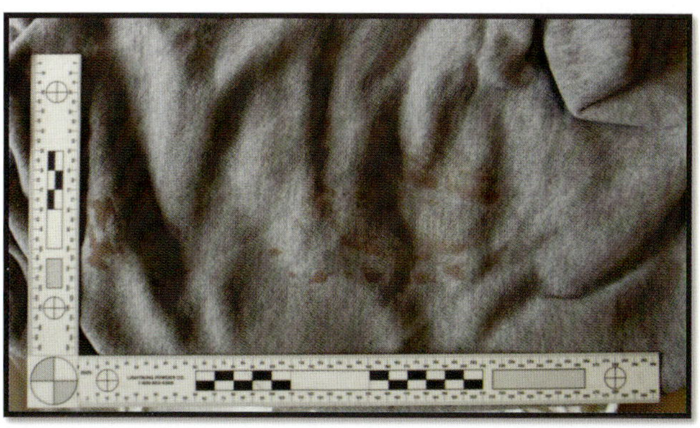

Figure 13.39 Closer view of the bloodstains on the decedent's sweatpants.

The decedent's slippers exhibit bloodstains on the tops and sides (Figure 13.40 and Figure 13.42). The dark discoloration on the soles comes from contact with blood (Figure 13.41 and Figure 13.43), which the wipes on the floor confirm as well.

The "stab-o-meter" was used to measure the force required both to penetrate and perforate (the length of the knife) a pig's neck (Figures 13.44 and 13.45). Although obviously not identical, the pig's skin and neck structures relevantly resemble the structure of human skin and muscle.

Measurements recording the force of both penetration and perforation (the length of the knife) were recorded for purposes of learning the maximum force required to penetrate

Figure 13.40 The decedent's slipper—top.

Figure 13.41 The decedent's slipper—bottom.

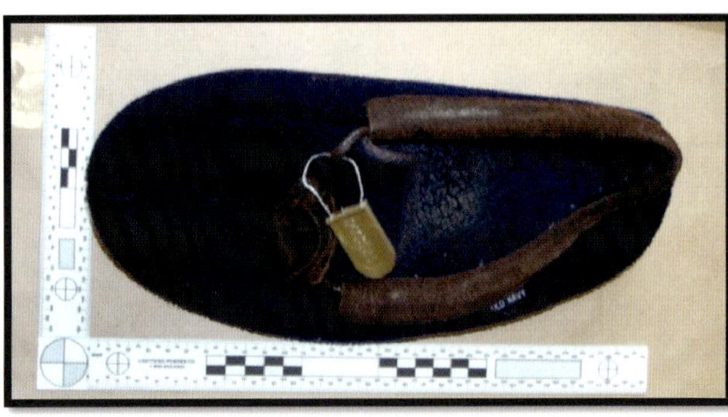

Figure 13.42 The decedent's slipper—top at a different angle.

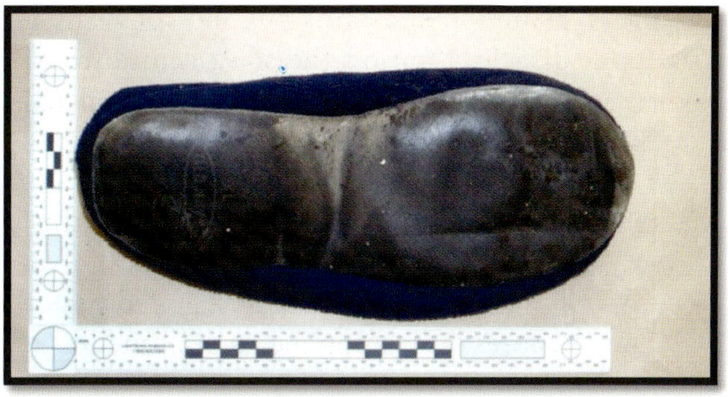

Figure 13.43 The decedent's slipper—bottom at a different angle.

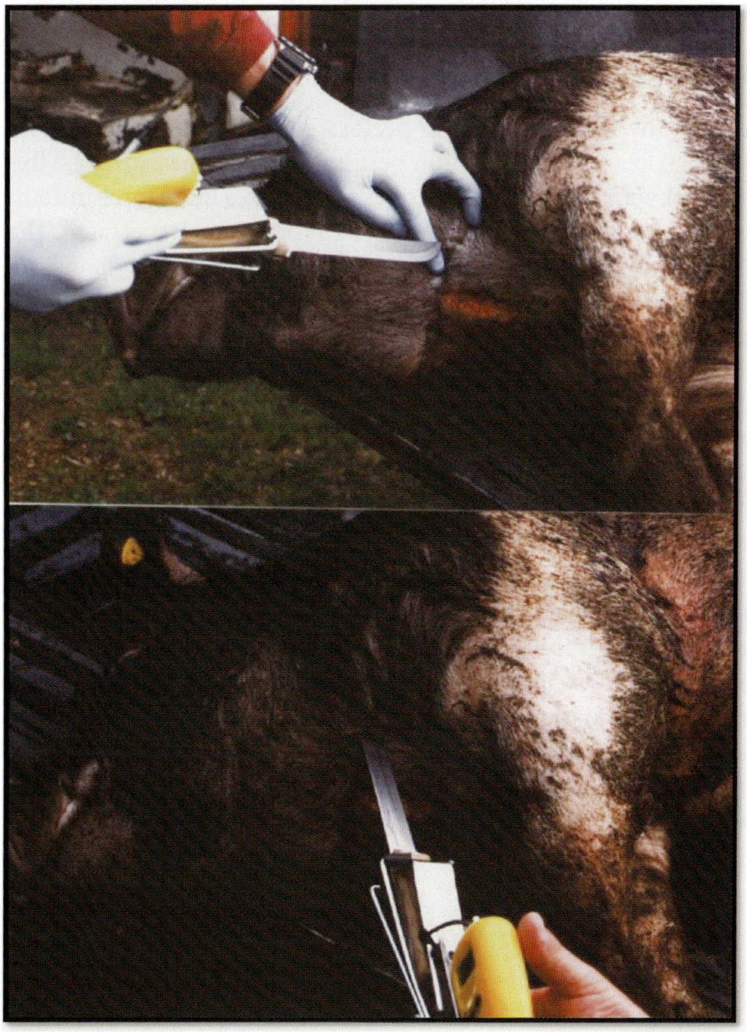

Figure 13.44 Using the stab-o-meter to measure forces required to penetrate the neck of a euthanized pig.

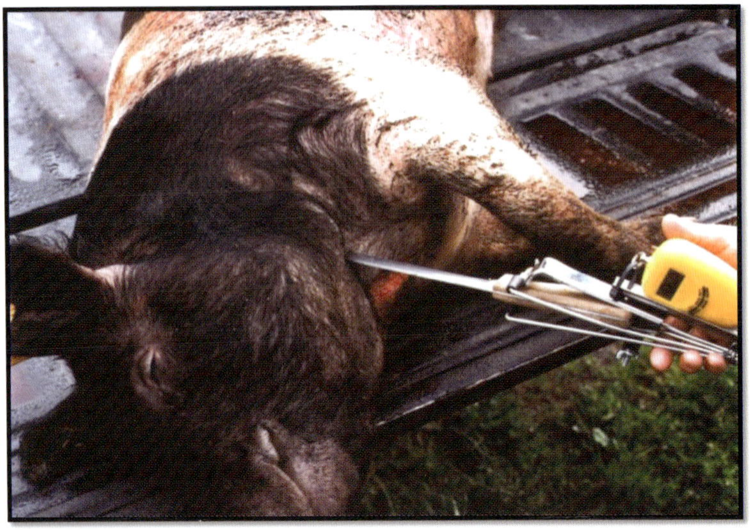

Figure 13.45 Another angle using the stab-o-meter on the euthanized pig.

the neck and then the force required to perforate the structure. Maximum penetration force measured 13.0 lb and the maximum perforation force measured 2.0 lb.

The experiment with the pig demonstrates that it takes more force initially to penetrate the tissues than it takes to move the knife through them once penetrated. That tells us that it took very little force for the knife to continue along its path to the victim's back. That could have happened when Raymond tried to help Bert stand up. But who knows? There are always elements of a reconstruction that remain missing, always questions that remain unanswered. Sometimes that is because we are not clever enough to develop some physical data to answer them. Sometimes it is because there just are no such data. When that is true, science has reached its limit.

Subjectively, I cannot help but think that both Raymond and Bert were a bit surprised by the volume of blood. Yet, this does not seem to be a case cloaked with the same sort of mystery as the dagger's travels into and out of Stephen Vaughn's pericardium (Chapter 9). The knife here was bent and the wound showed the knife's several movements. Certainly, some force was required to bend the knife. How was that force applied? Did the knife bend when Raymond plunged it into Bert and held its handle when Bert moved, trying to lift him or move him with it? Or did the knife move when Bert landed face down on the floor? It takes some force to bend the blade at the handle, in fact more force than I could measure within the limits of the calibrated scale of the stab-o-meter. Once the skin is penetrated, very little force is required to push the kitchen knife through the tissue of a human neck. How did the skin first come to be penetrated? This question, like the others, must remain forever unanswered. We just cannot get there from here. Sometimes "here" is a lonely place to be.

Slow But Sure
Testudines Chelonia

14

... pray you'll never know
The hell where youth and laughter go.

Siegfried Sassoon, "Suicide in the Trenches," *Counter-Attack and Other Poems* (1918)

Introduction

The three boys (Figure 14.1) left the bicycle in a field to go play by the bayou. They did not come home for supper. They never would. Their lifeless little bodies were found submerged in the water, face up, with their hands secured behind their backs, bound tightly with their own shoelaces. All of their clothes lay piled on the muddy bank. Under the opaque water, the hours turned into days before they were found.

Once the bodies were discovered, detectives arrived and drained the bayou to search for evidence (Figures 14.2 and 14.3). They recovered a muddy fingerprint on the bank, near the bodies, along with the standard miscellaneous items of as-yet-indeterminate probative value. It was obvious to wide-eyed detectives that the little bodies had stories to tell, and those stories developed loud and clear, amplified by the autopsies. Enter their main suspect, Lucifer, T. D., alias Mr. E, Satan, Master Brimstone, the master thug wanted for every crime against humanity since God created the Garden of Eden.

Some months before the gruesome discovery in the pond, local police officials had attended training on "Satanic Cults and Their Evil Practices." Detectives became almost hypervigilant in their pursuit of the forces of evil operating among them. When the boys' corpses were found floating in the pond, the evidence was as clear as a revelation from God himself: the boys had been tortured and sacrificed in an unholy satanic ritual. One lad's genitals were missing. Each of the three small bodies displayed innumerable sharp-force injuries, abrasions, and lacerations, exactly as the facilitator at the Satanism seminar had described. Evil lurked and worked among them in Arkansas.

The three suspects, quickly identified by their black clothing, garish makeup, assorted chains, and tortured tunes, lived rebelliously in the Goth tradition, popular among some conforming nonconformist teens in the 1980s. Their unholy music, their backtalk, and their contempt for authority confirmed their guilt in the eyes of the state and the public. They were arrested, tried, convicted, and sentenced for capital crimes. The alleged ringleader, Damian Echols, received the death penalty. He lingered on death row for 15 years before the case came to me.

This story will undoubtedly be told eloquently and often by others. HBO has already produced two specials on the case. There are thousands of details, with hundreds of work hours filled by diligent attorneys, private detectives, forensic scientists, and one dedicated

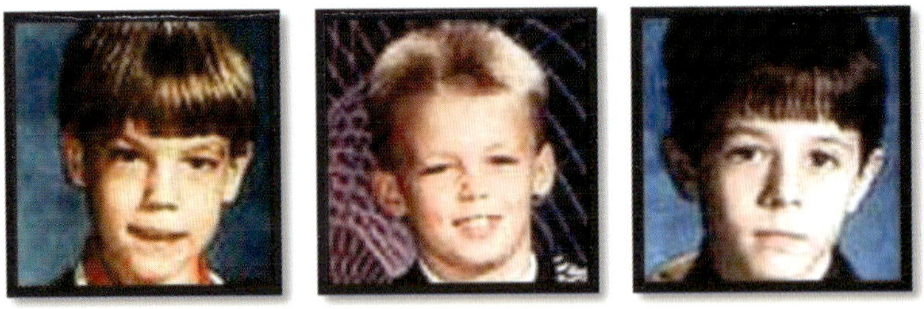

Figure 14.1 (From left) Christopher Byers, Steven Branch, and Michael Moore.

Figure 14.2 Wooded scene on the bank overlooking the water.

New York architect, Lori Davis.* She believed in the innocence of Echols and his codefendants, as did a number of talented, influential entertainers and musicians whose sense of justice had been inflamed by the television specials on the case. Although the case has been presented in various written, television, musical, and film formats, I focused on examining only the physical evidence.†

Despite the intervening years, many items of potential evidence were available for my examination. One victim's house remained unoccupied for many years. Another's house, while occupied, remained largely unchanged and available for my investigation. Private investigator Ron Lax, Inquisitor, Inc., was able to recover a truck owned and driven by the father of one of the victims at the time of the murders. None of these items had a high

* She later became Damian Echols's wife.

† Three films, *Paradise Lost: The Child Murders at Robin Hood Hills*, *Paradise Lost 2: Revelations*, and *Paradise Lost 3: Purgatory*, were directed by Joe Berlinger and Bruce Sinofsky. Metallica permitted its music to be used for the first time in a film production. Several books have been written about the case. *Blood of Innocents* by Guy Reel; *The Last Pentacle of the Sun: Writings in Support of the West Memphis Three*, edited by Brett Alexander Savory and M. W. Anderson; and *Devil's Knot* by Mara Leveritt. Damien Echols also wrote a book, *Almost Home, Volume 1*. Rock, hip hop, punk, and metal groups such as Black Flag and the Rollins Band–produced album *Rise Above: 24 Black Flag Songs to Benefit the West Memphis Three*. In 2002, Zao's album *Parade of Chaos* included a song "Free the Three." Investigative journalist Aphrodite Jones presented the case on her Discovery Network show *True Crime with Aphrodite Jones*. Web sites have also documented the case, keeping the issues in the case before the public.

Figure 14.3 Apparent friction ridge detail in the mud on the back at the scene, as recovered on the right.

Figure 14.4 Stains thought at one point to be blood were in reality cockroach scat.

probability of developing useful evidence in the homicides. But the old saying remains true: one does not know if one does not look. So I looked.

I methodically searched each house and the truck looking for bloodstain patterns that might lead us to develop additional evidence in the case (Figures 14.4 and 14.5). My search of the two houses, the truck (Figure 14.6), and the associated areas consumed many hours, weeks, and months.

Visiting the houses where the boys once lived gave rise to many questions: Are the stains we see here blood? If so, whose blood? When was it shed? The problem is seldom having too little to examine. There is usually too much to examine. Which is evidence and which is not? The coloring on the wall and the small specks seen in Figure 14.4 initially look like impact bloodstain patterns. The marks and the patterns are cockroach scat.

The searching and testing process fails to sustain any action-packed drama suitable for mainstream forensic television. The labor is no less difficult despite the lack of probative results. The results here were singularly disappointing. I looked. Ultimately, nothing developed from the searches.

A scientific reconstruction ultimately remains limited to the available data. If the data is lacking, or if additional relevant data cannot be developed, then the reconstruction goes only as far as that data and the associated sciences allow. This case presented that limitation. The results become limited, modest, and brief.

But modest results can present important insights, no matter how limited. The scene and autopsy photos, despite their paucity, spoke volumes when combined with information about the ecology of the area. During the original trial, the evidence that prosecutors

Figure 14.5 Luminol is very sensitive but not specific to blood. Its use to detect blood must be tempered with an understanding of the other mixtures that produce a similar reaction. Here, I used Blue Star and developed a positive reaction in the bathroom where one of the boys lived so long ago.

Figure 14.6 The recovered truck could still provide useful physical evidence even after so many years. That depends both upon what and how one seeks.

used against the teens seemed inadequate and undeveloped. In fact, it never even remotely supported bringing charges against the three in the first place.

The Evidence and the Trial

The three teens who were tried and convicted of the 1993 murders of the three little boys in West Memphis, Arkansas, were called "The West Memphis Three" by the press. The three were Damien Echols, judged to be the leader, Jessie Misskelley, Jr., and Jason Baldwin (Figure 14.7).

During the trial, the prosecution asserted that the three innocents had been murdered and tortured in a satanic ritual. The prosecution's case consisted of evidence heavily influenced by the detectives' original postseminar reactions to the death scene. The medical examiner apparently viewed the boys' injuries through the bias introduced by detectives. The medical examiner's explanation, however, should have depended solely on an unbiased interpretation of the boys' injuries.

One major thrust of the prosecution's argument stressed that the three suspects each wore black clothes and listened to heavy metal music containing references to Satanism.

Figure 14.7 (From left) Jessie Misskelley, Jr., Damien Echols, and Jason Baldwin.

The prosecution presented one witness who claimed to see Echols with muddy clothing on the night of the murders, and two teens who testified that they heard Echols confess. Defendant Jessie Misskelley confessed after a lengthy police interrogation and implicated both Echols and Baldwin, stating that they raped and murdered the three boys.[*]

The dearth of physical evidence included a knife recovered from behind Echols's home. The prosecution claimed that the knife was consistent with the marks on one of the bodies.[†] The other physical evidence is limited to fibers found in both Echols's and Baldwin's homes said to be similar to fibers found on the bodies.

Refutation of the physical evidence was equally telling. The knife blade in fact failed to match any marks on the victims' bodies, and had neither blood nor tissue present.[‡] The fiber evidence collected at the two suspects' homes also turned out to be dissimilar to fibers recovered from the scene.

The defense countered these assertions during the trial, arguing that simply wearing dark clothes and listening to heavy metal music is not equivalent to Satanism. In addition, all three teens offered documented alibis for the time the murders occurred. The witnesses exhibited a further weakness in the prosecution's case: the teens claiming to hear Echols's confession had a well-known history as liars and both disliked Echols for various reasons. Jessie Misskelley, who has an IQ of only 72, had been interrogated for 12 hours without benefit of either legal counsel or a parent or guardian present. His confession remained logically inconsistent with both facts among the police reports and with the medical evidence.

Each defendant pled not guilty. Each was convicted. On appeal, the Arkansas Supreme Court upheld the convictions. Once again, both truth and justice had apparently escaped courts of law. Postconviction appeals by various innocence projects throughout the country have shown by their success that wrongful convictions are not necessarily isolated incidents.

[*] Misskelley's confession was considered by jurors in their deliberations but never presented in court. The "confession" contained innumerable factual errors and inconsistencies with the evidence. For one example, no evidence indicated that the boys had been sodomized. Despite these lapses, it was accepted by detectives and prosecutors as a true and reliable account of events.

[†] "Consistent with" has become a typical catchphrase often used in courtroom testimony as well as in professional reports. The claim is actually very weak. In fact, every statement that is not in direct contradiction with a given statement is consistent with the claim. The issue remains: Does the evidence explain the injuries better than any other alternative explanation?

[‡] Later investigations proved that it had nothing whatever to do with the homicides.

The Injuries

A case of this complexity requires the efforts of many professionals working together to obtain justice for the victims. If the state has the incorrect solution, then a murderer remains free. This, many argue, is not the way to be tough on crime and criminals. In that event, the families of both the victims and the convicted become additional victims. If the evidence fails to support the convictions, then additional evidence must be discovered to solve the murders, one way or the other. The place to begin reworking the case was the evidence of the injuries suffered by three small boys. The autopsy offered the best evidence of the crime.

Incised Wounds, Lacerations, Contusions, and Abrasions

Incised wounds, or cuts, usually leave smooth margins and a relatively straight separation of tissues. Lacerations, or tears, usually have ragged margins with some evidence of their mechanism, such as bruises (called contusions) on the body or bloodstain patterns at the scene. The boys had what appeared to be both lacerations and scratches, called abrasions, without contusions.

An abrasion produces avulsed skin or other tissues, which pile up in the direction of the scratch. Abrasions are familiar to anyone who has ever fallen onto a sidewalk, scraping hands and knees. Once the three boys' wounds were catalogued with their types identified, the next step in this case involved determining when the injuries occurred. Then the search for alternative mechanisms of injury can begin.

Premortem, Paramortem, and Postmortem Injuries

One notable source of error involves misinterpreting an injury's type: for example, confusing an incised wound with a laceration or an abrasion. Another error involves misinterpreting the time of injury relative to the time of death. It remains obvious that many distinct errors contributed to the investigation's profound misdirection in this case. One of those errors involved confusing a premortem injury with a postmortem artifact. That issue belongs with forensic medicine.

The autopsies in this case were performed under the auspices of a state medical examiner, which had replaced the old coroner system. A coroner* is a political, administrative position reminiscent of medieval structures in not-so-merry old England, where, historically, sheriffs tended to collect taxes from residents of their shires without providing the king his legal cut. So the king appointed a "crowner" to represent his interests in tax collection. Hence the term *coroner* developed from the word *crowner*. Death, like everything else, involved a tax. Murderers, for example, paid a tax for their actions designed to compensate

* To qualify to be a coroner in most U.S. jurisdictions, one must be at least 18 years old, alive, and elected by the voters. Things are changing for the better. Some jurisdictions, like Arkansas, have eliminated the coroner position in favor of the medical examiner system. Medical examiners are usually board-certified forensic pathologists. The coroner's position is thought by its detractors to represent an unnecessary bureaucratic expense; autopsies must be performed by someone at least trained in medicine, and very few coroners have such training. Historically, some jurisdictions employed undertakers to perform autopsies. The potential conflict of interest should be clear. Similarly, some coroners also double as sheriffs or as district attorneys. Again, the conflict of interest should be transparent.

the victim's family. But, like all taxes, it eventually became just another way to support the sheriffs' and the king's extravagant lifestyles.* Suicide became a case of "self-murder" for tax purposes largely to raise more money for the sheriffs and the king.

That helps explain the legal status of suicide in many jurisdictions as an illegal action—self-murder is a form of murder. Murder, by definition, is wrongful killing. That results in taxes for the king. The more things taxable, the happier kings are. Homicide is neutral with respect to right and wrong—a homicide is simply one person killing another for acceptable, or perhaps for unacceptable, reasons. If it is unacceptable, then it is a murder and taxable for kingly purposes.

The autopsies on the three victims were performed by a state medical examiner with medical credentials. In Arkansas, the medical examiners are on the staff of the Arkansas State Crime Laboratory. Four of the six state medical examiners now working for the Arkansas State Crime Laboratory are certified by the American Board of Pathology. Frank J. Peretti, Jr., M.D., P.A., the associate medical examiner who performed the autopsies on all three murdered boys, is not.† That, of course, does not mean that Peretti is or is not a competent medical examiner. Only one's work can demonstrate competence.

Character assassination has no place in scientific evaluations applying critical intelligence to medical questions. In my review of autopsies, I focus on the work product and not on the worker. Whether a person is liked, admired, disliked, or reviled becomes irrelevant to the review of a document presenting allegedly reliable scientific findings. Questions about an autopsy's conclusions appropriately focus on the reasons provided in the report supporting a given interpretation of an injury, with supporting photographs, drawings, histology slides, and any other physical evidence collected at the autopsy. The work alone exhibits the presence or absence of scientific reliability. Any report stating conclusions based upon vast experience alone, absent supporting reasons, remains vacuous. Best for a scientist to leave the curriculum vitae (CV) battles, the nature of someone's training, and the quality of a person's work experience to others.

To answer questions put to the natural sciences requires knowing the reasons offered for the conclusions developed. This permits independent observation, additional or repeated tests, or the construction of creative experiments. The mechanisms that inform us about

* This is more an editorial comment on the history of taxation than a precise chronology. But it makes the point. One need only think of today's lotteries, which were originally designed and sold to the public as a way to reduce property taxes and to fund schools. Maybe some do. But most have structures that direct the profits to fund the bureaucracies generated to maintain the enterprise and to treat citizens who become addicted to gambling. Property taxes, despite the sales pitch, remain characteristically unaffected.

† According to Peretti's CV, he became a diplomate of the American Board of Forensic Medicine in 1999. This, unlike the American Board of Pathology, is a club almost anyone can join. According to the group's Web site, a member can elevate his or her status by becoming a diplomate. Requirements include holding a degree from an accredited university, successfully completing one of its credentialing programs, having at least five years of experience in a forensic-related field, and paying a fee of $250. Peretti became certified in 2005 by the American Association of Medical Review Officers, which apparently has similar standing among forensic pathologists, who populate a subspecialty in the American Board of Pathology, and forensic science professionals in general. The AAMRO describes itself as "a voluntary medical organization dedicated to establishing national standards for the certification of medical practitioners who perform drug and alcohol testing." It claims no connection with autopsy work, interpreting injuries, or establishing cause and manner of death. The requirements established for these organizations are not equivalent to board certified as used among all pathologists and all medical specialists, as well as among forensic medical specialties under the National Association of Medical Examiners (NAME) organization.

the reliability of the evidence applied to develop the cause, manner, and the mechanism of death must not go unchallenged.*

Although Peretti provided a clear cause of death (drowning in the three cases) documented in his autopsy report, he apparently did not consider observations distinguishing pre- and postmortem injuries. Nor did he consider any mechanisms for such injuries other than satanic, ritualistic torture with a sharp instrument, such as a knife. He appeared to accept the detectives' suggested satanic interpretations and was, perhaps, a bit too eager to please prosecutors. That suggestibility hampers scientific independence and scientists become advocates when they reach beyond the data into politics.

We have considered this suggestibility in the history of physics through a brief discussion of N-rays and the experimenter effect, which is the phenomenon of imposing inappropriate expectations onto the data. A form of the problem has even been described in the music of Simon and Garfunkel: "Still a man hears what he wants to hear and disregards the rest."†

When examining the autopsy report and studying photographs of the boys' injuries, unusual features appear. What are described as cuts are in fact irregular abrasions, also called scratches, with avulsed tissues piled up in the direction of the scratch. The injury described as a castration and surgically excising tissues is actually a complex laceration of the lower torso, which included ripping off one boy's genitalia. Multiple abrasions appear in various parts of the three boys' bodies. Furthermore, all of these injuries are postmortem: they occurred after the boys had drowned and were already dead.

The signs of postmortem trauma are unmistakable at both the macro and micro levels. For example, consider some of the signs showing that an injury has been inflicted before death. A bruise inflicted before death shows swelling and color changes depending upon the survival interval. An abrasion, or scratch, inflicted before death exhibits evidence of bleeding, the coagulation or clotting of blood, and a red area of inflammation with swelling around the margins. Evidence of bleeding, swelling, and inflammation will also appear in lacerations suffered before death. The tissues bleed and exhibit an inflammatory response since the blood continues to circulate during life.

Signs that an injury has occurred before death can also be seen at the cellular level. The signs of apoptosis were first studied on human skin. Histological samples taken of surrounding tissues will show signs of apoptosis, with red neurons visible under the microscope as part of the inflammatory response and the body's reaction to injury.‡ But if an injury has occurred postmortem, then none of these signs appear in the tissue samples. The tissue samples from the boys add an avenue to gather more data to establish the postmortem nature of the boys' external injuries.

* The challenge meant here must not reduce to an attack on the individual. It is meant to be a collegial, independent review of scientific work. Its purpose is to detect the strengths as well as the weaknesses of any scientific methodologies presented in a report based upon physical evidence. It is the kind of peer review central to the advancement of scientific knowledge. Such reviews form the basis for progress studied throughout the history of natural science. In fact, prior to 1846, when William Whewell coined the term *scientist*, the discipline was known as natural philosophy, or *philo sophia*, literally, the love of wisdom or knowledge. This scientific approach remains poorly understood by lawyers educated in adversarial techniques and approaches. The language of science and the language of law display a basic difference in approach at the root of both disciplines.

† Simon and Garfunkel, "The Boxer," *A Bridge Over Troubled Waters*, 1970.

‡ Apoptosis is one of the first signs of the body's inflammatory response to injury or assault. It only develops during life. It does not develop postmortem.

Histology Study

To help understand the potential role of histology in this case, it is useful to understand the basics of the wounding and healing process. While the following account is perhaps overly simple, it nonetheless helps clarify the potential significance of the microscopic results.

Chemicals Present in Cells and What They Can Tell Us

Normal human well-being, or "wellness," is called homeostasis. When injuries occur, the body initiates a series of chemical changes in response. The type of response depends upon the type of injury. The responses to injury have been classified into three types: inflammatory, proliferative, and remodeling. When injuries occur, a complicated set of biochemical processes start to repair the damage. For example, platelets called thrombocytes collect at the injury site to form a fibrin clot in order to control any active bleeding. The inflammatory phase removes foreign debris and dead cells with phagocytes, which release various proteins involved in the proliferative phase. Some of the released proteins include fibronectin, CD 3, and CD 15. Remodeling, then, results in hemostatis (the cessation of the blood-borne chemical responses to injury).

Although this verbal model greatly oversimplifies these processes, which, in fact, overlap, it remains a useful heuristic device. The main point to learn here is that this process occurs over time and that certain intercellular chemicals arrive at relatively predictable time intervals. Therefore, their presence or their absence can be quite useful in helping us understand the time between an injury and the time of death, or the cessation of these chemical processes.

To that end, tissue samples taken from the areas of injury suspected to involve satanic torture would receive H&E* and iron stains to address the relationship between the injury and time of death. These tissue samples, when suitably stained, can tell us more about the location of the injury, as well as the decedent's survival interval through the presence or absence of certain proteins such as fibronectin present in the tissues.

Microscopic slides are prepared from tissue samples taken from different locations. The tissues are sectioned into very thin layers and put onto glass slides. Staining is used to reveal the presence of various chemicals through reactions making the elements visible using transmitted light. Photographs, called photomicrographs, are then taken to show the results of these tests. This process of producing the slides and examining them is called histology.

Slides prepared from another case involving similar questions and similar injuries were appropriately stained in order to learn which, if any, of the proteins were present in the various tissues. When tissues taken from World Trade Center victims known to have jumped from the towers were examined, no such proteins appeared. This tells us that the interval between injury and death was not long enough to allow the aforementioned process described, called the *inflammatory response*, to develop fully enough to result in the presence of these proteins. Slides from the decedent in a case similar to the one in Arkansas were examined as follows. The slide's tissue sample has been cut with a microtome, floated, and specifically stained for fibronectin.

* Hematoxylin has an affinity for the nucleic acids in the cell nucleus and is used with eosin, an acidic dye that has an affinity for the cytoplasm in the cell. Other specialized stains are used to help visualize other specific elements in the cell.

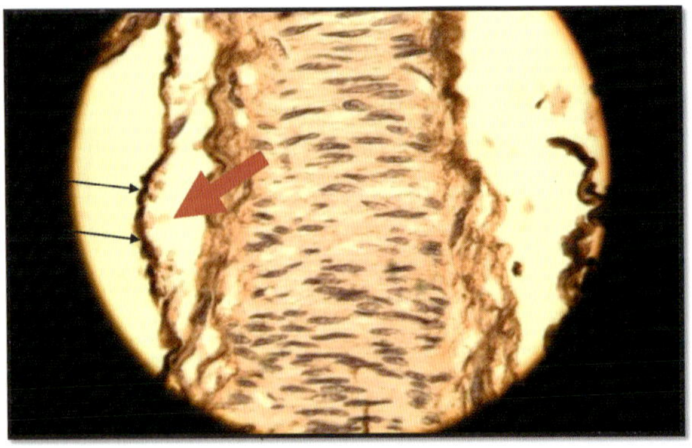

Figure 14.8 This slide from a different case involving trauma, shows fibronectin in a section taken from the pons, medulla, and cortex of the decedent's brain. An example of a red blood cell or erythrocyte is noted by the red arrow.

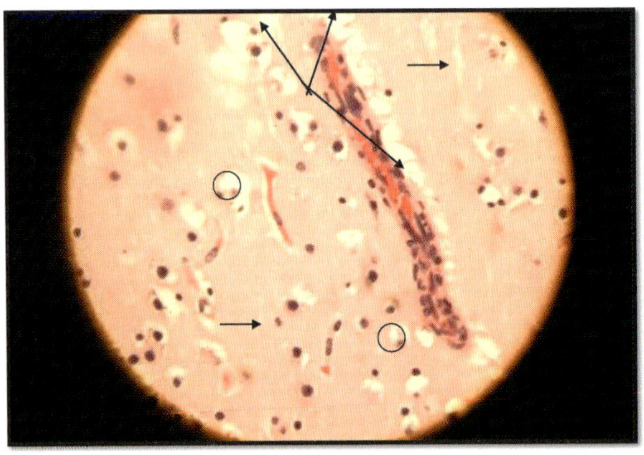

Figure 14.9 This slide from a different case, shows a sample from the corpus callosum. The circles indicate so-called red neurons. The conjoined arrows indicate signs of edema.

The slide in Figure 14.8 shows fibronectin in a section taken from the pons, medulla, and cortex of the decedent's brain. The slide shows vascular congestion and ischemic changes. The result is positive for fibronectin, a protein related to the body's inflammatory response to injury. An example of a red blood cell or erythrocyte is noted by the red arrow.

The slide in Figure 14.9 shows a sample from the corpus callosum. The slide reveals diffuse edema with ischemic changes, as discussed earlier. The circles indicate so-called red neurons, sometimes called ischemic neurons or eosinophilic neurons. The conjoined arrows indicate signs of edema. When the nucleus of a cell begins to shrink, it is a sign that cellular death and repair is in progress when seen along with other cellular activity. This cellular death, when present with evidence of the inflammatory response, is called apoptosis, a word that has come to mean cellular death, but which needs to be distinguished from cellular death called necrosis.

Apoptosis versus Necrosis

Distinguishing *apoptosis* and *necrosis* is not as simple as it might appear. Cell death can result from externally imposed events that leave no sign of cellular repair; it can result from externally imposed events that do leave signs of cellular repair; or it can result as the consequence of an active and ongoing cellular repair process. The former type of cell death is termed *necrosis*, the latter is called *apoptosis*.

Unlike necrosis, apoptosis plays a central role in developing and maintaining the health of an organism by eliminating old cells, unnecessary cells, and unhealthy cells. It is a type of cellular housekeeping. The human body replaces perhaps one million cells per second in this housekeeping process. Apoptosis describes this biological process of programmed cell death leading to predetermined and characteristic cellular changes.

Figure 14.10 shows a section of midbrain. Again, signs of ischemia appear along with evidence of apoptosis. Greater magnification (Figure 14.11) shows, within the yellow circle, an example of a red neuron, ischemic neuron, or dying, shrinking nuclei of a cell. The presence of fibronectin in cells from those tissues indicates the initiation of the inflammatory response and documents the extent of its progress before death ensued. Exact time estimates are difficult, if not impossible, to make. The evidence in these slides, however, shows the extent of a process taking time to develop. In this case, death was not immediate. In other words, when the injuries on this decedent occurred, the inflammatory response began at the site of the damaged tissues. Over time, ischemic changes overcame the brain's ability

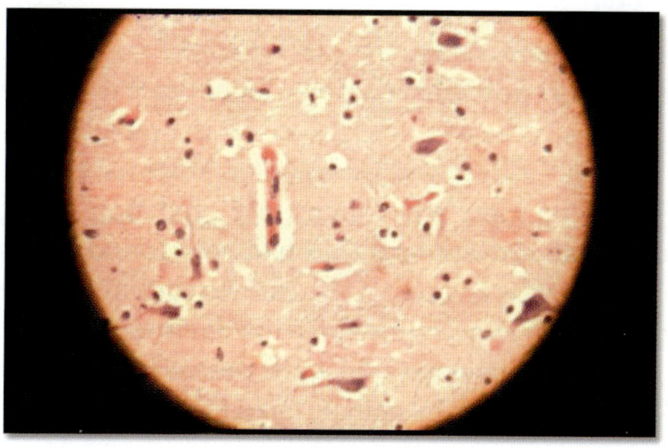

Figure 14.10 This slide from another different but similar type of death case, shows a section of midbrain. Again, signs of ischemia appear along with evidence of apoptosis.

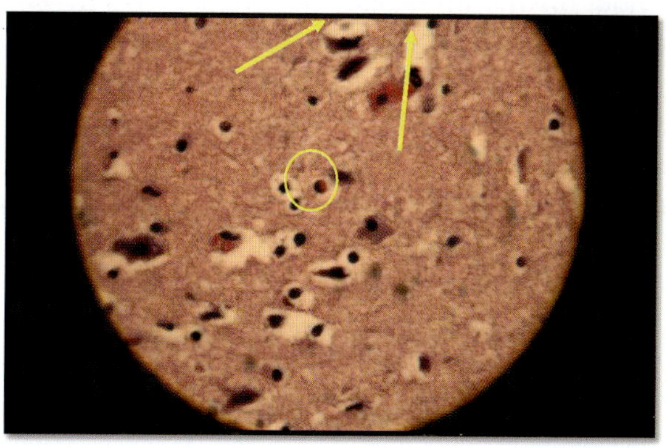

Figure 14.11 Greater magnification of a section from the slide shows, within the yellow circle, an example of a red neuron, ischemic neuron, or dying, shrinking nuclei of a cell. The arrows indicate the presence of ischemia and developing edema.

Figure 14.12 On the left, a sea turtle in Hawaii (not the type present at the scene). A turtle leaving the damage to the boys' bodies would have claws like the painted turtle on the right, not flipper-like legs as on the left. (The turtle on the right did not appreciate the help onto the fence post.)

to function and death ensued. Death is best understood as a process and not an event. But the histology in this other case shows that the injuries occurred before death. That is not what happened with the three boys in Arkansas. If appropriately examined, the histology samples from the boys' surrounding skin injuries would not have red neurons present.

This cellular response is part of the complex healing process that begins in life when an injury is suffered. All these signs are absent when the injuries are postmortem. None of these signs appear on the visible "satanic marks" on any of the three young victims. If only anyone would have looked. It would have saved a lot of heartache for everyone.*

The cause of the boys' deaths was determined by the medical examiner to be drowning, unequivocally. The boys were found underwater. Postmortem marks on their bodies, then, might be explained by some aquatic-creature predation. If their bodies rested on the ground for some time after death or if they had been found on the ground, one might expect postmortem predation by any relevant creatures common to the area. But it remains unlikely that they spent much time exposed to that environment. The cause of death and the recovery location support this point. Thus, the postmortem scratches, lacerations, and bruises were caused, at least in part, by aquatic bayou creatures. These might include crayfish, muskrats, turtles, and aquatic insects (Figures 14.12 and 14.13). Other nonaquatic creatures also might venture into the water for a quick meal. These include various insects, such as beetles and ants, or birds, and even some species of rats. Domestic dogs cannot be ruled out entirely without further investigation.

If postmortem animal interference is confused with injuries preceding death, investigations can go astray. Postmortem injuries caused by animal scavenging and the classification of the species involved can be performed by the forensic pathologist, usually assisted by a naturalist familiar with the area where the bodies were found. Knowledge of the shapes and patterns of injuries left by various scavenging creatures, including claw marks, tooth marks, jaw patterns, and beak marks, proves invaluable. Once some possible alternatives are recognized, then pattern matching and even experiments can begin to confirm the discovery. If the histology were properly prepared and examined, I expect that

* Today, DNA can also be used to test for animal activity. Of course, that possibility depends upon the type of samples taken and retained at the autopsies. A molecular biological investigation of the wounds may generate genetic evidence pointing toward the postmortem animal interference of a specific creature. A specific and sensitive cross-species multiplex polymerase chain reaction has been established including the species-specific animal markers, enabling the genetic identification of wounds caused postmortem by different animals on human remains.

Figure 14.13 The area supported a population of crayfish. They are also likely perpetrators of the visible postmortem injuries.

all this would have been unnecessary. But from a distance of many years, that is difficult to say. It depends upon the tissues that were retained in the autopsies.

When I learned that the area where the boy's bodies were recovered was called "turtle hill," I consulted with a naturalist familiar with the area. He looked at the autopsy photographs and said that the postmortem injuries showed typical patterns of turtle feeding. At any rate, I remain convinced that at the cellular level, if the proper tissues were retained as samples and if they were processed with the proper stains, we would see injury without cellular activity typical for apoptosis, or cellular repair. That is one more way to prove that the injuries visible on the boys were postmortem—one more bit of data that can help answer difficult questions through the applications of science. When samples are retained, we can usefully look to the future for further developments that may help us answer even more questions about this case and about other cases.

In this case, the evidence points strongly, even conclusively, toward postmortem animal scavenging—either by turtles or by other aquatic inhabitants populating the area. Although the murder of the three young boys is horrendous, no proof of satanic rituals or torture exists. Damage done to the bodies was just nature taking its course after they were murdered. Three boys were drowned in Arkansas. Three other young men lost years of their lives in prison before finally being released; they have not yet been exonerated. Compounded tragedy could have been prevented if detectives and investigators had not been so sure of what they were seeing when the bodies were pulled from the pond.

Appendix A: Error Rates, Certainty Factors, and Confidence Coefficients

The main objective of logic as a rational tool is to avoid error. This involves both recognizing and separating truth from falsity. We want to believe true statements and disbelieve false statements. So logic has already asserted itself: we note right away that to believe true statements is to disbelieve their negations. And so it begins.

Some scientists argue that error rates might only attach to operations performed by individuals, such as making a physical measurement, but do not properly attach to theoretically described relationships among various physical features of the subjects of one theory of the known universe; the subject of another theory in some predictable way. Of course whenever metrology (measurement *theory*) is applied to casework by making, say, a measurement under the microscope, I tend to think that there may be two sources for potential error: one obviously is the error introduced by any individual making the measurement, for example, by using a reticle or applying a photomicrographic software package. The error attaches to the limits of exactness when compared with some absolute standard (accuracy) and the frequency of varied results among many microscopists conducting similar measurements (precision).* So far so good. But some also think that there is a discoverable and measurable error built into the theory itself as it confers some significance to the measurement. Such error may arise in a complex and perhaps indirect way. Some argue that this indirect way works back to measurement error as just described. In keeping with good mathematics, let us call the above theory "Theory X" (or Tx for short).

If Tx depends upon any claims about size relationships among microscopic objects, for example, then that theory is open to an error rate calculation. Call it "acceptable variance" if you will, but a pig called Porky is still a pig. However, the discussion must develop at the philosophical level. It is both interesting and nontrivial. For example, I have discussed several theories in physics that require improvement by adding empirical approximations to correct what are recognized as "theoretical shortcomings."† And these are not the only theories that appear to have built-in errors that practitioners must both recognize and correct. These appear in the natural sciences in various ways as "correction factors," "ad hoc coefficients of X," and so on. One simple example occurs in theories of bloodstain pattern analysis.

* That is what validation of various tools and software packages require: any attempted validation of a method or a procedure. Can one develop validations without admitting theoretical error rates? That relates to the philosophical question raised earlier.

† See "How Approximations Take Us Away from Theory and Toward the Truth," Nancy D. Cartwright and Jon Nordby, *Pacific Philosophical Quarterly*, July 1983. Also, the issues are discussed in *How the Laws of Physics Lie*, Nancy Cartwright, with Jon Nordby coauthoring Essay #6, "For Phenomenological Laws," pp. 100–127; further reference to my work appears in the book's Introduction, pp. 1–20; published by Clarendon Press, Oxford University Press, New York, 1983.

Various theories model the point of origin and the point of convergence for bloodshedding events producing bloodstains on a flat surface.* Part of the theory, like for any theory, becomes the identification of the relevant features to consider and those appropriate to omit when building the model. For example, the colors of the socks worn by the researchers are not relevant, but the size of the stain measured in length and width is relevant (that "fact" builds the theory). In fact, several software approaches have been developed and refined over the years to aid in the determinations involved in locating the approximate point where the initial striking producing the bloodstains occurred. With such calculations performed on known points of origin (with points of origin documented while the stains were produced), the theory involved consistently results in errors of about 10 cm too high. The calculated result predicted by the theory gives a result that is consistently 10 cm higher than the actual origin of the bloodstains captured by documenting cameras. Since this "error" has been replicated many times and found by many researchers to be 10 cm, it does not do to call this a "measurement error." (There appear to be few if any relevant measurement errors occurring here.) Regardless, this 10-cm error is *not a measurement error*, despite the presence of any other errors that are such measurement errors. This must therefore be *an error built into the theory*.† As such, we should be able to calculate the error rate of any theory resulting in a prediction of a stain pattern's origin that is consistently 10 cm high. In this way, often our practice is an improvement on our theories.

Despite such rudimentary contributions to the discussion of error rates, this discussion moves us well beyond the scope of this brief appendix. To be more exact, in fact, no one has yet provided a coherent definition of just exactly what an error rate is in the practice of the forensic sciences, or how, once determined, it is supposed to inform us one way or another about a particular result. So when legal decisions‡ mandate the statement of error rates, many scientists rightly wonder just what is being requested. When an attorney asks for "known error rates," she is usually asking the scientist for a figure representing the times that the analyst has been mistaken in the past about the same matter currently at issue before the court. This is obviously impossible. First, it is unlikely that the scientist would ever know that he was mistaken—any mistakes remain independent of the case outcome, and he certainly believes that his results were correct; otherwise, he would not have offered them in the first place. So, how would one even discover such a mistake in this meaning of "error?" Further, how would the scientist discover them even if they were such mistakes? If this is the desired objective in asking the question, then it is the wrong question. That is why, rightly, scientists can remain puzzled by the attorney's request. This discussion remains ongoing but nonetheless important for the development of robust theories in the forensic sciences and the whole of the natural sciences.

What can be discussed here, however, involves the beliefs that we have about the truth of various statements. The cases discussed in this volume exemplify typical applications of an area called Bayesian statistics. Specifically, the inferences listed as "limited crime scene reconstructions" involve these inferences, which are possible to clarify. (Ironically, to

* See Appendix B for a brief meditation on bloodstain pattern analysis basics developed over the years to help lawyers and other nonscientists with some of the basic concepts.

† It may be an error of failing to recognize relevant explanatory factors affecting the property of bloodstain flight, modeling that mechanism, or translating it into geometrical rather than some alternative mathematical model.

‡ *Daubert v. Merrell Dow Pharmaceuticals*, Rule 702, *Kumho Tire Co. v. Carmichael*, as well as various state court decisions regarding scientific evidence.

clarify them we have to make them more technically complex! Simplification seems always to extract a price, after all.)

In the natural sciences, we often do not know the truth about a particular factual statement, but we may have good reasons for believing that it is true. Rational belief under conditions of uncertainty involves applications of the scientific methods discussed in the preceding chapters and forms one of the logical foundations for scientific crime scene reconstruction. Bayes' theorem is one of the main tools for manipulating probabilities of any kind; that is, it is applicable no matter what interpretation is being placed on the probabilities being manipulated. Bayesian inference is a formal approach to making statistical inferences in cases where some of the probabilities are interpreted as representing beliefs, or knowledge, rather than having a frequency-based interpretation.

This is particularly true for the so-called one-off cases that forensic scientists often confront. We do not have the luxury of dealing with large sample populations, which most people think of when they think of "natural science." We are not investigating claims about populations of patients and the clinical effects of a particular medication under certain specified conditions of administration.* Instead, in casework we have a suspect or suspects or no suspects, a victim or victims, witnesses or no witnesses, some injuries, scene or autopsy photographs, a proposed murder weapon, and some official reports.

Some of these cases do involve what are called prior probabilities: given beliefs about testimony, blood, as well as additional beliefs about the witnesses. This type of belief may also put us on our guard for examples of bias. That effort requires that we identify bias in all areas relevant to casework. Of course not all such beliefs are false or incorrect; we must simply be aware of what they are and what role they play in our inferences.

We also have beliefs about evidence. In the absence of any evidence, what do we believe about the victim, for example, or about a victim being shot in the head instead of the chest, or in the left leg as he sat in a chair, or in a vehicle seated behind the steering wheel, or seated beside the driver who is also dead? Or about vehicles hitting trees, or turning over, and so on?

Bayes' theorem (in inferences involving beliefs using Bayesian inference) can be stated as follows. Let *H* denote some hypothesis; that a certain statement of supposed fact is true or that a statistical parameter takes a certain value. Before observing data from a given experiment, one starts with some belief about whether the hypothesis, *H*, is true, expressed in the form of a probability, usually called the prior probability. Bayes' theorem is used to determine what one's probability for the hypothesis should be, once the outcome, *D*, from the experiment is known. Here, in lieu of experimental results, we can substitute the recognition of certain facts such as the presence of blood on the back surfaces of a victim's legs, the absence of blood on a vehicle's windows, or the corresponding locations of blood at some other location, such as the victim's bedroom.

The phrase "should be" is important here, as Bayes' theorem is a condensation of the rules that anyone should apply to updating beliefs, provided that they are acting according to the "reasonable rules" forming the requirements of rationality and consistency. The

* The notable exception, of course, is DNA evidence that must be based upon the probability of the uncommon alleles in the DNA chain occurring or not occurring among members of a relevant population group. These calculations usually make use of "relative frequency" models of probability calculation, since data has been collected about the frequency of alleles in various ethnic populations and are available for inclusion in casework.

probability of the hypothesis (once the outcome from the experiment is known) is called the posterior probability. In this way, the posterior probability for one hypothesis becomes the prior probability for another hypothesis. Thus, the requirement for rational consistency becomes part of the inferential structure moving from one premise or step of reasoning, to the next.

The posterior probability is proportional to the likelihood of the observed data, multiplied by the prior probability, and is given by Bayes' theorem as follows:

$$P(H \mid D) = \frac{P(D \mid H)P(H)}{P(D)}$$

where

P(*H*) is the prior probability of *H*; the probability that *H* is correct before the data *D* are seen.

P(*D*|*H*) is the conditional probability of seeing the data *D* given that the hypothesis *H* is true. This conditional probability is called the likelihood.

P(*D*) is the marginal probability of *D*.

P(*H*|*D*) is the posterior probability; the probability that the hypothesis is true, given the data and the previous state of belief about the hypothesis.

The quantity *P*(*D*) is the prior probability of witnessing the data *D* under all possible hypotheses, and it depends on the prior probabilities given to each of these other possible hypotheses. Given any exhaustive set of mutually exclusive hypotheses, H_i,

$$P(D) = \sum_i P(D, H_i) = \sum_i P(D \mid H_i)P(H_i)$$

Here *i* can be considered to index alternative instances, of which exactly one is really valid, and H_i is the hypothesis that instance *i* is valid. Then $P(D, H_i)$ is then the probability that both instance *i* is valid and that the data from the experiment or the observed fact from the research turn out to be what was observed. Since the set of alternative cases is assumed to be mutually exclusive and exhaustive, the above formula is a case of the *law of total probability*. In many cases, *P*(*D*), which is a normalizing constant, need not be evaluated. As a result, Bayes' formula is often simplified to

$$P(H \mid D) \propto P(D \mid H)P(H)$$

where $\propto$ denotes proportionality.

One of the crucial features of the Bayesian view is that a probability is assigned to a hypothesis, whereas under the frequentist view, a hypothesis is typically rejected or not rejected without directly assigning a probability.

With conditional probabilities, the standard confidence coefficient is .95 or 95%. That remains true in this rather direct application of Bayesian logical probability.

Error Types in Statistical Inferences

Type O Error (No Alarm): Failure to Recognize Relevant Explanatory Relationships

From the point of view of confidence intervals, getting it wrong is simply a matter of the population value being outside the confidence interval. These have been called Type O errors. The O stands either for outside (the confidence interval) or for zero (as opposed to errors of Type I and II).

Type I Error (False Alarm): Establishing Explanatory Relationships When There Are None

If we claim that an explanatory result is significant when there are no explanatory relationships among the *explanans* and *explanandum* among the population, then we have committed another error. In other words, it is the rate of false alarms or false positives. Such things happen, because some samples show a relationship just by chance. One example is thinking that vodka and water, scotch and water, and whiskey and water make for drunkenness; therefore since water is the common element, the water explains the drunkenness. Similar errors can result from undetected or uncorrected biases. Consider "a spouse always is the killer when the other spouse is murdered" may incorrectly be calculated to have a high prior probability, for example, due to imprecise formulations of just what the probability involves (like the water explaining the intoxication above, prior probabilities that may or may not apply to a case depend upon exactly what feature or set of facts is to be counted as relevantly prior, etc.).

Type II Error (Failed Alarm): Eliminating Explanatory Relationships When There Are Some

The other sort of error is the chance to miss an effect (i.e., declare that there is no significant effect) when it is actually present.

For most cases, standard confidence intervals remain in the 95% area with ±5% acceptable variance as the established error rate derived by most scientific laboratories. These "confidence intervals" may eventually be distinguished from error rates, but this seems to be the best that we can rationally provide. Each concept—error rates, certainty factors, and confidence coefficients—blurs into the next with no readily available distinction to help sort them out.

The confidence interval calculations produced for these individual laboratory ranges depend upon the practice of the given laboratory and are laboratory dependent: the necessary data includes the number of cases, the number of cases with given results, and the correctness of the results when tested for *accuracy* and *precision* as described in the above chapters. Such data development requires a great deal of effort and lots of patience.

The discussion of exactly what an error rate amounts to and just how we are to develop a formal mechanism for their exhibition still remains somewhat if not completely mysterious. That seems to be a symptom of our uncertainty about error rates. And here is yet another statement reflecting our uncertainty. And here we go again.

Appendix B: Bloodstain Pattern Basics

A. Introduction

Bloodstain pattern analysis involves the *scientific study* of the static consequences resulting from dynamic blood shedding events. A detailed study of bloodstain patterns at crime scenes often develops invaluable evidence. The distribution, size, and shape of bloodstains on a victim, on a suspect, or on the walls, floors, ceilings, or on objects at the scene can help reconstruct these blood-shedding events. Bloodstain pattern analysis can also help one evaluate the credibility of statements provided by a witness, a victim, or a suspect.

Physical evidence waits to be detected, preserved, evaluated, and analyzed. Herb McDonnell once said, to paraphrase, that in the course of a trial, both defense attorneys and prosecuting attorneys may lie, witnesses may lie, and the defendant certainly may lie. Only the evidence never lies. But if the evidence is not properly recognized, documented, preserved, and processed, all we have are attorneys, witnesses, and defendants as sources for any data. This poses obvious problems.

B. Characteristics of blood

In order to understand bloodstain evidence, one must understand some basic scientific principles.

1. When blood leaves the body as a drop, as spatter, or as a gushing flow from an artery, its behavior conforms to the laws of physics.
2. Understanding blood flows involves understanding the physical forces involved in blood-shedding events.
3. As a fluid, blood has physical properties similar to the properties of water. Physical forces affect blood much as physical forces affect water.
4. Unlike water, however, blood is a pseudoplastic non-Newtonian fluid.
5. Fluids lack the ability to support a shearing stress. This property distinguishes fluids from solids. A fluid will flow under the influence of a shearing stress, whereas a solid will not.
6. Viscosity is a form of internal friction in fluids. It results from frictional forces generated between layers of the fluid as they flow past each other. Viscosity, then, is a measure of a fluid's resistance to change in shape or flow. It is colloquially referred to as a fluid's "thickness" or "thinness." (Water is thinner, therefore less viscous, than motor oil.)
7. Viscosity results from the electrical forces attracting the fluid's molecules to each other. (Red blood cells concentrate sialic acid on their membranes producing a high negative electrical charge, giving blood a higher viscosity than water.)
8. Newtonian fluids maintain constant viscosity under shearing forces (ignore temperature and pressure changes for the moment). Water is a Newtonian fluid.

9. Non-Newtonian fluids do not maintain constant viscosity under shearing forces. Blood is a non-Newtonian fluid.
10. Blood's viscosity decreases with increases in blood's velocity gradient. Blood is a "shear thinning fluid."
11. Blood drops hold together in free flight through the forces of both cohesion and surface tension.
12. Cohesion is an electrical force attracting like molecules to each other. Surface tension is a force resulting from a fluid's molecules achieving the most stable low-energy configuration by minimizing exposed areas of the fluid. This phenomenon results in increased cohesive forces at the liquid's surface, in turn producing a "skin" or a "membrane" of cohesive force.
13. The forces of cohesion and surface tension cause blood drops to become and remain spherical once they fall free from their source. Cohesion and surface tension also cause the drops to resist breaking up even when striking targets such as floors, walls, body parts, or clothing.
14. Other forces affecting bloodstain patterns include adhesion and capillary action.
15. Adhesion is the force attracting unlike molecules to each other.
16. Capillary action is the force that results when the adhesive forces attracting a fluid to a surface are greater than the cohesive forces attracting that fluid to itself. When blood is drawn into any porous material it does so by capillary action. This force helps explain the concave meniscus in a test tube blood sample.
17. If the cohesive force of the liquid is greater than the adhesive force attracting the liquid to the container, then capillary action is reversed. Reverse capillary action, then, results in a convex meniscus.
18. Relative density (once called "specific gravity") compares the ratio of the density of a given substance to the density of water. (The density of water is 1 g/cm^3.) So any substance with a density lower than water will float on water; any substance with a density higher than water will sink in water.
19. Since relative density is a ratio, it has no units. Blood has a relative density of 1.060.
20. The maximum speed for a free-falling blood drop in air is 25.1 feet per second for a 0.05 ml drop. Little change in the blood drop appears from 0 to 25 feet. There will be no change in the blood drop after 25 feet because at that point, it has reached its terminal velocity. Only the target surface can change the shape or condition of the blood drop.

C. The forces of cohesion and surface tension in a blood drop must be overcome by some greater force(s) for the drop to spatter.
   1. Surface texture, not distance fallen, determines the degree of blood spatter.
      a. If the surface is smooth and clean, ruptures of the surface tension do not occur and the drop will not spatter.
      b. If the surface is rough, porous, or has protruding fibers, the surface tension will rupture, causing the blood to spatter.
   2. Blood striking an object at angles less than 90 degrees produces a teardrop shape. A 90-degree impact angle will yield a circular shape.
   3. In teardrop stains, the sharp end points in the direction of travel. This pointed end indicates the blood drop's forward direction of travel prior to striking the surface.

4. When blood forcefully strikes a surface at sharp angles, a smaller droplet is cast off from the larger parent drop much like a large breaking ocean wave issues smaller water volumes that travel up the beach as surf. The smaller "wave cast-off" droplet travels closely to the surface and in a very short distance, begins to streak the surface.

D. For blood drops to divide, energy must be applied to overcome cohesion and surface tension.
   1. Generally, the higher the energy, the greater the division and the finer the spatter.
   2. Generally, the lower the energy, the less the division and the bigger the spatter.

E. Blood dripping into blood
   1. Blood that drips into blood results in small round satellite spatters, usually from 0.1 to 1.0 millimeters in diameter, surrounding the center of the bloodstain.
   2. Commonly seen when a blood source continually drips into a pool of blood.

F. Blood transfer pattern
   1. Transfer pattern—Occurs when a wet bloody surface contacts a second unstained surface creating recognizable mirror image or at least a recognizable portion of the original surface.
   2. Examples of common transfer patterns
      a. Clothing patterns that reproduce fabric weaves
      b. Shoe print and hand print transfers
   3. Swipe pattern—The transfer of blood onto a surface not already contaminated with blood. One side is usually feathered, which indicates the direction of travel.
      a. One common pattern at scenes is a hair swipe, a long thin fine-line transfer.
      b. Another common pattern at scenes is a clothing swipe.
   4. Wipe pattern—Created when an object moves through blood that has not completely dried and moves, removes, or otherwise alters it.
   5. Smear pattern—A large volume of blood, at least 0.5 ml, which has been distorted so much that further classification is not possible.
   6. Smudge Pattern—Another reference to a bloodstain that has been distorted to such a degree that further classification is not possible.

G. Cast-off
   1. During a beating with an instrument that produces the bleeding, blood will not normally collect on the surface of the instrument from the first strike.
   2. On subsequent strikes at the same location, blood will adhere to the instrument since it now strikes a blood source. When the instrument is raised or swung backward, its movement allows small drops of blood to be released from its surface.
      a. Some of these small drops will strike a surface, often a ceiling, at a 90-degree impact angle.
      b. As the instrument continues to swing backward, the movement accelerates and additional blood droplets will be cast off the surface of the instrument. These drops will approximate a linear pattern.
      c. The small drops will strike at increasing acute angles and become elongated.
      d. When the swing of this bloody instrument is interrupted, the resulting patterns are called "cessation cast-off."

3. The movement and the number of swings can often be documented by examining the cast-off pattern.
    a. The instruments range from hammers, bats, tools, and hands, to just about anything that can be used as a weapon. The only requirement is that the object becomes bloodstained during multiple impacts.
    b. Cast-off bloodstains also can result from multiple stabbings.

H. Expirated bloodstains
1. This type of pattern usually occurs when blood is expelled through the mouth, nose, or some perforating airway injury.
    a. The victim may have had a gunshot wound that results in coughing up blood through his mouth and nose.
    b. The victim may also have internal injuries that cause blood to be coughed up or exhaled.
    c. CPR or other resuscitative efforts may also produce such stains.
2. The bloodstains may have tiny air bubbles in them, visible even after they dry. Magnification is often required to see these bubbles. They may not be seen.
3. This bloodstain pattern could be confused with other minute trace evidence, so it is important to closely examine the "bloodstains" and to perform presumptive tests where possible.

I. Projected bloodstains occur when a large volume of blood is propelled toward and strikes a surface.
1. This type of pattern, also called arterial spurting, usually occurs when an artery is damaged and the blood spurts or gushes from the wound in large volume pulses. It continues spurting as long as the heart continues beating.
2. Large drops striking a vertical surface decelerate from air resistance and produce a pattern without spines. The drops strike the surface and then characteristically drip or run downward due to their large volume.
3. This blood projection could also be created when a force acts upon a quantity of blood of approximately 0.10 ml or greater. For example, this occurs commonly at scenes where someone has stepped or stomped into a pool of blood, projecting the blood away from the blood pool.

J. Gunshot wounds
1. Forward spatter
    a. Forward spatter is blood that travels in the same direction as the source of the energy or force that produced the spatter.
    b. Forward spatter varies in size depending upon the injury. The size could vary from .1 millimeter or smaller with some larger stains obvious. The smaller bloodstains do not travel as far as the larger bloodstains.
2. Back spatter
    a. Back spatter is blood that is directed back toward the source of the energy producing the blood drops.
    b. Back spatter often results from a gunshot entrance wound to a body. The unimpeded blood spatters back toward the weapon.
3. Drawback effect
    a. The drawback effect sucks blood into the muzzle of a firearm immediately after its discharge due to the partial vacuum created by contracting discharge gasses.

   b. The drawback effect can be observed in contact gunshot wounds but the effect(s) of compensators, suppressors, and silencing devices as well as any other intervening items may alter the outcome.

K. Other impact spatter sources

Baseball bats, other long handled implements such as crowbars; high-speed machinery such as circular saws, band saws, drills; and many other possibilities

L. Determining bloodshed origin
   1. The location from which blood was shed can often be determined.
   2. Point of convergence
      a. This is the two-dimensional point establishing a bloodstain pattern's deposition direction. It is determined by tracing the long axis of well-defined bloodstains within a single pattern back to a point where the axes intersect or converge.
      b. This two-dimensional point is used to help determine the three-dimensional point of origin or the approximate position of the victim at the time blood was shed.
   3. Point of origin
      a. Measuring the length and width of blood drops and using trigonometry allows us to determine an approximate point of origin.
      b. The arc sin of the width of the bloodstain in millimeters, divided by the length of the bloodstain in millimeters, equals the impact angle.
      c. Only experienced analysts trained in this technique should perform these measurements.
      d. Strings can be placed over blood drops along the axes of the stains at the calculated impact angles, and a resulting point of origin can be visualized in three dimensions.
      e. Error rates are a hotly debated topic, but "error rate" for point of origin calculations can be thought of as an area about the size of a basketball.

M. Clothing
   1. Suspects, victims, and bystanders will often become spattered with blood during blood-shedding events. However, the absence of blood on a person's clothing does not by itself indicate that he or she is not the offender, or that he or she was not present when blood was shed.
      a. The clothing could be changed.
      b. No spatter may strike the suspect if the victim's body or another barrier is between the suspect and the source of the bloodshed.
      c. The suspect may be nude during the offense and so forth.
   2. Each item of bloodstained clothing may help indicate where and how the victim was injured when the bloodstains were deposited on the garment.

N. Summary
   1. Often at crime scenes, the physical evidence contradicts eyewitness testimony. Physical evidence, including bloodstain evidence, is more reliable than evidence based solely upon witness memory.
   2. To be presented accurately and usefully in court, bloodstain evidence must be recognized, documented, preserved, and correctly evaluated.
   3. Bloodstain pattern analysis is a valuable tool to help explain blood-shedding events.

Appendix C: Glossary

The following list of suggested terminology from the International Association of Bloodstain Pattern Analysts (IABPA) is offered as a simple reference. An updated terminology has been developed by the Scientific Working Group on Bloodstain Pattern Analysis (SWGSTAIN) and is to be made available at www.swgstain.org for use by bloodstain pattern analysts in an effort to help standardize terms. Some terms such as "high velocity," "medium velocity," and "low velocity" are so fraught with error that they have been abandoned entirely.

angle of impact: The acute angle formed between the direction of a blood drop and the plane of the surface it strikes

arterial spurting (or gushing) pattern: Bloodstain pattern(s) resulting from blood exiting the body under pressure from a breached artery

back spatter: Blood directed back toward the source of energy or force that caused the spatter

bloodstain: Evidence that liquid blood has come into contact with a surface

bubble rings: Rings in blood that result when blood containing air bubbles dries and retains the bubble's circular configuration as a dried outline

cast-off pattern: A bloodstain pattern created when blood is released or thrown from a blood-bearing object in motion

direction of flight: The trajectory of a blood drop that can be established by its angle of impact and directionality angle

directionality: The directionality of a bloodstain or pattern that indicates the direction the blood was traveling when it impacted the target surface; directionality of a blood drop's flight can usually be established from the geometric shape of its bloodstain

directionality angle: The angle between the long axis of a bloodstain and a predetermined line on the plane of the target surface that represents 0 degrees

drawback effect: Blood in the barrel of a firearm that has been drawn backward into the muzzle

drip pattern: A bloodstain pattern that results from blood dripping into blood

expirated blood: Blood that is blown out of the nose, mouth, or a wound because of air pressure and/or airflow that is the propelling force

flight path: The path of the blood drop, as it moves through space, from the impact site to the target

flow pattern: A change in the shape and direction of a bloodstain due to the influence of gravity or movement of the object

forward spatter: Blood that travels in the same direction as the source of energy or force that caused the spatter

impact pattern: Bloodstain pattern caused when blood received a blow or force resulting in the random dispersion of smaller drops of blood

impact site: That point where force encounters a source of blood

misting: Blood that has been reduced to a fine spray as a result of the energy or force applied to it

parent drop: A drop of blood from which a wave, cast-off, or satellite spatter originates

passive drop (bleeding): Bloodstain drop(s) created or formed by the force of gravity acting alone

perimeter stain: A bloodstain that consists of only its outer periphery, the central area having been removed by wiping or flaking after liquid blood has partially or completely dried

point (area) of convergence: The common point (area), on a two-dimensional surface, over which the directionality of several blood drops can be retraced

point (area) of origin: The common point (area) in three-dimensional space to which the trajectories of several blood drops can be retraced

projected blood pattern: A bloodstain pattern, such as arterial spurting, that is produced by blood released under pressure as opposed to an impact

ricochet: The deflection of blood after impact with one target surface that results in staining of a second target surface

satellite spatter: Small droplets of blood that are distributed around a drop or pool of blood as a result of the blood impacting the target surface

spatter: Blood that has been dispersed as a result of force applied to a source of blood; patterns produced are often characteristic of the nature of the forces that created them

spine: The pointed or elongated stains that radiate away from the central area of a bloodstain

swipe pattern: The transfer of blood from a moving source onto an unstained surface; the direction of travel may be determined by the feathered edge of the pattern

target: A surface upon which blood has been deposited

transfer/contact pattern: A bloodstain pattern created when a wet, bloody surface comes in contact with a second surface; a recognizable image of all or a portion of the original surface may be observed in the pattern

void: An absence of stains in an otherwise continuous bloodstain pattern

wave cast-off: A small blood drop that originates from a parent drop of blood due to the wavelike action of the liquid in conjunction with striking a surface

wipe pattern: A bloodstain pattern created when an object moves through an existing stain, removing and/or altering its appearance

Index

A

B

C

D

N

O

P

R

S

T